Lecture Notes in Computer Science　　12133

More information about this series at http://www.springer.com/series/7409

Khalid Saeed · Jiří Dvorský (Eds.)

Computer Information Systems and Industrial Management

19th International Conference, CISIM 2020
Bialystok, Poland, October 16–18, 2020
Proceedings

 Springer

Editors
Khalid Saeed (iD)
Bialystok University of Technology
Bialystok, Poland

Jiří Dvorský (iD)
VSB - Technical University of Ostrava
Ostrava, Czech Republic

ISSN 0302-9743   ISSN 1611-3349   (electronic)
Lecture Notes in Computer Science
ISBN 978-3-030-47678-6   ISBN 978-3-030-47679-3   (eBook)
https://doi.org/10.1007/978-3-030-47679-3

LNCS Sublibrary: SL3 – Information Systems and Applications, incl. Internet/Web, and HCI

This Springer imprint is published by the registered company Springer Nature Switzerland AG
The registered company address is: Gewerbestrasse 11, 6330 Cham, Switzerland

Preface

CISIM 2020 was the 19th in a series of conferences dedicated to computer information systems and industrial management applications. The conference was held during October 16–18, 2020, in Poland at the Białystok University of Technology.

62 papers were submitted to CISIM 2020 by researchers and scientists from a number of reputed universities around the world. These scientific and academic institutions belong to Bulgaria, Chile, Colombia, Czech Republic, India, Italy, Pakistan, Poland, Spain, and Tunisia. Most of the papers were of high quality, but only 55 of them were sent for peer review. Each paper was assigned to at least two referees initially, and the accept decision was taken after receiving two positive reviews. In case of conflicting decisions, another expert's review was sought for the respective papers. In total, about 130 reviews and comments were collected from the referees for the submitted papers. In order to maintain the guidelines of Springer's *Lecture Notes in Computer Science* series, the number of accepted papers was limited. Furthermore, a number of electronic discussions were held within the Program Committee (PC) chairs to decide on papers with conflicting reviews and to reach a consensus. After the discussions, the PC chairs decided to accept for publication in this proceedings book the best 40 of the total submitted papers. The main topics covered by the chapters in this book are biometrics, security systems, multimedia, classification and clustering, and industrial management. Besides these, the reader will find interesting papers on computer information systems as applied to wireless networks, computer graphics, and intelligent systems. We are grateful to the three esteemed speakers for their keynote addresses. The authors of the keynote talks were Prof. Rituparna Chaki, University of Calcutta, India; Prof. Marina Gavrilova, University of Calgary, Canada; Prof. Witold Pedrycz, University of Alberta, Canada; Prof. Danuta Rutkowska, University of Social Sciences in Lodz, Poland; and Prof. Michał Woźniak, Wroclaw University of Technology, Poland.

We would like to thank all the members of the PC and the external reviewers for their dedicated efforts in the paper selection process. Special thanks are extended to the members of the Organizing Committee both the international and the local ones and the Springer team for their great efforts to make the conference a success. We are also grateful to Andrei Voronkov, whose EasyChair system eased the submission and selection process and greatly supported the compilation of the proceedings.

We hope that the reader's expectations will be met and that the participants enjoyed their stay in the beautiful city of Białystok.

October 2020

Khalid Saeed
Jiří Dvorský

Organization

Conference Patron

Lech Dzienis   Białystok University of Technology, Poland

General Chair

Khalid Saeed   Białystok University of Technology, Poland

Conference Co-chairs

Marek Krętowski	Białystok University of Technology, Poland
Rituparna Chaki	University of Calcutta, India
Agostino Cortesi	Ca' Foscari University of Venice, Italy

International Organizing Committee

Zenon Sosnowski (Chair)	Białystok University of Technology, Poland
Pavel Moravec	VŠB-Technical University of Ostrava, Czech Republic
Jiří Dvorský	VŠB-Technical University of Ostrava, Czech Republic
Nabendu Chaki	University of Calcutta, India
Sławomir Wierzchoń	Polish Academy of Sciences, Poland

Local Organizing Committee

Pavel Moravec (Co-chair)	VŠB-Technical University of Ostrava, Czech Republic
Maciej Szymkowski (Co-chair)	Białystok University of Technology, Poland
Mirosław Omieljanowicz	Białystok University of Technology, Poland
Grzegorz Rubin	Lomza State University, Poland
Mariusz Rybnik	University of Białystok, Poland
Aleksander Sawicki	Białystok University of Technology, Poland
Marek Tabędzki	Białystok University of Technology, Poland

Program Committee

Chair

Khalid Saeed   Białystok University of Technology, Poland

Members

Waleed Abdulla	The University of Auckland, New Zealand
Adrian Atanasiu	Bucharest University, Romania
Aditya Bagchi	Indian Statistical Institute, India
Valentina Emilia Balas	University of Arad, Romania
Anna Bartkowiak	Wrocław University, Poland
Rahma Boucetta	National Engineering School of Gabes, Tunisia
Nabendu Chaki	University of Calcutta, India
Rituparna Chaki	University of Calcutta, India
Agostino Cortesi	Ca' Foscari University of Venice, Italy
Dipankar Dasgupta	University of Memphis, USA
Pierpaolo Degano	University of Pisa, Italy
Riccardo Focardi	Ca' Foscari University of Venice, Italy
Marina Gavrilova	University of Calgary, Canada
Jan Devos	Ghent University, Belgium
Andrzej Dobrucki	Wrocław University of Technology, Poland
Jiří Dvorský	VŠB-Technical University of Ostrava, Czech Republic
David Dagan Feng	The University of Sydney, Australia
Pietro Ferrara	IBM T. J. Watson Research Center, USA
Raju Halder	Ca' Foscari University of Venice, Italy
Christopher Harris	State University of New York, USA
Kauru Hirota	Tokyo Institute of Technology, Japan
Khalide Jbilou	Université du Littoral Côte d'Opale, France
Ryszard Kozera	The University of Western Australia, Australia
Tomáš Kozubek	VŠB-Technical University of Ostrava, Czech Republic
Marek Lampart	VŠB-Technical University of Ostrava, Czech Republic
Christoph Lange	Fraunhofer IAIS, Germany
Jens Lehmann	University of Bonn, Germany
Flaminia Luccio	Ca' Foscari University of Venice, Italy
Pavel Moravec	VŠB-Technical University of Ostrava, Czech Republic
Romuald Mosdorf	Białystok University of Technology, Poland
Debajyoti Mukhopadhyay	Maharashtra Institute of Technology, India
Yuko Murayama	Iwate University, Japan
Nobuyuki Nishiuchi	Tokyo Metropolitan University, Japan
Andrzej Pacut	Warsaw University of Technology, Poland
Jerzy Pejaś	WPUT in Szczecin, Poland
Marco Pistoia	IBM T. J. Watson Research Center, USA
Piotr Porwik	University of Silesia, Poland
Jan Pries-Heje	IT University of Copenhagen, Denmark
S. P. Raja	Vel Tech Institution of Science and Technology, India
Isabel Ramos	University of Minho, Portugal
Anirban Sarkar	National Institute of Technology Durgapor, India
Ewa Skubalska-Rafajłowicz	Wrocław University of Technology, Poland
Kateřina Slaninová	VŠB-Technical University of Ostrava, Czech Republic
Krzysztof Ślot	Lodz University of Technology, Poland

Václav Snášel	VŠB-Technical University of Ostrava, Czech Republic
Zenon Sosnowski	Białystok University of Technology, Poland
Jarosław Stepaniuk	Białystok University of Technology, Poland
Marcin Szpyrka	AGH Kraków, Poland
Andrea Torsello	Ca' Foscari University of Venice, Italy
Qiang Wei	Tsinghua University, China
Sławomir Wierzchoń	Polish Academy of Sciences, Poland
Michał Woźniak	Wrocław University of Technology, Poland
Sławomir Zadrożny	Polish Academy of Sciences, Poland

Additional Reviewers

Marcin Adamski	Białystok University of Technology, Poland
Paola Patricia Ariza Colpas	Universidad de la Costa, Colombia
Giancarlo Bigi	University of Pisa, Italy
Rituparna Chaki	University of Calcutta, India
Tomasz Grzes	Białystok University of Technology, Poland
Wiktor Jakowluk	Białystok University of Technology, Poland
Miguel A. Jimenez-Barros	Universidad de la Costa, Colombia
Wojciech Kwedlo	Białystok University of Technology, Poland
Marek Lampart	VŠB-Technical University of Ostrava, Czech Republic
Tomáš Martinovič	VŠB-Technical University of Ostrava, Czech Republic
Ireneusz Mrozek	Białystok University of Technology, Poland
Dionicio Neira Rodado	Universidad de la Costa, Colombia
Sabina Nowak	University of Gdansk, Poland
Miroslaw Omieljanowicz	Białystok University of Technology, Poland
Hugo Hernandez Palma	Universidad del Atlántico, Colombia
Nadia Pisanti	University of Pisa, Italy
Grzegorz Rubin	Lomza State University, Poland
Mariusz Rybnik	University of Białystok, Poland
Soharab Hossain Shaikh	BML Munjal University, India
Maciej Szymkowski	Białystok University of Technology, Poland
Marek Tabędzki	Białystok University of Technology, Poland
Amelec Viloria Silva	Universidad de la Costa, Colombia
Lukáš Vojáček	VŠB-Technical University of Ostrava, Czech Republic

Keynotes

NSP and Its Application Towards Increasing Patient Satisfaction in Assisted Living

Rituparna Chaki

University of Calcutta, India
rituchaki@gmail.com

Abstract. The domain of nurse scheduling is a well-researched one. Researchers have been focusing mainly to increase the efficiency of nurses' allocation while maintaining the nurse satisfaction at an acceptable level. However, most of the existing works focus on optimizing the utilization of nursing staff. In the age of IoT, as more and more researches are carried on in the domain of assisted living, it becomes more important to use NSP for maximizing patient recovery. In our bid to understand the challenging issues of adapting the convention solutions in the assisted living scenario, a thorough state-of-the-art study has been done. This study led us to note the following issues that need to be addressed:

- Nurse scheduling solutions mainly aim at assigning the nurses so as to maximize the utilization of available nurses. The solutions mostly lacked consideration for the perspective of patients, viz, their preferences and types of ailments.
- Matching nurses' expertise to patients' requirements, as well as patients' preferences with respect to nurse availability is an issue that needs to be better investigated.
- The problem of increased size of search space regarding increasing number of patients' requirements need to be focused in more specific way.

The problem is defined as the assignment of a set of available nurses N to a set of patients P, depending on a number of criteria. The identification of parameters and the definition of constraints (soft and hard) is to be considered in such a way so as to maximize patient satisfaction. The cost function also needs to be formulated in terms of these parameters and the goal is to keep the cost at an optimum level. In this talk, the focus will be on discussing some of the relevant techniques used for solving the nurse scheduling problem, including a novel solution specifically aimed to increase patient satisfaction.

Adaptive and Reliable Decision Making for Multi-modal Biometric Systems

Marina Gavrilova

University of Calgary, Canada
marina@cpsc.ucalgary.ca

Abstract. The area of biometrics, without a doubt, has advanced to the forefront of an international effort to secure societies from both physical and cyber threats. This keynote provides an overview of the state of the art in multi-modal data fusion and biometric system design, linking those advancements with real-world applications.

The rapid development of massive databases and image processing techniques has led over the past decade to the significant advancements in both fundamental biometric research and in a relevant commercial product development. Typical biometric applications include banking, border control, law enforcement, medicine, e-commerce, smart sensors, and consumer electronics. A variety of issues related to biometric system performance and analysis has been addressed previously. A high number of biometric samples, data variability, data quality, data acquisition, types of fusion and system architectures have been shown to affect an individual biometric system's performance. Addition of new types of behavioral data, based on social interactions, presents unique challenges and opportunities. This keynote reviews current trends related to design of adaptive and reliable multi-modal biometric systems, with the focus on issues of security and privacy of person data. It supports the theoretical developments with the practical examples on the use of multi-modal biometrics in industrial applications, including city planning, finance, medicine, and situation awareness systems.

Explainable AI: From Data to Symbols and Information Granules

Witold Pedrycz

University of Alberta, Canada
pedrycz@ee.ualberta.ca

Abstract. With the progress and omnipresence of Artificial Intelligence (AI), two aspects of this discipline become more and more apparent. When tackling with some important societal underpinnings, especially those encountered in strategic areas, AI constructs call for higher explainability capabilities. Some of the recent advancements in AI fall under the umbrella of industrial developments (which are predominantly driven by numeric data). With the vast amounts of data, one needs to resort themself to engaging abstract entities in order to cope with complexity of the real-world problems and delivers transparency of the required solutions. All of those factors give rise to a recently pursued discipline of *explainable* AI (XAI). From the dawn of AI, symbols and ensuing symbolic process have assumed a central position and ways of symbol grounding become of interest. We advocate that in the realization of the two timely pursuits of XAI, information granules and Granular Computing (embracing fuzzy sets, rough sets, intervals, among others) play a significant role. The two profound features that facilitate explanation and interpretation are about an accommodation of the logic fabric of constructs and a selection of a suitable level of abstraction. They go hand-in-hand with the information granules. First, it is shown that information granularity is of paramount relevance in building linkages between real-world data and symbols encountered in AI processing. Second, we stress that a suitable level of abstraction (specificity of information granularity) becomes essential to support user-oriented framework of design and functioning AI artifacts. In both cases, central to all pursuits is a process of formation of information granules and their prudent characterization. We discuss a comprehensive approach to the development of information granules by means of the principle of justifiable granularity. Here various construction scenarios are discussed including those engaging conditioning and collaborative mechanisms incorporated in the design of information granules. The mechanisms of assessing the quality of granules are presented. In the sequel, we look at the generative and discriminative aspects of information granules supporting their further usage in the AI constructs. A symbolic manifestation of information granules is put forward and analyzed from the perspective of semantically sound descriptors of data and relationships among data. With this regard, selected aspects of stability and summarization of symbol-oriented information are discussed.

Artificial Intelligence and Image Understanding

Danuta Rutkowska

University of Social Sciences in Lodz, Poland
darutko@gmail.com

Abstract. Applications of Artificial Intelligence (AI) have increased rapidly in recent years. A lot of interest is focused on Deep Learning and Big Data. We are witnesses of spectacular results presented by Google with regard to Image Recognition. Deep Learning has also been successfully applied in speech processing and much more.

However, this part of AI – that is data driven – reflects the process of learning from examples. In this case, an intelligent system, e.g. a Deep Learning network, achieves a result, e.g. an image recognition, by the learning ability that increases along with larger amount of data examples. In this way, the system solves the problem without an explanation concerning the result.

Although with regard to Image Recognition, the explanation is not always necessary (we see what we get), in other AI applications a user wants to know how and why the system come up with the result. This is very important, e.g., in recommender systems and medical applications (also referring to medical images).

On another side of AI are expert systems – that are knowledge based – and realize an inference process, with explanation facilities. The knowledge is usually represented by logical rules. Therefore, it is possible to explain how a result has been obtained by use of the rules.

Hybrid intelligent systems can reflect both aspects of intelligence: an inference based on the knowledge represented by rules and the learning ability. This means that the inference can be realized – when the rules are known, otherwise the knowledge of the form of the rules can be acquired from data (examples) during the learning process.

In contrary to the Deep Learning approach – that is viewed as a "black box" because of the lack of the explanation – we propose a rule based system to solve an Image Understanding problem. The rules considered with regard to the knowledge of this system are fuzzy rules or the rules generated within the rough set theory. The linguistic description of images are produced by the system, and then analyzed within the framework of databases and AI. The goal is to describe an image based on the color segmentation, and location of particular color granules, as well as their size and shape. Mutual relationships between the color granules are taken into account in order to explain the understandable description of an image.

Chosen Challenges of Imbalanced Data Classification

Michał Woźniak

Wrocław University of Science and Technology, Poland
michal.wozniak@pwr.edu.pl

Abstract. Imbalanced data classification is still a focus of intense research because most of the learning methods can work with a reasonably balanced data set. Still, many real-world applications have to face imbalanced data sets. A data set is said to be imbalanced when several classes are under-represented (minority classes) in comparison with others (majority classes). Learning from imbalanced data is among the contemporary challenges in Machine Learning, and multi-class imbalance, as well as an imbalanced data stream, stand out as the most challenging scenarios.

In binary imbalanced learning, the relationships between classes are easily defined: one class is the majority one, while the other is the minority one. However, in multi-class scenarios, this is no longer obvious, as the correlations among classes may vary, e.g., one class can be at the same time minority and majority, or one of the different classes. Therefore canonical methods designed for binary cases cannot be directly applied in such scenarios.

Another topic which we will discuss during the talk is imbalanced data stream classification because only a few of the authors distinguish the differences between the imbalanced data stream classification problem and a scenario where the prior knowledge about the entire data set is given. This discrepancy is a result of the lack of knowledge about the class distribution, and this issue is notably present in the initial stages of data stream classification. Another difficulty is the presence of the phenomenon called *concept drift*, which can usually lead to classifier quality deterioration. The concept drift may have different nature, but it causes the change of the probability characteristics of the decision task, e.g., it could lead to a shift in the prior probabilities, i.e., the frequency at which the objects appear in the examined classes. A typical example of such a case is the technical diagnosis in which the fault probability increases with utilization time, and it may be a result of material fatigue. Sometimes the relationship between the minority and majority classes changes in a way that the former becomes the majority class.

This talk will discuss the main problems of imbalanced data classification, as multi-class imbalanced data analysis or imbalanced data stream classification, with particular attention to the methods developed by the Machine Learning team from the Department of Systems and Computer Networks from Wroclaw University of Science and Technology.

Contents

Industrial Management and other Applications

Machine Learning and High Performance Computing

Modelling and Optimization

Biometrics and Pattern Recognition Applications

Transfer Learning Approach in Classification of BCI Motor Imagery Signal

Filip Begiełło[1], Mikhail Tokovarov[2] (ID),
and Małgorzata Plechawska-Wójcik[2(✉)] (ID)

[1] Smart Geometries Sp. z o.o., Lęborska 8/10/183, Warsaw, Poland
f.begiello@smartgeometries.pl
[2] Lublin University of Technology, Nadbystrzycka 36B, Lublin, Poland
{m.tokovarov,m.plechawska}@pollub.pl

Abstract. The paper presents application of a transfer learning-based, deep neural network classification model to the brain-computer interface EEG data. The model was initially trained on the publicly available dataset of motor imagery EEG data gathered from BCI experienced users. The final fitting was performed on the set of six participants for whom it was the first contact with a BCI system. The results show that initial training affects classification accuracy positively even in case of inexperienced participants. In the presented preliminary study five participants were examined. Data from each participant were analysed separately. Results show that the transfer learning approach allows to improve classification accuracy by even more than 10% points in comparison to the baseline deep neural network models, trained without transfer learning.

Keywords: Transfer learning · BCI · Motor imagery · Convolution networks

1 Introduction

A brain-computer interface (BCI) is understood as a system that performs measurements of activity of the brain using electroencephalography (EEG) and converts them into artificial stimuli that replace or strengthen the brain's natural activity in specific areas and thus change the interaction between the system and its external or internal environment [1, 2].

The goal of BCI design is to provide new or extend already existing user contact with the environment [3]. For this reason the interaction should be as natural as possible. That is why a BCI system must aim at fitting to the user's activity and his further training, usually related to identification of his intentional control. A typical BCI needs offline training to calibrate the system and select optimal features from multiple EEG channels [4].

The classical BCI design can be separated into several modules: a module responsible for signal registration, signal pre-processing module, feature extraction, classification module and the output device, providing feedback to the user [5].

The signal pre-processing phase covers cleaning and adaptation of the signal recorded and amplified by the measuring apparatus. The methods usually applied in this module include filtering algorithms: spectral, responsible for removing undesirable

K. Saeed and J. Dvorský (Eds.): CISIM 2020, LNCS 12133, pp. 3–14, 2020.
https://doi.org/10.1007/978-3-030-47679-3_1

signal components, and spatial, applied to isolate information about the region in which the signal was recorded [6].

Feature extraction is the next phase in classical BCI systems. The result of this process is a chosen set of key values describing the signal. Depending on the chosen BCI paradigm, the set of features relevant to system control might differ. Extracted features are usually frequency band power features or time point features [7]. The feature extraction process often precedes feature selection, which is a common approach to representing EEG signals in a compact and relevant manner [4] and to achieve fewer significant features with the largest meaning for further classification [8].

Signal classification is the last of the processing steps. The result of this process is to match a signal to one of the known patterns, which allows its interpretations. According to the classical BCI-related literature different learning techniques are used. The most common classification algorithms are linear classifiers such as LDA, SVM or MLP [4].

Alternative, high efficiency solutions applied to BCI are neural networks, in particular convolutional neural networks (CNNs). This approach gains more and more popularity. CNNs give the possibility of filtering the noise at the stage of initial analysis, so they require significantly less initial signal processing [9]. In the case of solutions based on neural networks, however, a problem arises of model degradation along with working time and the need for recalibration.

A convolutional neural networks is a type of deep learning neural networks, where features and the classifier are jointly learning directly from data [4]. The architecture of a deep learning model is based on a set of trainable feature extractor modules and layers. CNNs are so called feedforward neural networks with at least one convolutional layer, where information flows from the input through the hidden layers to the output [10]. Deep neural networks and CNNs are gaining popularity in all types of BCI systems. The great majority of the published results were offline studies. The first application of CNNs to BCI was presented in [11], where two convolutional layers were adapted in a P300-based BCI. There are also a few studies exploring motor imagery-based BCI with CNN classification. In [12] 84% performance was reached in a classification based on raw EEG signal. In [13] CNNs and stacked autoencoders (SAE) were applied to improve classification results on a BCI competition IV dataset. Other deep learning approaches to motor imagery-based BCI are based on deep belief networks (DBNs) and restricted Boltzmann machine (RBM) [14, 15].

In the last few years a new machine learning approach has been presented. It violates the basic hypothesis of machine learning, which is the classifier's training data, and its test data belong to the same feature space and follow the same probability distribution. This new approach, called transfer learning, is aimed at enhancing performance of a classifier initially trained on one task based on information gained while learning another task. The effectiveness of transfer learning depends on the level of relation between these two tasks [4]. Transfer learning turns out to be an effective technique for improving BCI classification performance. Motor imagery proved to be a paradigm, for which transfer learning application gained popularity. This technique allows to use pre-recorded sessions, from possibly different subjects, and analyse them jointly. However, it might encounter such problems as ensuring quality of the features and coping with data variability across subjects or sessions. In the literature the transfer learning approach is often related to the SVM [16–18] or LDA [19–22] classification methods.

The aim of the paper is to examine if the transfer learning approach allows to achieve higher classification performance even in the case of inexperienced BCI users who were never previously trained to generate motor imagery patterns in EEG signals.

The rest of the paper is structured as follows. Section 2 presents materials and methods used in the study including applied classification methods, data sets and the experiment details. Section 3 describes the data analysis procedure, Sect. 4 presents the results. Discussion and conclusion are covered in the Sect. 5.

2 Materials and Methods

2.1 Transfer Learning for CNNs

As mentioned in the introduction, a convolutional neural network is a type of deep neural network incorporating one or more convolutional layers. Each such layer passes its input through a series of filters performing simple mathematical operations on subsequent parts of the input. The aim of those operations is to extract so-called feature maps from the input signal, which in turn can be used as an input for another layer, or as a set of final key features on which classification can be performed. Over the course of training a CNN, both convolutional filters and dense layer weights can be adjusted, which results in a model precisely fit for extraction and classification of features present in the training dataset. While this approach is perfect for data domains in which key features are common and shared across all its elements, it can fall short when it comes to generalisation in domains with high inter-subject variability, such as EEG readings of motor imagery.

Transfer learning aims to improve the model's ability to generalise between two domains with different feature spaces. Of particular use, in the case of CNNs, can be feature-representation transfer learning, which assumes maximisation of classification accuracy by basing constructing the target domain feature space on already known feature spaces [23]. In other words, a model can be trained and fitted to data from a universal, large-source domain in order to develop a general feature extraction mechanism, and then further fit to a specific target domain by adjusting the weights and in turn overcome the problem of inter-subject variability, even if the target domain comprises only a small number of observations. In the case of CNNs the process would entail adjusting filters, fit for extracting general key features, to specific subject signal key features and thus improving its performance, without the need for a long training stage when used in a BCI.

2.2 Data Sets

Two datasets were used in the present research. A publicly available dataset published in [24] was applied for the initial stage of the experiment, when the models were pre-trained. The dataset was collected in accordance with the procedure described in [25].

The dataset contains the signals of 109 patients, 4 of them were rejected as outliers on the basis of their power density spectra. Only the parts of the signals corresponding to hand-movement related motor imagery patterns were chosen for the experiment. Each patient generated 20 observations of each class, i.e. left- or right-hand movement imagery, which led to the total number observations as high as 4200. The signal in [24] was sampled with the sampling frequency equal to 160 Hz. The electrode montage used was in accordance with the 10–10 system with exclusion of the following electrodes: Nz, F9, F10, FT9, FT10, A1, A2, TP9, TP10, P9, and P10. As for the second stage of training, i.e. final training on individual people's samples, it was conducted on the dataset which composed of 6 separate recordings, however one of them proved to be of insufficient quality, and each made for a different subject. As in the training dataset, only hand-movement imagery related parts were chosen. One recording contains 30 observations of each of 2 classes – left- or right-hand movement imagery, for the total of 60 observations per subject. Before starting transfer learning two major issues were to be dealt with:

- the frequency of the final-stage signal had to be decreased from 500 Hz down to 160 Hz, i.e. had to undergo the procedure of downsampling;
- the electrode order of the initial stage dataset had to be brought into accordance with the 10–20 system used for gathering of the second-stage signal, Fig. 1 presents the 10–10 and 10–20 systems. This issue was solved by rejecting the electrodes of the 10–10 system, which are not used in the 10–20 system. Another measure undertaken involved changing the electrode order.

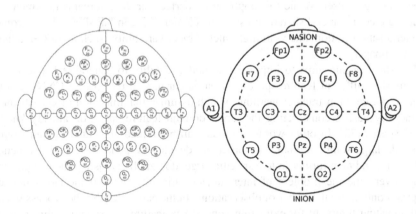

Fig. 1. EEG electrode montage systems: 10–10 (left) and 10–20 (right)

The signals of the initial-stage dataset were cut into 6 s long segments. Each segment contained 2 s of pre-stimulus and 4 s of post-stimulus signal, so each observation had the following dimensions 960 × 64, i.e. 6 s, 160 samples each, by 64 electrodes.

Observations in the final set were 4 s long - 1 s pre-stimulus and 3 s post-stimulus signal, which gives 2000 × 19 dimensions - 4 s, 500 samples each, by 19 electrodes.

2.3 Equipment Characteristic

Mitsar EEG 201 amplifier was used to record the EEG data. The data was transmitted to the computer with the EEG Studio software. The design motor imagery-based BCI was built in the Matlab software. A LabStreamingLayer library was applied to transfer data between the EEG Studio and Matlab.

The study used 19 electrodes arranged in accordance with the 10–20 EEG standard. Two reference electrodes (A1, A2) were placed on the ears. The ground electrode was placed in the middle part of the frontal lobe. The recording frequency was 500 Hz.

2.4 Experiment Construction

The study group for the final dataset included 5 subjects ranging from 20 to 35 years of age (on average: 25 years old), both male and female. For all participants it was their first contact with a motor imagery-based BCI and the BCI systems in general.

The BCI used in the experiment was constructed to represent 4 classes of motor imagery – left- and right-hand motions, tongue motion and feet motion. For each class a red arrow, pointing respectively left, right, up and down from the centre of the screen, was assigned. In downtime between stimuli a control image of a black dot centred in the middle of the screen was displayed.

Each participant was first introduced to the system and instructed as to how to interact with it. After preparation (seating, scalp cleanup, electrode montage) participants were once again given a brief overview of the experiment. The recording part of the experiment started with 60 s of calibration image being displayed to the participant. Next a sequence of stimuli followed by control images was displayed. Each pair consisted of 4 s of stimuli image and 2 s of control image. The entire sequence comprised 30 repetitions of every class of stimuli intermixed with one another as to avoid participants finding a pattern in the sequence.

3 Description of the Data Analysis Procedure

3.1 Preprocessing and Training Process

The signals in [24] are provided in unprocessed state, so initial processing included filtering the signals with a Butterworth filter of the 5-th order. The low and high-cuts were set as follows: low: 1 Hz, high: 45 Hz.

As it was described in 2.2, initial stage training observations had the following dimensions: 960 × 64. The training batch generator was defined in the way, that training samples had the length of 640 time samples, i.e. were 4 s long (approximately 1 s pre-stimulus and 3 s post-stimulus). In order to enrich the training set some uniformly distributed random shift was applied: the start of each training observation was shifted by the value obtained by pseudorandom number generator from the range [−10, 10]. By this approach one aspect of data augmentation was implemented. The Fig. 2 presents the idea of data augmentation by the means of pseudo-random shift.

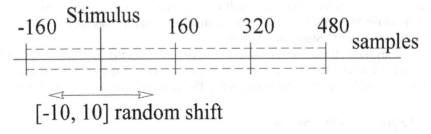

Fig. 2. Data augmentation by means of random window shift

Every training/testing observation was fed to the network as a *s*-x-*e* 2D array, where *s* is the number of samples in an observation and *e* is the electrode number. In order to deal with inter-subject and inter-session variation of conductivities, z-score standardization was performed for every separate electrode, so the signal of the *i*-th electrode after standardization was obtained in the way presented in Eq. (1).

$$x^i_{stand} = \frac{x^i - mean(x^i)}{std(x^i)} \tag{1}$$

Another approach to data augmentation applied in the research was based on adding a normally distributed weak noise. By the term weak the authors understand a normally distributed noise with a 0 mean and a small (around 0.1) standard deviation.

Combining the above methods of data augmentation with such a technique as application of maximal norm constraint allowed to decrease the variance of the classification model leading to better accuracy and lower overfitting.

3.2 Neural Network Architecture

Models used over the course of research were based on a feedforward CNN architecture consisting of pairs of convolution and pooling layers. In the initial tests an architecture comprising two pairs was chosen. Each of the pairs consisted of one 1D convolution layer and one 1D average pooling layer. Both convolution layers had penalties on layer parameters added in the form of L2 regularisers. After each of the pairs a dropout layer was added to further prevent overfitting. Presented on the Fig. 3. is the detailed model architecture with tensor shapes.

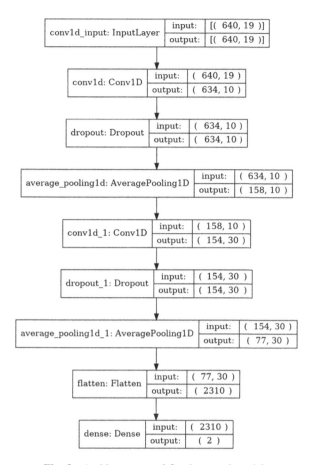

Fig. 3. Architecture used for the tested models

3.3   Comparison of Model Training Process with and Without Transfer Leaning Approach

For the purpose of comparison two approaches to training deep neural network models were tested: with and without application of transfer learning.

In the former case, models were trained only on samples from the final dataset. For subsets of each participant a set of 10 models was trained. The training process consisted of 10 epoch further split into 30 batches of 10 samples. Considering a relatively small size of the set before augmenting, a stochastic gradient descent optimiser with a learning rate of 0.001 was chosen to prevent overfitting. This approach scheme is given in Fig. 4.

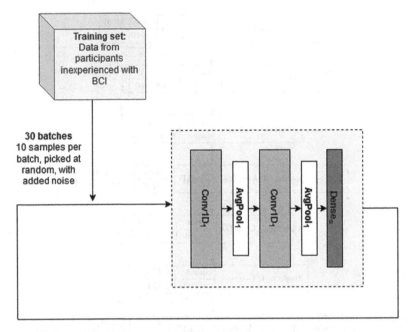

10 Epochs

Fig. 4. Model training process without transfer learning

The learning process in the transfer learning approach was separated into two stages. The first stage involved training models on the initial dataset for 10 epochs, 150 batches of 30 samples each. For this step a RMSprop optimiser with a learning rate of 0.001 was used as it yielded the most stable results. In the second stage pre-trained models from the previous stage were further trained on the final dataset for 10 epochs, 30 batches of 10 samples each. Similarly to the training process presented in Fig. 4, a stochastic gradient descent optimiser with a learning rate of 0.001 was applied in this step. As previously, models were trained in sets of 10 for the data from each participant. The entire process is presented in Fig. 5.

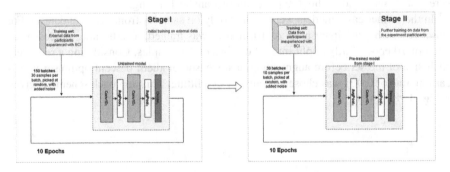

Fig. 5. Model training process with transfer learning

4   Results

The performance of a CNN model depends on a number of hyperparameters which need to be precisely adjusted to the task carried out. That is why a grid search technique was applied over a range of values for each of the hyperparameters and tested architectures. After initial tests, as mentioned in 3.2, the best performing architecture was chosen and a more thorough search was performed including multiple filter sizes, numbers of filters, activation functions etc. The resulting scores with hyperparameters of models trained in the search were then arranged prioritizing maximal accuracy and minimal loss on test set and ignoring entries with high (above 15% points) discrepancy between training and test sets as to avoid including overfitted models. This list served as a tool to find dependencies between scores and specific values of tested hyperparameters.

As a result, the set of hyperparameters ensuring the best performance of the CNN was chosen, which is presented in Table 1.

Table 1. Hyperparameters of the model

Hyperparameter		Value
Output filters	$Conv1D_1$	10
	$Conv1D_2$	30
Kernel size	$Conv1D_1$	7
	$Conv1D_2$	5
Stride length	$Conv1D_1$	1
	$Conv1D_2$	1
Activation function	$Conv1D_1$	ReLU
	$Conv1D_2$	ReLU
Pooling window size	$AvgPool1D_1$	4
	$AvgPool1D_2$	2
Dropout rate		0.3
Regularisers	Kernel	L2 ($\lambda = 0.02$)
	Bias	L2 ($\lambda = 0.01$)
Constraints	Kernel	MaxNorm (max = 1.2)
	Bias	MaxNorm (max = 1.2)

The results obtained for the separate participants are presented in Table 2. Classification accuracy was the main metric; additionally, values of the loss function are presented. The results confirm that the transfer learning approach allows achieving a higher accuracy in four of five cases. The improvement reached from 10 to 15% points. Only in the case of one participant (P5) the achieved result was lower for the transfer learning approach, although the losses for all five participants are notably (up to two times) lower.

Table 2. Comparison of classification metrics for models trained with and without application of transfer learning for separate participants

Patient No.	Accuracy (%)		Loss	
	Without transfer learning	Transfer learning	Without transfer learning	Transfer learning
P1	66.7	75	1.166	0.557
P2	58.3	66.7	1.149	0.742
P3	66.7	75	1.0111	0.579
P4	58.3	75	1.207	0.599
P5	75	66.7	1.151	0.672

5   Discussion and Conclusions

The main aim of the paper was to examine if the transfer learning approach allows to achieve higher classification performance even in the case of inexperienced BCI users who were never previously trained to generate motor imagery patterns in EEG signals. For initial stage of training a Physionet Motor Imagery dataset was used. This dataset is extensive enough and widely applied in a scientific community and thus serves as a good basis for initial model training.

The second stage involved training on the data gathered from novice BCI users. A low number of participants and samples was an additional obstacle for transfer learning to overcome. The transfer learning approach allowed to achieve higher classification accuracy compared to the baseline approach to deep neural network models training by the margin of at least 8.3% points: from 66.7 to 75. The mean accuracy improvement equalled 10.42% points (excluding P5). Participant P5 reported difficulties with generating motor imagery, and thus his results were less reliable. The maximum accuracy gain was equal to 16.7% points from 58.3 up to 75.

The commonly known fact is that deep models require large datasets in order to achieve satisfying performance, being prone to overfitting in the case of an insufficient amount of data. The results prove that the transfer learning approach might help to mitigate this problem, which can be concluded from the lower values of losses for this approach.

Even though the present study is preliminary, it provides promising results and further research on a larger number of participants might lead to even better performance.

Over the course of the present study several models based on different architectures were tested (2 Conv layer network, 3 Conv layer network, different combinations of convolution and pooling, different layer complexity, etc.). The result analysis reveals that the best performance is achieved by relatively simple models (as can be seen in the results of the grid search in Table 1. Furthermore, the obtained results correspond to the current trends presented in scientific literature.

References

1. Wolpaw, J.R., Wolpaw, E.W.: Brain–computer interfaces: something new under the sun. In: Wolpaw, J.R., Wolpaw, E.W. (eds.) Brain–Computer Interfaces, Principles and Practice, pp. 3–12. Oxford University Press Inc., New York (2012)
2. Clerc, M., Bougrain, L., Lotte, F.: Brain-Computer Interfaces 1: Foundations and Methods. Wiley, New York (2016)
3. Wolpaw, J.R.: Brain–computer interfaces. Handbook of Clinical Neurology **110**, 67–74 (2013)
4. Lotte, F., et al.: A review of classification algorithms for EEG-based brain–computer interfaces: a 10 year update. J. Neural Eng. **15**(3), 031005 (2018)
5. Arvaneh, M., Tanaka, T.: Brain–computer interfaces and electroencephalogram: basics and practical issues. In: Arvaneh, M., Tanaka, T. (eds.) Signal Processing and Machine Learning for Brain-Machine Interfaces, pp. 1–21. The Institution of Engineering and Technology, London (2018)
6. Ramoser, H., Muller-Gerking, J., Pfurtscheller, G.: Optimal spatial filtering of single trial EEG during imagined hand movement. IEEE Trans. Rehabil. Eng. **8**, 441–446 (2000)
7. Makeig, S., Kothe, C., Mullen, T., Bigdely-Shamlo, N., Zhang, Z., Kreutz-Delgado, K.: Evolving signal processing for brain–computer interfaces. Proc. IEEE **100**, 1567–1584 (2012)
8. Krusienski, D.J., McFarland, D.J., Principe, J.C.: BCI signal processing: Feature extraction. In: Wolpaw, J.R., Wolpaw, E.W. (eds.) Brain–Computer Interfaces, Principles and Practice, pp. 3–12. Oxford University Press Inc., New York (2012)
9. Cecotti, H.: Feedforward artificial neural networks for event-related potential detection. In: Arvaneh, M., Tanaka, T. (eds.) Signal Processing and Machine Learning for Brain-Machine Interfaces, pp. 173–192. The Institution of Engineering and Technology, London (2018)
10. LeCun, Y., et al.: Backpropagation applied to handwritten zip code recognition Neural Comput. **1**, 541–551 (1989)
11. Cecotti, H., Graser, A.: Convolutional neural networks for P300 detection with application to brain–computer interfaces. IEEE Trans. Pattern Anal. Mach. Intell. **33**, 433–445 (2011)
12. Schirrmeister, R.T., et al.: Deep learning with convolutional neural networks for EEG decoding and visualization. Hum. Brain Mapp. **38**(11), 5391–5420 (2017)
13. Tabar, Y.R., Halici, U.: A novel deep learning approach for classification of EEG motor imagery signals. J. Neural Eng. **14**, 016003 (2016)
14. Sturm, I., Lapuschkin, S., Samek, W., Müller, K.R.: Interpretable deep neural networks for single-trial EEG classification. J. Neurosci. Methods **274**, 141–145 (2016)
15. Lu, N., Li, T., Ren, X., Miao, H.: A deep learning scheme for motor imagery classification based on restricted Boltzmann machines IEEE. Trans. Neural Syst. Rehabil. Eng. **25**, 566–576 (2017)
16. Lotte, F., Guan, C.: Regularizing common spatial patterns to improve BCI designs: unified theory and new algorithms. IEEE Trans. Biomed. Eng. **58**, 355–362 (2011)
17. Kang, H., Nam, Y., Choi, S.: Composite common spatial pattern for subject-to-subject transfer. IEEE Signal Process. Lett. **16**, 683–686 (2009)
18. Morioka, H., et al.: Learning a common dictionary for subject-transfer decoding with resting calibration. NeuroImage **111**, 167–178 (2015)
19. Fazli, S., Popescu, F., Danóczy, M., Blankertz, B., Müller, K.R., Grozea, C.: Subject-independent mental state classification in single trials. Neural Netw. **22**, 1305–1312 (2009)

20. Cho, H., Ahn, M., Kim, K., Jun, C.S.: Increasing session-to-session transfer in a brain–computer interface with on-site background noise acquisition. J. Neural Eng. **12**, 066009 (2015)

21. Jayaram, V., Alamgir, M., Altun, Y., Scholkopf, B., Grosse-Wentrup, M.: Transfer learning in brain-computer interfaces. IEEE Comput. Intell. Mag. **11**, 20–31 (2016)

22. Kang, H., Choi, S.: Bayesian common spatial patterns for multi-subject EEG classification. Neural Netw. **57**, 39–50 (2014)

23. Azab, A., Toth, J., Mihaylova, L.S., Arvaneh, M.: A review on transfer learning approaches in brain–computer interface. In: Arvaneh, M., Tanaka, T. (eds.) Signal Processing and Machine Learning for Brain-Machine Interfaces, pp. 173–192. The Institution of Engineering and Technology, London (2018)

24. Schalk, G., McFarland, D.J., Hinterberger, T., Birbaumer, N., Wolpaw, J.R.: BCI2000: a general-purpose brain-computer interface (BCI) system. IEEE Trans. Biomed. Eng. **51**(6), 1034–1043 (2004)

25. Goldberger, A.L., et al.: PhysioBank, PhysioToolkit, and PhysioNet: components of a new research resource for complex physiologic signals. Circulation **101**(23), 215–220 (2000)

Time Removed Repeated Trials to Test the Quality of a Human Gait Recognition System

Marcin Derlatka[(⊠)] [iD]

Bialystok University of Technology,
Wiejska Street 45C, 15-351 Bialystok, Poland
m.derlatka@pb.edu.pl

Abstract. The field of biometrics is currently an area that is both very interesting as well as rapidly growing. Among various types of behavioral biometrics, human gait recognition is worthy of particular attention. Unfortunately, one issue which is frequently overlooked in subject-related literature is the problem of the changing quality of a biometric system in relation to tests that are repeated after some time. The present article describes tests meant to assess the accuracy of a human gait recognition system based on Ground Reaction Forces in time removed repeated trials. Both the initial testing as well as the repeated trials were performed with the participation of the same 40 people (16 women and 24 men) which allowed the recording of nearly 1,600 stride sequences (approximately 800 in each trial). Depending on the adopted scenario correct recognition ranged from 90.4% to 100% of cases. These results indicate that the biometric system had greater problems with recognition the longer the period of time which passed since the first trials. The present article also analyzed the impact of footwear change in the second series of testing on recognition results.

Keywords: Human gait recognition · Ground Reaction Forces · Behavioral biometrics · Repeated trials · Ensemble learning

1 Introduction

The science of biometrics concerns the identification and verification of a person on the basis of his or her physical (fingerprints, vein patterns, hand geometry) or behavioral (voice, gait, signature) characteristics. The function of biometric systems is based on the assumption that every person is unique with respect to physical and behavioral traits that do not change over time or when that change is relatively small. It is necessary to mention that physical attributes ensure a greater degree of correct recognitions but are more prone to falsification. Behavioral biometric systems, on the other hand, do not provide as high a rate of recognition but the possibility to imitate them is either very difficult or impossible.

Among types of behavioral biometrics special attention should be directed at human gait. Human gait is the most natural form of locomotion and that is the result of coordinated interaction of the skeletal, muscle and nervous systems. It is accepted that

© Springer Nature Switzerland AG 2020
K. Saeed and J. Dvorský (Eds.): CISIM 2020, LNCS 12133, pp. 15–24, 2020.
https://doi.org/10.1007/978-3-030-47679-3_2

after maturity movement patterns remain practically unchanged. In reality, the way a person moves depends on a number of factors, including:

– type of footwear being worn [5],
– physical features or circumstances such as walking speed and load bearing [10, 18, 20],
– physical and emotional state [22, 24],
– time [15].

Classification of works concerning human gait recognition in relation to the signal measured to describe this phenomenon was completed in [4]. The following ways of human gait recognition were identified within this work:

– methods using pictures from video cameras or a Kinect controller [1, 13],
– the measurement of pressure exerted by a person's foot on the ground [7, 12, 16],
– accelerometers and other wearable devices [17, 27],
– audio [8].

Subjects connected to human gait recognition which are most commonly addressed in literature include the following issues:

– the quality of a biometric system understood as its highest accuracy [9],
– preprocessing methods [14],
– forensics [19, 23].

It is worth pointing out that despite the growing scientific and utilitarian significance of biometrics as well as an increase in the number of publications dealing with this subject matter, articles connected with the effect of time on the quality of gait recognition are rare. One of the very few publications dealing with this notion is [15]. This work includes a database with several elapsed periods (0, 1, 3, 4, 5, 8 and 9 months) between gallery and probe instances. The study involved 25 people whose gait was measured in a special tunnel using 12 video cameras. Over time, the recognition rate fell approximately 5% with a portion of that decrease in the quality of a biometric system being explained through different clothes worn by participants in some series of experiments.

Pataky et al. [21] recorded the plantar pressure data of 104 people with 10 of those participants taking part in a other session which occurred from 1.5 years to 5 years before or after testing of all subjects had been completed. During the classification of this subgroup, correct recognition rate reached a level of 94% and with respect to the entire set of data from all 104 participants it was at 99.6%.

Vera-Rodriguez et al. [25] also used underfoot pressure information including GRF in the recognition of 40 people whose data was recorded in various sessions occurring over a period of 16 months. The utilization of an SVM classifier with RBF as a kernel allowed the achievement of equal error rates (EER) at levels ranging from 15.5% (with GRF only) to 2% (fusion data). Unfortunately the authors did not perform an analysis of the impact of time on results.

The aim of the present work is the study of the influence of time passage on human gait recognition based on ground reaction forces.

2   Materials and Method

2.1   Materials

Research conducted as part of the present study was carried out at the Bialystok University of Technology. The study included 40 people, 16 women and 24 men. Participants' age spanned a range of 21.3 ± 0.79 years. Their body weight was at 71.98 ± 14.45 kg and body height at 174.16 ± 9.03 cm. Their gait was recorded once more after the passage of 2 to 55 weeks and these participants can be divided into two subgroups:

– (I) those where the time between testing consisted of a few or several weeks (26 people, 9 women, 17 men) and
– (II) those where repeated tests were performed after approximately a year (14 people, 7 women and 7 men).

Changes in body weight between the two cycles of testing were slight and amounted to 0.33 ± 1.82 kg for the entire group.

In all trials, both primary and repeated, the participants walked along a measuring path along which two Kistler-made force plates were concealed. Subjects were not informed about the location of the plates. In cases where a person did not tread appropriately on the force plates, the test was repeated with a slight modification of its starting point.

Each test subject walked in their own sports shoes. For repeat testing, participants were asked to bring the same footwear which they used during the first series of the experiment. However, this turned out to be impossible for 7 people (all from subgroup II) because the footwear in question had been destroyed as a result of everyday use.

During each of the two series of tests, study participants walked along the measuring path numerous times allowing the recording of 16–25 stride sequences for each of them. In every case, after approximately 10 such walks, a short, 1–2 min break was made to avoid fatigue. In total, measurement data consisted of 1,589 stride sequences where 789 gait cycles were captured in the first series of the experiment and 800 in its second part.

2.2   Method

A method that has been described in detail in the works of [5, 6] has been used to fulfill the aims of the present article so only a brief outline of this method will be presented in this paper (Fig. 2).

1. Measurement signals obtained from force plates consisted of the three components of Ground Reaction Forces (GRF): vertical, anterior-posterior and lateral (Fig. 1). GRF is the force with which the ground acts on the lower limb of a person during the support phase of a gait. Maximum values for the vertical component Fy (Fig. 1a) correspond to the moments of: transferring the entire body weight onto the analyzed limb (first maximum—maximum of the overload phase) and the load of the forefoot (the heel is not in contact with the ground) right before the toes off (the

second maximum—maximum of propulsion). In a typical gait these maximum
values reach approximately 120% of body weight. Half way through the supporting
phase the entire active surface of the foot is in contact with the ground. This is a
period of unloading (minimum of the unloading phase) and the force value decrease
below 100% of body weight. The anterior-posterior Fx component (Fig. 1b) consists of two phases. During the first its value is negative when it is opposite to the
direction of movement. It is the result of the deceleration of the analyzed lower
limb. Similarly, during the second phase the anterior-posterior component shows
positive values. It is then that the process of acceleration begins concluded by
pushing off the ground with the toes. The value of the lateral Fz component
(Fig. 1c) depends on the limb being analyzed. Assuming that movement occurs in
the direction determined by the orientation of the Fx force than the values of the Fz
component will be positive for the right leg and negative for the left leg. The value
of the Fz force depends on the manner in which the test subject places his feet. The
values of these forces are about 10% of the body weight of the test subject.

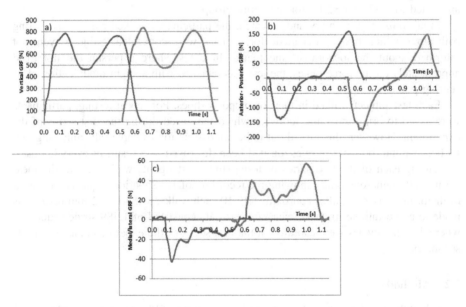

Fig. 1. The components of Ground Reaction Forces in a) vertical, b) anterior/posterior and c)
medial/lateral direction during the support phase of the left lower limb (blue line) and of the right
one (red line). (Color figure online)

2. Each component was then divided into sub-phases in accordance with divisions
   utilized in the biomechanics of human gait [26]: Loading response (LR), Mid-
   Stance (MSt), Terminal Stance (TSt) and Pre-swing (PSw).
3. To identify a person being subjected to the testing using GRF measurements, a
   similarity between that person and those contained within a data base was calculated using a dynamic time warping algorithm. Assuming that $\rho_{v,s}$ signifies the

distance between two time sequences describing GRF in the v-phase of the gait cycle for lower limb s. This distance was determined using the formula:

$$\rho_{v,s} = \sum_{m=1}^{M} DTW_m \tag{1}$$

where: DTW_m designates the distance between the two time sequences calculated for the m component of GRF. M – the number of GRF components taken into account in determining the distance, in this work $M = 3$ (all GRF components). Additionally, the total stride distance without division into phases and individual limbs is also determined. So, as a result 9 distances were obtained: $\rho_{LR,L}$, $\rho_{MSt,L}$, $\rho_{TSt,L}$, $\rho_{PSw,L}$, $\rho_{LR,R}$, $\rho_{MSt,R}$, $\rho_{TSt,R}$, $\rho_{PSw,R}$, ρ_{Stride}.

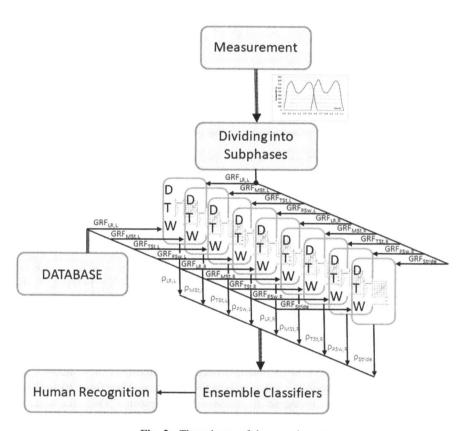

Fig. 2. The scheme of the experiment

4. The classification was carried out using ensemble classifiers with k-nearest neighbors as base classifiers. A decision for a set of classifiers was made through a weighted vote with weights based on rank order. The decision of the ensemble classifiers is a class label with the largest sum of weights:

$$cl = \arg\max\left(\sum\nolimits_{j=1}^{9} w_j \cdot d_{j,i}\right) \tag{2}$$

where: cl – class label; $d_{j,c}$ – decision of the j-th classifier, which indicates the k nearest neighbors, $d_{j,i} \in \{0,1\}$, if j-th classifier chooses class i then $d_{j,i} = 1$ otherwise $d_{j,i} = 0$; $w_j = [w_1, \ldots, w_R, \ldots, w_k]$ - weights, which are calculated from the following formula:

$$w_R = \frac{k+1-R}{k} \tag{3}$$

where: R – indicates the rank for j-th classifier, $R = \{1, 2, \ldots, k\}$;

5. The final decision was equal to 'NONE' if at least two labels had the same sum of their weights or the obtained weighted sum was lower than the assumed threshold. Within the present study, the assumed threshold value was equal to the maximum value.

3 Results and Discussion

The present work assumes the occurrence of the following four scenarios:

- scenario (a) – both the learning sequence (prototype points of kNN classifier), as well as the testing sequence, contain data recorded during the first test;
- scenario (b) – the learning sequence contains data from the first test while the testing sequence is constructed from data registered solely during the second test; this is the most realistic scenario;
- scenario (c) – both the learning sequence, as well as the testing sequence, contain data from both tests;
- scenario (d) – similar to (b) but only with respect to those people who wore the same footwear during the second test as during the primary testing. In this case, the impact of footwear change on the functioning of a biometric system is eliminated by emphasizing the influence of time on the results.

The number of gait cycles in a testing set varied and depended on the number of people considered in individual tests.

To facilitate the comparison of obtained results with those of other authors the results for a varying number of randomly selected people ranging from 10 to 40 in increments of 5 (10, 15, 20 and so forth) have been presented. Only for scenario (d) where the number of people amounted to 33 no tests for 35 and 40 people were conducted. In order to reduce the impact of randomness on study results, the tests were repeated 10 times for every group. On the basis of preliminary results, the number of considered k-nearest neighbors was equal to 9. The results presented below assume the acceptance of the most liberal strategy. Data presented in tables contain the Correct Classification Rate, the False Rejected Rate and the False Accepted Rate.

Table 1. Correct Classification Rate (CCR), False Rejected Rate (FRR) and False Accepted Rate (FAR) for the reference scenarios: (a)–(d)

Num of subj.		10	15	20	25	30	35	40
Scenario (a)	CCR	100%	100%	100%	100%	100%	100%	100%
	FRR	0%	0%	0%	0%	0%	0%	0%
	FAR	0%	0%	0%	0%	0%	0%	0%
Scenario (b)	CCR	97.88%	93.91%	92.67%	92.45%	92.24%	90.19%	90.40%
	FRR	0%	0.04%	0.03%	0.07%	0.04%	0.09%	0.07%
	FAR	2.12%	6.05%	7.30%	7.47%	7.72%	9.71%	9.52%
Scenario (c)	CCR	99.56%	99.52%	99.24%	99.19%	99.30%	99.58%	99.20%
	FRR	0%	0%	0%	0.02%	0%	0%	0.01%
	FAR	0.44%	0.48%	0.76%	0.79%	0.70%	0.42%	0.79%
Scenario (d)	CCR	96.44%	94.67%	95.91%	94.70%	94.65%	94.83%*	–
	FRR	0.03%	0.04%	0.07%	0.04%	0.07%	0.11%*	–
	FAR	3.52%	5.29%	4.03%	5.25%	5.28%	5.06%*	–

*33 subjects

Results obtained in scenario (a) are surprisingly good. Regardless of the number of people considered in the calculations the Correct Classification Rate was always 100%. This result caused the simulations for this scenario to be repeated twice but each time the effect was the same. Similar research was presented in [5] where the rise in the number of study participants caused the CCR to decrease and for 40 people it reached a level of 98.53%. However, with respect to that work, the study sample consisted of only women making the set of data more uniform. This uniformity, which causes greater difficulty in the correct identification of people, can be seen, for example, in bodyweight which in that study was 61.90 ± 11.07 kg. Although the uniformity of data in the selection of study subjects in the research conducted as part of the present work was also quite large, the variability of the body weight component was higher since, as mentioned above, its standard deviation was equal to nearly 14.5 kg. It must be mentioned, however, that this fact does not diminish the significance of presented research because the results obtained in [5] could be treated as underestimated in comparison to studies done on a standard group of participants.

The results of scenario (b) show a relatively fast decrease in the CCR that is proportional to the rise in the number of people considered in the experiment. Although this decrease is quite large it is not as drastic as that occurring after a change of footwear from sports shoes to high heels for women in [5]. On this basis, it should be concluded that psychophysical factors play an important role in the way a person moves but not as large as a change in footwear.

A more detailed result analysis allowed the observation that CCR decreases along with the expansion of the time period between the two series of testing. In respect to subgroup (I), for which time between the primary series of testing and the repeated testing was at most only several weeks, CCR reached 93.79%. However, for subgroup (II) where the two series of testing were separated by about a year, CCR was 83.56% (Fig. 3a). It is worth mentioning that if only those people who were retested within a month of primary tests would be considered then a rise in CCR to 96.71% would be seen provoking the conclusion that when it comes to a working biometric system it is necessary to systematically update its database of patterns.

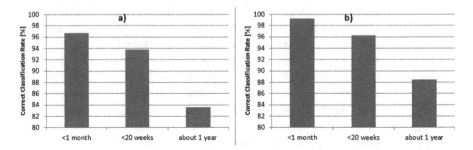

Fig. 3. Correct Classification Rate depend on time removed repeated trials for: a) subgroup (I), b) subgroup (II)

The confirmation of this conclusion can be found in the results of scenario (c) in which the learning sequence data contained recorded stride sequences of people from both series of the experiment. In this case, a very high level of correct recognitions exceeding 99% was achieved. This value falls slightly along with a rise in the number of considered people.

In the description of people taking part in the study, attention was drawn to the fact that not all people who participated in the second set of tests occurring after approximately a year wore the same footwear. Undoubtedly, the fact that different shoes were used impacts the results of the identification [2, 3, 11]. Thanks to this, a decrease in the FAR of nearly half was observed while the CCR rose to 94.83% with respect to a group of 33 people. It is worthy of notice (Fig. 3b) that in this scenario, similar to scenario (b), the CCR value falls along with the passage of time between the two series of tests. It should be mentioned that the obtained results confirm those of other works which, despite the use of different signals to describe the stride of a person, also saw a fall in the CCR of about 5–6% [15, 21]. With respect to people who were tested the second time within 1 month of the primary tests the accuracy of recognition was as high as 99.23% coming considerably closer to the result reached in scenario (a).

It is also worth noticing that FAR values (Table 1) are relatively high as a result of the assumption of a high threshold value (see point 5, Subsect. 2.2). By decreasing this value a lower FAR value will be seen but it will also result in higher FRR and a fall in the CCR.

4   Conclusions

This article presents the quality of a human gait recognition system based on GRF signals generated during time removed repeated gait trials for the same group of people. Repeat testing was carried out after the passage of several weeks. Generally, the obtained results are very good and show the high potential of biometric systems of this type. It has also been shown that the passage of time impacts results to a lesser degree than a change of footwear that a participant wears during testing. It has been concluded that in practical applications it is necessary to update pattern databases describing the manner in which a particular person moves with newly recorded and correctly identified patterns.

Further work within this scope can be carried out in two directions. First, they should focus on increasing the number of people who take part in the experiments which, in consequence, will facilitate a more detailed analysis of results. Additionally, testing of the same people should be repeated several times and not only done twice.

Acknowledgments. This work is supported by research grant no. WZ/WM-IIB/1/2020 of the Institute of Biomedical Engineering, Bialystok University of Technology.

References

1. Bari, A.H., Gavrilova, M.L.: Artificial neural network based gait recognition using kinect sensor. IEEE Access **7**, 162708–162722 (2019)
2. Connie, T., Goh, M., Ong, T.S., Toussi, H.L., Teoh, A.B.J.: A challenging gait database for office surveillance. In: Proceedings of the 2013 6th International Congress on Image and Signal Processing (CISP), Hangzhou, China, vol. 3, pp. 1670–1675 (2013)
3. Connor, P.C.: Comparing and combining underfoot pressure features forshod and unshod gait biometrics. In: Proceedings of the 2015 IEEE International Symposium on Technologies for Homeland Security (HST), Waltham, MA, USA, pp. 1–7 (2015)
4. Connor, P., Ross, A.: Biometric recognition by gait: a survey of modalities and features. Comput. Vis. Image Underst. **167**, 1–27 (2018)
5. Derlatka, M., Bogdan, M.: Recognition of a person wearing sport shoes or high heels through gait using two types of sensors. Sensors **18**(5), 1639 (2018)
6. Derlatka, M., Bogdan, M.: Combining homogeneous base classifiers to improve the accuracy of biometric systems based on ground reaction forces. J. Med. Imaging Health Inf. **5**(8), 1674–1679 (2015)
7. Derlatka, M., Bogdan, M.: Ensemble kNN classifiers for human gait recognition based on ground reaction forces. In: 2015 8th International Conference on Human System Interaction (HSI), Warsaw, pp. 88–93. IEEE (2015)
8. Geiger, J.T., Kneißl, M., Schuller, B.W., Rigoll, G.: Acoustic gait-based person identification using hidden Markov models. In: Proceedings of the 2014 Workshop on Mapping Personality Traits Challenge and Workshop, Istanbul, Turkey, pp. 25–30. ACM, New York (2014)
9. Guan, Y., Li, C.T., Roli, F.: On reducing the effect of covariate factors in gait recognition: a classifier ensemble method. IEEE Trans. Pattern Anal. Mach. Intell. (T-PAMI) **37**, 1521–1528 (2015)

10. Jordan, K., Challis, J.H., Newell, K.M.: Walking speed influences on gait cycle variability. Gait Posture **26**(1), 128–134 (2007)
11. Kim, M., Kim, M., Park, S., Kwon, J., Park, J.: Feasibility study of gait recognition using points in three-dimensional space. Int. J. Fuzzy Log. Intell. Syst. **13**, 124–132 (2013)
12. Li, Y., et al.: A convolutional neural network for gait recognition based on plantar pressure images. In: Zhou, J., et al. (eds.) CCBR 2017. LNCS, vol. 10568, pp. 466–473. Springer, Cham (2017). https://doi.org/10.1007/978-3-319-69923-3_50
13. Lv, Z., Xing, X., Wang, K., Guan, D.: Class energy image analysis for video sensor-based gait recognition: a review. Sensors **15**, 932–964 (2015)
14. Mason, James Eric, Traoré, Issa, Woungang, Isaac: Machine Learning Techniques for Gait Biometric Recognition. Springer, Cham (2016). https://doi.org/10.1007/978-3-319-29088-1
15. Matovski, D.S., Nixon, M.S., Mahmoodi, S., Carter, J.N.: The effect of time on gait recognition performance. IEEE Trans. Inf. Forensics Secur. **7**(2), 543–552 (2012)
16. Moustakidis, S.P., Theocharis, J.B., Giakas, G.: Subject recognition based on ground reaction force measurements of gait signals. IEEE Trans. Syst. Man. Cybern. B Cybern. **38**, 1476–1485 (2008)
17. Muaaz, M., Mayrhofer, R.: Accelerometer based gait recognition using adapted gaussian mixture models. In: Proceedings of the 14th International Conference on Advances in Mobile Computing and Multi Media, pp. 288–291 (2016)
18. Nakajima, K., Mizukami, Y., Tanaka, K., Tamura, T.: Footprint-based personal recognition. IEEE Trans. Biomed. Eng. **47**(11), 1534–1537 (2000)
19. Nixon, M.S., Imed, B., Arbab-Zavar, B., Carter, J.N.: On use of biometrics in forensics: gait and ear. In: European Signal Processing Conference, Aalborg, pp. 1655–1659 (2010)
20. Qu, X., Yeo, J.C.: Effects of load carriage and fatigue on gait characteristics. J. Biomech. **44**(7), 1259–1263 (2011)
21. Pataky, T.C., Mu, T., Bosch, K., Rosenbaum, D., Goulermas, J.Y.: Gait recognition: highly unique dynamic plantar pressure patterns among 104 individuals. J. R. Soc. Interface **9**, 790–800 (2011)
22. Roether, C.L., Omlor, L., Christensen, A., Giese, M.A.: Critical features for the perception of emotion from gait. J. Vis. **9**(6), 15 (2009)
23. Seckiner, D., Mallett, X., Maynard, P., Meuwly, D., Roux, C.: Forensic gait analysis— morphometric assessment from surveillance footage. Forensic Sci. Int. **296**, 57–66 (2019)
24. Sloman, L., Berridge, M., Homatidis, S., Duck, T.: Gait patterns of depressed patients and normal subjects. Am. J. Psychiatry **139**(1), 94–97 (1982)
25. Vera-Rodríguez, R., Mason, J.S.D., Fierrez, J., Ortega-Garcia, J.: Comparative analysis and fusion of spatiotemporal information for footstep recognition. IEEE Trans. Pattern Anal. Mach. Intell. **35**(4), 823–834 (2013)
26. Whittle, M.W.: Gait analysis: an introduction. Butterworth-Heinemann, Edinburgh (2014)
27. Yang, G., Tan, W., Jin, H., Zhao, T., Tu, L.: Review wearable sensing system for gait recognition. Cluster Comput. **22**(2), 3021–3029 (2019)

Spiral-Based Model for Software Architecture in Bio-image Analysis: A Case Study in RSV Cell Infection

Margarita Gamarra[1]([envelope]) [ID], Eduardo Zurek[2] [ID], Wilson Nieto[2] [ID],
Miguel Jimeno[2] [ID], and Deibys Sierra[3]

[1] Universidad de la Costa, CUC, Barranquilla, Colombia
mgamarra3@cuc.edu.co
[2] Universidad del Norte, Barranquilla, Colombia
[3] Universidad del Magdalena, Santa Marta, Colombia

Abstract. The advancement in biological and medical image acquisitions has allowed the development of numerous investigations in different fields supported by image analysis, from cell to physiological level. The complexity in the treatment of data, generated by image analysis, requires a structured methodology for software development. In this paper we proposed a framework to develop a software solution with a Service-Oriented Architecture (SOA) applied to the analysis of biological images. The framework is completed with a novel image analysis methodology that would help researchers to achieve better results in their image analysis projects. We evaluate our proposal in a scientific project related to cell image analysis.

Keywords: Spiral methodology · Bio-image informatics · Cell image processing · Respiratory Syncytial Virus

1 Introduction

Current researches in biology include both quantitative and qualitative analysis, which are important for clinical applications and biological research. In these studies, the amount of information and metadata contained in a single sample is large.

The introduction of new models, measurements and methods has produced an everyday increasing amount of data using image-based evidence [1]. However, the raw data and the extracted information of an image are difficult to organize, manage, process and analyze. As a natural extension of the existing biomedical image analysis field, the Bio-image informatics develops and uses various image data analysis and informatics techniques to extract, compare, search and manage the biological knowledge of the respective images [1].

There are many software tools useful to the researchers, which include several image processing techniques and data analysis. In the biological image case some problems arise: techniques are limited and do not work well in all situations, the

© Springer Nature Switzerland AG 2020
K. Saeed and J. Dvorský (Eds.): CISIM 2020, LNCS 12133, pp. 25–38, 2020.
https://doi.org/10.1007/978-3-030-47679-3_3

amount of images and the large data and metadata generated require computational resources with high processing capacity, the parameters selection are unknown for biologist and it is necessary to create a software program to improve the established techniques (it requires an expert in image processing). In addition, although the stages in the image processing are well defined (acquisition, segmentation, feature extraction and classification), implementing a software development methodology is necessary in order to create a framework and improve the results according to the specific issue. Furthermore, the joint work between biologist and software developers is fundamental to obtain a product with high quality and requirements fulfillment.

Our primary aim in this paper is to propose a software methodology which meets the needs of biologists, scientists and engineers to extract useful and novel knowledge from biological images, using advanced algorithms and a service oriented software [2]. Subsequently, we proposed a framework for software development, applied to bio-image analysis, which included a Spiral-based model with a Service Oriented Architecture, which was exposed in [3]. This complete framework for Bio-image informatics has not been proposed in the literature. In this way the main features of our proposal are:

A Spiral-Based Methodology for Software Development: This model is adequate for large projects, like cell image informatics, and its flexibility allows changes to be implemented at several stages of the project. This methodology allows incorporating the step of the spiral methodology to the stages of digital image processing. The main advantage of this methodology is that the client (biologists, scientists or laboratory technicians), who will be involved in the development of each segment, retains control over the direction and implementation of the project.

Service-Oriented Architecture: It allows a configurable and elastic hosting in the web and it can be deployed as a Software-as-a-Service. SOA allows defining integration architectures based on the concept of a service for Bio-image informatics. An advantage that SOA offers is its design, based on layers, which allows the improvement of certain parts of the system without affecting the rest [4]. It also allows modularity, and if correctly implemented, allows scalability of their parts as the project requires.

Furthermore we expose a cell image analysis use case: The bio-image informatics research scenario we chose is cell image analysis, with a specific application to analyze the cells infected by Respiratory Syncytial Virus (RSV), research supported by the National Institutes of Health under award number R01AI110385. This field is widely studied because of the information obtained of a cell image is highly significant in the disease research.

2   Related Work

Many biological image management tools have been built for specialized domains. We will focus on the platforms and software related with cell image analysis, software development models and architectures in bio-image informatics.

The authors of [5] have an interesting review about open software dedicated to analysis of scientific images. They part from the NIH Image and ImageJ as the first and most popular ones, from which many have proposed new open applications. ImageJ [6] is a Java based cross-platform tool for biomedical image processing and measurement. Two of the reasons that helped its popularity were the support for 32 different image formats and the fact that it was cross-platform.

Some of the popular open image processing tools are described now, some of which have evolved from ImageJ. A number of image analysis toolboxes were developed by various research groups. ITK [7] is an open-source, cross-platform system that provides developers with an extensive suite of software tools for image analysis. ITK employs leading-edge algorithms for registering and segmenting multi-dimensional data, and focuses on 3D medical data segmentation and registration algorithms.

CellProfiler [8] is a free open-source software designed to enable biologists without training in computer vision or programming to quantitatively measure phenotypes from thousands of images automatically. Authors in [9] developed an image analysis plat-form named YeastQuant to simplify data extraction by offering an integrated method to turn time-lapse movies into single cell measurements. The database is connected to the engineering software Matlab®, which allows extracting the desired information by automatically segmenting and quantifying the microscopy images. The authors claimed that their main contributions were the ideas of integrating different segmentation methods, enabling automated data extraction, and annotating and storing the results in a single database for future reference.

Some Platform-as-a-Service (PaaS) has been developed, integrating cell image processing with web services. For example, the UCSB Bisque [10] system provides an integrated online environment for users to upload, search, edit and annotate images. Although it is not specialized in cell images, it includes a few analysis and visualization modules for this field.

Authors in [11] propose a science-oriented model named Butterfly Model. It consists of four wings: Scientific Software Engineering, Human Computer Interaction, Scientific Methodology and Scientific Application. Moreover it leads to continuous improvement. The achievements translate the goals into software. The Butterfly model implementation mechanism has a three-layer architecture: Gray (the abstract layer), Yellow (the basis for design and development), and Green (the implementation and testing by the user). In this paper some case studies are presented.

Another approach of a methodology applied to bioinformatics is presented in [12]. In this study, the application of Aspect-Oriented Programming (AOP) methodology has been investigated in the development of Bioinformatics Software – Bioseqsearch, using Eclipse-AJDT environment. The study concludes that AOP methodology in Eclipse-AJDT environment is highly useful in design and implementation of efficient, cost-effective and quality bioinformatics software projects.

The research developed in [13] suggests the use of Service Oriented Architecture (SOA) to integrate biological data from different data sources. This work shows that SOA solves the problems of facing integration process and accessing biological data in an easier way. The Microsoft .Net Framework was used to implement the proposed architecture.

Authors in [14] presented a web services based platform focused on the cell counting problem. Using OpenCF, a web services development framework, authors integrate in a single platform services oriented to image processing and classifying, cell counting based on a set of parameters, and data post-processing (plot generation, datasheets, etc.). A GUI added to the platform helps to launch jobs with image sets, and the execution of different tasks from a web service based client.

The software developed in [15] LOBSTER (Little Objects Segmentation & Tracking Environment), an environment designed to help scientists design and customize image analysis workflows to accurately characterize biological objects from a broad range of fluorescence microscopy images, including large images exceeding workstation main memory.

Despite the wide software development in the bio-image processing and analysis, to the best or our knowledge, no one methodology has been proposed for the software engineering applied to biological image analysis. Our proposal improves the project management, as it introduces the use of the modified spiral methodology, which ensures that the right algorithms are used in the most appropriate way.

The proposed Spiral-based methodology allows the incorporation of the different stages of image processing to the phases that are defined in this methodology: analysis, evaluation, modeling and development. This is why it can be used in multiple image processing projects; this model has been successfully applied in the image analysis of infected cells by RSV (Respiratory Syncytial Virus).

3 Proposed Framework

In this paper we have proposed a framework for software development in Bio-image informatics. In this proposal not only processes are handled but this framework includes three main components:

- A robust methodology with layers for abstraction, design, implementation and quality control. A spiral-based methodology is merged with the basic stages of image processing and adapted to Bio-image informatics.
- A contextualized architecture divided by layers and service-oriented.
- Techniques and artifacts specified for each phase in the process.

In Fig. 1 the framework is summarized with the three components, which are related to each other. The methodology and the architecture are exposed in the next sections.

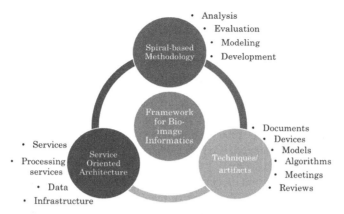

- Analysis
- Evaluation
- Modeling
- Development

Spiral-based Methodology

Framework for Bio-image Informatics

- Services
- Processing services
- Data
- Infrastructure

Service Oriented Architecture

Techniques/ artifacts

- Documents
- Devices
- Models
- Algorithms
- Meetings
- Reviews

Fig. 1. Framework for software development in bio-image informatics.

4   Adapted Methodology for Biological Image Analysis

In this section we present a Progressive Spiral Methodology (PSM) modified to analyze images adaptively according to the research requirements. This model reflects the underlying concept that each cycle involves a progression that addresses the same sequence of steps, for each portion of the product and for each of its levels of elaboration, from an overall concept of operation document down to the coding of each individual program [16].

An important feature of the spiral model is that each cycle is completed by a review involving the primary people or organizations concerned with the product. Another advantage is the ease to complete the process in any stage, this means that the final objective is not the classification; it could be only improving quality, segmentation, cells counting, extracting features to realize statistical analysis or finally a classification. Additionally, the applied methodology is progressive due to the fact that the process is developed gradually or in stages, proceeding step by step. Then, the proposed Spiral-based methodology is suitable for software development in biological imaging allowing the incorporation of the different stages of image processing to the phases that are defined in this methodology.

4.1   Stages for the Biological Image Processing Component

The main stages in a digital image processing are:

- Acquisition: Before any image processing, an image must be captured by a device and converted into a manageable entity. The image acquisition process consists of three steps: energy reflected from the object of interest, an optical system which focuses the energy and finally a sensor which measures the amount of energy [17].
- Pre-processing stage, which usually implies several steps, as explained by [18], like filtering, image enhancement and segmentation.

- Feature extraction deals with extracting attributes that result in some quantitative information of interest or are basic for differentiating one class of objects from another [18, 19].
- Classification is the process that assigns a label to an object based on its descriptors [18]. Features like computational cost, hit rate and evaluation time are compared between the classifiers.

All the stages are included in the proposed methodology. If they are correctly followed, the image analysis process is going to be successful. The purpose of the methodology is helping inexperienced researchers, and researchers willing to get better results of their analysis.

We have adapted the spiral model to our proposal, merging the spiral form with the stages for image processing. In the center of the spiral is the core of the project, which includes the definition of the main goal, a global vision of the entire plan, restrictions and requirements, related works and resources. This preceding phase can avoid delays or going back to a previous stage during the project. The proposed methodology is shown in Fig. 2, and it is composed by 4 main phases:

(1) Analysis: in this phase the development team exposes the objectives, alternatives and constrains. Each stage of the image processing has a different analysis and it is important the discussion between the development team and the expert biologists.
(2) Evaluation: in this phase the possibilities and constrains exposed in the previous phase are evaluated, the risks are identified and the suitable solution is selected. Likewise, the expert opinion is significant.
(3) Modeling: in this phase the objects of study (images, pixels, features, and vectors) are the inputs to the model, which is a mathematical representation to perform the processing in each stage (acquisition, preprocessing, feature extraction and classification). The verification is important in this phase because the modeling leads to the implementation.
(4) Development: in this phase the algorithms or the techniques are implemented. In many cases a machine with a high processing capacity is required, because of the size and the amount of the images to process and the complex mathematical operations. This is a strong reason to employ cloud computing for cell images processing. Once again, the verification and assessing of the health or biology expert is essential in this phase.

According to the CMMI® (Capability Maturity Model® Integration), the verification and validation process must be included in the software development projects. The purpose of Verification is to ensure that selected work products meet their specified requirements [20]. Verification is inherently an incremental process because it occurs throughout the development of the product and work products, beginning with verification of requirements, progressing through the verification of evolving work products, and culminating in the verification of the completed product [20].

In our proposal, for each phase, an expert verification and assessing process is added, it aims to guarantee the results and to involve the expert judgment. In the modeling phase, only a process of verification is necessary, because the mathematical methods used for processing do not involve health or biology experts. Furthermore, when each stage has finished, a validation process is performed in order to guarantee compliance of objectives.

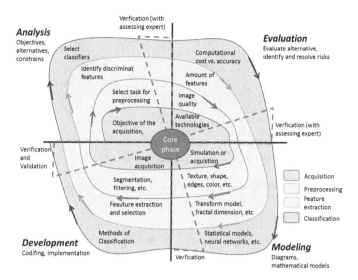

Fig. 2. Progressive Spiral Methodology (PSM) adapted to bio-image analysis.

5   Service-Oriented Architecture

The proposed architecture is presented in Fig. 3. The top layer corresponds to the Services layer. These services are divided in two types: user and internal services. The User Services facilitates the functionalities offered by the platform related with the image management and selection of options. The Internal Services are used by the development and management team. In this layer we have included an Enterprise Service Bus (ESB) to enable a layer of communication between the services. An ESB typically allows an easy integration between different components used in the architecture. This is useful if for example the implementers of this architecture decide to use a multiplicity of providers for the listed services.

The second layer is an extension of the first layer, specifying the processing services, corresponding to the stages of the image processing. It is possible that a user needs only one of these services or requires a group of these, aiming to analyze a set of cell images.

The Data layer corresponds to the needed information to perform the image analysis. These data are provided by the user or are taken of our database when the user requires it. This is a very interesting component of the architecture. If the cloud is chosen as the base for the implementation, the data could be shared with multiple open image analysis projects stored elsewhere. It is even possible to replace the layer with a Big Data implementation with important resources like metadata, patterns, models and image files. The bottom layer shows the two main components of the infrastructure that supports our platform: a database available for users and the Platform-as-a-Service deployed in the cloud.

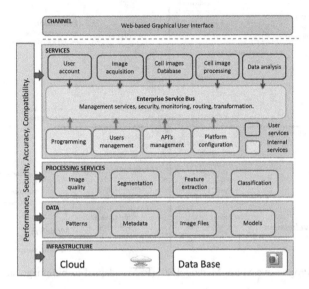

Fig. 3. Proposed multilayered architecture.

The scheme shows on top of the layers the channel used to offer the services to the users. A web-based graphical user interface is the best option when a non-expert in image processing requires the services [21]. There are several methods to implement SOA but web service is the most popular method [13]. A web-based environment enables scientists to more effectively define a task, perform the task at a desired time, monitor the execution status, and view the results [22]. Additionally the SOA covers the requirement of usability, accessibility, configurability and scalability.

Furthermore, these layers are affected by the non-functional requirements: security with the treatment of the images, high performance to process the number of images and data, accuracy in the results and compatibility with the format of the data.

Finally, in the scheme shown in Fig. 3, it can be clearly seen that this architecture is closely related with the stages proposed in the Spiral-based methodology: the Processing Services layer and the Services layer are adjusted with the processing stages (acquisition, preprocessing, feature extraction and classification). The Data layer exposes all the data used during the different phases and stages in the methodology. The infrastructure and the channel support the four phases proposed for the methodology (Analysis, Evaluation, Modeling and Development), since these phases can be performed in the cloud, using a data base and through a web-based graphical user interface.

6 Case Study: Analysis of Cell-to-Cell Variability in RSV Infection

Figure 4 shows the domain model for the image analysis component of the project. It is necessary to highlight three main classes from the model: the Preprocessor, the Feature Extractor, and the Classifier, which correspond to the main steps of the image analyzing process. Those are interfaces which are implemented by other classes. Another interesting point to highlight is the implementation of multiple types of classifiers that utilize the Classifier interface. This enables the implementation of multiple options in the image analysis methodology and gives the researcher options that are not easily found in other open image analyzing tools in the literature.

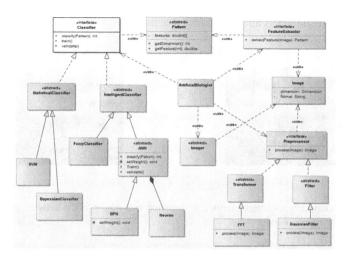

Fig. 4. Domain model for image processing projects.

The proposed methodology is applied to the study of human cells infected with RSV (Respiratory Syncytial Virus). This research aims to develop a morphological characterization and statistical analysis of image cell under laboratory conditions, to relate the population context features with the probability of infection in an individual cell, which contribute to the understanding of mechanisms of entry of RSV to the body human. The used image database correspond to Human epithelial type 2 (HEp-2) cells infected with Respiratory Syncytial Virus (RSV). The set of images contains different density of cells seeded, which allows a more complete analysis of the viral infection. The image database is a key point for the design of the SOA infrastructure, but it is also a point where the cloud architecture can show its advantages. The cloud can escalate up or down the database services as the image database increases or decreases. If a well-funded image research project is started, services could be escalated and optimized as needed.

In the initial Core phase of the project the project manager and the development team had a meeting with the experts in virology: a medical virologist, a biologist and the laboratory technicians. They exposed the goals, the restrictions (equipment, devices, cell samples, amount and quality of the images, time and available budget) and requirements (security in the platform and data management, image quality, big-data processing, usability of the platform, accuracy for the results, availability database, compatibility). In this stage the researches give a global approach about the development of the project and it is necessary a review of the state-of-the-art.

The four phases described in Sect. 4.1 are applied to the RSV research. The spiral-based methodology is applied in each stage of the project. The global vision at the start of the project and the recommended techniques give a benchmark for the development team. The final selection is defined by the performance of the algorithms.

In Fig. 5 an example of the acquired images is shown. For this first stage a fluorescence microscope (Reference: Axio Observer, Zeiss) was used and the goal was reached (visualize infected and non-infected cells).

a)   b)

Fig. 5. a) Infected cells in green channel. b) Infected and non-infected cells in green and red channels. (Color figure online)

In this case of study, the preprocessing phase was performed by the segmentation algorithm, SM-Watershed, developed in [23]. The method combines the Marker-Controlled Watershed algorithm with a two-step method based on Watershed, Split and Merge Watershed (SM-Watershed): in the first step, or split phase, the algorithm identifies the clustered cells using inherent characteristics of the cell, such as size and convexity, and separates them using watershed transform. In the second step, or the merge stage, it identifies the over-segmented regions using proper features of the cells and eliminates the divisions. The flow chart of the SM-Watershed is shown in Fig. 6. The results of the algorithm were satisfactory, as it achieve a high performance in the cell segmentation for the dataset Hep-2.

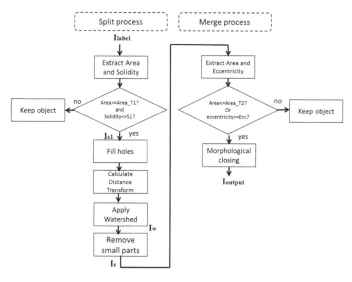

Fig. 6. Flowchart of the cell segmentation algorithm.

In a meeting between the development team and the biologist experts, five different cell context features were selected to associate them with the virus infection. These features are related with microenvironment and cell state:
Microenvironment:

(1) Population Size: (POP.SIZE) the number of cells in each spot was counted based on the segmentation results.
(2) Edge: The location of cells within a local population (EDGE) (center or edge of a local cell population).
(3) Local cell density (LCD) was estimated using a Gaussian kernel density estimator based on nuclei centers.

Cell morphology:

(4) Size (SIZE) of each cell was directly computed from the cell segmentation method. It is the number of pixels in the region.
(5) Circularity: (CIRC) this is a measurement for eccentricity: the ratio of the distance between the foci of the ellipse and its major axis length.

These five features are representative of the population context and the shape of the cells. The feature extraction required of the segmentation stage to obtain characteristics cell-by-cell. Other internal features of the cells could affect the virus infection, but they are not visible with the fluorescence microscopy technology used in this study.

Finally, the last stage was the statistical analysis. The objective of the statistical analysis in this research was to determine the relation, between the features (microenvironment and cell state) extracted from each cell and the Syncytial Respiratory Virus infection. A logistic regression analysis was conducted to analyze the relationship between the RSV infections in Hep-2 cells with the five predictors.

With the aim to consolidate the segmentation algorithms and the feature extraction for the final user, we developed software named ALICIA® (Assistant for Laboratory Investigations in Cell Image Analysis). This software was developed in Matlab® platform and it includes a graphical user interface, image import, several algorithms for segmentation, feature extraction, data exportation, data organization and search, and statistical description.

7 Conclusions

Projects of software development applied to biological image analysis require a proper methodology as frequently their duration is long, the results are unknown and the communication between the engineering team and the biologist or medical team is essential. These factors make that the spiral methodology is appropriate.

Although the image-based diagnostic is finally carried out by human experts, it is necessary to use automatic processing due to the large data that the images contain. Then, software to process and analyze the images is necessary in the biomedical field. In this article we proposed a methodology for software development of projects in bio-image analysis. The proposed spiral-based methodology provide a guide for the development of bio-images researches, where the joint work between computer sciences and life sciences is fundamental for the success of the project. The methodology is based on the four main stages of the digital image processing (acquisition, preprocessing, feature extraction and classification) which follow four phases (analysis, evaluation, modeling and development). This spiral-based methodology was adapted to the RSV project and the results were satisfactory with the application of the four phases proposed in the methodology: analysis, evaluation, modeling and development.

Additionally a SOA was proposed as a core infrastructure from which a bio-image analysis project could build a scalable base. As a continuation of this work, ALICIA will include the whole set of proposed services and specialized algorithms for cell image processing. Other platforms in the cloud could be used to support big data and high processing requirements.

References

1. Peng, H.: Bioimage informatics: a new area of engineering biology. Bioinformatics **24**, 1827–1836 (2008). https://doi.org/10.1093/bioinformatics/btn346
2. Yao, J., Zhang, J., Chen, S., Wang, C., Levy, D., Liu, Q.: A mobile cloud with trusted data provenance services for bioinformatics research. In: Liu, Q., Bai, Q., Giugni, S., Williamson, D., Taylor, J. (eds.) Provenance and Data Management in eScience. SCI, vol. 426, pp. 109–128. Springer, Heidelberg (2013). https://doi.org/10.1007/978-3-642-29931-5_5
3. Gamarra, M., Zurek, E., Nieto, W., Jimeno, M., Sierra, D.: A service-oriented architecture for bioinformatics: an application in cell image analysis. In: Rocha, Á., Correia, A.M., Adeli, H., Reis, L.P., Costanzo, S. (eds.) WorldCIST 2017. AISC, vol. 569, pp. 724–734. Springer, Cham (2017). https://doi.org/10.1007/978-3-319-56535-4_71

4. Zorrilla, M., García-Saiz, D.: A service oriented architecture to provide data mining services for non-expert data miners. Decis. Support Syst. **55**, 399–411 (2013). https://doi.org/10.1016/j.dss.2012.05.045
5. Schneider, C.A., Rasband, W.S., Eliceiri, K.W.: NIH Image to ImageJ: 25 years of image analysis. Nat. Methods **9**, 671–675 (2012). http://www.ncbi.nlm.nih.gov/pubmed/22930834. Accessed 11 Sept 2017
6. Abramoff, M.D., Magalhães, P.J., Ram, S.J.: Image processing with ImageJ. Biophotonics Int. **11**, 36–42 (2004). http://dspace.library.uu.nl/handle/1874/204900. Accessed 12 Apr 2016
7. Yoo, T.S., Ackerman, M.J., Lorensen, W.E., Schroeder, W., Chalana, V., Aylward, S., et al.: Engineering and algorithm design for an image processing API: a technical report on ITK– the Insight Toolkit. Stud. Health Technol. Inform. **85**, 586–592 (2002)
8. Carpenter, A.E., Jones, T.R., Lamprecht, M.R., Clarke, C., Kang, I.H., Friman, O., et al.: CellProfiler: image analysis software for identifying and quantifying cell phenotypes. Genome Biol. **7**, R100 (2006). https://doi.org/10.1186/gb-2006-7-10-r100
9. Pelet, S., Dechant, R., Lee, S.S., van Drogen, F., Peter, M.: An integrated image analysis platform to quantify signal transduction in single cells. Integr. Biol. (Camb). **4**, 1274–1282 (2012). https://doi.org/10.1039/c2ib20139a
10. Kvilekval, K., Fedorov, D., Obara, B., Singh, A., Manjunath, B.S.: Bisque: a platform for bioimage analysis and management. Bioinformatics **26**, 544–552 (2010). https://doi.org/10.1093/bioinformatics/btp699
11. Ahmed, Z., Zeeshan, S., Dandekar, T.: Developing sustainable software solutions for bioinformatics by the "Butterfly" paradigm. F1000Research **3**, 71 (2014). https://doi.org/10.12688/f1000research.3681.2
12. Sharma, A., Vidyapeeth, J.R.N.R.: Application of AOP methodology in eclipse-AJDT environment for developing bioinformatics software (n.d.). http://www.ijcaonline.org/icwet/number15/SE243.pdf. Accessed 11 Sept 2017
13. Al-Otaibi, N.M., Noaman, A.Y.: Biological data integration using SOA. Int. J. Comput. Electr. Autom. Control Inf. Eng. **5**, 74–79 (2011)
14. Castillo, J.C., Almeida, F., Blanco, V., Ramírez, M.C.: Web services based platform for the cell counting problem. In: Lopes, L., et al. (eds.) Euro-Par 2014. LNCS, vol. 8805, pp. 83–92. Springer, Cham (2014). https://doi.org/10.1007/978-3-319-14325-5_8
15. Tosi, S., Bardia, L., Filgueira, M., Calon, A., Colombelli, J.: LOBSTER: an environment to design bioimage analysis workflows for large and complex fluorescence microscopy data. Bioimage Inform. **36**(8), 2634–2635 (2019). https://doi.org/10.1093/bioinformatics/btz945
16. Boehm, B.W.: A spiral model of software development and enhancement. Comput. (Long. Beach. Calif.). **21**, 61–72 (1988). https://doi.org/10.1109/2.59
17. Moeslund, T.B.: Image Acquisition, pp. 7–24 (2012). https://doi.org/10.1007/978-1-4471-2503-7_2
18. Gonzalez, R.C.: Digital Image Processing. Pearson Education, Upper Saddle River (2009)
19. Oliveira, R.B., Papa, J.P., Pereira, A.S., Tavares, J.M.R.S.: Computational methods for pigmented skin lesion classification in images: review and future trends. Neural Comput. Appl., 1–24 (2016). https://doi.org/10.1007/s00521-016-2482-6
20. CMMI® for Development, Version 1.3—CMMI Institute. CMMI Institute (2010)
21. González-Castaño, D.M., Pena, J., Gómez, F., Gago-Arias, A., González-Castaño, F.J., Rodríguez-Silva, D.A., et al.: eIMRT: a web platform for the verification and optimization of radiation treatment plans. J. Appl. Clin. Med. Phys. **10**, 2998 (2009)

22. Xiang, X.: Service-oriented architecture for integration of bioinformatic data and applications. University of Notre Dame (2007)
23. Gamarra, M., Zurek, E., Escalante, H.J., Hurtado, L., San-Juan-Vergara, H.: Split and merge watershed: a two-step method for cell segmentation in fluorescence microscopy images. Biomed. Signal Process Control (53) (2019). https://doi.org/10.1016/j.bspc.2019.101575

Artificial Intelligence System for Drivers Fatigue Detection

Waldemar Karwowski(✉) , Przemysław Reszke , and Marian Rusek

Institute of Information Technology, Warsaw University of Life Sciences—SGGW,
ul. Nowoursynowska 159, 02-776 Warsaw, Poland
{waldemar_karwowski,marian_rusek}@sggw.pl

Abstract. Driver drowsiness is one of major causes of growing number of road accidents. To combat this problem car makers install their proprietary and expensive driver alert systems. In this paper an analogous system based on an opensource machine learning library is presented. The proper eye aspect ratio for closed eyes is discussed and estimated. The accuracy of closed eyes recognition is tested on a basis of several public available face libraries.

Keywords: Driver alert · Drowsiness detection · Closed eyes

1 Introduction

Driver drowsiness and fatigue cause many road accidents and have had a significant impact on the safety of all road users. According to Polish police annual report, in 2018, 533 road accidents in which 76 people were killed were caused by drivers fatigue or drivers falling asleep[1]. The same is true in other countries of the world, for example in 2015, 2.3% (824) of the fatalities that occurred on U.S. roadways are reported to have involved drowsy driving. In 2015, the total number of fatalities increased by 7% compared to 2014[2]. Many accidents could have been avoided if the drowsy driver had been warned in proper time. To do that it is necessary to monitor the physical and mental state of driver. In other words driver fatigue detection method is required. Nowadays some modern cars have advanced driver-assistance systems (ADAS), which help the vehicle driver during driving or parking. Designed with a proper human-machine interface they are intended to increase car safety and more generally road safety. The probably best known system of this kind is Tesla Autopilot, which contains an adaptive cruise control, lane-keep assistance and forward collision warnings. Very advanced but less known is Super Cruise, the highly automated system developed by General Motors. In this system the driver should observe the road while driving, otherwise when the driver looks away or falls asleep, the system triggers a sound

---

[1] http://statystyka.policja.pl/st/ruch-drogowy/76562,Wypadki-drogowe-raporty-rocz ne.html.

[2] https://crashstats.nhtsa.dot.gov/Api/Public/ViewPublication/812446.

© Springer Nature Switzerland AG 2020
K. Saeed and J. Dvorský (Eds.): CISIM 2020, LNCS 12133, pp. 39–50, 2020.
https://doi.org/10.1007/978-3-030-47679-3_4

alarm. Other solution of this kind is for example BMW's Active Driving Assistant. Together with Attention Assistant it analyses driving behaviour and, if necessary, advises the driver to rest. Such fatigue detection systems also appear in more popular car models. Skoda's iBuzz Alert advises the driver on the basis of information about his steering behaviour. In order to do so it evaluates data from the power steering sensors to detect any driver fatigue behaviour and warns the driver to take a break. Volkswagen's Driver Alert System identifies failing concentration of the driver and provides him with a five-second acoustic warning as well as an optical indication on the instrument cluster and recommends to take a break. If the driver does not take a break within 15 min, the warning is repeated[3]. However, these solutions are still rather expensive and available only on the latest models, more accessible solutions would be useful for most popular older cars. Of course there are cheaper solutions like Anti Sleep Pilot[4]—device that can be fitted to any vehicle, uses a combination of accelerometers and reaction tests, but its operation is still imperfect.

Different people react differently to fatigue and generally awareness of the driver depends highly on his physiology. There are many factors which indicate that the person driving the vehicle is falling asleep like for example eyelid closure, yawning, head tilt or heart rate. It is possible to monitor the driver with sensors to record electrical activity of the brain (EEG) or measure electrodermal activity (EDA). Both the biological condition of the driver's body, as well as vehicle behaviour, can be used for driver's drowsiness detection. In addition to observing the driver's condition it is possible using sensors to monitor the seat movements, steering velocity occurrences, hands-on-wheel frequency, behaviour of the car position in relation to the centre of the lane. Research on driver drowsiness has a long history and has been of interest to both researchers and institutions interested in road safety for years. One of the fundamental works is [5] where it was found that one of the most important indicators of drowsiness is the PERCLOS—percentage of time that the eyes are 80% to 100% closed. Despite criticism, eyelid closure is one of the primary ways to study drowsiness researchers propose to combine this measure with others [6] reports on a project aimed at integrating PERCLOS with other drowsiness metrics. A comprehensive survey on drowsiness detection techniques is presented in [11] where generally drowsiness detection methods are classified into three main categories: behavioral parameter-based techniques, physiological parameters-based techniques, and vehicular parameters-based techniques. Moreover the pros and cons of the diverse method were discussed and a comparative study of them was presented. In [3] a computer vision-based system is described that keeps track of the eyes and detects the sleep onset of fatigued drivers. The proposed system uses template matching for detecting the state of the eyes. The method for real-time video monitoring with a 3D convolutional neural network, providing early warning signals to a drowsy driver is presented in [20]. On the basis of this review, we can conclude that eyes analysis is non-intrusive and should be easy

---

[3] https://www.volkswagen-newsroom.com/en/driver-alert-system-3932.

[4] https://www.stopsleep.co.uk/Anti-Sleep.solutions.html.

to implement using available face detection algorithms. In this paper we analyze the performance and accuracy of its possible implementation.

This paper is organized as follows. In Sect. 2 major face detection and landmark discovery algorithms and their implementations are presented. In Sect. 3 open and closed eyes are analysed numerically on face photos from several image datasets. The pseudocode of an sample application is shown in Sect. 4. We finish with some comments and remarks in Sect. 5.

2 Face Detection and Landmark Discovery

An important step in facial analysis is the determination of the characteristic points of the face. The shape and size of the nose, mouth or jaw varies from person to person. Another characteristics like the distance between eyes or mouth width are also unique to individuals. Their identification of selected facial landmarks points can help in many tasks connected with face analysis. They are facial recognition, age estimation, gender classification, facial expression analysis, and other similar tasks even in the detection of certain diseases. The face detection process is the first step of an automated landmarking and significantly affects the whole process performance. In general, most object detection methods can be applied to this problem, however two foundational methods of face detection applies here. The first method is the Viola-Jones method which is a kind of Haar casacade classifier [19]. The Viola-Jones method consists of three main steps. First is the image representation, called the "Integral Image", which is based on the value of simple features not the pixels. The simple features used are reminiscent of Haar basis functions. Next step is AdaBoost learning algorithm, which selects a small number of critical visual features. Last step is combining classifiers in a "cascade", which allows background regions of the image to be quickly discarded. The second method is Histogram of Gradients (HOG) [2]. In this method the image is converted into a series of histograms of image gradient orientations in a dense grid. This process is based on the orientation and magnitude of pixel gradients within the image. Finally well-normalized local histograms are evaluated with an SVM classifier used to identify the face within the image. Nowadays there are many more sophisticated face detectors based on convolutional neural networks (CNN) [9]. Many modifications of CNN method are performed for example in [10]. Despite newer methods, Viola-Jones and HOG are still used due to the speed of operation. These methods nowadays are not state of the art but achieve good accuracy and processing speed. This is the reason that they are in many "off-the-shelf" implementations such as in OpenCV and Dlib. Those two "off-the-shelf" frontal face detectors performance were tested in [7], authors concluded that the Dlib, HOG-based face detector outperformed all of the OpenCV variants with greater accuracy and fewer false positives.

In order to assess and compare the various facial landmarking methods, it was necessary to establish the common criteria for evaluation. In [14] authors recognized the variation in annotation schemes between public datasets and described methods and resources for landmarking performance comparison. Moreover they proposed a semi-automatic landmarking tool [15] to provide the same a 68 points annotation schema for most known datasets. Generally automatic landmarking methods are based on earlier work of annotator experts. Annotators signed the positions on the nose, eyes, the outer border of the lips, eye brows and line of the mouth and on this basis, the tools can be trained. A comparative study of face landmarking techniques was completed in [21]. Very comprehensive review of the current state of automated image-based facial landmarking processes and a comparison of their performance is presented in [7]. Authors presented selection of publicly available facial landmarking databases and reviewed current facial landmarking methods. Generally landmarking methods can be divided into three categories: generative methods, discriminative methods and methods combining the two, producing statistical methods. Main examples of generative methods are active appearance models (AAM) which are an improvement of active shape models (ASM) introduced in [1]. In ASM a statistical-based representation of the face using the shape information provided by the landmarks is build, in AAM additionally the texture is utilized. These techniques have been modified by researchers for many years and were successfully used to determine the landmarks. Simple but very effective method is presented in [17], this method is implemented in OpenCV. Discriminative methods requires training a regression functions which maps image values to facial landmark coordinate. Example of this kind of methods is presented in [12] where a set of local binary features, and a locality principle are used to learn a linear regression for the final output. Similarly in [8] is showed how an ensemble of regression trees can be used to estimate the face's landmark positions directly from a sparse subset of pixel intensities, achieving realtime performance and high quality predictions. The latter two methods are implemented in OpenCV. There are many extensions of the cascade shape model, in [4] is described a multi-view, multi-scale, and multi-component cascade shape model (M3CSR). Recent research in facial landmarking is dominated by deep learning techniques and the use of convolutional neural networks. In [18] the use of a cascade of Neural Net regressors to increase the accuracy of the estimated facial landmarks is presented together with short review of the latest convolutional neural network landmarking models. As we mentioned above in OpenCV due to performance and not very high computing requirements methods presented in [8,12,17] are implemented but a trained model is available only for [12]. In Dlib facial landmarking is implemented only via method presented in [8].

3   Open and Closed Eyes Analysis

To build a drivers drowsiness detector, we need to discriminate between open and closed eyes. To do so we will be using a metric called the eye aspect ratio (EAR) [16]. In facial landmark among 68 points 12 describe eyes, left eye points are numbered from 37 to 42, and right eye points are numbered from 43 to 48.

Using appropriate facial point annotations[5] [13–15] the values of these coefficients for the left and right eyes read as

$$\text{EAR}_{L,R} = \frac{|p_{38,44} - p_{42,48}| + |p_{39,45} - p_{41,45}|}{2|p_{37,43} - p_{40,46}|} \tag{1}$$

Each eye is described by two vertical lines (described by points $p_{38,44}$ and $p_{42,48}$ or $p_{39,45}$ and $p_{41,45}$) and a horizontal line (described by points $p_{37,43}$ and $p_{40,46}$). The value of EAR from Eq. (1) can be understood as a ratio of the average length of the vertical lines (height) to the length of the horizontal line (width). For a typical open oval-shape eye the value of the EAR coefficient is about 0.5 (twice as wide as tall). For a completely closed eye the EAR coefficient becomes 0.

Let us now study the probability distribution function of EAR coefficient for face images downloaded from the Closed Eyes In The Wild (CEW) dataset[6]. This dataset contains 2423 subjects, among which 1192 subjects with both eyes closed are collected directly from Internet, and 1231 subjects with eyes open are selected from the Labeled Face in the Wild[7] (LFW) database. Using HOG algorithm implemented in Dlib we were able to detect 1025 faces with closed eyes and 1114 faces with open eyes. For CNN algorithm form Dlib these numbers are 1169 and 1150 respectively. Thus as expected CNN detects faces better. For each face detected 68 landmarks p_i were discovered using DLib implementation of Ensemble of Regression Trees (ERT) method [8] and average EAR coefficient for both eyes computed. Their histograms are presented in Fig. 1.

It is seen from inspection of Fig. 1 that the maxima of the probability density functions of EAR corresponding to closed and open eyes are well separated but their tails do not. To estimate this overlap in Fig. 2 we have plotted the cumulative distribution functions for densities from Fig. 2. Note, that in the case of open eyes the distribution is inverted as one minus distribution (or the integration limits are changed from $\int_{-\infty}^{\text{EAR}}$ to $\int_{\text{EAR}}^{+\infty}$). Both distributions cross at EAR $\simeq 0.23$. This means that for EAR > 0.23 the chance that the eyes are closed is about 20%, and for EAR < 0.23 the chance that the eyes are open is also about 20%.

We would like to estimate a threshold value of EAR for our drowsiness detection system. The camera would periodically track the driver's face, the EAR coefficient would be computed. If it is lower than the threshold for a prolonged period of time an alarm would be activated. To better estimate this threshold in Fig. 3 we have plotted the sum of the distribution functions from Fig. 2. This sum has a clear minimum at EAR $\simeq 0.2$. For EAR > 0.2 the chance that the eyes are closed is less than 30%, and for EAR < 0.2 the chance that the eyes are open is less than 10%. Therefore lowering the threshold value of EAR to 0.2 would decrease the chance that the alarm is activated when the eyes are open at a slight decrease of safety (not all cases with closed eyes are properly detected).

---

[5] https://ibug.doc.ic.ac.uk/resources/facial-point-annotations/.
[6] http://parnec.nuaa.edu.cn/xtan/data/ClosedEyeDatabases.html.
[7] http://vis-www.cs.umass.edu/lfw/#download.

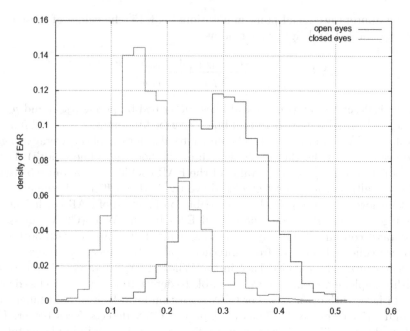

Fig. 1. Histograms of average EAR coefficients for faces with closed and open eyes from the CEW database.

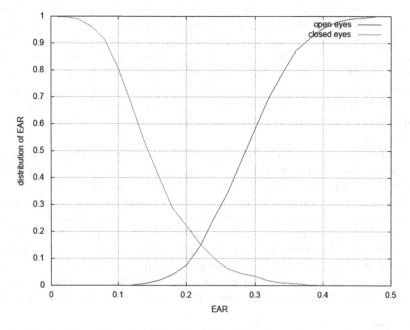

Fig. 2. Distributions of average EAR coefficients for faces with closed and open eyes from Fig. 1. In the case of open eyes the distribution is inverted.

It is interesting to mention that 0.2 is also the mean value of the probability distribution of EAR coefficient corresponding to closed eyes plus standard deviation.

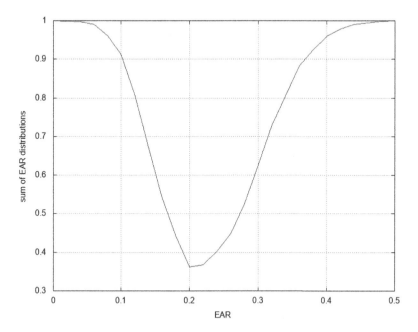

Fig. 3. Sum of distributions of average EAR coefficients for faces with closed and open eyes from the CEW database from Fig. 2.

Now we would like to verify our choice of the threshold value of EAR using the face photos from the LFW database. Using CNN we were able to detect all 13233 faces and compute the EAR coefficients for them: 1622 (12%) are below threshold 0.2 and 11611 (88%) above threshold. The histogram of these coefficients is plotted in Fig. 4 and compared with the histogram for all (both open and closed) eyes from the CEW database. Note that the latter histogram (resulting from "merging" the histograms from Fig. 1) still contains a dip around the value 0.2. The histogram for the LFW database is symmetric about the value 0.3.

We looked very carefully at more than 2558 faces with $0.24 \leq EAR < 0.28$ and found 4 of them with closed eyes. They are shown in Fig. 5. Therefore the threshold proposed is by no means perfect but works well for randomly chosen faces.

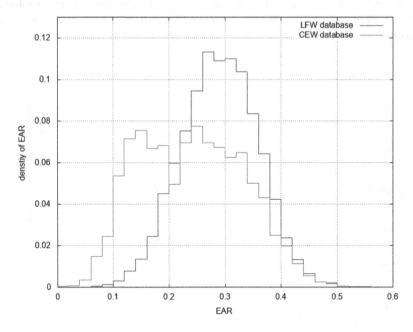

Fig. 4. Histograms of average EAR coefficients for all faces from the LFW and CEW databases.

Fig. 5. Faces with closed eyes from the LFW database with EAR coefficients above the threshold. From left to right they belong to David Surrett, Aiysha Smith, Sun Myung Moon, and Ariel Sharon.

4   Sample Application Pseudocode

The main idea of the application is to read consecutive frames from the camera. Than for each frame, a face landmark or the fact of non-recognition of the face is determined. It should be noted that the absence of a recognized face means that the driver's head is facing downwards or falling. If the EAR coefficient is less than ratio threshold value, or face is not recognized, within a few seconds, the system initiates an alarm. In the application the EAR threshold value was set to 0.21, but it is possible to give a different value. To realize this idea we define two constants, EYE_CLOSED_SECONDS and NO_FACE_SECONDS that will determine after how many seconds the alarm will be turned on when the eyes are closed or when the camera does not locate the face. Then, after multiplying both variables by frame rate, we get appropriate values EYE_FRAMES and NO_FACE_FRAMES in the form of frame rates. In the application main loop size of captured frame is reduced and converted into gray scale to increase the speed of action. Next, faces are detected. If no face is identified, the COUNTER_NO_FACE is increased, and it is checked that the critical value has not been exceeded. If so alarm is turned on. If faces are recognised (in our case, it is one face), then for each face landmark points the EAR ratio is counted according to Eq. (1). If ratio is less than the previously defined RATIO_THRESHOLD value, the COUNTER_EYE_CLOSED is increased, and it is checked that the critical value has not been exceeded. If so alarm is turned on. To break testing user can press the ESC key at any time. The whole algorithm is listed below:

```
 1: EYE_CLOSED_SECONDS ← 2.0
 2: NO_FACE_SECONDS ← 5.0
 3: fps ← GetFramesPerSecond()
 4: EYE_FRAMES ← fps * EYE_CLOSED_SECONDS
 5: NO_FACE_FRAMES ← fps * NO_FACE_SECONDS
 6: COUNTER_EYE_CLOSED ← 0
 7: COUNTER_NO_FACE ← 0
 8: ALARM_ON ← false
 9: ESC_KEY ← 27
10: loop
11:    read next frame.
12:    resize frame and change it into gray scale image.
13:    faces ← Detector(image).
14:    if len(faces) = 0 then
15:       if COUNTER_NO_FACE ≥ NO_FACE_FRAMES then
16:          ALARM_ON ← true
17:       else
18:          COUNTER_NO_FACE ← COUNTER_NO_FACE + 1
19:       end if
20:    else
21:       COUNTER_NO_FACE ← 0
```

22:      **for all** $face \in faces$ **do**
23:          $shape \leftarrow Predictor(face)$.
24:          $ratio \leftarrow Count(shape)$.
25:          **if** $ratio <$ RATIO_THRESHOLD **then**
26:              COUNTER_EYE_CLOSED $\leftarrow$ COUNTER_EYE_CLOSED + 1
27:              **if** COUNTER_EYE_CLOSED $\geq$ EYE_FRAMES **then**
28:                  ALARM_ON $\leftarrow$ **true**
29:              **end if**
30:          **else**
31:              COUNTER_EYE_CLOSED $\leftarrow$ 0
32:          **end if**
33:      **end for**
34: **end if**
35: **if** $WaitKey =$ ESC_KEY **then**
36:      break
37: **end if**
38: **end loop**

The aforementioned libraries used for estimating a threshold value of EAR, Dlib, and OpenCV, provide a convenient interface to Python. Thus the sample application shown above has been implemented in this language. Note, that OpenCV library was used only to manage the video stream from the camera. Faces are detected with Dlib HOG detector which, although less accurate, works much faster than CNN. Landmark points are determined with the help of Dlib ERT method. Application was tested on a laptop with standard camera. The tests were conducted on the authors of the application and were not too exhaustive but each longer eyelid closure was captured.

5 Summary

A software side of a drowsiness detection system was created and tested on large datasets of still images of faces with open and closed eyes: Closed Eyes In The Wild and Labelled Faces in the Wild. Together with low cost cameras that can provide relatively high-quality image, this indicates a possibility of a simple and cheap driver alert systems. In this paper only space analysis of an eye was performed. A metric called Eye Aspect Ratio together with a threshold value was used to distinguish between an closed and an open eye. Threshold value was chosen based on the analysis of the distributions of aspect ratio coefficients for eyes known to be open and closed. It was shown that it depends on the properties of the probability density of coefficients describing the closed eyes only. This method is not perfect but works quite well for randomly chosen faces. Additional study should consider the behaviour of an eye in time on a database consisting of face videos like, e.g., Driver Drowsiness Detection Dataset[8]. This will be done in a following paper.

---

[8] http://cv.cs.nthu.edu.tw/php/callforpaper/datasets/DDD/.

References

1. Cootes, T.F., Taylor, C.J., Cooper, D.H., Graham, J.: Active shape models-their training and application. Comput. Vis. Image Underst. **61**, 38–59 (1995)
2. Dalal, N., Triggs, B.: Histograms of oriented gradients for human detection. In: 2005 IEEE Computer Society Conference on Computer Vision and Pattern Recognition (CVPR 2005), vol. 1, pp. 886–893 (2005)
3. De Rubira, T.T.: Automatic fatigue detection system (2009)
4. Deng, J., Liu, Q., Yang, J., Tao, D.: M3 CSR: multi-view, multi-scale and multi-component cascade shape regression. Image Vis. Comput. **47**, 19–26 (2016)
5. Dingus, T.A., Hardee, H.L., Wierwille, W.W.: Development of models for on-board detection of driver impairment. Accid. Anal. Prev. **19**(4), 271–283 (1987)
6. Hanowski, R.J., Bowman, D., Alden, A., Wierwille, W.W., Carroll, R.: PERC-LOS+: development of a robust field measure of driver drowsiness. In: 15th World Congress on Intelligent Transport Systems and ITS America's 2008 Annual Meeting, New York, NY (2008)
7. Johnston, B., de Chazal, P.: A review of image-based automatic facial landmark identification techniques. EURASIP J. Image Video Process. **2018**(1), 1–23 (2018). https://doi.org/10.1186/s13640-018-0324-4
8. Kazemi, V., Sullivan, J.: One millisecond face alignment with an ensemble of regression trees. In: Proceedings of the IEEE Conference on Computer Vision and Pattern Recognition, pp. 1867–1874 (2014)
9. Li, H., Lin, Z.L., Shen, X., Brandt, J., Hua, G.: A convolutional neural network cascade for face detection. In: 2015 IEEE Conference on Computer Vision and Pattern Recognition (CVPR), pp. 5325–5334 (2015)
10. Peng, C., Bu, W., Xiao, J., Wong, K.C., Yang, M.: An improved neural network cascade for face detection in large scene surveillance. Appl. Sci. **8**(11), 2222 (2018)
11. Ramzan, M., Khan, H.U., Awan, S.M., Ismail, A., Ilyas, M., Mahmood, A.: A survey on state-of-the-art drowsiness detection techniques. IEEE Access **7**, 61904–61919 (2019)
12. Ren, S., Cao, X., Wei, Y., Sun, J.: Face alignment at 3000 fps via regressing local binary features. In: 2014 IEEE Conference on Computer Vision and Pattern Recognition, pp. 1685–1692 (2014)
13. Sagonas, C., Antonakos, E., Tzimiropoulos, G., Zafeiriou, S., Pantic, M.: 300 faces in-the-wild challenge: database and results. Image Vis. Comput. **47**, 3–18 (2016)
14. Sagonas, C., Tzimiropoulos, G., Zafeiriou, S., Pantic, M.: 300 faces in-the-wild challenge: the first facial landmark localization challenge. In: Proceedings of the IEEE International Conference on Computer Vision Workshops, pp. 397–403 (2013)
15. Sagonas, C., Tzimiropoulos, G., Zafeiriou, S., Pantic, M.: A semi-automatic methodology for facial landmark annotation. In: Proceedings of the IEEE Conference on Computer Vision and Pattern Recognition Workshops, pp. 896–903 (2013)
16. Soukupová, T., Cech, J.: Eye blink detection using facial landmarks. In: 21st Computer Vision Winter Workshop, Rimske Toplice, Slovenia (2016)
17. Tzimiropoulos, G., Pantic, M.: Optimization problems for fast AAM fitting in-the-wild. In: 2013 IEEE International Conference on Computer Vision, pp. 593–600 (2013)
18. Valle, R., Buenaposada, J.M., Baumela, L.: Cascade of encoder-decoder CNNs with learned coordinates regressor for robust facial landmarks detection. Pattern Recogn. Lett. (2019). https://doi.org/10.1016/j.patrec.2019.10.012

19. Viola, P., Jones, M., et al.: Robust real-time object detection. Int. J. Comput. Vision 4(34–47), 4 (2001)
20. Wijnands, J.S., Thompson, J., Nice, K.A. et al.: Real-time monitoring of driver drowsiness on mobile platforms using 3D neural networks. Neural Comput. Appl. 1–13 (2019). https://doi.org/10.1007/s00521-019-04506-0
21. Çeliktutan, O., Ulukaya, S., Sankur, B.: A comparative study of face landmarking techniques. J. Image Video Proc. **2013**, 13 (2013). https://doi.org/10.1186/1687-5281-2013-13

Automatic Marking of Allophone Boundaries in Isolated English Spoken Words

Janusz Rafałko[1]([⊠]) [iD] and Andrzej Czyżewski[2] [iD]

[1] Faculty of Mathematics and Information Science,
Warsaw University of Technology, Warsaw, Poland
j.rafalko@mini.pw.edu.pl
[2] Faculty of Electronics, Telecommunications, and Informatics, Gdańsk
University of Technology, Gdańsk, Poland

Abstract. The work presents a method that allows delimiting the borders of allophones in isolated English words. The described method is based on the DTW algorithm combining two signals, a reference signal and an analyzed one. As the reference signal, recordings from the MODALITY database were used, from which the words were extracted. This database was also used for tests, which were described. Test results show that the automatic determination of the allophone limits in English words is possible with good accuracy. Tests have been carried out to determine the error of particular allophones borders marking and to find out the cost of matching the given allophone to the reference one. Based on this cost, a coefficient has been introduced that allows for determining in percentage how much the automatically marked allophone is similar to the reference one. This coefficient can be used for an assessment of the correctness of the pronunciation of the allophone. The possibilities of further research and development of this method were also analyzed.

Keywords: Speech recognition · Speech analysis · Phoneme · Allophone

1 Introduction

This work is a part of a project devoted to the multimodal signal analysis of allophones, where an allophone can be defined as a variant of a particular phoneme [5, 8]. The project is based on combining audio and visual modalities [3]. The combination of these two modalities leads to improved accuracy of allophone transcription and recognition.

The applied algorithm is intended for the automatic marking of allophone boundaries in isolated English words based on the reference speech base obtained in the result of the preparation of the audio-video corpus [4, 17]. Although the corpus contains multimodal material, this work employs a single modality only, namely sound. This is the first part of the work that will answer the question if we can mark the limits of allophones in a continuous speech based on the sound signal only, with good accuracy. Further research, on the other hand, will allow answering the question of whether adding a second modality will improve this accuracy.

© Springer Nature Switzerland AG 2020
K. Saeed and J. Dvorský (Eds.): CISIM 2020, LNCS 12133, pp. 51–62, 2020.
https://doi.org/10.1007/978-3-030-47679-3_5

The problem of marking the borders of allophones in continuous speech is a very important issue in speech technology. It is related to such subjects as, for example, speech recognition and transcription, speech synthesis, or learning a foreign language. Speech transcription and recognition are described extensively in the literature, as well as speech synthesis, which is also covered broadly. For example, the paper of Szpilewski et al. [16] shows the approach to the concatenative TTS (Text-to-Speech) system based on allophones in the context of multilingual synthesis.

Another area where the appropriate marking of the allophone boundaries is an important task is the field of foreign language learning. Appropriate marking allophones in English has high development potential, because nowadays according to David Cristal, author of the "English as a Global Language" [2], non-native English speakers are three times as many as native speakers. That is why algorithms related to the correct delimitation of allophones can be very helpful in learning the correct pronunciation.

In Sect. 2, the speech recording process done in the project will be described, as well as the reference speech database used for this work to mark the allophone boundaries is described.

Section 3 is devoted to the description of the determination of the allophone boundaries algorithm in continuous speech. The algorithm is based on the DTW (Dynamic Time Warping) method. The modification of this method allows to marking of the boundaries of allophones. The parameters of the method are also presented.

In Sect. 4, a coefficient related to the correctness of the allophone boundaries and the correctness of the pronunciation of the allophone will be defined.

Section 5 contains an evaluation of the results. The test will be described that were carried out for various recordings, i.e., speech of different people.

Section 6 contains a summary and conclusions, as well as a discussion on planned further research.

2 Database

2.1 Phonetic Material

Audio recordings from the Modality [4] corpus were used as a referenced corpus in this study as well as a data source to carry out the tests (www.modality-corpus.org). The Modality corpus material consists of spoken numbers, names of months and days, and a set of verbs and nouns mostly related to controlling computer devices. It was presented to speakers as a list containing a series of consecutive, isolated words, and sequences of continuous speech. The corpus includes recordings of 35 speakers. The gender composition is 26 male and 9 female speakers. The corpus is split between native and non-native English speakers. Approximately half of the isolated words represented some typical command-like sentences, while the rest was formed into isolated word sequences. Every speaker participated in 12 recording sessions. Half of the sessions were recorded in quiet conditions, and the second half includes noise. Reference noise-only recording sessions were performed in order to enable a precise calculation of SNR (signal-noise ratio) for every sentence spoken by the speaker. The audio-visual material was collected in an acoustically adapted room. The video material was recorded using two Basler ace 2000-340kc cameras. The cameras were set up to capture video streams at 100 frames per second, in 1080 × 1920 resolution. The audio material was collected from an array of 8 B&K measurement microphones placed in different distances from the speaker. The audio data were recorded using 16-bit samples at 44.1 kSa/s sampling rate with PCM encoding. The setup was completed by loudspeakers placed in the corners of the room, serving as noise sources. The average SNR calculated in the 300–3500 Hz of frequency range was 36 dB for the quiet condition, and 17.2 dB for noisy conditions.

2.2 Reference Database Description

The reference base used for the presented system contains words from the Modality database stored in separate files. In addition, each word contains the boundaries of the allophones that it consists of. The marking of these boundaries was performed manually. Figure 1 shows an example word "clever" of the male native speaker, IPA notation: ˈklevəʳ, from the reference database with the allophone boundaries marked manually by the author of this work. Recordings of twenty people, eight women, and eleven men were selected for the reference database, of which two female and six male voices belonged to native speakers. The speakers were between the ages of 22 and 58. One of these voices was always used as a reference voice in the system, and the others were used to mark allophones in them. Only recordings without additional noise have been selected from the Modality database, which is only with ambient noise.

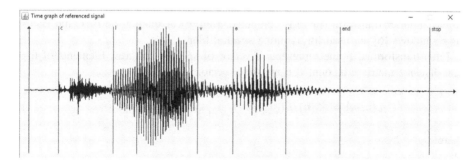

Fig. 1. Reference word "clever" with manually marked allophone borders

Because of the need to compare a homogenous material, there was not possible to mark the allophone boundaries automatically in any speech signal, but only in such recordings for which there is an equivalent in the reference database. Based on the marked allophones in the word from the reference database, it is possible to mark the boundaries of the allophones in the same word also from outside the base.

For the automatic system building, these boundaries are stored in a separate text file in the format: "allophone, sample number".

3 Algorithm for Automatic Marking of Allophones

The algorithm for automatic marking of allophone boundaries is based on the DTW method [7, 9, 15]. A system using a similar approach, but designed to create bases of acoustic units in speech synthesis, was presented in the author's earlier paper [12]. The approach presented in this paper differs from the previous one in several fundamental issues. First of all, it concerns the English language, instead of Polish. Secondly, in the previous system, the bases of acoustic units were created employing a much larger, redundant recording corpus, which allowed for averaging the results and selecting the best one unit. In the previous work, the databases were created for the concatenation speech synthesis, in which only one realization of a given allophone was needed, hence it was necessary to find the best one. Here we should approach each allophone individually because e.g. we want to determine whether the given allophone was pronounced correctly or not. In this case, the point is not to find one allophone in the whole corpus. In this case, we want to define limits for each allophone in the spoken word, to determine how correctly the given allophone has been pronounced. The consecutive steps leading to this aim are described in subsequent subchapters.

3.1 Combining Reference and Analyzed Signals

The algorithm of automatic segmentation of the speech signal is based on the DTW method [1, 9–11] however, unlike the classic DTW, it does not rely on the signal in the time domain, but on the frequency domain representation. The referenced speech signal (spoken word) and the analyzed signal (the same word pronounced by another person) in the time domain are divided into frames that can overlap each other. In each frame, the Fast Fourier transform (FFT) is computed. Before calculating the FFT, the Hamming windows [6] are used for avoiding spectral blur.

Each transformed frame represents a vector of spectral features. Elements of the local distance matrix are counted using these vectors as:

$$c(n,m) = \|S(n), E(m)\| = \sum_{k=1}^{K} |S(n,k) - E(m,k)| \tag{1}$$

where:

$S(n)$ - vector of spectral features of the referenced signal in the n-th frame
$E(m)$ - vector of spectral features of the natural signal in the m-th frame
K - length of the spectral features vector

The reference signal frame is combined with the analyzed signal frame, and then the distance between these vectors is calculated using the signal spectrum in the frame. In formula 1, the distance is calculated according to the Manhattan metric, being used for the algorithm.

The global distance matrix calculated is shown in Fig. 2. It presents speech signals to which 256-sample frames were applied, employing the Hamming window and no overlapping. The reference signal is shown in the form of a spectrogram drawn vertically on the left side of the drawing. The analyzed signal is also presented in the form of a spectrogram but at the bottom of the drawing. Both signals combined are identical, and in the considered case, there are of the voice of a male native speaker. The square area in the center of Fig. 2 shows the global distance matrix. The bright areas indicate small values of the distance between the signal frames that are the signals spectrally similar, while the dark areas indicate a large distance between frames, which are signals that differ in spectral features.

The warping path is also shown there, which is going through the areas of the lowest cost. Because in this case, both signals are identical, the warping path is going alongside the anti-diagonal of the global distance matrix.

Having the optimal warping path and the boundaries of the allophones in the reference signal, it is possible to determine the allophone boundaries in the analyzed signal, as is shown in Fig. 2. A vertical spectrogram on the left-hand side shows a reference signal in which allophone boundaries are known because they have been manually marked by a phonetician expert. In the figure, they are marked with horizontal lines, which additionally go into the area of the global distance matrix and end in the optimal warping path. The points of intersection of these lines and the warping path determine the boundaries of the allophones in the analyzed signal, as it is represented in the figure by vertical lines going down from the warping path. These lines pass to the spectrogram of the analyzed signal, as it is shown in Fig. 2, and they determine the limits of the allophones in this signal. In this example, both phrases are identical, therefore the warping path is a straight line.

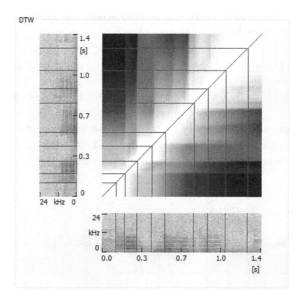

Fig. 2. Global distance matrix of "keep moving" phrase and determination of allophone boundaries in the analyzed signal

In case when we set boundaries in a word spoken by a different speaker than the one who produced the reference one, the matching path will no longer be a straight line, as is shown in Fig. 3. The reference signal belongs to a native speaker, whereas the analyzed signal to a non-native speaker. The parameters for determining the path are the same as before.

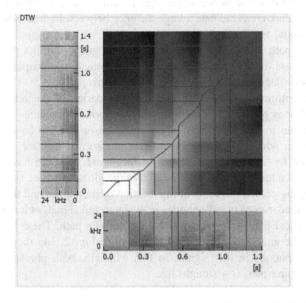

Fig. 3. The warping path and the allophone boundaries in the "keep moving" phrase of speaker different to the reference speaker

The allophone boundaries set in this way can be compared with manually marked boundaries, as is seen in Fig. 4. The same "keep moving" phrase is shown on both graphs. The upper graph shows the boundaries of allophones determined manually, while the lower graph - boundaries determined automatically. In Fig. 4 we can visually evaluate the location, and thus roughly the quality of automatically defined boundaries. As is seen, the differences, in this case, are minimal. In Sect. 5 the statistical assessment will be presented computed on the reference set discussed earlier.

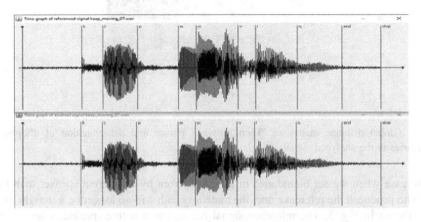

Fig. 4. Comparison of boundaries marked manually and automatically

The boundaries of allophones obtained in this way are not always correct. The reason may be incorrect pronunciation or errors in the automatic determination of the border. Errors resulting from algorithm inaccuracies can be corrected. A discussion on this topic is presented in previous papers [13, 14]. The work [13] refers to the correction of allophone boundaries in acoustic databases used in Polish speech synthesis, where only one allophone cut out from a set of many redundant allophones of the same type is subject to correction. In [14], an approach to the correction of all allophones in a word is presented.

4 The Cost of Allophone Matching

While determining the boundaries of allophones in the analyzed word, we use a reference word spoken by another speaker. By juxtaposing both words, the warping path is determined. The values on the path are the cost of matching, and they are derived from the global distance matrix. These values determine the similarity of both signals. Having determined allophone limits manually in the reference signal, besides determining the allophone limits in the analyzed signal, we can also calculate the cost of matching of individual allophones. The result represents the cost of matching the allophone designated on the warping path in terms of a difference in the value of points on the path (formula 2):

$$C_M(allophone) = C_{end}(allophone) - C_{start}(allophone) \qquad (2)$$

where:

$C_M(allophone)$ – the matching cost of allophone
$C_{end}(allophone)$ – the cost of matching of the end of the allophone in the point of the intersection of its border with the warping path
$C_{start}(allophone)$ – the cost of matching of the beginning of the allophone in the point of the intersection of its border with the warping path

The above cost (formula 2), determined when two identical signals are juxtaposed, will have a value of 0. Such a situation will take place, as is shown in the example presented in Fig. 2. In case of determining the allophone limits in a different signal than the reference one, the cost will be greater than zero. Table 1 shows the cost of matching of allophones in the word "clever" determined using the previously described method. The table also includes the limits of the beginning of allophones marked manually and automatically. The fourth column presents the absolute boundary determination error calculated in relation to the length of allophones marked manually. To obtain this error, allophones should be manually marked in the analyzed word, too.

The absolute error applies only to the location of the boundary, while the cost of matching determines the similarity of the allophones. For example, allophones 4 (v) and 5 (e): the error of defining the border for the allophone "v" is larger than that of the allophone "e". However the cost of matching is presented in the opposite way:

for the allophone "v" the cost is less than for "e", which means that "v" is more frequency-related to the reference allophone than "e" to the reference "e".

Table 1. Error and cost of allophone matching

Allophone	Manual borders	Automatic borders	Error	Cost of matching	Correctness
c	22960	21999	7,10%	508.0	0,01
l	27568	27633	7,91%	406.1	0,13
e	29611	29779	5,53%	1440.8	0,04
v	33812	33536	7,13%	222.8	0,16
e	36831	36755	1,33%	549.2	0,17
r	42753	42389	6,22%	131.7	0,10

The specified match cost can be used for calculating the coefficient determining the correctness of the automatically marked allophone. An absolute error cannot play the role of such a factor, because it requires correctly, manually set the limits of the allophones, thus it can be used only for testing purposes. In order to obtain such a factor, the average cost of determining the allophone at a given reference signal should be calculated. Furthermore, this average cost should be set for correctly pronounced allophones, which is for the voices of native speakers. With this average cost, we can associate the coefficient of similarity of the given allophone to the reference one as shown in formula 3. This coefficient can be used to determine the correctness of the pronunciation of the allophone.

$$S_{al} = \left| \frac{C_{al} - C_{avr}}{C_{avr}} \right| \tag{3}$$

where:

S_{al} – allophone similarity to the referenced one coefficient
C_{al} – the matching cost of the allophone
C_{avr} – the average matching cost of the allophone

The smaller this ratio is, the more allophone is similar to the reference one. An example illustrating this procedure for the word "clever" is presented in Table 1.

5   Tests and Evaluation of Results

The tests were carried out using the referenced database described in Sect. 2. Table 2 presents the parameters of the example allophones marked automatically. The average error is presented related to the marking of the allophone together with the standard deviation of this error for the case of determining this allophone in the words of native speakers and non-native speakers. Similarly, the average cost of matching the allophone on the warping path of the global distance matrix of the DTW algorithm is given. Similarly, as in the case of a marking error, the average cost and its standard deviation are calculated separately for native and non-native speakers. In order to achieve these results, 20 different words were used uttered by 20 speakers described earlier. The upper part of the table shows male voices, the lower part - female voices. In both cases, the reference signal was a native speaker's voice. The signal processing parameters are 256-samples frame, with the Hamming window overlapped by 50%.

Table 2. The average error of the allophone marking and the average cost of matching

Male	Non-native		Native		Non-native		Native	
	Av. Error %	Err. St. Dev.	Av. Error %	Err. St. Dev.	Av. Cost	Cost St. Dev.	Av. Cost	Cost St. Dev.
s	7,90	5,03	7,57	7,64	2221	591	2065	102
e	5,21	3,16	6,15	6,11	1918	1122	1443	203
t	28,86	19,07	7,08	8,93	1057	211	908	44
ə	32,57	23,32	10,00	5,14	570	235	379	30
n	11,09	12,91	8,23	5,74	697	217	633	119
Female	Non-native		Native		Non-native		Native	
	Av. Error %	Err. St. Dev.	Av. Error %	Err. St. Dev.	Av. Cost	Cost St. Dev.	Av. Cost	Cost St. Dev.
s	9,32	7,19	7,69	5,16	4117	496	3583	237
e	6,61	2,73	3,62	2,15	2544	898	2052	240
t	15,49	12,50	9,70	7,84	1899	362	1042	178
ə	25,24	11,77	17,46	10,02	2765	725	1287	52
n	20,36	17,89	2,58	1,61	1129	513	882	218

Depending on the allophone, average errors in their marking can range from about 2% to about 20% for native speakers, and from about 5% to even several dozen for non-native-speakers. However, in cases where the pronunciation is correct, allophones marking errors of non-native speakers are similar to native ones. Similarly, the standard deviation for non-native speakers is larger but does not differ significantly from native speakers.

When the marking error exceeds the value of about 20%, it means that the allo- phone limit has been determined incorrectly and that it was significantly shifted in relation to the manually determined one. It should be remembered that this error is determined in relation to the correctly, manually marked border. This case is illustrated in Fig. 5, where for the non-native speaker, the beginning of allophone "c" has been incorrectly designated. The upper part of the drawing shows the boundaries marked manually, the lower part shows borders marked automatically. Such cases where the error exceeds 50% and the allophone is correctly pronounced have not been included in the calculation of the average errors in Table 2.

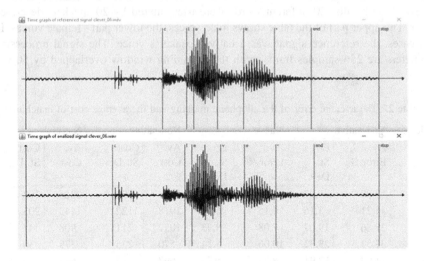

Fig. 5. Error in determining the border

The considerably high error in determining the border does not mean, however, that the allophone was pronounced wrongly. To determine the correctness of the spoken allophone, the cost of matching it to the reference one can be used. A slightly different situation is for this cost parameter, as Table 2 shows. The average cost for non-native speakers in the case of each allophone is higher, but also the standard deviation is noticeably higher. This average cost of matching can be used as a correctness coefficient, as is defined by the formula 3. Table 3 shows this coefficient with the error of the allophone limit marking for words belonging to different categories. The reference signal is a male native speaker. On its basis, the boundary for the voice of a male native speaker is marked, as well as for the male non- native speaker and for the female non-native voice.

Table 3. The correctness coefficient of the allophone

Reference male	Native male		Non-native male		Non-native female	
	Error	Correctness	Error	Correctness	Error	Correctness
s	2,23	0,22	7,30	0,22	23,75	0,04
e	4,16	0,17	4,20	0,03	6,55	0,55
t	0,15	0,03	36,23	0,54	16,51	0,98
ə	11,03	0,34	44,34	0,62	38,83	4,15
n	1,29	0,01	11,20	0,30	40,41	0,27

When the correctness factor has a value close to zero, it means that the given allophone is similar to the reference one, and that also means that it is correctly pronounced. For correctly pronounced words of voices of the same gender speakers, this coefficient is below 1. However, in the case of voices significantly differing in laryngeal frequency, very often, it can be as high as about 5, as is shown in Table 3, where on the basis of the male voice, the boundaries in the female voice are determined. For the allophone "ɔ", the correctness factor is 4.15, and also the error of determining the border is high, over 38%. The same situation occurs in the opposite case of determining the boundaries in the male voice on the basis of the female voice.

It can also be noticed another case here, that is a wrong set boundary does not mean that the allophone was pronounced wrongly. For the non-native male speaker, allophones "t" and "ɔ" are marked wrong in automatic mode, and errors exceed 30%, but similarity coefficients remain below 1, which means that these allophones were pronounced not quite wrongly.

6 Conclusions

The presented approach allows for determining the borders of allophones in isolated English words with satisfactory accuracy. Knowing determined average matching costs for every allophone, the correctness coefficient reflecting the similarity of the marked allophone to its reference counterpart is introduced. Using this factor, we can determine how much the spoken allophone is similar to the referenced allophone. It can be used e.g., for learning the correct pronunciation of a given allophone. In this way, it is also possible to determine compatibility with the correct pronunciation of words or phrases.

The results presented in this work allows for stating that allophone boundaries are determined correctly in the majority of cases. The correct determination of these boundaries in combination with the coefficient determining similarity to the referenced allophone can be used to improve the efficiency of recognizing specific allophones in speech, and thus to improve the quality of speech recognition.

The presented method can be further developed. In the presented work, all tests were carried out using the reference signal of one speaker, while the boundaries of allophones were determined for the words uttered by other speakers, with their voices produced employing different laryngeal tone. The presented method could be modified in such a way that the frequency of the laryngeal tone in the reference signal is changed so that for voiced allophones, it will be closer to the frequency of the analyzed signal. In addition, in the presented solution, the limits of allophones are determined only for words that have equivalents in the reference database. Meanwhile, the research and the solution can be extended towards determining boundaries in any words or phrases. It could be done, for example, by using a speech synthesizer, which would synthesize an appropriate signal from the reference database for which the boundaries in the analyzed voice will then be determined.

Acknowledgments. Research sponsored by the Polish National Science Centre, Dec. No. 2015/17/B/ST6/01874.

References

1. Bellman, R., Kalaba, R.: On adaptive control processes, automatic control. IRE Trans. **4**(2), 1–9 (1959)
2. Crystal, D.: English as a Global Language, 2nd edn. Cambridge University Press, Cambridge (2003)
3. Czyżewski, A., Ciszewski, T., Kostek, B.: Methodology and technology for the polymodal allophonic speech transcription. J. Acoust. Soc. Am. **139**(4), 2017 (2017)
4. Czyżewski, A., Kostek, B., Bratoszewski, P., Kotus, J., Szykulski, M.: An audio-visual corpus for multimodal automatic speech recognition. J. Intell. Inf. Syst. **49**(2), 167–192 (2017)
5. Gafos, A.: The Articulatory Basis of Locality in Phonology. Routledge Taylor & Francis Group, Abingdon (1999)
6. Harris, F.J.: On the use of windows for harmonic analysis with the discrete fourier transform. Proc. IEEE **66**(1), 51–84 (1978)
7. Keogh, E.J., Pazzani, M.J.: Derivative dynamic time warping. In: the 1st SIAM International Conference on Data Mining, Chicago, IL, USA (2001)
8. Kiritani, S., Itoh, K., Hirose, H., Sawashima, M.: Coordination of the consonant and vowel articulations—X-ray microbeam study on Japanese and English. Ann. Bull. Res. Inst. Logoped. Phoniatry **11**, 31–37 (1977)
9. Müller, M.: Information Retrieval for Music and Motion. Springer, Heidelberg (2007). Part I, chapter 4, Dynamic Time Warping, pp. 69–74
10. Myers, C.S., Rabiner, L.R.: A comparative study of several dynamic time-warping algorithms for connected word recognition. Bell Syst. Tech. J. **60**, 1389–1409 (1981)
11. Rabiner, L.R., Rosenberg, A., Levinson, S.: Considerations in dynamic time warping algorithms for discrete word recognition. IEEE Trans. Acoust. Speech Signal Process. **26**, 575–582 (1978)
12. Rafałko, J.: The algorithms of automation of the process of creating acoustic units databases in the polish speech synthesis. In: Atanassov, K.T., et al. (eds.) Novel Developments in Uncertainty Representation and Processing. AISC, vol. 401, pp. 373–383. Springer, Cham (2016). https://doi.org/10.1007/978-3-319-26211-6_32
13. Rafałko, J.: Algorithm of allophone borders correction in automatic segmentation of acoustic units. In: Saeed, K., Homenda, W. (eds.) CISIM 2016. LNCS, vol. 9842, pp. 462–469. Springer, Cham (2016). https://doi.org/10.1007/978-3-319-45378-1_41
14. Rafałko, J., Czyżewski, A.: Adjusting automatically marked voiced English allophone borders. In: Signal Processing: Algorithms, Architectures, Arrangements, and Applications (SPA), Poznan, 18–20 September 2019. https://doi.org/10.23919/spa.2019.8936805
15. Salvador, S., Chan, P.: FastDTW: toward accurate dynamic time warping in linear time and space. In: KDD Workshop on Mining Temporal and Sequential Data, pp. 70–80 (2004)
16. Szpilewski, E., Piórkowska, B., Rafałko, J., Lobanov, B., Kiselov, V., Tsirulnik, L.: Polish TTS in multi-voice slavonic languages speech synthesis system. In: SPECOM'2004 Proceedings, 9th International Conference Speech and Computer, Saint-Petersburg, Russia, pp. 565–570 (2004)
17. Modality Corpus. http://www.modality-corpus.org. Accessed 26 Mar 2019

Computer Information Systems and Security

Combined State Splitting and Merging for Implementation of Fast Finite State Machines in FPGA

Adam Klimowicz[✉] [iD]

Bialystok University of Technology, Bialystok, Poland
a.klimowicz@pb.edu.pl

Abstract. A new method of the synthesis of finite state machines is proposed. In this method, the speed of FSM is taken into account already at the early stage of synthesis process. The method is based on sequential merging and splitting two internal states regarding to speed of FSM. This parameter may decrease with reduction of internal states, but splitting internal states leads to decrease of number of variables in logic functions which describe combinational part of FSM. This parameter has a great influence on a critical delay path. The results of experiments showing efficiency of proposed approach are also presented.

Keywords: Finite state machines · Logic synthesis · Speed optimization · State minimization · State splitting · FPGA

1 Introduction

The high-speed performance of electronic projects is important in such spheres as computers, robotics, telecommunication, embedded systems, wired and wireless networks, transport, military and etc. For this reason the problem of increase of speed of electronic equipment becomes especially actual.

In the general case, the digital system can be represented as a combination of combinational schemes and finite state machines (FSM). The FSMs are also widely used as separate nodes as control devices. Usually, when creating a new project, the engineer has to re-create the original state machines every time. It is obvious that the parameters of the FSMs included in the digital system largely determine the success of the entire project. There are several approaches to increase the performance of the electronic equipment: the technological - using the elements with a high-speed performance, the system - using a piping and multi-core processors, the circuitry - increase of a supply voltage and the logical - using the synthesis methods allowing building FSMs and control units with the maximum high-speed performance, etc.

Today the most of scientific researches is connected to the first three directions (especially to the first) while the fourth direction is not sufficiently developed. At the same time, the methods of a logic synthesis can be used for any technological basis, can be applied together with system methods, and do not depend on supply voltage.

At present, field programmable gate arrays (FPGA) are widely used for constructing digital systems. Since modern FPGAs have a large number of look-up table

K. Saeed and J. Dvorský (Eds.): CISIM 2020, LNCS 12133, pp. 65–76, 2020.
https://doi.org/10.1007/978-3-030-47679-3_6

(LUT) logical elements, the implementation cost has not been a critical parameter. Recently, the most relevant optimization criteria are speed and power consumption.

2 State of the Art

One of the first attempts to solve speed optimization problem for implementation of FSMs on FPGA was presented in paper [1]. This work shows a new technique for improving the speed of a synchronous circuit configured as a look-up table based FPGA without changing the initial circuit configuration; only the register location is altered. It improves clock speed and data throughput at the expense of latency.

In [2] a new sequential circuit synthesis methodology is discussed that targets FPGAs and reconfigurable SoC (System-on-Chip) platforms. The methodology is based on the information-driven approach to circuit synthesis, general decomposition and the previously developed theory of information relationship measures. The paper [3] proposes a timing optimization technique for a complex finite state machine that consists of random logic and data operators. The proposed technique adds a functionally redundant block to the considered circuit, which includes a fragment of combinational logic and several registers. In result, the timing critical paths can be split into stages. In [4], a method for encoding the internal states of FSM realized on FPGA is proposed which allows optimizing parameters such as implementation cost, speed and power consumption. The paper [5] is concerned with the problem of state assignment and logic optimization of high speed FSMs. The method is oriented for PAL-based CPLD (Complex Programmable Logic Device) implementation.

New methods for the synthesis of high-speed asynchronous finite state machines were developed in [6–8]. In [6], a method for the synthesis of asynchronous finite state machines with a local synchronization signal is proposed, which makes it possible to implement an asynchronous FSM on any FPGA. In addition, the method allows using standard procedures of synchronous circuits minimization. In [7], a modification of the feedback of asynchronous finite state machines and convergent state coding was proposed. It allows an asynchronous FSM to be realized as easily as synchronous. In [8], the XBM (Extended Burst-Mode) architecture was presented, based on local synchronization signals, which allows for the synthesis of asynchronous FSMs using approaches for synthesizing synchronous machines.

An increase of the speed of finite automata by splitting the internal states is considered in [9, 10]. In [9], a method is proposed for reducing the number of arguments for transition functions, which does not depend on how the finite state machine is implemented and can be used for both FPGA/SoC and ASIC. In [10], split algorithms of FSM internal states for the synthesis of high-speed FSMs are described which are oriented to the realization of a FSM on FPGA. The parametric method of the minimization of finite state machines is proposed in [11]. In this method, based on state merging such optimization criteria as the critical delay path and also possibility of merging other states are taken into account already at the early stage of minimization of internal FSM states.

The analysis of available studies showed that there are no works in which the state splitting and state merging procedures are simultaneously used to achieve the reduction of critical path delay of FSM. In the present paper, a heuristic method for the synthesis of FSMs is proposed that makes it possible to optimize the FSM performance already at the stage of state minimization combined with state splitting. The proposed approach is intended for the implementation of FSMs on FPGA.

3   Idea of the Approach

The discrete optimization task is defined as searching for an FSM with minimal critical delay path (a minimization problem). It can be formulated as follows.

A function $f : U \rightarrow V$ is given, where U is a set of different forms of the same FSM (equivalent forms of FSM), V – a set of values corresponding to the length critical delay path. An element (FSM) $u_{min} \in U$ is sought, such as $f(u_{min}) \leq f(u)$ for all $u \in U$.

The idea of the approach to above formulated task is performing sequential operations of splitting states or joining states, if possible. In this way, new equivalent FSMs are created, whose implementation can give different results in terms of speed (critical path delay). To calculate the critical path length and then implement the machine, we must first do the state assignment using the selected method. To know which pair of states to choose or which state to split, the trial merging and trial splitting operations are performed. From the obtained results, the action to be performed (merging or splitting) and a state (for splitting) or a pair of states (for merging) is selected, for which the estimated critical path length is lowest (after the transformation of the FSM). Concerning the above considerations, the general synthesis algorithm can be described as follows.

At the beginning of Algorithm 1 an initial FSM representation is saved as the best one (line 1). Then, the procedures of searching pairs for merging and states (set G) for splitting (set D) are executed (lines 2–3). If there are no states to merge and split, the algorithm stops, otherwise the trial merging and trial splitting is performed in a similar way: first, the current FSM must be saved, then merging or splitting is performed, then codes of states are assigned and the speed of FSM is calculated (lines 6–26). The estimation procedure for speed parameter (function *CriticalPath*()) will be precisely described in Sect. 5. Among all solutions, the one is selected for which the critical path length is minimal. Then the final merging or splitting is executed and choosing states for merging or splitting is performed again (lines 27–38). The final FSM representation is the one with the highest speed from all considered equivalent representations (lines 39–40).

The splitting procedure may be divergent, so the stop condition for splitting has to be introduced. It is done in lines 30–38 of Algorithm 1, where the overall and average critical path length of currently splitting FSM is compared to the overall and average critical path length of the last performed splitting. If splitting does not lead to a further increase in speed – it is not performed.

Algorithm 1: General algorithm for FSM synthesis

```
1: bestFSM ← FSM
2: G ← FindMergePairs(FSM)
3: D ← FindSplitStates(FSM)
4: WHILE G ≠ ∅ and D ≠ ∅ DO
5:     S_st ← ∞
6:     WHILE G ≠ ∅ DO
7:         Save(FSM)
8:         FSM ← Merge(FSM, (a_s, a_t) ∈ G)
9:         Encode(FSM)
10:        IF CriticalPath(FSM) < S_st THEN
11:            S_st ← CriticalPath(FSM)
12:            SelectedPair ← (a_s, a_t)
13:        END IF
14:        Restore(FSM)
15     END WHILE
16:    S_i ← ∞
17:    WHILE D ≠ ∅ DO
18:        Save(FSM)
19:        Encode(FSM)
20:        FSM ← Split(FSM, a_i ∈ D)
21:        IF CriticalPath (FSM) < S_i THEN
22:            S_i ← CriticalPath (FSM)
23:            SelectedState ← (a_i)
24:        END IF
25:        Restore(FSM)
26:    END WHILE
27:    IF S_st ≤ S_i THEN
28:        FSM ← Merge(FSM, SelectedPair)
29:    ELSE
30:        IF CriticalPath(CurrentSplit) ≤ CriticalPath(LastSplit) THEN
31:            IF AverCriticalPath(CurrentSplit) < AverCriticalPath(LastSplit) THEN
32:                FSM ← Split(FSM, SelectedState)
33:            ELSE
34:                NoSplit ← TRUE
35:            ELSE
36:                NoSplit ← TRUE
37:        END IF
38:    END IF
39:    IF CriticalPath (FSM) < CriticalPath (bestFSM) THEN
40:        bestFSM ← FSM
41:    END IF
42:    G ← FindMergePairs(FSM)
43:    IF NoSplit = FALSE THEN
44:        D ← FindSplitStates(FSM)
45:    ELSE
46:        D ← ∅
47: END WHILE
48: END
```

4   State Merging Procedure

The proposed approach is based on the method for the reduction of the number of internal states proposed in [12] for Mealy machine and in [13] for Moore machine. The idea of this method is to sequentially merge two states. For this purpose, the set G of all pairs of internal states of the FSM satisfying the merging condition is found at each step. Then, for each pair in G, a trial merging is done. Next, the pair that leaves the maximum possibilities for merging other pairs in G is chosen for real merging.

Two FSM states a_s and a_t can be merged, i.e. replaced by one state a_{st}, if they are equivalent. Equivalency of two FSM states means that FSM behavior does not change when these states are merged in one. FSM behavior does not change after states a_s and a_t merge if the transition conditions from the states a_s and a_t that lead to different states are orthogonal. If there are transitions from states a_s and a_t that lead to the same unique state, then the transition conditions for such transitions should be equal. Moreover, the output vectors that are generated at these transitions should not be orthogonal.

The procedures of selecting pairs of states to merge (line 2 of Algorithm 1) and merging of the chosen states a_s and a_t (lines 8 and 26 of Algorithm 1) are precisely described in [12].

5   Estimation of Speed of FSM

Let us denote by L the number of FSM input variables of a set $X = \{x_1, \ldots, x_L\}$, by N the number of FSM output variables of a set $Y = \{y_1, \ldots, y_N\}$, by M the number of FSM internal states of a set $A = \{a_1, \ldots, a_M\}$, and by R the minimal number of bits required to encode internal states, where $R = \text{int}(\log_2 M)$.

To estimate the optimization criteria, all pairs of states in G and all states in set D are considered one after another. For each pair of states (a_s, a_t) in G, a trial merging is performed and for all states a_i in D – trial splitting is performed. After merging and splitting the internal states are encoded using one of the state assignment methods and the system of Boolean functions W corresponding to the combinational part of the FSM is built. Next, for the every pair (a_s, a_t), maximum critical delay path (speed) S_{st} is estimated. After trial merging, for the each state a_i after trial splitting the maximum critical delay path S_i is also determined. After selection of the optimal solution, the final merging or splitting can be done.

In the general case, the architecture of modern FPGAs can be represented as a set of logic elements based on functional LUT generators. A feature of LUT functional generators is that they can realize any Boolean function but with a small number of arguments (typically, 4–8). In the case when the number of arguments of functions to be realized exceeds the number of LUT inputs n, the Boolean function must be decomposed with respect to the number of arguments [14]. Among the great number of decomposition methods for Boolean functions with respect to the number of arguments, linear and parallel decomposition methods are most popular. Figure 1 and Fig. 2 present block diagrams of linear and parallel decomposition, where FB_i and FB_{ij} are functional FPGA blocks, and X_i are subsets of the set of input variables X.

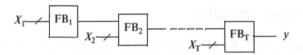

Fig. 1. Linear decomposition

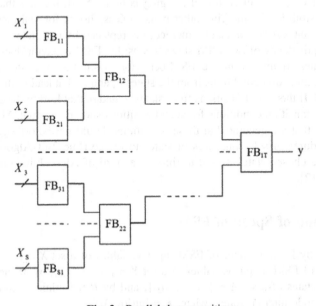

Fig. 2. Parallel decomposition

Let us denote by $B(a_i)$ – a set of states, which have transitions to state a_i and $X(a_i)$ – a set of input variables which initiate transitions to state a_i. In general, a number of arguments r_i of transition function to state a_i can be defined as:

$$r_i = |X(a_i)| + R \tag{1}$$

where operator $|A|$ returns a cardinality of set A.

In case when we apply one-hot encoding, which is most common used for FPGA FSM implementations, every FSM state corresponds to single memory element (flip-flop). If flip-flop state is logical one – the FSM is in corresponding state. In this case, a number of arguments r_i of transition function to state a_i can be defined as in [10]:

$$r_i = |X(a_i)| + |B(a_i)| \tag{2}$$

The upper boundary r^* of the number of arguments of the transition functions when splitting the FSM states for binary encoding type is defined as:

$$r^* = \max(|X(a_m, a_s)|) + R, \; m, s = 1..M, \tag{3}$$

but for one-hot encoding style is defined as in [10]:

$$r^* = \max(|X(a_m, a_s)|) + 1, \; m, s = 1..M, \tag{4}$$

where $X(a_m, a_s)$ is a set of FSM input variables, whose values initiate the transition from state a_m to state a_s.

As it was mentioned above, there are two most commonly used approaches to decomposition of Boolean functions: linear and parallel. In case of linear decomposition application the number of logic levels for transition function to state a_i, can be defined as:

$$L_i = 1 + \text{int}((r_i - n)/(n - 1)). \tag{5}$$

In case of using parallel decomposition the number of logic levels for transition function to state a_i, can be defined as:

$$L_i = \text{int}(\log_n r_i), \tag{6}$$

where L_i cannot be less than 1. If $r_i = 1$ then $L_i = 1$.

The average critical length for transition functions is the arithmetic mean of above parameter L_i for all states:

$$L_{av} = \left(\sum_{i=1}^{M} L_i \right) / M \tag{7}$$

The speed of operation of a whole FSM is determined by the length of the critical path of its combinational part, which is equal to the FPGA logic elements involved in the critical path. After state encoding and creating transition functions we can determine the maximum number L_{max} of arguments of the functions realized by the combinational part of the FSM.

$$L_{\max} = \max_{w_i \in W} |L(w_i)| \tag{8}$$

where $L(w_i)$ is a set of arguments of the function $w_i \in W$.

If the FSM is implemented on the basis of FPGA, the length of the critical path is determined only based on the maximum number of arguments. In case of use linear decomposition it can be defined as:

$$S = 1 + \text{int}((L_{\max} - n)/(n - 1)). \tag{9}$$

In case of using parallel decomposition it can be defined as:

$$S = \text{int}(\log_n L_{max}). \tag{10}$$

6   State Splitting Procedure

Splitting of internal FSM states is an operation of equivalent conversion of an FSM that does not change the algorithm of its functioning. During splitting of FSM states, the FSM type (Mealy, Moore) and general structure of the FSM does not change. For that reason, adding splitting procedure to synthesis method when implementing FSM on FPGA has practical usage and can be easily added to the process of digital system design.

The state splitting procedure is based on method from paper [10], but is modified to use not only one-hot but also binary state assignment. In distinction from paper [10], in this work the main strategy consists in finding the set D of all the FSM states satisfying the splitting conditions:

$$\exists a_i \in A, |B(a_i)| > 1 \tag{11}$$

$$\exists a_j \in B(a_i), \ r_j \leq r^* \tag{12}$$

Then for each state from D the trial splitting is carried out. Every state $a_i \in D$ is divided into two states. First state is corresponding to transitions from state $a_j \in B(a_i)$ where r_j = max. Second state is corresponding to rest of transitions to state a_i.

Finally a state a_i for splitting is selected that best satisfies the optimization criteria in terms of speed (minimization of values of critical path lengths: overall - S and average - L_{av}). The given process repeats as long as overall and average critical path length of currently splitting FSM is less than the overall and average critical path length of the last performed splitting.

7   Experimental Results

The method of minimization of finite state machines was implemented as a part of CAD system called ZUBR. To examine the efficiency of the presented method MCNC FSM benchmarks [15] were used. The experiments were performed using IntelFPGA Quartus Prime version 18.1 EDA tool.

Three parameters were taken from report files for further analysis: Maximum Clock Frequency - Fmax (F) in MHz, Core Dynamic Power (P) in mW and Total Logic Elements (C). For an implementation the EP4CE115F29I8L device – a popular low cost FPGA from the Cyclone IV E family was chosen. Two state assignment methods were chosen for FSM implementation: binary and one-hot encodings. All benchmarks in four cases (without minimization, minimized with STAMINA [16], synthesized with proposed method using speed estimation with linear decomposition and parallel decomposition) were implemented using identical Quartus Prime optimization parameters. The STAMINA was chosen to show the relation of speed of FSMs synthesized by the proposed method to speed of fully minimized FSMs.

The experimental results for binary and one-hot encodings are presented in Table 1 and Table 2, where C_0, F_0 and P_0 are, respectively, the number of used logic elements, maximum working frequency and dissipated dynamic power of the initial FSM (without minimization); C_1, F_1 and P_1 are, respectively, the number of used logic elements, maximum working frequency and dissipated dynamic power after minimization using STAMINA. Finally, parameters C_2, F_2 and P_2 are, respectively, the number of internal states, number of logic elements, maximum frequency and power after synthesis using proposed method with speed estimation using linear decomposition and C_3, F_3 and P_3 are, respectively, the number of internal states, number of logic elements, maximum frequency and power after synthesis using proposed method with speed estimation using parallel decomposition. The parameters shown in Table 2 (one-hot encoding) have subscript "O" instead of "B" in the Table 1 (binary encoding). *Mean* row contains the average values.

Table 1. The experimental results for binary encoding

Name	C_{OB}	F_{OB}	P_{OB}	C_{1B}	F_{1B}	P_{1B}	C_{2B}	F_{2B}	P_{2B}	C_{3B}	F_{3B}	P_{3B}
BBSSE	70	239.69	0.22	142	139.66	0.24	51	320.62	0.22	70	239.12	0.22
EX1	297	108.86	0.43	297	108.86	0.43	310	105.71	0.31	198	126.81	0.28
S1488	491	98.38	0.87	491	98.38	0.87	452	98.81	0.8	468	100.84	0.79
S1494	468	100.11	0.81	468	100.11	0.81	477	101.58	1.02	426	115.06	0.83
S208	145	124.05	0.24	145	124.05	0.24	175	115.79	0.27	122	134.55	0.23
S420	151	128.14	0.25	151	128.14	0.25	210	103.09	0.28	133	136.31	0.25
S820	247	100	0.34	262	118.91	0.4	243	120.55	0.3	216	121.94	0.3
S832	297	115.33	0.35	255	113.11	0.31	231	120.64	0.34	222	115	0.29
SAND	328	100.83	0.47	328	100.83	0.47	262	131.86	0.32	266	129.08	0.33
SSE	70	239.69	0.22	142	139.66	0.24	51	320.62	0.22	70	239.12	0.22
TMA	118	215.56	0.23	118	180.31	0.23	105	225.23	0.23	105	225.23	0.23
Mean	243.82	142.79	0.40	254.45	122.91	0.41	233.36	160.41	0.39	208.73	153.01	0.36

The analysis of Table 1 shows that after the synthesis, the average benefit in the speed of the FSM using linear decomposition makes 1.08 times, and on occasion (example *bbsse*) 1.34 times. When the parallel decomposition was used, average gain on the speed parameter was 1.09 times and on occasion (*sand*) 1.28 times. In comparison to STAMINA the average increase of the speed of the FSM makes 1.19 times, and on occasion (*bbsse*) 2.3 times using linear decomposition and 1.24 times for parallel decomposition (occasionally almost 2 times). The above gains were calculated as an relation of speed of FSMs designed with proposed method to speed of FSMs designed with reference method. An average benefit is a geometric mean of these relations for all considered FSMs. In 7 of 11 cases for the linear and 9 of 11 cases for parallel decomposition the number of used logic elements (cost) is also reduced. Accordingly, the number of used logic elements was lower in 7 of 11 examples for linear decomposition and in all cases for parallel decomposition in comparison to STAMINA. Dissipated power is in almost all cases a bit lower comparing to initial FSM and STAMINA minimized FSM.

Table 2. The experimental results for one-hot encoding

Name	C_{00}	F_{00}	P_{00}	C_{10}	F_{10}	P_{10}	C_{20}	F_{20}	P_{20}	C_{30}	F_{30}	P_{30}
BBSSE	37	282.09	0.22	116	184.09	0.24	40	386.55	0.22	41	366.3	0.22
EX1	216	157.23	0.32	216	157.23	0.32	157	187.76	0.26	220	175.78	0.31
S1488	364	168.8	0.44	364	168.8	0.44	335	175.44	0.46	341	166	0.49
S1494	328	162.84	0.4	328	162.84	0.4	313	159.34	0.38	345	166	0.44
S208	150	171.94	0.27	150	171.94	0.27	115	181.79	0.25	115	185.7	0.25
S420	135	173.55	0.26	135	173.55	0.26	127	193.5	0.25	127	193.5	0.25
S820	227	143.25	0.34	238	152.74	0.38	203	157.06	0.32	219	159.03	0.34
S832	247	144.26	0.36	228	151.56	0.34	212	161.08	0.33	224	150.99	0.34
SAND	221	164.1	0.34	221	164.1	0.34	180	185.46	0.32	246	184.95	0.36
SSE	37	332.01	0.21	116	184.09	0.24	40	386.55	0.22	41	366.3	0.22
TMA	84	393.39	0.23	96	241.31	0.24	85	384.62	0.23	105	394.79	0.24
Mean	186.00	208.50	0.31	200.73	173.84	0.32	164.27	232.65	0.29	184.00	228.12	0.31

The analysis of Table 2 shows that after the synthesis, the average gain on the speed of the FSM using linear decomposition makes 1.11 times, and occasionally (*bbsse*) 1.37 times. In case of the parallel decomposition the average gain on the speed parameter was 1.09 times and in best case (*bbsse*) 1.37 times. In comparison to STAMINA the average increase of the speed of the FSM makes 1.26 times, and on occasion (*bbsse*) 2.1 times using linear decomposition and 1.24 times for parallel decomposition (occasionally almost 2 times). In 8 of 11 cases for the linear and in 5 of 11 cases for parallel decomposition the cost of implementation is also reduced. Accordingly, the number of used logic elements was lower in 6 of 11 examples for parallel decomposition and in all cases for linear decomposition in comparison to STAMINA. There was also observed a little decrease of power consumption in case of application of proposed method comparing to initial FSM and STAMINA minimized FSM.

The comparison of average results for all state assignment methods is presented in Table 3. The analysis shows that the average results obtained using presented approach are better than results obtained using initial FSM in both styles of encoding used. The one-hot encoding style was the least area and power consumable and also fastest solution for all synthesis styles used. The average results obtained using presented approach are in all cases better than results obtained after minimization using the STAMINA in both styles of encoding used.

Table 3. The comparison of average results for all considered parameters

Parameter	Encoding	Initial FSM	STAMINA	Linear decomposition	Parallel decomposition
Speed [MHz]	Binary	142.79	122.91	160.41	153.01
	One-hot	208.50	173.84	232.65	228.12
Power [mW]	Binary	0.40	0.41	0.39	0.36
	One-hot	0.31	0.32	0.29	0.31
Cost [LC]	Binary	243.82	254.45	233.36	208.73
	One-hot	186.00	200.73	164.27	184.00

Comparing results from proposed method application in 7 of 11 cases speed estimation using parallel decomposition was better or equal to linear decomposition for binary encoding. On the other hand, in 9 of 11 cases speed estimation using linear decomposition was better or equal to parallel decomposition for one-hot encoding. The conclusion is that the parallel decomposition is more suitable method of speed estimation for binary encoding and its modification, but linear decomposition is a better way to estimate speed when using one-hot encoding.

8 Conclusion

In this paper, the new method for the synthesis of FSMs is proposed. In this method, the speed parameter is taken into account already in the first stage of the synthesis. It is shown that the speed is increased due to the combined splitting-merging procedure. The presented method allows increasing speed of FSM in almost all cases, also simultaneously with area and power decrease. Additionally the method allows in few cases to reduce the number of internal states of FSM. Two methods of speed estimation were presented: linear decomposition more suitable to estimate speed when using one-hot encoding and the parallel decomposition which was better method for binary encoding and its modification. The method can be further elaborated by considering other optimization criteria (e.g. power and area) and by improving optimization algorithms.

Acknowledgements. The work was supported by the grant from Bialystok University of Technology and funded with resources for research by the Ministry of Science and Higher Education in Poland.

References

1. Miyazaki, N., Nakada, H., Tsutsui, A., Yamada, K., Ohta, N.: Performance improvement technique for synchronous circuits realized as LUT-based FPGA's. IEEE Trans. Very Large Scale Integr. VLSI Syst. **3**(3), 455–459 (1995)
2. Jozwiak, L., Slusarczyk, A., Chojnacki, A.: Fast and compact sequential circuits through the information-driven circuit synthesis. In: Proceedings of the Euromicro Symposium on Digital Systems Design, Warsaw, Poland, 4–6 September 2001, pp. 46–53 (2001)
3. Huang, S.-Y.: On speeding up extended finite state machines using catalyst circuitry. In: Proceedings of the Asia and South Pacific Design Automation Conference (ASAP-DAC), Yokohama, January–February 2001, pp. 583–588 (2001)
4. Jóźwiak, L., Ślusarczyk, A., Gawlowski, D.: Multi-objective optimal FSM state assignment. In: Proceedings of the 9th Euromicro Conference on Digital System Design (DSD 2006), Dubrovnik, Croatia. IEEE (2006)
5. Czerwiński, R., Kania, D.: Synthesis method of high speed finite state machines. Bull. Pol. Acad. Sci. Tech. Sci. **58**(4), 635–644 (2010)
6. Oliveira, D.L., Bompean, D., Curtinhas, T., Faria, L.A.: Design of locally-clocked asynchronous finite state machines using synchronous CAD tools. In: Proceedings of 4th Latin American Symposium on Circuits and Systems (LASCAS), Cusco, Peru. IEEE (2013)

7. Pedroni, V.A.: Introducing deglitched-feedback plus convergent encoding for straight hardware implementation of asynchronous finite state machines. In: Proceedings of the IEEE International Symposium on Circuits and Systems (ISCAS), May, pp. 2345–2348. IEEE (2015)

8. Barbosa, F.T.D.F., De Oliveira, D.L., Curtinhas, T.S., de Abreu Faria, L., Luciano, J.F.D.S.: Implementation of locally-clocked XBM state machines on FPGAs using synchronous CAD tools. IEEE Trans. Circuits Syst. I Regul. Pap. **64**(5), 1064–1074 (2017)

9. Solov'ev, V.V.: Splitting the internal states in order to reduce the number of arguments in functions of finite automata. J. Comput. Syst. Sci. Int. **44**(5), 777–783 (2005)

10. Salauyou, V.: Synthesis of high-speed finite state machines in FPGAs by state splitting. In: Saeed, K., Homenda, W. (eds.) CISIM 2016. LNCS, vol. 9842, pp. 741–751. Springer, Cham (2016). https://doi.org/10.1007/978-3-319-45378-1_64

11. Klimowicz, A.: Performance targeted minimization of incompletely specified finite state machines for implementation in FPGA devices. In: Proceedings of Euromicro Conference on Digital System Design, Vienna, pp. 145–150 (2017)

12. Klimovich, A.S., Solov'ev, V.V.: Minimization of mealy finite-state machines by internal states gluing. J. Comput. Syst. Sci. Int. **51**(2), 244–255 (2012). https://doi.org/10.1134/S1064230712010091

13. Klimovich, A.S., Solov'ev, V.V.: A method for minimizing Moore finite-state machines by merging two states. J. Comput. Syst. Sci. Int. **50**(6), 907–920 (2011). https://doi.org/10.1134/S1064230711040113

14. Zakrevskij, A.D.: Logic Synthesis of Cascade Circuits. Nauka, Moscow (1981). (in Russian)

15. Yang, S.: Logic synthesis and optimization benchmarks user guide. Version 3.0. Technical report. North Carolina. Microelectronics Center of North Carolina (1991)

16. Rho, J.-K., Hachtel, G., Somenzi, F., Jacoby, R.: Exact and heuristic algorithms for the minimization of incompletely specified state machines. IEEE Trans. Comput. Aided Des. **13**, 167–177 (1994)

Securing Event Logs with Blockchain for IoT

Mateusz Kłos⬤ and Imed El Fray$^{(\boxtimes)}$⬤

Faculty of Computer Science and Information Technology,
West Pomeranian University of Technology, Szczecin, Poland
{mateusz.klos,ielfray}@zut.edu.pl

Abstract. The Internet of Things (IoT) is growing in popularity in recent years. With the increasing use, security threats are also becoming a bigger concern, especially considering new challenges, like limited computational power, low storage capacity and unprecedented number of independent, uncoordinated hardware device manufacturers. Unfortunately, modern attacks are more and more sophisticated and some attackers may even use anti-forensics techniques to hide any evidence of their malicious activity. As a result, digital forensics will be crucial in investigating crimes committed against IoT devices. One of the challenges is to create secure, lightweight and tamper-proof event log for the IoT system. Proposed solution relies on a blockchain to store event logs, guaranteeing the integrity of data. Event logs from IoT devices are being sent to multiple servers using multiple channels of communication to ensure logs availability and to move most of computational effort from the IoT device to the server. Logs are (optionally) encrypted to provide confidentiality of stored data. A set of security and performance tests were performed to prove effectiveness of proposed solution.

Keywords: IoT · Blockchain · Event log · IoT devices · IoT security

1 Introduction

The term, Internet of Things (IoT), referring to uniquely identifiable objects, 'things' and their virtual representations in an internet-like structure, has received much attention in recent years. It provides an integration of objects and sensors that can communicate directly with one another without human intervention. The IoT system includes various physical devices, monitors and gathers all types of data on machines and social life [1]. Concept was first proposed in 1998 [2] and, with growing popularity, it faces new challenges, also in the security subject because of following reasons:

1. the IoT extends the traditional internet with mobile network and sensor network.
2. every 'thing' will be connected to the Internet, and 3) all these devices will communicate with each other. Security concerns are even bigger when the IoT system will be applied to the crucial areas, such as medical service, healthcare, and intelligent transportation [3].

Many of IoT-related security challenges are variations of issues previously seen in technology but several key differences exist between the IoT and conventional wireless networks in term of dealing with security and privacy. The deployment of the IoT is

© Springer Nature Switzerland AG 2020
K. Saeed and J. Dvorský (Eds.): CISIM 2020, LNCS 12133, pp. 77–87, 2020.
https://doi.org/10.1007/978-3-030-47679-3_7

different compared to the traditional Internet, e.g. devices are set up on the Low power and Lossy Networks (LLNs), while others have extremely dynamic topologies relying on the application. LLNs are characterized by great data losses due to node impersonation. They are strained by dynamism, memory and processing power [4]. These aspects are not considered for the standard Internet. Moreover, sensor nodes in the IoT have limited computational power and low storage capacity. It makes frequency hopping communication application and public key encryption to secure the IoT devices impossible so lightweight encryption technology, including lightweight cryptographic algorithm is required [5].

1.1   Motivation

The IoT network is prone to man-in-the-middle and counterfeit attacks, in the network layer. These attacks may result in capturing from or sending fake information to communicating nodes in the network [6]. Another important issue involves an unprecedented number of independent, uncoordinated hardware device manufacturers. For many of these device manufacturers, the IoT-specific components and functionality are not areas of expertise. It results in creating a soft target [7]. In the era of the IoT, digital forensics will be crucial in investigating crimes committed against IoT devices [8, 9]. However, modern attacks are more and more advanced and anti-forensics techniques are used by attackers to hide any evidence that can be traced back to them [10]. One of the challenges is to create secure, lightweight, tamper-proof event log for the IoT system.

2   Related Works

Logs are the most important source of information used to discover what has happened in the systems and networks. They are used not only for troubleshooting and performance optimisation, but also in forensics, when some malicious actions against system or network were performed. Logs may be used as an evidence in court against malicious actors. However, it may be difficult if not impossible to tell the difference between authentic and tampered log records [11].

During operation, computer systems often generate event logs. Secure system should verify an integrity of the event logs to ensure event data is accurate, including individual events within the event logs. and to verify that the event logs have not been altered [12]. The most basic functions that a log system needs to cover are log collection, log storage, and log query analysis.

Moreover, in a distributed environment, e.g. in IoT systems, log files of different origins and format types should be collected by the log management system, and then unified, centralized storage and management should be performed on them. Best logging systems are characterised by flexible log collections capabilities, secure persistent storage capabilities, a highly available distributed architecture design, secure and efficient log queries function, and the ability to self-repair errors. The log storage may be increased when it is needed. Unfortunately, most log systems used today are not designed adequately addressing security issues, so when the system is attacked, it is possible for malicious actors to cover up their intrusions by deleting or tampering log records without much effort [8].

There are four distinctive categories of the secure log system implementations:

3. Encryption-based solutions utilise symmetric or asymmetric encryption to preserve the logs confidentiality. Techniques based on symmetric algorithms are less computationally expensive but strict protection of shared secret key is required. Asymmetric algorithms resolve the problem of key management but they require a specialized infrastructure (PKI) and they are characterised by worse performance which is a meaningful disadvantage in the IoT context. The solution based on forward integrity was proposed in 1997 [13]. It uses the Message Authentication Codes (MAC) of received logs and add them sequentially. Symmetric key is updated at regular intervals. Even if attacker compromise one key, he must modify past entries, what is difficult task. This solution preserves the logs integrity but it does not protect their confidentiality and availability. A similar solution was proposed by Schneier and Kelsey [14]. It also utilises chain of log MACs in order to protect them. In this solution, the secret key is pre-shared between the logging devices and updated after each log entry. Another encryption-based solution relies on FssAgg (Forward-Secure Sequential Aggregate) [15]. FssAgg is a scheme that relies on aggregated chain of signatures in order to achieve public verification and to prevent attacks aiming to delete logs. Every device having the public key is able to verify the integrity of the received log. BAF (Blind-Aggregate-Forward) [16] and FI-BAF (Fast-Immutable BAF) are extensions of the concept proposed in FssAgg. Both these schemes are suitable for distributed systems. BAF can produce publicly verifiable forward secure and aggregate signatures with near-zero computation, signature storage, and signature communication overheads for the loggers, without any online Trusted Third Party (TTP) support [17]. The downside of BAF is necessity to verify all log entries in order to verify one of them. Another problem is that, if the verification fails, it is not possible to know which entry has been modified. FI-BAF extends BAF by allowing selective verification of log entries without compromising the security of BAF as well as retaining its computational efficiency [18]. Different encryption-based approach was proposed in 2009 [19], utilising Merkle trees for tamper-evident logging. This scheme relies on a history tree, enabling the efficient verification of each log event. The downside of this solution is its requirement of logarithmic complexity in order to generate proof of integrity. The downside of encryption-based solutions is their inability to ensure log availability if end devices are compromised.

4. There are known hardware solutions proposed for secure logging problem. Some of them are based on the Software Guard Extensions (SGX) introduced by Intel in 2015, with the sixth generation of Intel Core microprocessors. SGX involves encryption by the CPU of a portion of memory. The enclave is decrypted on the fly only within the CPU itself, and even then, only for code and data running from within the enclave itself [20]. An example of SGX-based solution for secure logging is LogSafe proposed in 2018 [21]. It provides the Confidentiality, Integrity and Availability of logs and is designed for IoT systems. LogSafe stores the device logs on one or more cloud nodes running SGX. It ensures confidentiality by applying the

AES (Advanced Encryption Standard) algorithm. The integrity is provided by hash chaining. The availability is ensured by backups performed on a different node and a different location to mitigate impact of DoS (Denial of Service) attacks. The downside of this solution is limited availability protection, given that a backup is done on only one device and presents a storage overhead if the data is duplicated. Moreover, there are known attacks on SGX [22–24] and secure hardware solutions are limited by requirement of specialised hardware (cost) and data availability ensured by saving multiple copies of the same data, which introduces high communication and storage overhead.

5. A cloud-based log was proposed in 2013 [25]. It introduces logging-as-a-service by pushing logs into the cloud. However, it does not guarantee availability unless the backup is done.

6. Blockchain can be used as a part of the secure logging system. In 2019, distributed, blockchain based storage was proposed [27]. Authors have created decentralized storage with the verification system for IoT digital evidence. In this solution, the metadata and hash are saved in Cyber Blockchain Trust (CBT). The downside of this solution is that the privacy of evidence is not well protected, as it is saved in a permissioned ledger built on top of the HyperLedger fabric. Moreover, the performance of this solution makes it inappropriate for limited IoT device. This issue concern most of blockchain solutions, as handling numerous devices may result in large ledger sizes. In consequence, high computation and a lot of resources is required.

In view of analyzed solutions and challenges, authors are proposing highly scalable and configurable blockchain-based model. Most of blockchain-based solutions are not suitable for IoT devices because of their limited resources and computational capabilities. There is lacking of flexible solution giving balance between performance and high security. The details of proposed model are presented below.

3   Blockchain-Based Secure Log Solution for IoT

Blockchain received a lot of attention recently, not only from scholars but also in mass media. Blockchain is a distributed, continuously growing list of records (called blocks) which are linked and secured using cryptography. The nodes (peers) are part of the P2P network and every node is able to verify the integrity of the whole block. Each peer can create a block. The block of the first successful peer is propagated to the rest of peers. The new block can be properly created when the verification of the previous block was successful. It means that the reliability of the entire block is increasing with the length of the block. The correctness of the past blocks is verified by the comparison of hash values [26].

Below is a sample architecture and implementation of a secure, lightweight and tamper-proof event log for a blockchain-based IoT system.

3.1   The Proposed Blockchain-Based Secure Log Solution for IoT

The proposed model utilises blockchain to provide secure event log for IoT devices. It is highly scalable solution, sufficient for both smart home and enterprise purpose.

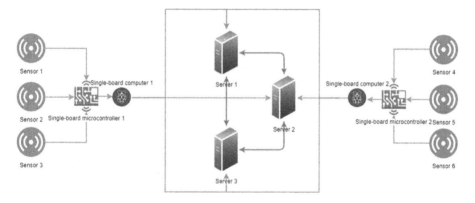

Fig. 1. Architecture of blockchain-based secure log solution for IoT.

The proposed architecture presented in Fig. 1 consists of two single-board microcontrollers (Arduino Mega 2560 R3), each of these microcontrollers gather data from three sensors (ultrasonic sensors) and send it to single-board computers 1 and 2 (Raspberry Pi 2 B) via USB serials. It is possible to replace Arduino + Raspberry Pi setup with ESP32, since the calculation is done on a virtual server. It reduces cost and allows to avoid latency between Arduino and Raspberry Pi (however, this latency was excluded from performance tests). Data is processed by single-board computers and every event is sent via HTTPS to each remote server as a blockchain transaction. Servers (nodes) are responsible for all blockchain operations and all nodes are communicating with each other via encrypted connection (HTTPS). This solution can be used with numerous, various sensors, depending on its purpose and one needs. It works with at least one working server but distributed nature of the blockchain implies using more for increased security. The more nodes, the better security but it may affect performance when the system is configured to send each event to every node (security over performance). The most secure configuration requires using multiple, independent channels of communication between micro-computers and each node to mitigate risk of attacks on connection between them, leading to whole module being cut off from the servers. Optionally, event logs may be encrypted using AES-256 Password Based Encryption on the server side, before they will be added to the blockchain.

3.2   Implementation

Sensors are connected to the single-board microcontroller, Arduino Mega 2560 R3 which is programmed in C++ and is sending all sensor data to the single-board microcomputer- Raspberry Pi 2 B programmed in Python 3.4.2, using Raspbian Jessie.

Microcomputer has a list of all servers available in the system. The first time the device is connected to the network, it generates the pair of keys. The public key is sent over HTTPS to each server known by the device. The microcomputer interprets sensor data, searching for valid events (e.g. "room entrance" event from ultrasonic sensor or "overheating" event from a temperature sensor). Microcomputer sends data, signed with its private key, in json format. Sent data includes: including date, time, device ID, event, and severity level (Emergency, Alert, Critical, Error, Warning, Notice, Informational, Debug). Exemplary event sent by microcomputer:

{"date": "12/01/2020", "time": 21:23:30, "device_id": 12345, "event": "Entrance", "level": "Informational"}

Data is sent via HTTPS POST requests (using Requests Python library) to each node so communication between microcomputer and each server is encrypted. Servers are listening on port 5000 for incoming data (Python Flask framework is used) registered as a transaction for another blockchain's block. Another server module, called miner.py, is trying to mine the new block every 10 s and if it finds the awaiting transaction (new registered event) it starts the mining process as shown in Fig. 2.

```
@app.route('/mine', methods = ['GET'])
def mine():

    if (len(blockchain.current_transactions) > 0):
        # We run the proof of work algorithm to get the next proof
        last_block = blockchain.last_block
        last_proof = last_block['proof']
        proof = blockchain.proof_of_work(last_proof)

        # Forge the new Block by adding it to the chain
        previous_hash = blockchain.hash(last_block)
        block = blockchain.new_block(proof, previous_hash)

        response = {
            'message': "New Block Forged",
            'index': block['index'],
            'transactions': block['transactions'],
            'proof': block['proof'],
            'previous_hash': block['previous_hash'],
        }
        return jsonify(response), 200
    else:
        response = {
            'message': "No transactions"
        }
        return jsonify(response), 200
```

Fig. 2. Implementation of mining function

This solution utilises proof of work algorithm, computing a number p' such that hash(pp') contains leading 4 zeroes, where p is previous p', so:

- p is the previous proof
- p' is the new proof

The algorithm is looking for a number p that when hashed with the previous block's proof, gives a hash with 4 leading zeroes.

SHA-256 is used for hashing blocks. Difficulty may be increased by configuring blockchain to compute hashes until the one containing leading 5 zeroes is found but it extremely affects overall performance and it is not suitable for IoT. When the proof is found, the block is added to the local copy of blockchain by each node. Miner module working on each node is resynchronizing blockchain in 60 s intervals by sending requests (HTTPS) to all known nodes (each node is configured with the list of other nodes IPs) and performing consensus algorithm when conflicts are encountered. A conflict is when one node has a different blockchain to another node. This implementation of consensus algorithm assumes that the longest, valid chain on the network is authoritative.

However, before a node accepts new blockchain and replace its own (old blockchain), it verifies following criteria:

1. the new blockchain must be longer than the old blockchain,
2. the new blockchain must have all transactions (log events) that are stored in the old blockchain, in correct order,
3. if criteria 1 and 2 are met, the node is verifying hashes of new blocks (blocks that are not present in the old blockchain),
4. if hashes are correct, the old blockchain is replaced with the new blockchain.

Only blocks with transactions are mined and every event is logged by each node. In result, the system is working even when all but one nodes are shutdown (e.g. as a result of malicious actions). After they are reconnected to the network, nodes will be resynchronized with the working one without any further actions so it is an efficient solution for recoveries after potential server failures. Each block has an index, a timestamp (in Unix time), a list of transactions (events), a proof and the hash of the previous block. Exemplary blockchain, containing first two blocks- genesis block and a block with single transaction, is shown below in Fig. 3.

```
{
    "chain": [
        {
            "index": 1,
            "previous_hash": 1,
            "proof": 100,
            "timestamp": 1585160312.0635436,
            "transactions": []
        },
        {
            "index": 2,
            "previous_hash": "32080f36107d35105b207b4740a7ff78e62e0401eaa11e4f3b7787c1115e2172",
            "proof": 35293,
            "timestamp": 1585160350.0799255,
            "transactions": [
                {
                    "date": "25/03/2020",
                    "device_id": 12345,
                    "event": "Entrance",
                    "level": "Informational",
                    "time": "18:19:05"
                }
            ]
        }
    ],
    "length": 2
}
```

Fig. 3. First two blocks of an exemplary blockchain

3.3 Results and Discussion

The performance of proposed solution was measured in local transactions-per-second, eliminating the network delay factor. The delay between Arduino and Raspberry Pi was excluded from measurements, because proposed solution can be also implemented on the single board (e.g. ESP32), so the outcome of this paper will be valid for this case. In the first run, the script written in Python was sending requests for 10 s to one of nodes. Node has been running on virtual machine with Ubuntu 18.04.4 and 2 GB RAM. 1090 transactions were registered by the node and mined, giving the result of 109 TPS. It is not the fastest solution because, e.g. "Blockchain Based Data Logging And Integrity Management System For Cloud Forensics" proposed by J.H. Park, J.Y. Park and E.N. Huh reached the number of 166.67 TPS. However, proposed model has better performance than well-known cryptocurrencies like Bitcoin (6.41 TPS), Ethereum (15.65 TPS) or Zcash (unshielded: 26.67 TPS and shielded: 6.67 TPS). This performance should be sufficient for most of use cases. Another test runs were performed with 2 and 3 nodes in the network. Each transaction were sent to all nodes.

Number of nodes	1	2	3
TPS	109	48.9	32

As expected, performance is lower with increasing number of nodes awaiting every transaction. It is the cost of higher security. For most purposes, sending each event to only one of available nodes may be reasonable, giving the best performance. For the most sensitive systems, requiring higher security level, sending data to more nodes, using multiple, independent channels of communications may be better solution.

Additionally, a security test (sniffing and denial-of-service attack) have been performed to solve the problem. The results presented in Fig. 4 show that the proposed solution is resistant to sniffing attacks, because transmission between microcomputers and nodes are encrypted (HTTPS).

Proposed model mitigates the risk of Denial-of-service attacks on individual nodes. The test with shutting down two of three working nodes was performed. The network works without any issues with only one, last node and the blockchain is propagated to other nodes automatically when they are available in the network again.

It is also resistant to Man-in-the-Middle attacks because of HTTPS and usage of certificates signed by a trusted Certificate Authority, which are stored on each node. It also gives protection against Replay attacks. Eavesdropping is not possible as shown in Fig. 4 and any trial of altering or breaking the communication failed because, regardless of attacks, correct data was sent via other, not attacked channels of communication.

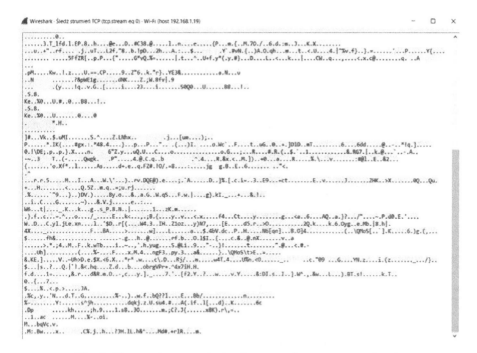

Fig. 4. Encrypted communication between microcomputer and node

4   Conclusion

In this paper, the tamper-proof, immutable event log, suitable for IoT devices, was proposed. It is flexible and scalable. Most of storage and computationally expensive tasks were delegated to the servers, taking into account limited performance of IoT devices. This solution may be used e.g. in smart homes but also in the most security demanding areas, requiring dedicated modules and custom configuration. This solution may be helpful for digital forensics in investigating crimes committed against IoT devices.

- However, there is still room for improvement. The risk of Denial-of-service attacks on microcomputers can be mitigated by checking their availability from nodes, so if the microcomputer is not responding then relevant event should be logged. Since the private keys are stored on devices, the possibilities of protection against private key being stolen should be considered as well as the ways to improve overall performance.

References

1. Yan, Z., Zhang, P., Vasilakos, A.V.: A survey on trust management for Internet of Things. J. Netw. Comput. Appl. **42**, 120–134 (2014)
2. Weber, R.H.: Internet of Things – new security and privacy challenges. Comput. Law Secur. Rev. **26**, 23–30 (2010)
3. Suo, H., Wan, J., Zou, C., Liu, J.: Security in the Internet of Things: a review. In: Proceedings of the IEEE International Conference on Computer Science and Electronics Engineering (ICCSEE), vol. 3, pp. 648–651 (2012)
4. Lu, C.: Overview of Security and Privacy Issues in the Internet of Things. http://www.cse.wustl.edu/~jain/cse574-14/ftp/security.pdf
5. Alaba, F.A., Othman, M., Hashem, I.A.T., Alotaibi, F.: Internet of Things security: a survey. J. Netw. Comput. Appl. **88**(Suppl. C), 10–28 (2017)
6. Zhao, K., Ge, L.: A survey on the Internet of Things security. In: Proceedings of the 9th International Conference on Computational Intelligence and Security, CIS 2013, pp. 663–667 (2013)
7. van Oorschot, P.C.: Internet of Things security: is anything new? IEEE Secur. Priv. **16**, 3–5 (2018)
8. Yaqoob, I., Hashem, I.A.T., Ahmed, A., Kazmi, S.M.A., Hong, C.S.: Internet of Things forensics: recent advances, taxonomy, requirements, and open challenges. Future Gener. Comput. Syst. **92**, 265–275 (2019)
9. Chernyshev, M., Zeadally, S., Baig, Z., Woodward, A.: Internet of Things forensics: the need, process models, and open issues. IT Prof. **20**(3), 40–49 (2018)
10. Noura, H.N., Salman, O., Chehab, A., Couturier, R.: DistLog: a distributed logging scheme for IoT forensics. Ad Hoc Netw. (2019). Article 102061
11. Blockchain Backed Log Assurance. https://guardtime.com/solutions/blockchain-backed-log-assurance. Accessed Jan 2020
12. Roberts, C.L., Windley, J.: Storing and verifying event logs in a blockchain. International Business Machines Corporation, US20180157700A1 (2018)
13. Bellare, M.: Forward integrity for secure audit logs. Technical report (1997)
14. Schneier, B., Kelsey, J.: Secure audit logs to support computer forensics. ACM Trans. Inf. Syst. Secur. **2**(2), 159–176 (1999)
15. Ma, D., Tsudik, G.: A new approach to secure logging. Trans. Storage **5**(1), 2:1–2:21 (2009)
16. Yavuz, A.A., Ning, P.: Baf: an efficient publicly verifiable secure audit logging scheme for distributed systems. In: 2009 Annual Computer Security Applications Conference, pp. 219–228. IEEE (2009)
17. Sohraby, K., Minoli, D., Znati, T.: Wireless Sensor Networks -Technology, Protocol and Applications, 2nd edn. (1991)
18. Shibin, D., Blessed Prince, P.: Survey on efficient and forward secure schemes for unattended WSNs. Int. J. Innov. Technol. Explor. Eng. (IJITEE) **2**(3), 54–57 (2013). ISSN 2278–3075
19. Crosby, S.A., Wallach, D.S.: Efficient data structures for tamper-evident logging. In: USENIX Security Symposium, August 2009
20. Researchers Use Intel SGX To Put Malware Beyond the Reach of Antivirus Software - Slashdot. it.slashdot.org. Accessed Jan 2020
21. Nguyen, H., Ivanov, R., Phan, L.T.X., Sokolsky, O., Weimer, J., Lee, I.: LogSafe: secure and scalable data logger for IoT devices. In: IEEE/ACM Third International Conference on Internet-of-Things Design and Implementation (IoTDI), pp. 141–152. IEEE (2018)

22. Chirgwin, R.: Boffins show Intel's SGX can leak crypto keys. The Register, 7 March 2017. https://www.theregister.co.uk/2017/03/07/eggheads_slip_a_note_under_intels_door_sgx_can_leak_crypto_keys/. Accessed 1 May 2017
23. Sample code demonstrating a Spectre-like attack against an Intel SGX enclave. github.com/lsds/spectre-attack-sgx. Accessed Jan 2020
24. Schwarz, M., Weiser, S., Gruss, D.: Practical enclave malware with Intel SGX. arXiv:1902.03256 [cs.CR], 08 February 2019
25. Zawoad, S., Dutta, A.K., Hasan, R.: SecLaaS: secure logging-as-a-service for cloud forensics. In: Proceedings of the 8th ACM SIGSAC Symposium on Information, Computer and Communications Security, ASIA CCS 2013, pp. 219–230. ACM, New York (2013)
26. Park, J.H., Park, J.Y., Huh, E.N.: Block chain based data logging and integrity management system for cloud forensics. In: International Conference on Computer Science, Engineering & Applications, pp. 149–159 (2017)
27. Bellini, A., Bellini, E., Gherardelli, M., Pirri, F.: Enhancing IoT data dependability through a blockchain mirror model. Future Internet **11**(5), 117 (2019)

Securing Data of Biotechnological Laboratories Using Blockchain Technology

Krzysztof Misztal[1,2]([⊠]) [iD], Tomasz Służalec[1,2] [iD],
and Aleksandra Kubica-Misztal[2] [iD]

[1] Jagiellonian University, Kraków, Poland
{krzysztof.misztal,t.sluzalec}@uj.edu.pl
[2] diCELLa Ltd., Kraków, Poland
a.kubica-misztal@dicella.com

Abstract. A few years ago blockchain technology was used in cryptocurrency. Nowadays, a variety of diverse areas are seeing the benefits of applying this technological approach to their needs. One way transactions without reverse mode is making blockchain a desirable platform for maintaining data. The authenticity, transparency and authorization of it make it ideal for healthcare or laboratory data systems. Research data should be authorized by a specific employee and never changed. The information collected should be never forged or falsified. Designing a blockchain system with access and permission rules is ideal in such a situation. In this article, we present the adaptation of blokchain technology to medical research laboratories and diagnostics. Afterwards, an RSA signature user can store information of any type and size. The new SHA-3 hash function is used to bind blocks together. This technological path makes laboratory workflow more efficient and fulfills restriction on medical laws.

Keywords: Blockchain · Laboratories · Database

1 Blockchain System

Blockchain system. In 2008, Satoshi Nakomoto published a paper [5] describing bitcoin, which is now referred to as the first cryptocurrency in the world. The prefix *crypto* is derived from the word cryptography. The bitcoin is a complete and complex system which provides the ability to perform anonymous and cashless transactions over the Internet. Bitcoin is based on a decentralized settlement book, which can be managed by any of the system's users. Such users are called miners because their reward for the management is the newly mined currency. The miners gather information about transactions and validate whether they can be performed. Then they choose the transactions to store in the settlement book. This book is known as a blockchain. Thanks to the used cryptography

K. Saeed and J. Dvorský (Eds.): CISIM 2020, LNCS 12133, pp. 88–96, 2020.
https://doi.org/10.1007/978-3-030-47679-3_8

algorithms, it is impossible to make any transaction as another user. Moreover, if a transaction is stored in the blockchain, no one can change or delete it.

It is worth noticing that while all bitcoin transactions are stored in the blockchain unencrypted, it is possible to keep encrypted data in the blockchain.

1.1 Bitcoin Technology

There are several implementations of blockchain and we will focus on the well known precursor bitcoin. For the more information the reader is reffered to [2]. The sellers or payers that want to make a transaction must submit their action to blockchain. Everyone that is connected to the blockchain system receive the information about the transaction. Receivers are called miners and are the clients which validate the action by algorithm. The validation by cryptographic algorithm is compound with two computation steps. Miners provide the computation power. Software for mining is free and simple. Moreover no certification is required so everyone can volunteers with their computers. Miner must validate existence of the bitcoin and that can be used for transaction. Then he checks the form and accepts the transaction. If the miner validate the whole block he is rewarded with a bitcoin. Block is has the detailed information about validation processes of transactions. The block is stored in the blockchain after it is validated by a fixed number of miners. What is more within every block are timestamps and a mathematically generated complex variable sums called hashes that secure the interruption in data block. These cryptographic hashes of blocks are compounded of preceding and current block. This distinct and unique sum make up the security signature that combine blocks into blockchain. The irreversibility of hash code makes blockchain immutable. This way we obtain digital trust. Decentralized validation of actions and saving them to the history makes mediator entity like banks useless to perform transaction. Example:

1. John wishes to pay bitcoins to Alice.
2. John's transaction is made.
3. Transactions are combined into a new block.
4. Blocks are broadcasted to miners.
5. Miners validates and accepts the transactions.
6. New block is encrypted with hash of last block.
7. New block become a part of blockchain.
8. Alice receives bitcoins.

1.2 Sensitive Data in Blockchain

Decentralized system can be widely used. The greatest benefit is that there is no need of third parties. As we have seen before, the decentralized data may be secured and immutable. Nowadays, most organizations collect our personal data in order to optimize their services. They want to predict our needs and create our digital personal image thus personal information in every aspect is a valuable resource in today economy. Although there are many benefits to a personalized system, the concern for our privacy is doubtful. What is more, now more often organizations have little control over the personal and sensitive data. That is why these information should not be handled by third-parties. The centralized system is vulnerable to hacker attacks, as well asphishing. The solution is for the user to own and be responsible for providing its information. We can do this by applying blockchain storage solution proposed by [8]. Authors proposed personal data management system that a advantage of blockchain and off-blockchain storage. This requires several key aspects: data ownership, data transparency audibility and fine-grained access control. The proposed platform devided to users that own and control their personal information and the guest services. Every user has information about collocated data and how is it accessed by organization. Services have delegated premissions from users to access their data. What is more, the owner of the data is able to make changes to the set of permissions to deny the access to data. This solution is similar to that of mobile apps. It will not have an impact on the user-interface because the access-control policies are stored in a blockchain. Only users are able to make changes in the blockchain (Fig. 1).

Fig. 1. Simplified schema of the above grant-access data blockchain.

Some ideas about blockchain apllication in health care can be seen in [2].

2   Laboratory

The results obtained in GMP and GLP laboratories have to be reliable. Any modification of partial results can make the final result fail. It is the reason why the lab records have to be collected precisely and without any forgery. It is also important to remember every the author of every record. The authorship has to be beyond any doubt, which means that nobody can create a record on behalf of another user and no user can deny his own authorship of any of his records. Moreover, the documentation has to be immutable so any stored data cannot be modified nor deleted. These requirements make users keep exact records, which objectively increases the reliability of the laboratory.

Solving this problem required us to cooperate with the employees of biotechnological laboratories and diagnostics. It required a great amount of time to translate the real life problem to algorithmic technological approach.

There were several key aspects for lab employees:

- constant access to warehouse- with restriction and access rules,
- rules of processing the production - to improve the process,
- rules of gathering lab results - the authorship of obtained data,
- rules of reporting everyday work - to avoid forgery and falsification,
- communication with clients.

All these aspects have a significant impact on the lab workflow.

2.1 Blockchain Application

The author of bitcoin, [5], stated that there can be a central authority which produces coins and validates the transaction. It brings up the problem that the authority would have infinite knowledge regarding every transaction. This problem can be solved in a very tricky way. Before we attempt to explain the solution to this let us focus on transaction that we define in blockchain for GMP and GLP laboratories. We want every data information to be unmodified and has its author. In our platform the transaction of a single information would represent a coin in bitcoin system.

Let us consider a scenario in which user A has created a note about some topic. Our system needs to remember the authorship of that note as well as the fact that user A cannot deny any part of that note. That is why the user certifies the note by his digital signature. The different B user stated that the note has faulty so he want to make a correction. He can do this but not in the obvious way. User B takes the note, adds the calculated hash code from that note, add a correction to the note. After his digital signature he submits the information. In this way every user can validate which part is A's and which is B's. We see a compatibility between bitcoin, crypto-coin and our data note. The correction mechanism made by B is similar to case of spending same cryptpo-coin twice. We observe that system witch verifies folding signatures would solve this problem, the same as in bitcoin scenario.

A different aspect is when a need to create information arises. Coins can be made in a specific regularized way. The information stored in our platform should be for a every authorized user. That is why we need a second type of coin that would represent the permission access for users. That is why every authorized laboratory user would have the ability to add a digital key of his coworker allowing him to make transitions in the blockchain. As well as every authorized user it the same way can deny specific user to access the system. To sum up, in the bitcoin system the information about from who to whom the coin is sent is stored. In out platform we store the information about:

- users *digital keys* with permission information - we track if the user is allowed to signature the transaction,
- data information - we track authorship of the laboratory data; this solution is general that any type of lab data can be stored.

2.2   Technological Details

Programming Language and Libraries. We choose specific cryptographic packages and programming language Python3 so we can crate application on a any technology. Although python apps are not as efficient as in C language but they are more portable. What is more professional programmers that are working with cyber-security chooses also python. In the prototype we used Django framework and cryptography python package becasue of the SHA-3 implemented hash function. It allows to simple and effective programming and is extending rapidly. After the theoretical analysis ensures us that these algorithms are sufficient to our needs.

Block Hash Function. The base and essential ingredient for the blockchain is a well-defined hash function. The National Institute of Standards and Technology (NIST) published two documents [1] and [3]. Where they certified only three hash functions:

- SHA-1,
- SHA-2,
- SHA-3.

The SHA-3 is technologically completely different. The SHA-1 and SHA-2 is based on the same mathematical mechanism. In 2017 Google announced [7] that SHA-1 is broken and should not be used for security. In comparison to SHA-1 and SHA-2, chosen SHA-3 is new and free of bugs algorithm.

User Digital Signature. Next important step in creating the blockchain is the algorithm for digital signatures and its validation. Here, the key solution is an asymmetric one-way cryptograhpic function. The National Institute of Standards and Technology (NIST) in [4] have defined three cyprographic algorithms that used digital signatures:

- RSA,
- DSA,
- ECDSA.

Due to the controversy with elliptic curves proposed by cryptographic authority Bruce Schneier [6] we chose the RSA algorithm with a key size of 4096 bits (Fig. 2).

Blocks Implementation Details. To avoid the future problems we designed the blocks in the form of dictionary with a predefined set of keys. Those objects can be easily serialized and deserlialized to the JSON format. Listing 1.1 shows seven keys for dictionary-blocks.

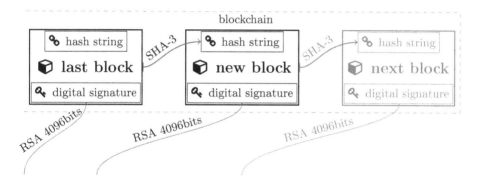

Fig. 2. Hash linking and RSA signature in blocks.

```
class BlockField(Enum):
    BLOCK_TYPE = 0
    TIMESTAMP = 1
    AUTHOR = 2
    SIGNATURE = 3
    OBJECT = 4
    PREVIOUS_HASH = 5
    NONCE = 6
```
Listing 1.1. Enum describing all parameters in block object.

We also defined five types of blocks shown at Listing 1.2.

```
class BlockType(Enum):
    GENESIS = 0
    INSERT_BLOCK = 1
    UPDATE_BLOCK = 2
    INSERT_KEY = 3
    REVOKE_KEY = 4
```
Listing 1.2. Enum describing possible types of blocks in the chain.

The *genesis block* is a first one in the chain. There can be only one *genesis block* in the blockchain. It contains the system administrator key and it cannot be rejected. Public key is controlled by admins managing the software, not in the hands of laboratory employees.

2.3   Platform

We have ended the work of implementation of blockchain technology prototype, moreover, we have made an unconventional modification of blockchain technology in the way that is adapted to the objectives given by our theoretical analysis.

We validated time resources that were saved by our system. There was a significant improvement. System was effective to time-consuming and resource-consuming problems. The designed platform was verified in laboratory conditions. We observed that the designed solution had an impact on the laboratory workflow (Fig. 3).

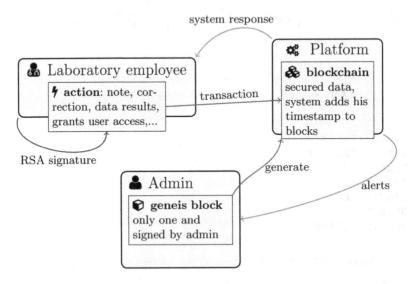

Fig. 3. Proposed laboratory blockchain platform usage. Platform is validating and storing signed and secured transaction. Additional lab-supporting algorithms perform analyses for the users basing on the blockchain data. Alerts are send when needed.

We have confirmed the effectiveness in every aspect of the use cases, as well as in the performance of our system. The processing time and memory usage by blockchain mechanism were not computation demanding.

Our designed template platform with small customization provides benefits that are widely desired in the laboratories. Below is the list of advantages that comes from proposed blokchain system.

1. Data are secured against unauthorized changes.
2. Blockchain security of data makes them authentic and repudiation.
3. Data stored in blockchain are transparent.
4. Due to GMP CFR 21 part 11 system allows to electronic documentation conducting.
5. System fulfils the acts of biobank laws.
6. System helps realization in tissue and cell procurement in order with doctor and dentist's law.
7. System fulfills the acts of pharmaceutic laws in terms of production advanced medicinal products.

8. System satisfies UE documents regarding production and medical therapy (transplanting cells, tissue and organs).
9. System additional algorithms archive the statistics improving the lab workflow.
10. System helps in stable managing the labolatory storage magazine.
11. System build in analytical algorithms save time for employees - list demands verification, production.

2.4 System Stability

The author in [5] introduce the *timestamp server* idea which binds blocks chronologically. In our proposed system there is only one *timestamp server*. The web or physical access to the server is limited to minimum. System takes the created and edited data in laboratory as well as key of users whose can submit these data. *Timestamp server* will sort transactions in the chronological order. When the new data is added system is validating its propriety. For example:

- if the note was made before,
- if edited content of the note corresponds to the last one in blockchain,
- if a user is allowed to add a note - system verifies if user *digital key* is active and permitted to add a content.

We have made stability list to determine if the blockchain is corrupted. Below is the checklist that is validated by system when inserting a block to blockchain:

1. the fist block is not a *genesis block*,
2. more then one block is *genesis block*,
3. *prev_hash* filed for *genesis block* is different then expected,
4. *timestamp* from name of *block* is not corresponding to the content of block,
5. *prev_hash* value of the *block* is not compatible with its *previous block*,
6. *hash string* do not begins with fixed length of zeros,
7. user *digital key* is added twice,
8. user *digital key* is deleted twice,
9. *genesis key* is deleted,
10. block added by unauthorized user - due to no *digital key*,
11. block added by unauthorized user - due to *digital key* rejection,
12. lack of mandatory filed in *block*.

We tried to hack our system. We used brute-force attacks but luckily without any success. The cryptographic packages the we have chosen in the carrying out of our system are dedicated to this kind of unwanted hacking access. Moreover the SHA-3 algorithm is used as the hash function is the one of the safest, most secure and the most efficient hash algorithms. In the case of non-quantum computers the SHA-3 algorithm should not be broken in the 20–30 years.

3   Conclusions

The blockchain technology is widely used in accounting and banking, but not exclusively in those fields. There are many projects adapting this technology in many different areas of daily lives. Our project, by using blockchain technology in the widely understood area of healthcare, is among them.

The requirements for the documentation in biotechnological labs cannot be enforced when records are being kept on paper. If one read the above descriptions carefully, one can notice that the requirements for lab records are satisfied by the main features of blockchain. In our talk we would like to present our innovative system for securing data of biotechnological laboratories using blockchain technology. Our presentation will show what data is stored in our blockchain and how the correctness of newly added *transactions* is checked.

Acknowledgement. This work was supported by the Regional Operational Programme for the Małopolska Region 2014–2020 as research project: Development of an innovative data security system in biotechnology and diagnostic laboratories based on blockchain technology. Contract number: RPMP.01.02.01-12-0183/18-00.

References

1. Secure hash standard: Federal Information Processing Standard (FIPS) 180-4 (2012)
2. Angraal, S., Krumholz, H.M., Schulz, W.L.: Blockchain technology: applications in health care. Circul. Cardiovasc. Qual. Outcomes **10**(9), e003800 (2017)
3. Dworkin, M.J.: SHA-3 standard: permutation-based hash and extendable-output functions. Federal Information Processing Standards (NIST FIPS)-202 (2015)
4. Gallagher, P.: Digital signature standard (DSS). Federal Information Processing Standards Publications, volume FIPS, p. 186–3 (2013)
5. Nakamoto, S.: Bitcoin: a peer-to-peer electronic cash system (2019)
6. Schneier, B.: The NSA is breaking most encryption on the internet. blog post, September 2013
7. Stevens, M., et al.: Announcing the first SHA1 collision. Google Security Blog (2017)
8. Zyskind, G., Nathan, O., et al.: Decentralizing privacy: using blockchain to protect personal data. In: 2015 IEEE Security and Privacy Workshops, pp. 180–184. IEEE (2015)

The Synthesis Method of High-Speed Finite State Machines in FPGA

Valery Salauyou$^{(\boxtimes)}$, Damian Borecki, and Tomasz Grzes

Bialystok University of Technology, Wiejska 45A, 15-351 Bialystok, Poland
valsol@mail.ru, pan.d.borecki@o2.pl, t.grzes@pb.edu.pl

Abstract. The synthesis method of high-speed finite state machines (FSMs) in field programmable gate arrays (FPGAs) based on LUT (Look Up Table) by internal state splitting is offered. Estimations of the number of LUT levels are presented for an implementation of FSM transition functions in the case of sequential and parallel decomposition. Split algorithms of FSM internal states for the synthesis of high-speed FSMs are described. The method can be easily included in designing the flow of digital systems in FPGA. The experimental results showed a high efficiency of the offered method. FSM performance increased by 1.73 times. In conclusion, the experimental results were considered, and prospective directions for designing high-speed FSMs are specified.

Keywords: Synthesis · Finite state machine · High-speed · High performance · State splitting · Field programmable gate array · Look up table

1 Introduction

The speed of a digital system and functional blocks depends directly on the speed of their control devices. The mathematical model for the majority of control devices and controllers is a finite state machine (FSM). Because of this, the synthesis methods of high-speed FSMs are necessary for designing high-performance digital systems. In this work we consider the synthesis of high-speed FSMs in field programmable gate arrays (FPGA) based on LUT (Look Up Table).

In [1], a technique for improving the performance of a synchronous circuit configured as an FPGA-based look-up table without changing the initial circuit configuration is presented. Only the register location is altered. In [2], the methods and tools for state encoding and combinational synthesis of sequential circuits based on new criteria of information flow optimization are considered. In [3], the timing optimization technique for a complex FSM that consists of not only random logic but also data operators is proposed. In [4, 5], the styles of FSMs description in VHDL language and known methods of state assignment for the implementation of FSMs are researched. In [6], evolutionary methods are applied to the synthesis of FSMs. In [7], the task of state assignment and optimization of the combinational circuit at implementation of high-speed FSMs is considered. In [8], a novel architecture that is specifically optimized for implementing reconfigurable FSMs, Transition-based Reconfigurable FSM (TR-FSM), is presented. The architecture shows a considerable reduction in area, delay, and power consumption compared to FPGA architectures. In [9], a new model of the automatic

© Springer Nature Switzerland AG 2020
K. Saeed and J. Dvorský (Eds.): CISIM 2020, LNCS 12133, pp. 97–107, 2020.
https://doi.org/10.1007/978-3-030-47679-3_9

machine named the virtual finite state machine (Finite Virtual State Machine - FVSM) is offered. FVSM implemented on new architecture have an advantage on high-speed performance compared with traditional implementation of FSMs on storage RAM. In [10], an implementation of FSMs in FPGA with the use of integral units of storage ROM is considered. Two pieces of FSMs architecture with multiplexers on inputs of ROM blocks which allow reducing the area and increasing high-speed FSM performance are offered. In [11], the reduction task of arguments of transition functions by state splitting is considered; this allows reducing an area and time delay in the implementation of FSMs on FPGA. This paper also uses splitting of FSM states, but the purpose of splitting is an increase of FSMs performance in LUT-based FPGA. The offered synthesis method of high-speed FSMs in FPGA is aimed at practical usage and can be easily included in the general flow of digital system design.

2 Estimations for the Number of LUT Levels for Transition Functions

Let $A = \{a_1, \ldots, a_M\}$ be the set of internal states, $X = \{x_1, \ldots, x_L\}$ be the set of input variables, $Y = \{y_1, \ldots, y_N\}$ the set of output variables, and $D = \{d_1, \ldots, d_R\}$ the set of transition functions of an FSM.

A one-hot state assignment is traditionally used for the synthesis of high-speed FSMs in FPGAs. Thus, each internal state $a_i (a_i \in A)$ corresponds to a separate flip-flop of FSM's memory. A setting of this flip-flop in 1 signifies that the FSM is in the given state. The data input of each flip-flop is controlled by the transition function $d_i, d_i \in D$, i.e. any internal state $a_i (a_i \in A)$ of the FSM corresponds with its own transition function $d_i, i = \overline{1, M}$

Let $X(a_m, a_i)$ be the set of FSM input variables, whose values initiate the transition from state a_m to state $a_i (a_m, a_i \in A)$. To implement some transition from state a_m to state a_i, it is necessary to check the value of the flip-flop output for the active state a_m (one bit) and the input variable values of the $X(a_m, a_i)$ set, which initiates the given transition. To implement the transition function d_i, it is necessary to check the values of the flip-flop outputs for all states, such that transitions from which lead to state a_i, i.e. $|B(a_i)|$ values, where $B(a_i)$ is the set of states from which transitions terminate in state a_i, where $|A|$ is the cardinality of set A. Besides, it is necessary to check the values of all input variables, which initiate transitions to state a_i, i.e. $|X(a_i)|$ values, where $X(a_i)$ is the set of input variables, whose values initiate transitions to state a_i, $X(a_i) = \cup_{a_m \in B(a_i)} X(a_m, a_i)$.

Let r_i be a rank of the transition function d_i, where

$$r_i = |B(a_i)| + |X(a_i)|. \tag{1}$$

Let n be the number of inputs of LUTs. If the rank r_i for transition function $d_i (i = \overline{1, M})$ exceeds n, there is a necessity to decompose the transition function d_i and its implementation on several LUTs.

Note that by splitting internal states it is impossible to lower the rank of the transition functions below the value

$$r^* = \max(|(Xa_m, a_s)| + 1, m = \overline{1, M}, s = \overline{1, M}. \tag{2}$$

In this method, the value r^* is used as an upper boundary of the ranks of the transition functions in splitting the FSM states.

It is well-known that there are two basic approaches to the decomposition of Boolean functions: sequential and parallel. In the case of sequential decomposition, all the LUTs are sequentially connected in a chain.

The n arguments of function d_i arrive on inputs of the first LUT, and the $(n-1)$ arguments arrive on inputs of all remaining LUTs. So the number l_i^s of the LUT's levels (in the case a sequential decomposition of the transition function d_i having the rank r_i) is defined by the expression:

$$l_i^s = int\left(\frac{r_i - n}{n - 1}\right) + 1, \tag{3}$$

where int(A) is the least integer number more or equal to A.

In the case of parallel decomposition, the LUTs incorporate in the form of a hierarchical tree structure. The values of the function arguments arrive on LUTs inputs of the first level, and the values of the intermediate functions arrive on LUTs inputs of all next levels. So the number of LUT's levels (in the case parallel decomposition the transition function d_i having the rank r_i) is defined by the following expression:

$$l_i^p = int(\log_n r_i). \tag{4}$$

It is difficult to predict what type of decomposition (sequential or parallel) is used by a concrete synthesizer. The preliminary research showed that, for example, the Quartus Prime design tool from Intel simultaneously uses both sequential and parallel decomposition. The number l_i levels of LUTs in the implementation on FPGA transition function d_i with the rank r_i can be between values l_i^s and l_i^p, $i = \overline{1, M}$.

Let k be an integer coefficient ($k \in [0,10]$) that allows adapting the offered algorithm in defining the number of LUT's levels for the specific synthesizer. In this case the number l_i of LUT's levels for the implementation of the transition function d_i having the rank r_i will be defined by following expression:

$$l_i = \left(\frac{l_i^s - l_i^p}{10} k + l_i^p\right). \tag{5}$$

If $k = 0$, we have $l_i = l_i^p$, i.e. the number l_i of levels corresponds to the fastest parallel decomposition, and if $k = 10$, we have $l_i = l_i^s$, i.e. the number l_i of levels corresponds to the slowest sequential decomposition. The specific value of coefficient k depends on the architecture of the FPGA and the used synthesizer.

The following problem is the answer to the question: when is it necessary to stop splitting the FSM states? In this algorithm, the process of state splitting is finished, when the following condition is met:

$$l_{max} \leq int(l_{mid}), \qquad (6)$$

where l_{max} is the number of LUT levels, which is necessary for the implementation of the most "bad" function having the maximum rank; l_{mid} is the arithmetic mean value of the number of LUT levels for all transition functions.

3   Method for High-Speed FSM Synthesis

According to the above discussion, the algorithm of state splitting for high-speed FSM synthesis is described as follows.

Algorithm 1

1. The coefficient $k(k \in [0, 10])$ is determined, which reflects the method used by the synthesis tool for the decomposition of Boolean functions.
2. According to (1) ranks r_i $(i = \overline{1, M})$ for all FSM transition functions are defined.
3. On the basis of (3), (4), and (5), for each transition function d_i the number l_i of LUT levels is defined.
4. The values l_{max} and l_{mid} are determined. If condition (6) is met, then go to step 7, otherwise go to step 5.
5. The state a_i, for which r_i = max, is selected. If there are several such states, from them the state for which $|A(a_i)|$ = min is selected.
6. The state a_i (which was selected in step 5) is split by means of algorithm 2 on the minimum number H of states $a_{i\_1}, \ldots, a_{i\_H}$ so that for each state $a_{i\_h}(h = \overline{1, M})$ was fulfilled $r_{i\_h} \leq r*$, where $r*$ is defined according to (2); go to step 2.
7. End.

For splitting some a_i state, $i = \overline{1, M}$, which is executed in step 6 of algorithm 1, Boolean matrix W is constructed as follows. Let $C(a_i)$ be the set of transitions to state a_i. Rows of matrix W correspond to the elements of set $C(a_i)$. Columns of matrix W are divided on two parts according to types of arguments of transition function d_i. The first part of matrix W columns correspond to set $B(a_i)$ of FSM states, the transitions from which terminate in state a_i, and the second part of matrix W columns correspond to set $X(a_i)$ of input variables, whose values initiate the transitions in state a_i. A one is put at the intersection of row t $(i = \overline{t, T}, T = |C(a_i)|)$ and column j of the first part of matrix W if the transition $c_t(c_t \in C(a_i))$ is executed from state $a_j(a_j \in B(a_i))$. A one is put at the intersection of row t and column j of the second part of matrix W if input variable $x_j(x_j \in X(a_i))$ accepts a significant value (0 or 1) on transition $c_t(c_t \in C(a_i))$. Now the task is reduced to a partition of matrix W on a minimum number H of row minors $W_1, \ldots, W_H$ so that the number of columns, which contain ones in each minor $W_h(h = \overline{1, H})$, do not exceed value $r*$ defined according to (2). The rows of each minor W_h will define transitions in state $a_{i\_h}(h = \overline{1, H})$.

Let w_t be some row of matrix W. For finding the row partition of matrix W on a minimum number H of row minors $W_1, \ldots, W_H$, the following algorithm can be used.

Algorithm 2

1. Put $h := 0$.
2. Put $h := h + 1$. A formation of minor W_h begins. The row w_t, which has the maximum number of ones, is selected in minor W_h as a reference row. The row w_t is included in minor W_H and the row w_t is eliminated from further reviewing, put $W_h := \{w_t\}, W := W \backslash \{w_t\}$.
3. The rows are added in minor W_h. For this purpose, among rows of matrix W, the row w_t is selected, for which the next inequality is satisfied $|W_h \cup \{w_t\}| \leq r^*$, where $|W_h \cup \{w_t\}|$ is the total number of ones in the columns of minor W_h and the row w_t after their joining on OR. If such rows can be selected from several among them, row w_t is selected, which has the maximum number of common ones with minor W_h, i.e. $|W_h \cap \{w_t\}| = max$. The row w_t is included in minor W_h and row w_t is eliminated from further reviewing, put $W_h := W_h \cup \{w_t\}, W := W \backslash \{w_t\}$.
4. Step 3 repeats until at least a single row can be included in minor W_h.
5. If in matrix W all the rows are distributed between the minors, then go to step 5, otherwise go to step 2.
6. End.

We show the operation of the offered synthesis method in the example. It is necessary to synthesize the high-speed FSM whose state diagram is shown in Fig. 1.

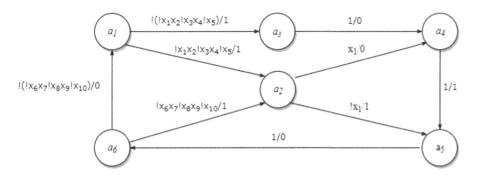

Fig. 1. State diagram of the initial FSM

This FSM represents the machine Moore, which has 6 states $a_1, \ldots, a_6$, 10 input variables $x_1, \ldots, x_{10}$, and one output variable y. The transitions from states a_3, a_4, and a_5 are unconditional, therefore the logical value 1 is written on these transitions as a transition condition. The values of sets $B(a_i)$ and $X(a_i)$, and also ranks r_i of the transition functions for the initial FSM are presented in Table 1, where Ø is an empty set. Since for this example we have $max(|X(a_m, a_s)|) = 5$, then (according to (2)) the value $r^* = 6$. Let it is necessary to construct the FSM on FPGA with 6-input LUT, i.e. we have n = 6.

According to (3) and (4), the values l_i^s and l_i^p are defined for each state (they are presented in the appropriate columns of Table 1). We do not know how the compiler performs a decomposition of Boolean functions, therefore we assume the sequential decomposition (a worst variant) and the value of coefficient k in expression (5) is equal to 10. As a result, the number of LUT levels (which are necessary for the implementation of each transition function) is defined by the value $l_i = l_i^s$. Thus, for our example we have $\mathrm{int}(l_{mid}) = \mathrm{int}(8/6) = 2$.

For this example, we have $l_{max} = l_2^s = 3$, i.e. the condition (9) does not meet for state a_2, since $l_{max} = l_2^s = 3 > \mathrm{int}(l_{mid}) = 2$. For this reason, state a_2 is split by means of algorithm 2. Matrix W is constructed for splitting state a_2 (Fig. 2).

	a_1	a_6	x_1	x_2	x_3	x_4	x_5	x_6	x_7	x_8	x_9	x_{10}
w_1	1	0	1	1	1	1	1	0	0	0	0	0
w_2	0	1	0	0	0	0	0	1	1	1	1	1

Fig. 2. Matrix W for splitting state a_2

Matrix W has two rows. Row w_1 corresponds to the transition from state a_1 to state a_2, and row w_2 corresponds to the transition from state a_6 to state a_2. The execution of algorithm 2 leads to a partition of rows of matrix W into two subsets: $W_1 = \{w_1\}$ and $W_2 = \{w_2\}$. So, state a_2 is split into two states a_{2_1} and a_{2_2}, as shown in Fig. 3.

Table 1. Values of $B(a_i), X(a_i), r_i, l_i^s$ and l_i^p for the initial FSM

State	$B(a_i)$	$X(a_i)$	r_i	l_i^s	l_i^p
a_1	$\{a_6\}$	$\{x_6, x_7, x_8, x_9, x_{10}\}$	6	1	1
a_2	$\{a_1, a_6\}$	$\{x_1, x_2, x_3, x_4, x_5, x_6, x_7, x_8, x_9, x_{10}\}$	12	3	2
a_3	$\{a_1\}$	$\{x_1, x_2, x_3, x_4, x_5\}$	6	1	1
a_4	$\{a_2, a_3\}$	$\{x_1\}$	3	1	1
a_5	$\{a_2, a_4\}$	$\{x_1\}$	3	1	1
a_6	$\{a_5\}$	$\varnothing$	1	1	1

Table 2. Values of $B(a_i), X(a_i), r_i, l_i^s$ and l_i^p after splitting state a_2

State	$B(a_i)$	$X(a_i)$	r_i	l_i^s	l_i^p
a_1	$\{a_6\}$	$\{x_6, x_7, x_8, x_9, x_{10}\}$	6	1	1
a_{2_1}	$\{a_1\}$	$\{x_1, x_2, x_3, x_4, x_5\}$	6	1	1
a_{2_2}	$\{a_6\}$	$\{x_6, x_7, x_8, x_9, x_{10}\}$	6	1	1
a_3	$\{a_1\}$	$\{x_1, x_2, x_3, x_4, x_5\}$	6	1	1
a_4	$\{a_{2\_1}, a_3\}$	$\{x_1\}$	3	1	1
a_5	$\{a_{2\_2}, a_4\}$	$\{x_1\}$	3	1	1
a_6	$\{a_5\}$	$\varnothing$	1	1	1

The new values of $B(a_i), X(a_i), r_i, l_i^s$ and l_i^p are presented in Table 2. Now we have $l_{max} = l_{mid} = 1$ and (according to (6)) running of algorithm 1 is completed.

Thus, for the given FSM by splitting state a_2 we reduced the number of LUT levels from 3 to 1, in the case of sequential decomposition, and from 2 to 1, in the case of parallel decomposition.

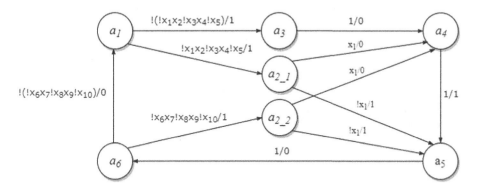

Fig. 3. State diagram of the FSM after splitting state a_2

4 Experimental Results

The presented method was evaluated against the FSM benchmarks, MCNC [13]. For this purpose, the considered synthesis method was applied to each benchmark of the FSM. Both finite state machines, the initial FSM and synthesized FSM, were described in the Verilog language. Then, standard implementation of FSMs in FPGA by means of CAD Quartus Prime was fulfilled. The number n of the LUT inputs has been set by 4. The offered method allowed to reduce the rank of transition functions for 21 benchmarks of 48, i.e. in 43.75% of cases.

Table 3 shows results of synthesis of high-speed FSMs by means of presented method, where FSM is the name of benchmark; Ns is the number of the splitted states; M, P, and $lmax$ are the number of FSM's states, FSM's transitions, and the maximum rank of transition functions before application of the synthesis method; $M*$, $P*$, and l_{max}^* are the number of FSM's states, FSM's transitions, and the maximum rank of transition functions after application of the synthesis method; $M*/M$, $P*/P$, and $l_{max}/l_{max}^* l_{max}^*$ are relations of the corresponding parameters; mid is the mean parameter value; max is the maximum parameter value.

Table 3 shows that for synthesized benchmarks the average number of splits is equal 3.00, and maximum – 11; the number of FSM states are increased by a factor of 1.68 on the average and maximum by a factor of 3; the number of FSM transitions are increased by a factor of 1.63 on the average and maximum by a factor of 3.36. The use of the synthesis method allows to reduce the rank of transition functions by a factor of 2.40 on the average, and maximum by a factor of 6.33. Thus, the presented method allows considerably to reduce the maximum rank of transition functions by despite increasing the number of FSM states and the transitions.

Table 3. Results of the experimental researches of the synthesis method of high-speed FSMs for benchmarks

FSM	Ns	M	$M*$	$M*/M$	P	$P*$	$P*/P$	l_{max}	$l*_{max}$	$l_{max}/l*_{max}$
BBSSE	11	16	48	3,00	56	188	3,36	7	3	2,33
CSE	1	16	23	1,44	91	154	1,69	8	5	1,60
EX1	1	18	22	1,22	233	379	1,63	8	5	1,60
EX2	1	19	34	1,79	72	72	1,00	6	2	3,00
EX3	1	10	17	1,70	36	36	1,00	3	2	1,50
EX5	1	9	16	1,78	32	32	1,00	3	2	1,50
EX7	1	10	18	1,80	36	36	1,00	4	2	2,00
KEYB	1	19	26	1,37	170	261	1,54	8	4	2,00
PLANET	2	48	51	1,06	115	120	1,04	3	2	1,50
PMA	8	24	39	1,63	73	106	1,45	5	2	2,50
S208	2	18	43	2,39	153	332	2,17	9	5	1,80
S298	1	218	246	1,13	1096	1236	1,13	27	19	1,42
S386	2	13	18	1,38	64	99	1,55	7	3	2,33
S420	2	18	43	2,39	137	296	2,16	9	5	1,80
S820	3	25	36	1,44	232	324	1,40	14	5	2,80
S832	3	25	38	1,52	245	412	1,68	14	5	2,80
S1488	3	48	63	1,31	251	341	1,36	19	3	6,33
S1494	3	48	63	1,31	250	364	1,46	19	3	6,33
SAND	3	32	39	1,22	184	325	1,77	6	5	1,20
SSE	11	16	48	3,00	56	188	3,36	7	3	2,33
STYR	2	30	41	1,37	166	259	1,56	9	5	1,80
mid	3,00			1,68			1,63			2,40
max	11			3			3,36			6,33

The benchmarks were realized on the following families FPGA Arria II, Cyclone V, MAX II, and Stratix V. The results of researches are given in tables, where LE and $LE*$ are the number of the logical elements (area), which necessary for realization of the initial FSM and the synthesized FSM; F and $F*$ are the frequency of functioning of the initial FSM and the synthesized FSM; $LE*/LE$ and $F*/F$ are the relations of the corresponding parameters; other parameters have former value.

Tables 4 show that the offered method allows to increase the frequency of FSMs for various FPGA families by a factor of 1.08–1.13 on the average and by a factor of 1.52–1.73 maximum. Using of the presented method also increases the area by a factor of 1.27–1.35 on the average.

Table 5 compares the considered method to the known university programs for FSM state assignment JEDI [14] and NOVA [15], where *FJ*, *FN*, and *F** are the frequency of the FSM when using the program JEDI, NOVA, and the offered method respectively. Table 5 shows that the offered method allows to increase the maximal frequency of FSMs by a factor of 2.03–2.33 on the average in comparison with the JEDI program and by a factor of 3.68–4.47 on the average in comparison with the NOVA program. The maximum increase in frequency of FSMs makes 3.11 times in comparison with the JEDI program and 8.31 times in comparison with the NOVA program (Fig. 4).

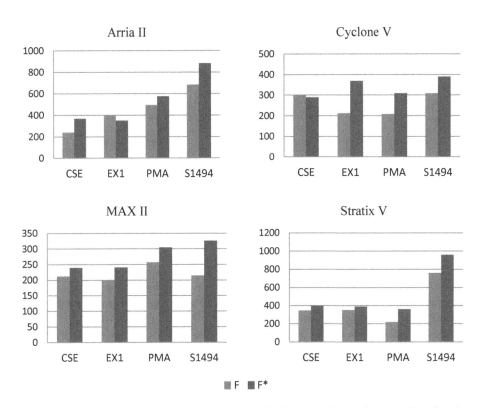

Fig. 4. The frequency of functioning of the initial FSM and the synthesized FSM for the selected examples

Table 4. The implementation results of FSM benchmarks in the Arria II, Cyclone V, MAX II, and Stratix V FPGA families

Family	*LE*/LE*		*F*/F*	
	mid	*max*	*mid*	*max*
Arria II	1.31	2.07	1.10	1.54
Cyclone V	1.35	2.14	1.13	1.73
MAX II	1.33	2.03	1.19	1.52
Stratix V	1.27	1.89	1.08	1.64

Table 5. Comparing of the offered method with the JEDI and NOVA programs in case of FSM implementation in the Cyclone V and MAX II FPGA families

FSM	Cyclone V					MAX II				
	FJ	FN	F*	F*/FJ	F*/FN	FJ	FN	F*	F*/FJ	F*/FN
BBSSE	402	119	937	2,33	7,87	311	163	937	3,01	5,75
CSE	300	221	370	1,23	1,67	222	161	370	1,67	2,30
EX1	267	156	352	1,32	2,26	217	129	352	1,62	2,73
EX2	333	188	787	2,36	4,19	302	175	787	2,61	4,50
EX3	280	294	629	2,25	2,14	348	219	629	1,81	2,87
EX5	422	298	587	1,39	1,97	340	206	587	1,73	2,85
EX7	398	308	587	1,47	1,91	320	193	587	1,83	3,04
KEYB	360	264	876	2,43	3,32	282	152	876	3,11	5,76
PLANET	435	215	1030	2,37	4,79	391	124	1030	2,63	8,31
PMA	224	198	575	2,57	2,90	267	116	575	2,15	4,96
S386	408	285	928	2,27	3,26	336	192	928	2,76	4,83
SSE	402	119	937	2,33	7,87	311	163	937	3,01	5,75
mid				2,03	3,68				2,33	4,47
max				2,57	7,87				3,11	8,31

5 Conclusions

Using of the offered method allows considerably to lower the maximum rank of transition functions, however increase a performance of the FSMs is watched not in all cases. It is explained by complexity of the synthesis task of the fast FSMs, for example, in comparison with the task of area reduction. The matter is that performance of the FSMs is influenced by not only results of a logical synthesis, but also results of placement and routing. Besides, performance of FSMs, except transitions functions, is also influenced by complexity of output functions.

The offered method can be also applied to creation of high-speed FSMs in ASIC chips. For this purpose it is enough to define estimates (3) and (4) the number of the circuit levels for specific architecture of ASIC.

Further development of synthesis methods of high-speed FSMs can go on the way of accounting of output function complexity, use of special FSM structural models, architectural FPGA properties, special control of FSM synchronization, etc.

The present study was supported by a grant S/WI/3/2018 from Bialystok University of Technology and founded from the resources for research by Ministry of Science and Higher Education.

References

1. Miyazaki, N., Nakada, H., Tsutsui, A., Yamada, K., Ohta, N.: Performance improvement technique for synchronous circuits realized as LUT-Based FPGA's. IEEE Trans. Very Large Scale Integr. (VLSI) Syst. **3**(3), 455–459 (1995)

2. Jozwiak, L., Slusarczyk, A., Chojnacki, A.: Fast and compact sequential circuits through the information-driven circuit synthesis. In: Proceedings of the Euromicro Symposium on Digital Systems Design, Warsaw, Poland, 4–6 September 2001, pp. 46–53 (2001)

3. Huang, S.-Y.: On speeding up extended finite state machines using catalyst circuitry. In: Proceedings of the Asia and South Pacific Design Automation Conference (ASAP-DAC), Yokohama, January-February, 2001, pp. 583–588 (2001)

4. Kuusilinna, K., Lahtinen, V., Hamalainen, T., Saarinen, J.: Finite state machine encoding for VHDL synthesis. Comput. Dig. Techniques, IEE Proc. **148**(1), 23–30 (2001)

5. Rafla, N.I., Davis, B.A.: Study of finite state machine coding styles for implementation in FPGAs. In: Proceedings of the 49th IEEE International Midwest Symposium on Circuits and Systems, San Juan, USA, 6–9 August 2006, vol. 1, pp. 337–341 (2006)

6. Nedjah, N., Mourelle, L.: Evolutionary synthesis of synchronous finite state machines. In: Proceedings of the International Conference on Computer Engineering and Systems, Cairo, Egypt, 5–7 November 2006, pp. 19–24 (2006)

7. Czerwiński, R., Kania, D.: Synthesis method of high speed finite state machines. Bull. Polish Acad. Sci. Techn. Sci. **58**(4), 635–644 (2010)

8. Glaser, J., Damm, M., Haase, J., Grimm, C.: TR-FSM: transition-based reconfigurable finite state machine. ACM Trans. Reconfigurable Technol. Syst. (TRETS) **4**(3), 231–2314 (2011)

9. Senhadji-Navarro, R., Garcia-Vargas, I.: Finite virtual state machines. IEICE Trans. Inf. Syst. **E95D**(10), 2544–2547 (2012)

10. Garcia-Vargas, I., Senhadji-Navarro, R.: Finite state machines with input multiplexing: a performance study. IEEE Trans. Computer-aided Design Integr. Circ. Syst. **34**(5), 867–871 (2015)

11. Solov'ev, V.V.: Splitting the internal states in order to reduce the number of arguments in functions of finite automata. J. Comput. Syst. Sci. Int. **44**(5), 777–783 (2005)

12. Yang, S.: Logic synthesis and optimization benchmarks user guide. Version 3.0. Microelectronics Center of North Carolina (MCNC). North Carolina, USA (1991)

13. Lin, B., Newton, A.R.: Synthesis of multiple level logic from symbolic high-level description languages. In: Proceedings of the International Conference on VLSI, 1989, Munich, pp. 187–196 (1989)

14. Villa, T., Sangiovanni-Vincentelli, A.: Nova: state assignment of finite state machines for optimal two-level logic implementation. IEEE Trans. Comput. Aided Des. Integr. Circuits Syst. **9**(9), 905–924 (1990)

Industrial Management and other Applications

A Framework of Business Intelligence System for Decision Making in Efficiency Management

Daniela Borissova[1,2(✉)] ⓘ, Petya Cvetkova[1] ⓘ, Ivan Garvanov[1] ⓘ,
and Magdalena Garvanova[1] ⓘ

[1] University of Library Studies and Information Technologies,
1784 Sofia, Bulgaria
pcvetkova@abv.bg,
{i.garvanov, m.garvanova}@unibit.bg
[2] Institute of Information and Communication Technologies at the Bulgarian
Academy of Sciences, 1113 Sofia, Bulgaria
dborissova@iit.bas.bg,

Abstract. The business decisions at different levels require processing different kinds of information. In this regard, the usage of suitable tools will contribute to making effective business decisions. The described framework of the business intelligence system aims to support such decisions in an effective way. The core of the proposed decision support system relies on several modules with a different database. One of them contains the input data of the particular problem, second include multi-criteria design analysis models, while the next contains optimization models to support decision-making. These optimization models are the focus of the current article. Two single and one multi-objective optimization models are formulated to express different situations and to support business decisions via reasonable solutions. Depending on the particular purpose, one of the models can be used to determine the best or compromise decision, which contributes to the effectiveness in business management. The applicability of the proposed models and respectively the core of the framework of business decision-making in efficiency management is illustrated in public street lights renovation. The obtained results show that all models are practically applicable in the determination of corresponding decisions in accordance with the selected goal. As the essences of the proposed framework are the optimization models this proves the effectiveness of optimization models in decision making to support efficient management.

Keywords: Business intelligence · Decision support framework · MCDA · MCDM · Optimization models

1 Introduction

With the growing economy nowadays it is difficult to be successful without taking business decisions at different levels (Shalamanov 2017). To gain the maximum advantage, input information is to be properly processed to make effective business decisions (Borissova et al. 2019). Such suitable kind of tool is decision support systems. They refer to a broad range of interactive systems that utilize data and models to solve

© Springer Nature Switzerland AG 2020
K. Saeed and J. Dvorský (Eds.): CISIM 2020, LNCS 12133, pp. 111–121, 2020.
https://doi.org/10.1007/978-3-030-47679-3_10

different problems. Instead of using many different tools for the variety aspects of information management, the more suitable decision is the usage of integrated approach by custom business intelligence tools with ability to implement artificial intelligence in some cases (Pedrycz et al. 2008). There are different approaches to design of custom software depending of the specific application areas and using of optimization software tools (Borissova & Mustakerov 2016; Mustakerov & Borissova 2009) or by using other appropriated tools in MS Excel (Borissova et al. 2018; Borissova 2008) or web-based applications (Mustakerov & Borissova 2014; Borissova et al. 2013). In some situations the complex interconnected systems could be modeled by means of Mamdani fuzzy networks with feedforward rule bases presented as an equivalent Mamdani fuzzy system by linguistic composition of its nodes (Gegov et al. 2016). A rule base simplification method for fuzzy systems could be also integrated when data processing to get the recommended decision (Gegov et al. 2017). The advantages of such decision support systems are the ability end-users to manipulate the data without necessity to have technical skills. Thus, the obtained solutions make possible to create flexible and easy-to-use analysis that serve as base to make reasonable decisions. All of these charac-teristics of business intelligent systems could be implemented in custom software to support decision making in the efficiently management.

The major challenge nowadays is the problem with climate change that imposes the reduction of energy usage. In this regard, some local agreements put the emphasis on supporting the implementation of energy-saving measures via knowledge-sharing networks (Cornelis 2019). Two distinguish directions could be determined—one direction is related to using renewable energy resources, while the second emphasizes the utilization of devices with less power consumption (Lee et al. 2013; Ivanov 2013). Some authors show that used energy for public lighting has the greatest impact on energy consumption of a municipality up to 54% of total energy consumption and 61% of electricity consumption of municipal facilities (Elejoste et al. 2013). In this sense, some main practices used worldwide in terms of energy efficiency of urban public lighting could be evaluated to determine the best options considering diffident multi-criteria decision-making methods to tackle with multi-attribute decision-making problems (Salvia et al. 2019). It is shown, that energy efficiency effect on the public street lighting could be realized by using LED light replacement (Sudarmono et al. 2018). On the other hand, using emerging LED technology enables intelligent street lighting based on sensing of individual vehicles and dimming street lights that lead to considerable energy reduction exceeding 50% on roads with low traffic (Nefedov et al. 2014). The energy efficiency could be realized by a multi-objective optimization approach based on the formulation of the bi-criterion problem (Guliashki et al. 2019).

The planning of public street lights is highly dependable on location, streets, side-walks, travel patterns, behavioural characteristics, safety, security, etc. The unique in their placement is the fact that street lights can be located anywhere along a street. Some of the main outputs in street lighting are the parameters related to luminaire spacing, height, and illuminance, which can be obtained using the proposed multi-objective evolutionary algorithms (Rabaza et al. 2013). To cope with the problem of public street lighting design, an optimization discrete integer programming model capable of deriving the minimum number of street lights is proposed (Murray & Feng 2016). In contrast this model, the current article describes three models able to simulate three different scenarios for different goals in determination of the best suitable LED lamps for renovation of lights street.

The rest of the paper is organized as follow: Sect. 2 provide problem description and given input data; Sect. 3 presents the framework of business intelligence decision making and several optimization models for effective management, Sect. 4 describes some preliminary test results about the applicability of proposed models for intelligent business decision making; while conclusions and future investigations are drawn in Sect. 5.

2 Problem Description

The investigated problem concerns the determination of the suitable LED lamp type and determination of the needed number for the renovation of several villages. The selection of a particular LED lamp type should meet the requirement about the distance between each street light pole while minimizing the overall costs for the renovation of the entire length of the road. It should be noted that in some streets where more illumination is needed the goal is to determine such an LED type that will provide needed luminance. Determination of proper LED lamp type selection is realized by using predefined suitable LED lamps for street luminance as shown in Table 1.

Table 1. LED lamps parameters

#	Luminous flux, lm	Luminaire Wattage, W	Pole distance, m	Mounting height, m	Overhang, m	Price, BGL
L-1	1928	13.9	40.0	7.5	0.3	150
L-2	3772	27.3	36.0	8.0	0.2	180
L-3	4735	34.1	31.0	8.0	0.8	204
L-4	2064	15.5	32.0	7.0	0.9	160
L-5	2170	17.0	30.0	6.3	0.0	160
L-6	3062	25.4	28.0	8.0	0.0	150
L-7	2394	18.1	30.0	7.0	0.2	135
L-8	1755	13.2	36.0	8.5	0.2	135

There are three essential parameters influencing to the selection: 1) distance between each street light pole that will influence on the needed number to cover the road length; 2) the luminous or luminaire wattage that influence on the energy consumption and respectively on the illumination; and 3) costs per unit.

To solve the described above problem, two basic approaches can be integrated–multi-criteria design analysis and single or multi-objective optimization. These two approaches represent the core of the proposed framework to support the business decisions in efficiency management.

3 Framework of Business Intelligence System for Decision Making in Efficiency Management

In the context of decision support systems, different techniques and methods that contribute to fulfilling the goal could be integrated. Two basic approaches can be involved to support management inefficiency way – multi-attribute decision making and single or multi-objective decision making (Borissova 2015). The multi-attribute decision-making techniques rely on a subjective point of view toward evaluation criteria and could involve different utility functions. The single or multi-objective decision making is based on the formulation of an optimization model with one or several criteria. The framework of a business intelligence system for decision making can be realized of different modules and sub-modules depending on the flow of collecting and processing data (Borissova and Mustakerov 2012, 2013). Despite this variety of modules there are three basic components that have always to present in such systems such as 1) user interface, 2) databases and 3) analytical tools with visualization as shown in Fig. 1.

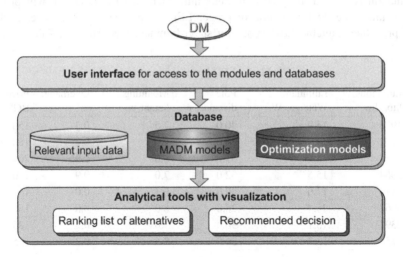

Fig. 1. Framework of business intelligence system for decision making

The user interface should be designed in such a way to provide easy access to all resources required the active participation of decision-makers. The second essential com-ponent is consists of different databases. One of them will contain the needed input data for the particular domain area, while the rest two are designed to contain multi-attribute decision-making models and optimization models. There are different and well-known models for MADM that can be used for ranking a predefined set of alternatives in accordance to the decision-maker preferences. More important is the second database where different optimization criteria could be formulated and used as objective function/s. The advantage of this approach is the fact that the obtained solution is surely optimal for the given objective function and used restrictions. One database is intended for single and multi-objective models as shown because the solution methods for multi-objective problems are transformed into a single optimization problem. Regardless of which approach will be used, the analytical tools for a ranking list of alternatives or recommended decision should be properly visualized.

3.1 Optimization Models to Support Decision Making in Efficiency Management

While multi-attribute methods aim to provide estimation of alternatives performance toward given attributes, the optimization methods make possible to determine design variables for the system as whole and they are often are used to eliminate the clearly unsatisfactory design variants. Using optimization for a preliminary estimation is the primary way to estimate effectiveness of designed system and operating policies.

For the goal of helping business decisions of efficiency management in a reasonable way, three optimization models with three different objective functions are proposed. These models enable to obtain optimal solution in determining the best suitable type of LED lamps for renovation of street lighting. The models differs from each other to the objective functions used to guide the searching the optimal solution. The first model (M-1) aims to determine such type of LED lamp that provides the minimum costs for renovation of entire length of road. In this case the optimization model has the following formulation:

$$min(N * P) \tag{1}$$

subject to

$$N = (RL/D) \tag{2}$$

$$P = \sum_{i=1}^{M} x_i P_i \tag{3}$$

$$L = \sum_{i=1}^{M} x_i L_i \tag{4}$$

$$D = \sum_{i=1}^{M} x_i D_i \tag{5}$$

$$\sum_{i=1}^{M} x_i = 1 \tag{6}$$

where N express the number of needed LEDs and is calculated as relation between road length RL considered as given in advance value and distance between LED lamps denoted by D and determined by the relation (5). The price P of the selected type of LED lamps is expressed by the relation (3) and is determined as task solution. The parameter L is the luminance of the selected type of LED lamp expressed by the relation (4). An essential role in the formulated model (1)–(6) plays the used binary decision variables x_i determined by (6). The particular choice of the LED lamp type is realized by relation (6) where the corresponding decision variable is associated with particular type of LED lamp used in determination of the best selection.

The second model (M-2) aims to determine such type of LED lamp that provides the maximum luminance in renovation of entire length of road. In this case the optimization model has the following expression:

$$max(N * L) \tag{7}$$

subject to

$$N = RL/D \tag{8}$$

$$P = \sum_{i=1}^{M} x_i P_i \tag{9}$$

$$L = \sum_{i=1}^{M} x_i L_i \tag{10}$$

$$D = \sum_{i=1}^{M} x_i D_i \tag{11}$$

$$\sum_{i=1}^{M} x_i = 1 \tag{12}$$

All parameters are the same as the formulated above model, but here the luminance of the selected type of LED lamp L is determined as task solution.

The third multi-objective model (M-3) aims to determine such type of LED lamp that simultaneously provides the minimum costs and maximum luminance in renovation of entire length of road. In this situation, the optimization model has the following formulation:

$$\begin{cases} min(N * P) \\ max(N * L) \end{cases} \tag{13}$$

subject to the same restrictions (8)–(12).

It should be noted, that formulated above model with multi-objective function (13) s. t. (8)–(12). requires selecting a proper method for it solving. These methods rely on: 1) a priori aggregation of preference information, 2) posteriori aggregation of preference information or 3) interactive articulation of preference information depending when decision maker express his preferences. The well-known and easy to perform are the weighted sum and lexicographic method (Marler & Arora 2004). Both of these methods can be used for a priori and posteriori articulation of preferences. In contrast to the lexicographic method, the weighted sum requires scalarization technique to transforms the original multi-objective problem into weighted linear sum of the normalized objective functions (Genova et al. 2013).

The described above three different optimization models contribute in reasonable decision making in efficient management of plan for business strategy.

4 Result Analysis and Discussion

The formulated three different models aiming to support the business decision in efficiency management are used to formulate three different optimization problems. The input data for these optimization problems are used the parameters of LED lamps shown in Table 1. For the simplicity of numerical testing of the applicability of the proposed models, the overall road length for renovation is considered equivalent for all models and is equal to 5600 m. The obtained recommended decisions toward the selected LED lamp types are illustrated in Fig. 2.

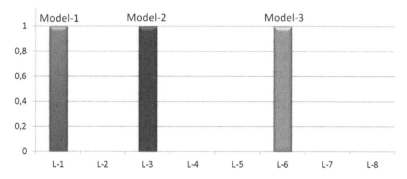

Fig. 2. The selected LED lamp type by three different models

The solution from model M-1 (1)–(6) aims to determine such type of LED lamp that provides the minimum costs for renovation of entire length of road. In this situation, the best solution when minimizing the overall costs is achieved at value for the objective function equal to 21000. The value of decision variable x_1 is equal to 1, which corresponds to the choice of L-1 LED type. The needed number of LED lamps is equal to 140.

When the model M-2 (7)–(12) is used, the best solution is achieved at objective function value concerning maximum luminance is equal to 6160. Here, the value of decision variable x_3 is equal to 1 that corresponds to the choice of L-3 LED type. In this case, the needed number of LED lamps determined as result of task solution is equal to 180.6452 that means not an integer and has to be considered equal to 181.

For solving model M-3 (13) s. t. (8)–(12) by weighted sum method using, the transformation of the objective function (11) is as follows:

$$max\{w_1 f_1^* + w_2 f_2^*\} \tag{14}$$

subject to (8)–(12) plus the following additional restrictions

$$w_1 + w_2 = 1 \tag{15}$$

$$f_1^* = \left(\frac{L - L^{min}}{L^{max} - L^{min}} \right) \tag{16}$$

$$f_2^* = \left(\frac{P^{max} - P}{P^{max} - P^{min}} \right) \tag{17}$$

where w_i express the coefficients for the objective importance and their sum are to be equal to 1, and the f_1^* and f_2^* are the normalized objective functions concerning to the luminance L and price P for overall road length for renovation, and L^{min}, L^{max}, P^{min}, P^{max} express the minimum and maximum values each objective could take.

When model M-3 is used, the obtained solution determines that L-6 LED type is the compromise choice when considering both objective functions with equal importance, i.e. $w_1 = w_2 = 0.5$. In contrast to the previous models, where the objective function expresses the costs or the luminance for the renovation of entire roads length, the objective function value of model M-3 is dimensionless. This is due to the used normalization of both objective functions within the range of $(0, 1)$ and the obtained value for (14) is equal to 0.5879729. The task solution defines that the number of LED lamps is to be exactly 200 to cover the given length of 5600 m.

The obtained results about the renovation of street lights concerning roads length of 5600 m are summarized in Table 2.

Table 2. Solution results for the entire length road of 5600 m

Model	Objective function values	Decision variable equal to 1	LED lamp type	Number of LED lamps	Luminaire Wattage	Costs
M-1	21000	x_1	L-1	140	13.9	150
M-2	6160	x_3	L-3	181	34.1	204
M-3	0.588	x_6	L-6	200	5080	30000

The values of used binary decision variables (x_i) determine the corresponding type of LED lamps from Table 1. The values for the renovation costs when using model M-1 and model M-2 are of the same range (150 and 204) in contrast to the model-3 where they are considerably increased to the 30000 (Table 2). This could be explained by the selected type L-6 that requires less separation distance equal to 28 m compared to the 40 m and 31 m and corresponding price per unit. Except the costs, the illumination for the entire length is increased too.

In such way, depending on the goal it is possible to use one of the formulated optimization models, which solutions determine the best selection of LED lamps for lights street renovation. These models are the core of the described framework for business intelligence system for decision making in efficiency management. The numerical testing of these models proves their applicability in determining the most appropriate solution (Fig. 2). In conclusion, from the predefined set of 8 LED lamps

(Table 1) and by using 3 different strategies only 3 LED lamps type could be considered as potential pretenders for lights street renovation, namely L-1, L-3 and L-6 (Table 2).

Once the selection of the proper LED lamp type is done, it is needed to determine how the lights will be controlled. The easy and promising way is to use some kind of wire-less technologies that provide remote control of LEDs. This imposes to take into ac-count the specific characteristics of the electromagnetic propagation environment to avoid conflict situations with radio communication systems. Besides this, the possibility for the usage of already installed smart technologies is to be considered also to prevent some possible psychological and physical impacts on the human body in urban areas. That means some additional input data about the surround signals environment are to be collected and used to determine the suitable remote control system. These activities are planned as the future directions of the investigations.

5 Conclusion

The article deals with problems related to the development of an integrated framework for effective management of business decision making. The basic structure with the needed modules of such a framework is proposed. The most essential part of the framework for effective management of business decision making is the database with optimization models. Three different optimization models are formulated to express different situations and to support the business decisions by reasonable solutions. Depending on the particular purpose, one of the models can be used to determine the best or compromise decision, which contributes to the effectiveness in business management.

The problem with the renovation of lights street is used to demonstrate the applicability of the formulated optimization models that support decision-making. The used input data for numerical testing are the parameters of a predefined set of suitable LED lamps type and a total length of roads for renovation. Based on the proposed models, the corresponding optimization tasks using the input data are formulated. The obtained results show that all three models are practically applicable in the determination of corresponding decisions in accordance with the selected goal. As the user interface and visualization are subject to the particular problem, the essence of the proposed framework of the decision support system in the effective management of business decisions is the optimization models. Therefore, numerical testing proves the basic functionality of the proposed framework of the decision support system in the effective management of business decisions.

Future developments are related to extending the formulated optimization models with additional restrictions to help in the determination of selection and placement of remote control for LED lamps. Thus, the managers will be supported in an efficient way to make the best business decisions.

Acknowledgment. This work is supported by the Bulgarian National Science Fund, Project title "Synthesis of a dynamic model for assessing the psychological and physical impacts of excessive use of smart technologies", KP-06-N 32/4/07.12.2019.

References

Borissova, D., Mustakerov, I.: A framework for designing of optimization software tools by commercial API implementation. Int. J. Adv. Eng. Manage. Sci. **2**(10), 1790–1795 (2016)

Borissova, D.: Single- and multicriteria models and algorithms for optimal design, planning and management of engineering systems. Abstract Dissertat. IICT-BAS **4**, 1–60 (2015)

Borissova, D., Korsemov, D., Mustakerov, I.: Multi-attribute group decision making considering difference in experts knowledge: an Excel application. In: 12th International Management Conference – Management Perspectives in the Digital Era, pp. 387–395. Bucharest, Romania (2018)

Borissova, D., Korsemov, D., Mustakerov, I.: Multi-criteria decision making problem for doing business: comparison between approaches of individual and group decision making. In: Saeed, K., Chaki, R., Janev, V. (eds.) CISIM 2019. LNCS, vol. 11703, pp. 385–396. Springer, Cham (2019). https://doi.org/10.1007/978-3-030-28957-7_32

Borissova, D., Mustakerov, I., Bantutov, E.: Web-based architecture of a system for design assessment of night vision devices. Int. J. Inf. Sci. Eng. **7**(7), 62–67 (2013)

Borissova, D., Mustakerov, I.: A concept of intelligent e-maintenance decision making system. In: International Symposium on Innovations in Intelligent Systems and Applications, IEEE, Albena, Bulgaria (2013). https://doi.org/10.1109/inista.2013.6577668

Borissova, D., Mustakerov, I.: An integrated framework of designing a decision support system for engineering predictive maintenance. Int. J. Inf. Technol. Knowl. **6**(4), 366–376 (2012)

Borissova, D.: Optimal scheduling for dependent details processing using MS Excel Solver. Cybernet. Inf. Technol. **8**(2), 102–111 (2008)

Cornelis, E.: History and prospect of voluntary agreements on industrial energy efficiency in Europe. Energy Policy **132**, 567–582 (2019)

Elejoste, P., et al.: An easy to deploy street light control system based on wireless communication and LED technology. Sensors **13**, 6492–6523 (2013)

Gegov, A., Sanders, D., Vatchova, B.: Aggregation of inconsistent rules for fuzzy rule base simplification. Int. J. Knowl. Based Intell. Eng. Syst. **21**, 135–145 (2017)

Gegov, A., Sanders, D., Vatchova, B.: Mamdani fuzzy networks with feedforward rule bases for complex systems modelling. J. Intell. Fuzzy Syst. **30**(5), 2623–2637 (2016)

Genova, K., Kirilov, L., Guljashki, V.: New reference-neighbourhood scalarization problem for multiobjective integer programming. Cybernet. Inf. Technol. **13**(1), 104–114 (2013)

Guliashki, V., Marinova, G., Groumpos, P.: Multi-objective optimization approach for energy efficiency in microgrids. IFAC-PapersOnLine **52**(25), 477–482 (2019)

Ivanov, V.: On the approach for automatic generation of small embedded PicoBlaze system. In: 13th International Conference on Application of Concurrency to System Design, pp. 257–260. Barcelona (2013)

Korsemov, Ch., Toshev, H., Mustakerov, I., Borissova, D., Grigorova, V.: An optimal approach to design of joinery for renovation of panel buildings. Int. J. Sci. Eng. Investigat. **2**(18), 123–128 (2013)

Lee, X.-H., Moreno, I., Sun, Ch.-Ch.: High-performance LED street lighting using microlens arrays. Opt. Express **21**(9), 10612–10621 (2013)

Marler, R.T., Arora, J.S.: Survey of multi-objective optimization methods for engineering. Struct. Multidisciplinary Optim. **26**, 369–395 (2004)

Murray, A.T., Feng, X.: Public street lighting service standard assessment and achievement. Socio-Econ. Plan. Sci. **53**, 14–22 (2016)

Mustakerov, I., Borissova, D.: A web application for group decision-making based on combinatorial optimization. In: 4th International Conference on Information Systems and Technologies, pp. 46–56. Valencia (2014)

Mustakerov, I., Borissova, D.: LINDO API using for development of customized applied optimization software. In: International Conference "Automatics and Informatics", pp. I-113–I-116. Sofia, Bulgaria (2009)

Nefedov, E., et al.: Energy efficient traffic-based street lighting automation. In: 23rd International Symposium on Industrial Electronics (ISIE), pp. 1718–1723. IEEE, Istanbul (2014)

Pedrycz, W., Ichalkaranje, N., Phillips-Wren, G., Jain, L.: Introduction to the computational intelligence for decision making. In: Phillips-Wren, G., Ichalkaranje, N., Jain L.C. (eds.) Intelligent Decision Making: An AI-Based Approach. Studies in Computational Intelligence, vol. 97, pp. 79–96. Springer, Heidelberg (2008). https://doi.org/10.1007/978-3-540-76829-6_3

Rabaza, O., Pena-Garcia, A., Perez-Ocon, F., Gomez-Lorente, D.: A simple method for designing efficient public lighting, based on new parameter relationships. Expert Syst. Appl. **40**, 7305–7315 (2013)

Salvia, A.L., Brandli, L.L., Filho, W.L., Kalil, R.M.L.: An analysis of the applications of Analytic Hierarchy Process (AHP) for selection of energy efficiency practices in public lighting in a sample of Brazilian cities. Energy Policy **132**, 854–864 (2019)

Shalamanov, V.: Institution building for IT governance and management. Inf. Secur. Int. J. **38**, 13–34 (2017)

Sudarmono, P., Deendarlianto, Widyaparaga, A.: Energy efficiency effect on the public street lighting by using LED light replacement and kwh-meter installation at DKI Jakarta Province, Indonesia. Journal of Physics: Conference Series 1022(1), Article id. 012021 (2018). https://doi.org/10.1088/1742-6596/1022/1/012021

Generalized Approach to Support Business Group Decision-Making by Using of Different Strategies

Daniela Borissova[1,2(✉)] , Dilian Korsemov[1] ,
and Nina Keremedchieva[1]

[1] Institute of Information and Communication Technologies at the Bulgarian
Academy of Sciences, 1113 Sofia, Bulgaria
dborissova@iit.bas.bg, dilian_korsemov@abv.bg,
ninakeremed@gmail.com
[2] University of Library Studies and Information Technologies,
1784 Sofia, Bulgaria

Abstract. The recent advances in ICT and increased market competition make
the problem of business decisions more significant and more complex. This is
related to the performance evaluation of a variety of business decisions that have
multi-level and multi-factor features. In this regard, the current article aims to
propose a generalized flexible approach to support group decision making by
using different strategies. The different decision-making strategies aim to provide the most preferable alternative; several good alternatives simultaneously, or
ranking of all given alternatives. These different strategies are realized via
corresponding optimization models that are capable to consider differences in
the knowledge and expertise of the group experts. The contribution of the
descried approach is focused on the aggregation stage of the known simple
multi-attribute rating technique. The applicability of the proposed approach and
formulated optimization models are demonstrated in the determination of
preferable offer/s for printing a book considering several different evaluation
parameters. The numerical results show that imposing requirements for different
strategy realization it is quite helpful to get the group decision. Furthermore, the
final group decision is modelled in such a way to reflect the particular background in expertise of each group' member.

Keywords: Group decision-making · Business decisions · Different strategies ·
Simple multi-attribute rating technique · Weights of experts

1 Introduction

Due to globalization and a growing economy, today is difficult to make successful
business decisions without the assistance of experts at different levels [1]. Such decisions often involve business intelligence to gain maximum information from available
data in order to make effective business decisions [2]. The expansion of digital media,

© Springer Nature Switzerland AG 2020
K. Saeed and J. Dvorský (Eds.): CISIM 2020, LNCS 12133, pp. 122–133, 2020.
https://doi.org/10.1007/978-3-030-47679-3_11

entertainment, advertising, retail, and transport to the digital economy is due to the transformation of business models to the e-commerce, and software-as-a-service [3]. The e-media can be successful if a proper business model is implementing to meet the company goal [4]. Some good practices related to the similar service-based management processes are to be considered involving cyber resilience too [5, 6]. Business intelligence relies on analytics experts with capabilities to provide believable information to help in effective and high-quality business decision-making. The core of any decision support system is to help the higher management in different decision problems. Increasing competition and developing ICT makes the problem of business decisions more significant and more complex. The complexity is related to the essence of the selection process that involves various quantitative and qualitative criteria. For such problems the multi-criteria decision making (MCDM) methods can be used to tackle with different and conflicting criteria. The performance evaluation and selection of the most appropriate alternative have multi-level and multi-factor features. This defines essential difficulties in making trade-offs between evaluation criteria to get the best compromise solution. For classification and forecasting, artificial neural networks could be used, but they take considerable consuming time for training. To overcome this drawback, a permutation of the neurons activation function in order the convergence speed-up to be achieved seems to be useful [7].

The latest achievements in ICT reflect across different areas including the publishing sector. Each of these publishers has his own business models that include different parameters of the processes for publishing. Due to the digital publishing revolution, research institutions and academic publishers have begun to examine the possibilities of working together to promote academic digital publishing [8]. All of this, along with advances in digital printing is prerequisites of a variety of offers for manuscripts publishing. Therefore, the main question is how to select or how to determine a restricted set of good offers for manuscripts publishing. During the selection, different parameters are to be considered as price per unit, the discount rate per circulations, quality of paper and covers, printing mode (offset, digital), lead time delivery, a minimum allowed number for printing, way of delivery, samples, guarantee when a large number of defective items are available (who will bear the expense) and last but not least e-book and print on demand.

When modeling real-world decision problems three basic problem formulations can be distinguished: choice, ranking and sorting problems [9]. In respect to this classification, the current article deals with the problems of choice and ranking. An appropriate quantitative tool for group decision making (GDM) that can be applied in such problems is the multi-attribute utility theory (MAUT) [10]. MAUT seeks to over-come different dimensions via one dimension values followed by aggregation through a weighted linear average. The simple additive weighting (SAW) method is a common aggregation approach in MAUT. A simple multi-attribute rating technique (SMART) is another form of MAUT with the possibility to be applied to any type of weight assignment technique like absolute, relative, etc. [11, 12]. It should be noted that different methods of MAUT could lead to different solutions [13].

The selection problem could be viewed as multi-attribute group decision making (MAGDM) problems as often the executives have no confidence in an untested supplier and to overcome this is needed to take different indicators to guarantee objectives performance [14]. There are different approaches proposed to tackle with the problem of selection [15]. For example, an integrated approach is proposed to select using AHP-ARAS-MCGP methodology [16]. Other authors emphasize the usage of multi-criteria intuitionistic fuzzy TOPSIS method [17]. A group decision-making approach by extended SMART is used in the evaluation and ranking of students [18]. To cope with different competencies of the group members modifications of simple additive weighting and weighted product models are proposed [19]. These modifications are implemented in the spread sheet as a tool to support group decisions [20]. The comparison between approaches of individual and group decision making toward multi-criteria decision problems shows that individual point of view of each expert significantly influences to the final decision [21]. In this respect, to express more transparently the expertise of each group member a two-component function of objective and subjective part seems to be useful [22].

In processes of selection and ranking, the GDM could be applied to find a solution – a single alternative or set alternatives according to the preferences provided by the authorized group of experts. While the most publications considered the problem for handle different expressions of experts' preferences by utility-based individual preferences [23], fuzzy relations [24, 25], decision Markov logic [26], fuzzy logic based on multi-criteria evaluation [27], multidimensional preference [28], linguistic information [29], etc., the current paper emphasizes on the evaluation and aggregation phase of the group decision-making process.

In contrast to the published results, the current article proposes a generalized approach to support business decisions incorporating different strategies to make better decisions by analysing the existing alternatives. The different strategies aim to determinate the most preferable alternative, or to determinate several good alternatives simultaneously, or to rank of all given alternatives. Furthermore, the described approach takes into account the experts' knowledge and experience when forming the final group decision. Each strategy is based on the formulated optimization model.

The rest of the paper is organized as follows: Sect. 2 describes the proposed algorithm, Sect. 3 provides a description of the numerical application in the publishing sector where several offers are to be evaluated, Sect. 4 contains results analysis and conclusions are drawn in Sect. 5.

2   Generalized Approach to Support Business Group Decision Making by Three Different Strategies

The possibility to incorporate a wide variety of quantitative and qualitative criteria makes SMART a popular and widely used method. The originally SMART describes the whole process as rating the alternatives and weighting of attributes [11]. The proposed approach to support the business by group decision making with three different strategies is based on SMART. The basic stages of the proposed modification of SMART are graphically illustration is presented in Fig. 1.

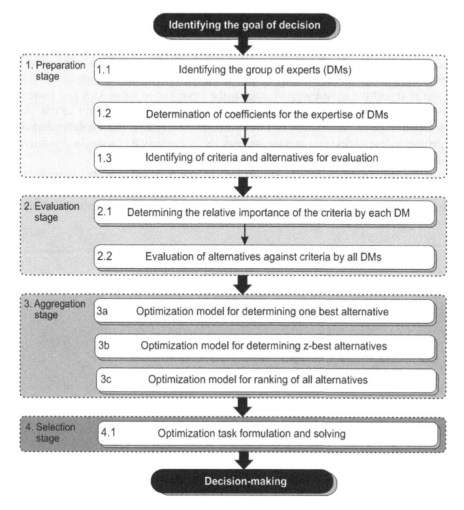

Fig. 1. Generalized approach to support group decision-making by three different strategies

The *preparation stage* 1 of the proposed approach starts with the identification of the main goal of group decision making that includes: 1.1) identification of evaluation criteria and alternatives; 1.2) identification of needed experts (DMs) for group decision making; and 1.3) determination of corresponding weights for each DM in accordance to its knowledge and experience. This stage starts with the determination of alternatives and suitable evaluation criteria. Next, the group of experts responsible to evaluate different aspects of the decision goal is to be selected. According to Yetton & Botter [30], groups of five experts are the most effective in group decision making closely followed by a group of seven experts. Weights of DMs play a very important role in multiple-attribute group decision-making and in the proposed algorithm the corresponding weights for each expert reflect their expertise.

The *evaluation stage* 2 is composed of two steps: 2.1) determination of weighted coefficients for the evaluation criteria by DMs; and 2.2) evaluation of the given alternatives against the criteria by each DM. The weighted coefficients for criteria importance and alternatives evaluations are expressed by scores derived by the experts where the higher value means better performance.

The next and the most essential stage is the *aggregation stage* 3. A key issue in MAGDM is the aggregation of individual preferences. In the current article, the aggregation stage can be realized by three different optimization models depending on the particular goal of decision-making strategy for: 3a) selection of a single alternative; 3b) defining of z best alternatives; 3c) ranking of all alternatives.

The first decision-making strategy of the aggregation stage (Fig. 1) illustrates the situation where the selection of single alternative is to be done considering the alternatives performance and the corresponding formulation of the combinatorial optimization problem is as follows:

$$\text{maximize } \sum_{q=1}^{K} \lambda^q \left(\sum_{j=1}^{N} \sum_{i=1}^{M} w_j^q p_{ij}^q x_i \right) \tag{1}$$

subject to

$$\sum_{q=1}^{K} \lambda^q = 1, \ \lambda_q \in (0,1) \tag{2}$$

$$\sum_{j=1}^{N} w_j^q = 1, \ q = \{1, 2, \ldots, K\} \tag{3}$$

$$\sum_{i=1}^{M} x_i = 1, \ x_i \in \{0,1\} \tag{4}$$

where λ^q is weighting coefficient for q-th expert; weight of n-th criterion accordingly the q-th expert is denoted by w_n^k; the performance of i-th alternative in respect to j-th by q-th expert is denoted by p_{ij}^q where $0 \leq p_{ij}^q \leq 1$ and x_i are binary integer variables assigned to each alternative that indicate whether item i-th is selected or not.

The objective function (1) seeks to maximize the overall performance of alternatives in accordance to the evaluation criteria, weights of these criteria assigned from each DM and corresponding weighted coefficients for DMs expertise. Constraint (3) provides the selection of only one alternative.

The second decision-making strategy of the aggregation stage (Fig. 1) aims to determine several good alternatives by replacing the relation (4) with the following statement [31, 32]:

$$\sum_{i=1}^{M} x_i = z, \, x_i \in \{0, 1\}, \, 1 < z < M \tag{5}$$

where z is integer number bigger than 1 and less then number of alternatives (M).

The third decision-making strategy of the aggregation stage of the proposed group decision-making approach makes possible to rank all alternatives using the following algorithm for ranking as shown in Fig. 2.

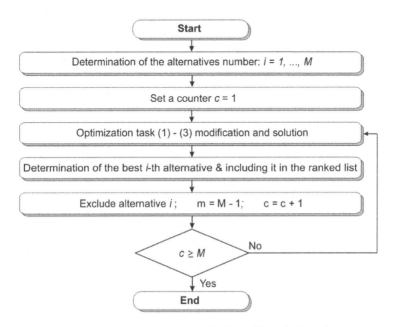

Fig. 2. Flowchart of the algorithm for ranking of alternatives

This algorithm for ranking of all alternatives is based on the sequential solution of multiple tasks of the type (1) – (4). The solution on the first iteration provides the best alternative which goes on the top of the ranked list. On the second iteration, this alternative is excluded from the set of alternatives and the modified optimization task (1) – (3) is solved again to get the next alternative in the ranked list. This process is repeated until only one alternative remains.

During the *selection stage* 4, the optimization tasks corresponding to the simulation of three different decisions making scenarios (strategies) are to be formulated and solved. Depending on the obtained solution for one best alternative, several good alternatives or ranking of all alternatives, the higher management could make the most appropriate choice from the three scenarios. The higher management could make a decision based on the obtained solution from stage 3 taking into account also the defined goal and policy of the company.

3 Numerical Application

The numerical example is from the publishing sector where several offers are given, each of them with different parameters and one of them is to be selected. The five different offers for printing a book are considered that correspond to five alternatives (A-1, A-2, A-3, A-4, and A-5). The decision should be taken based on the evaluations of a group of five experts: production manager (E-1), R&D manager (E-2), sales manager (E-3), marketing manager (E-4) and salesman (E-5). The offers' evaluation criteria to be estimated involve ten parameters as follows: 1) unit price (C-1); 2) discount rate per circulations (C-2); 3) quality paper and covers (C-3); 4) printing mode – offset/digital (C-4); 5) lead time delivery (C-5); 6) minimum allowed number for printing (C-6); 7) way of delivery (C-7); 8) samples (C-8); 9) guarantee – when a large number of defective items are available, who will bear the expense (C-9); and 10) e-book and print on demand (C-10).

The evaluations of five potential offers for printing a book with respect to the described ten criteria from five experts are shown in Table 1.

Table 1. Evaluations of potential alternative offers from the experts

Experts	Alter-natives	Criteria/Weights/Evaluations									
		C-1	C-2	C-3	C-4	C-5	C-6	C-7	C-8	C-9	C-10
E-1		*0.04*	*0.13*	*0.14*	*0.08*	*0.15*	*0.15*	*0.08*	*0.1*	*0.07*	*0.06*
	A1	0.2	0.9	0.9	1	0.3	0.4	0.8	0.6	0.5	0.4
	A2	0.8	0.9	1	0.9	0.7	0.8	0.9	0.8	0.6	0.7
	A3	0.8	0.3	0.6	0.8	0.3	0.5	0.7	0.5	0.6	0.8
	A4	0.2	0.6	0.2	0.4	0.3	0.5	0.7	1	0.4	0.4
	A5	0.4	0.8	0.7	0.8	0.6	0.8	0.6	0.3	0.2	0.8
E-2		*0.05*	*0.14*	*0.16*	*0.05*	*0.09*	*0.06*	*0.08*	*0.15*	*0.16*	*0.06*
	A1	0.4	0.8	0.2	0.6	0.1	0.5	0.9	0.4	0.6	1
	A2	0.5	0.8	0.5	0.6	1	0.5	0.7	0.9	1	0.8
	A3	0.4	0.7	0.5	0.8	0.7	0.9	0.4	1	0.5	1
	A4	1	0.5	0.8	0.1	0.6	0.4	0.8	0.9	0.4	0.2
	A5	0.8	0.9	0.1	0.2	0.5	1	0.9	0.8	0.6	0.8
E-3		*0.1*	*0.1*	*0.1*	*0.1*	*0.11*	*0.1*	*0.08*	*0.1*	*0.12*	*0.09*
	A1	0.6	0.7	0.8	0.4	0.9	0.3	0.4	0.6	0.8	0.8
	A1	0.8	0.7	0.7	0.8	0.7	0.7	1	0.8	0.6	0.8
	A3	0.6	0.5	0.7	0.8	0.2	0.2	0.6	0.6	0.6	0.6
	A4	0.8	0.6	0.2	0.2	0.2	0.2	0.1	1	0.1	0.7
	A5	0.7	0.8	0.7	0.8	0.8	0.6	0.5	0.4	0.3	0.6
E-4		*0.11*	*0.11*	*0.06*	*0.08*	*0.14*	*0.06*	*0.11*	*0.11*	*0.14*	*0.08*
	A1	0.1	0.7	0.7	0.4	0.1	0.5	0.2	0.3	0.4	0.1
	A2	0.9	0.5	1	0.8	0.7	0.8	0.7	0.9	0.8	0.7
	A3	0.8	0.1	0.8	0.6	0.6	0.7	0.7	0.7	0.5	0.6
	A4	0.5	0.7	0.6	0.3	0.5	0.6	0.4	0.7	0.4	0.3
	A5	0.7	0.8	0.6	0.7	0.7	0.6	0.8	0.5	0.1	0.5

(continued)

Table 1. (*continued*)

Experts	Alter-natives	Criteria/Weights/Evaluations									
		C-1	C-2	C-3	C-4	C-5	C-6	C-7	C-8	C-9	C-10
E-5		*0.07*	*0.1*	*0.1*	*0.08*	*0.15*	*0.08*	*0.1*	*0.1*	*0.15*	*0.07*
	A1	0.1	0.6	0.6	0.4	0.2	0.2	0.2	0.2	0.2	0.1
	A2	0.9	0.6	0.9	0.8	0.7	0.7	0.5	0.6	0.7	0.6
	A3	0.8	0.2	0.6	0.7	0.9	0.7	0.6	0.8	0.7	0.9
	A4	0.5	0.8	0.5	0.7	0.6	0.6	0.1	0.5	0.9	0.2
	A5	0.6	0.7	0.6	0.6	0.8	0.8	0.7	0.4	0.3	0.6

In accordance with stage 2 of the proposed generalized group decision-making algorithm (Fig. 1), all potential offers for printing the book are evaluated in respect to given criteria (attributes) from a group of experts. Furthermore, the relative importance between criteria from the particular expert' point of view has to be done by corresponding weights as shown in Table 1.

Three different cases, representing three scenarios about the experts' knowledge and experience are used to demonstrate the importance of experts in group decision making. Case-1 expresses the situation where the knowledge and experience for all experts in the group are equal, while Case-2 and Case-3 make a difference between involved experts. Case-2 considers expert E-5 as the most trusted expert, while the dominant role in Case-3 is assigned on expert E-4. The results of the solution of optimization tasks, formulated on step 3.1a are shown in Table 2.

Table 2. Solution results of tasks on step 3

Cases	Weighting coefficients of experts					The best alternative	Two good alternatives	Ranking of alternatives
	E-1	E-2	E-3	E-4	E-5			
Case-1	0.20	0.20	0.20	0.20	0.20	A-2	A-2 & A-5	A-2, A-5, A-3, A-1, A-4
Case-2	0.10	0.15	0.18	0.22	0.35	A-3	A-3 & A-5	A-3, A-5, A-2, A-1, A-4
Case-3	0.16	0.13	0.11	0.44	0.17	A-5	A-3 & A-5	A-5, A-3, A-2, A-4, A-1

4   Results and Discussion

The proposed approach to support business by group decision-making with different strategies is applied for: 1) determination of the best alternative, 2) determination of 2 good alternatives and 3) ranking of all alternatives. According to the input data from Table 1 and three different cases for the weights of experts from Table 2, the obtained solutions for selection and ranking are illustrated in Fig. 3.

The main idea of the proposed approach for group decision-making is to aggregate different experts' evaluation for the alternatives into a final group decision. The flexibility of the modification of SMART is realized by formulation and solving of combinatorial optimization tasks for determination of a single alternative, for several good alternatives or for ranking of all alternatives in accordance with the selected strategy. Regardless of the strategy used, all of the optimization models (respectively tasks) take into account the knowledge of the experts by corresponding weighting coefficients.

The numerical application when using strategy for selection of one alternative, determines as the best option alternative A-2 when all experts are considered with equal level of competency and knowledge. For the same strategy, but considering the differences in experts' knowledge and experience, the tasks solution determine respectively alternative A3 (Case-2) and alternative A5 (Case-3) as the best decision.

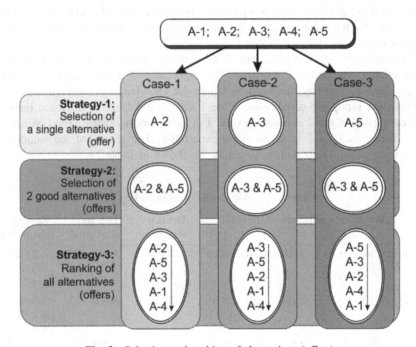

Fig. 3. Selection and ranking of alternatives (offers)

In the second strategy, there is no information which of the selected alternatives A2 & A5 (respectively A3 & A5, A3 & A5) is better than the other, but both alternatives are certainly better than the others. It should be noted that determined several good alternatives is obtained simultaneously as a result of solution via single run of the task. The usage of third strategy allows ranking of all alternative according to the point of view of all experts and taking into account the level of the expertise of group members. The obtained solutions under different strategies and three different cases of experts' expertise are shown in Fig. 3. The implementation of described approach allows simulation of different strategies for decision making that provides the flexibility in the achievement of the main goal of group decision making process. The results of all strategies implementation determine one alternative, several good alternatives or ranking of all alternatives that can be serving as base from which the executive managers could form the final best decision.

The described generalized approach to support business decision by using different strategies can helps companies to make better decisions by analysing the existence alternatives within business context. The essence of business intelligence is to provide performance and competitor benchmarks to make the organization run more efficiently. In this regard, the proposed generalized group decision making approach with incorporated three different strategies make clear basis on which the alternatives are being evaluated taking into account many different points of view about alternatives decisions which is particularly important in group decision making situations. Such business model contributes to make the most suitable business decisions in accordance to the goal and vision of the company. In such way, the executive managers could easier to understand data without the technical know-how to dig into the data themselves.

5   Conclusion

The article describes a group decision-making modeling approach to support business decisions. This is realized by the proposed flexible algorithm that implements three different decision-making strategies. The described flexible algorithm is based on SMART and includes also usage of corresponding weights for each group' expert in accordance to his knowledge and experience, and formulation of three combinatorial optimization models corresponding to three different decision-making strategies. These strategies are realized on the aggregation stage and are related to the determination of a single best alternative, determination of several good alternatives or determination of the ranked list of all alternatives. The implementation of each strategy relies on the proposed combinatorial models that are used to formulate and solve the corresponding optimization tasks.

The main contribution of the present article is the flexibility of the proposed approach with different decision-making strategies that make possible the highest executive managers to make the most reasonable decision based on the aggregated opinions of the group of experts. The described modeling approach could be applied for any other kind of problems that can be represented with a certain set of criteria about given alternatives.

Numerical testing demonstrates the practical applicability for all described decision-making scenarios. Because of the well-structured weighted decision matrix, the described combinatorial optimization modeling approach can be coded as a software tool to support managers without specific mathematical background.

References

1. Shalamanov, V.: Organizing for IT effectiveness, efficiency and cyber resilience in the academic sector: National and regional dimensions. Inf. Secur. Int. J. **42**, 49–66 (2019)
2. Borissova, D., Mustakerov, I., Korsemov, D.: Business intelligence system via group decision making. Cybern. Inf. Technol. **16**(3), 219–229 (2016)
3. Zaheer, H., Breyer, Y., Dumay, J.: Digital entrepreneurship: an interdisciplinary structured literature review and research agenda. Technol. Forecast. Soc. Change **148**, 119735 (2019). https://doi.org/10.1016/j.techfore.2019.119735
4. Andreev, R., Borissova, D., Shikalanov, A., Yorgova, T.: Model-driven design of emedia: virtual technology transfer office. In: Lugmayr, A., Stojmenova, E., Stanoevska, K., Wellington, R. (eds.) Information Systems and Management in Media and Entertainment Industries. ISCEMT, pp. 279–298. Springer, Cham (2016). https://doi.org/10.1007/978-3-319-49407-4_14
5. Shalamanov, V.: Institution building for IT governance and management. Inf. Secur. Int. J. **38**, 13–34 (2017)
6. Shalamanov, V.: Towards effective and efficient IT organizations with enhanced cyber resilience. Inf. Secur. Int. J. **38**, 5–10 (2017)
7. Balabanov, T., Atanasova, T., Blagoev, I.: Activation function permutation for multilayer perceptron training. In: Proceedings of International Conference on Big Data, Knowledge and Control Systems Engineering, 21–22 November, Sofia, Bulgaria, pp. 9–14 (2018)
8. Wen-Qi, F., Mei, Z., Ling-Yan, Y.: Academic e-book publishing in china: an investigation of current status and publishers' attitudes. J. Acad. Librariansh. **44**, 15–24 (2018)
9. Roy, B.: Multicriteria Methodology for Decision Aiding. Springer, Boston (1996). https://doi.org/10.1007/978-1-4757-2500-1
10. Rao, R.V.: Introduction to multiple attribute decision-making (MADM) methods. In: Decision Making in the Manufacturing Environment, Springer Series in Advanced Manufacturing, pp. 27–41. Springer, London (2007). https://doi.org/10.1007/978-1-84628-819-7_3
11. Barfod, M.B., Leleur, S. (eds.) Multi-Criteria Decision Analysis for Use in Transport Decision Making. DTU Lyngby: Technical University of Denmark, Transport (2014)
12. Edwards, W.: How to use multiattribute utility measurement for social decision making. IEEE Trans. Syst. Man Cybern. **7**(5), 326–340 (1977)
13. Korsemov, D., Borissova, D., Mustakerov, I.: Group decision making for selection of supplier under public procurement. In: Kalajdziski, S., Ackovska, N. (eds.) ICT Innovations 2018, Engineering and Life Sciences, ICT 2018, Communications in Computer and Information Science, vol. 940, pp. 51–58 (2018)
14. Borissova, D., Atanassova, Z.: Multi-criteria decision methodology for supplier selection in building industry. Int. J. 3-D Inf. Model. **7**(4), 49–58 (2018)
15. Wetzstein, A., Hartmann, E., Benton Jr., W.C., Hohenstein, N.O.: A systematic assessment of supplier selection literature – state-of-the-art and future scope. Int. J. Prod. Econ. **182**, 304–323 (2016)
16. Fu, Y.-K.: An integrated approach to catering supplier selection using AHP-ARAS-MCGP methodology. J. Air Transp. Manag. **75**, 164–169 (2019)
17. Memari, A., Dargi, A., Jokar, M.R.A., Ahmad, R., Rahim, A.R.A.: Sustainable supplier selection: a multi-criteria intuitionistic fuzzy TOPSIS method. J. Manuf. Syst. **50**, 9–24 (2019)

18. Borissova, D., Keremedchiev, D.: Group decision making in evaluation and ranking of students by extended simple multi-attribute rating technique. Cybern. Inf. Technol. **18**(3), 45–56 (2019)
19. Korsemov, D., Borissova, D.: Modifications of simple additive weighting and weighted product models for group decision making. Adv. Model. Optim. **20**(1), 101–112 (2018)
20. Borissova, D., Korsemov, D., Mustakerov, I.: Multi-attribute group decision making considering difference in experts knowledge: an excel application. In: Proceedings of 12th International Management Conference – Management Perspectives in the Digital Era, 1–2 November, Bucharest, Romania, pp. 387–395 (2018)
21. Borissova, D., Korsemov, D., Mustakerov, I.: Multi-criteria decision making problem for doing business: comparison between approaches of individual and group decision making. In: Saeed, K., Chaki, R., Janev, V. (eds.) CISIM 2019. LNCS, vol. 11703, pp. 385–396. Springer, Cham (2019). https://doi.org/10.1007/978-3-030-28957-7_32
22. Borissova, D.: A group decision making model considering experts competency: an Application in personnel selections. Comptes rendus de l'Academie Bulgare des Sciences **71**(11), 1520–1527 (2018)
23. Huang, Y.S., Chang, W.C., Li, W.H., Lin, Z.L.: Aggregation of utility-based individual preferences for group decision-making. Eur. J. Oper. Res. **229**, 462–469 (2013)
24. Peneva, V., Popchev, I.: Aggregation of fuzzy relations using weighting function. Comptes rendus de l'Academie bulgare des Sciences **60**(10), 1047–1052 (2007)
25. Peneva, V., Popchev, I.: Models for fuzzy multicriteria decision making based on fuzzy relations. Comptes rendus de l'Academie bulgare des Sciences **62**(5), 551–558 (2009)
26. Sgurev, V.: Decision Markov logic. Comptes rendus de l'Academie bulgare des Sciences **67**(2), 181–188 (2014)
27. Buyukozkan, G.: Multi-criteria decision making for e-marketplace selection. Internet Research **14**(2), 139–154 (2004)
28. Zhang, S., Zhu, J., Liu, X., Chen, Y.: Regret theory-based group decision-making with multidimensional preference and incomplete weight information. Inf. Fusion **31**, 1–13 (2016)
29. Javier, F., Herrera-Viedma, C.E., Pedrycz, W.: A method based on PSO and granular computing of linguistic information to solve group decision making problems defined in heterogeneous contexts. Eur. J. Oper. Res. **230**(3), 624–633 (2013)
30. Yetton, P., Botter, P.: The relationships among group size, member ability, social decision schemes, and performance. Organ. Behav. Hum. Perform. **32**, 145–159 (1983)
31. Mustakerov, I., Borissova, D.: A combinatorial optimization ranking algorithm for reasonable decision making. Comptes rendus de l'Academie bulgare des Sciences **66**(1), 101–110 (2013)
32. Borissova, D.: Group decision making for selection of k-best alternatives. Comptes rendus de l'Academie bulgare des Sciences **69**(2), 183–190 (2016)

A Generic Materials and Operations Planning Approach for Inventory Turnover Optimization in the Chemical Industry

Jairo R. Coronado-Hernández[1]([⊠]) ⓘ, Alfonso R. Romero-Conrado[1] ⓘ,
Olmedo Ochoa-González[2], Humberto Quintero-Arango[2], Ximena Vargas[3],
and Gustavo Gatica[3] ⓘ

[1] Universidad de la Costa, Barranquilla, Colombia
{jcoronad18,aromero17}@cuc.edu.co
[2] Universidad Tecnológica de Bolívar, Cartagena, Colombia
[3] Universidad Andrés Bello, Santiago, Chile
ggatica@unab.cl

Abstract. Chemical industries usually involve continuous and large-scale production processes that require demanding inventory control systems. This paper aims to show the results of the implementation of a mixed-integer programming model (MIP) based on the Generic Materials and Operations Planning Problem (GMOP) for optimizing the inventory turnover in a fertilizer company. Results showed significant improvements for Inventory Turnover Ratios and overall costs when compared with an empirical production planning method.

Keywords: Inventory turnover · Production planning · GMOP · Fertilizers · Chemical industry · Optimization

1 Introduction

The chemical industry involves highly complex production systems, with production planning scenarios that include multilevel product structures, multi-period planning horizons, and co-production environments (processes share raw materials and have common outputs) [7,23].

Production planning studies for the chemical industry have been mainly focused on lot sizing and scheduling problems considering batch processes in large-scale plants. The reduction of holding and setup costs has been one of the main objectives for these applications [3,4,8,15,16,18,19].

In practice, the presence of batches and co-production parameters increases the complexity of inventory systems for controlling quantities of raw materials, work in progress, and final products. Some implementations for inventory control systems in chemical plants have involved mainly a "push" approach for a continuous process, based on ERP platforms [6,22,24–26].

© Springer Nature Switzerland AG 2020
K. Saeed and J. Dvorský (Eds.): CISIM 2020, LNCS 12133, pp. 134–145, 2020.
https://doi.org/10.1007/978-3-030-47679-3_12

The inventory turnover ratio measures how many times the plant has sold and replaced inventory during a given period and gives a useful perspective for commercial, manufacturing, inventory and purchase decisions [5,17], improving efficiency and reducing costs.

This paper aims to show the implementation results of a mixed integer programming model (MIP) based on the Generic Materials and Operations Planning Problem (GMOP) to optimize the inventory turnover ratio in a fertilizer company. Two approaches for production planning will be compared: an empirical method based on the planner experience, and an exact method based on generic materials and operations. The exact method approach has been widely proved in similar production planning scenarios [9,12–14,21] and allows considering the complex product structures inside an NPK fertilizers plant.

The paper is organized as follows: Sect. 2 shows the generalities and basic concepts for the two planning approaches, Sect. 3 explains the case study in a fertilizers company, Sect. 4 shows the implementation results (Costs and Inventory turnover), and finally, main conclusions and further research are stated in Sect. 5.

2 Methodology

Using a quantitative approach, this paper aims to compare inventory and cost performance results of an empirical production planning methodology with the results obtained from a well documented MIP model, parameterized with demand forecasts.

2.1 Inventory Turnover as Operational Efficiency Measure

Chemical industries usually involve continuous and large-scale production processes that require demanding inventory control systems. Generally, lead times for chemical raw materials are considerably high, and production planning turns into a complex task that has to deal with finding a balance between holding costs and raw material availability [11].

The inventory turnover (IT) is a key indicator that allows measure performance and operational efficiency [10]. IT is an estimation of how fast the company is replacing its inventory: a higher inventory turnover is a sign of a high synchronization between purchases, production, and sales. As shown in Eqs. 1, 2, 3, and 4, the inventory turnover can be calculated according different analysis approaches.

$$Turnover\ (Raw\ Materials) = \frac{Total\ Cost\ of\ Raw\ Materials}{Average\ Inventory\ of\ Raw\ Materials} \quad (1)$$

$$Turnover\ (Work\ in\ Progress) = \frac{Total\ Production\ Costs}{Average\ Inventory\ of\ Work\ in\ Progress} \quad (2)$$

$$Turnover\ (Final\ Products) = \frac{Total\ Sales\ Cost}{Average\ Inventory\ of\ Final\ Products} \quad (3)$$

$$Inventory\ Turnover = \frac{Sum\ of\ product\ inventory}{Average\ Inventory} \quad (4)$$

2.2 An Empirical Production Planning Method

The production planning method for the fertilizer company in this case study is considered as empirical and considerably subjective (is based on the experience of the production planner). The planning process is supported by an ERP database and a spreadsheet, following a heuristic procedure similar to the described in Fig. 1.

Production Planning starts with a sales forecasts provided by the commercial department. Forecasts are made for a mid-term horizon, a year usually, segmented in months. Estimated sales are included in the spreadsheet and distributed along the planning horizon for NPK fertilizers and Calcium Nitrate.

The company has five different plants for the production of Ammonia, Nitric Acid, Ammonium Nitrate, Potassium Chloride, and Calcium Nitrate. The plants operate 24 h a day, with 3 work shifts of 8 h each. Programmed maintenance is carried out every certain amount of produced metric tons. This information is used for determining the possible inventory levels each month and then, proposing changes to the purchase orders made by the commercial department. Once inventory levels are adjusted, the Master Planning Schedule (MPS) is prepared and socialized.

For turnover ratio calculation purposes, data for inventory levels, purchases and costs were collected during a complete year of operation.

2.3 The Generic Materials and Operational Planning Model (GMOP)

As cited by [21], the Generic Materials and Operational Planning model was proposed by García Sabater [9] and Maheut et al. [13], as an alternative for modeling the existing relations between the processes and the materials needed for the elaboration of a product. This lot-sizing model focuses on the planning of operations (Strokes). A Stroke is defined as any activity or operation that allows for the transformation of a set of products or Stock Keeping Units (SKUs) into another set of SKUs, using or immobilizing a certain amount of resources. A Stroke can contain the following attributes [9]:

- Outputs (Stroke Outputs): The product or set of products obtained from the stroke execution.
- Inputs (Stroke Inputs): The product or set of products consumed at the execution of the stroke.
- Lead times
- Operation times and costs

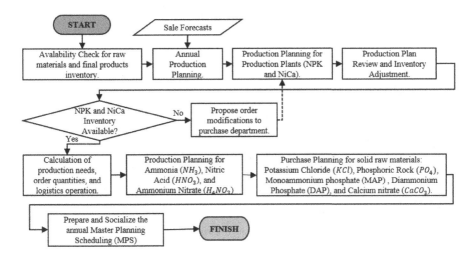

Fig. 1. Empirical production planning method

- Set-up times and costs
- Resource usage: Resources can be, for example, machinery, the workforce, and so on.

The GMOP model can easily include capacity constraints, as well as direct, inverse, and alternate bill of materials, multi-site production, resource requirements, by-products, transportation modes, and packaging processes [9,21].

The problem is presented as a mixed-integer programming model, whose parameters and variables are shown in Table 1. The objective function (5) aims to minimize the total planning cost Z, which includes storage, operation, and set up costs generated by the execution of strokes.

$$Z = Min \sum_{t} \sum_{i} (h_{i,t} * X_{i,t}) + \sum_{t} \sum_{k} (CS_{k,t} * \delta_{k,t} + CO_{k,t} * z_{k,t}) + \sum_{t} \sum_{i} (CB_{i,t} * w_{i,t}) \quad (5)$$

Equation (6) represents the inventory constraint. It considers the stock levels from the previous period, demand requirements, purchased items, and the quantities generated and consumed by the strokes in every period.

$$X_{i,t} = X_{i,t-1} - D_{i,t} + w_{i,t} - \sum_{k} (SI_{i,k} * z_{k,t}) + \sum_{k} (SO_{i,k} * z_{k,t-LT_k}) \forall (i,t) \quad (6)$$

Equation (7) ensures the inclusion of a setup cost when a stroke is used: If $z_{k,t}$ is larger than zero, then $\delta_{k,t}$ must be 1 in order to satisfy the constraint.

$$z_{k,t} - M * \delta_{k,t} \le 0 \forall (k,t) \quad (7)$$

Equation (8) is a capacity constraint that limits the use of resources by considering both setup and operations times.

$$\sum_{k} (TS_{k,r} * \delta_{k,t}) + \sum_{k} (TO_{k,r} * z_{k,t}) \le KAP_{r,t} \forall (r,t) \quad (8)$$

Table 1. Parameters and decision variables for the GMOP model [9,21].

Symbol	Description
i	Index set of products (includes product, packaging, and site)
t	Index set of planning periods
r	Index set of resources
k	Index set of strokes
$D_{i,t}$	Demand of product i for period t
hi, t	Cost of storing a unit of product i in period t
$CO_{k,t}$	Cost of stroke k in period t
$CS_{k,t}$	Cost of the setup of stroke k in period t
$CB_{i,t}$	Cost of purchasing product i in period t
$SO_{i,k}$	Number of units i that generates a stroke k
$SI_{i,k}$	Number of units i that stroke k consumes
LT_k	Lead time of stroke k
$KAP_{r,t}$	Capacity availability of resource r in period t (in time units)
M	A sufficiently large number
$TO_{k,r}$	Capacity of the resource r required for performing one unit of stroke k (in time units)
$TS_{k,r}$	Capacity required of resource r for setup of stroke k (in time units)
$z_{k,t}$	Amount of strokes k to be performed in period t
$\delta_{k,t}$	=1 if stroke k is performed in period t (and 0 otherwise)
$w_{i,t}$	Purchase quantity for product i in period t
$X_{i,t}$	Stock level of product i on hand at the end of period t
Z	Total Planning Cost

Finally, Eqs. (9, 10, 11 and 12) define the range and domain of the decision variables.

$$X_{i,t} \geq 0 \tag{9}$$

$$w_{i,t} \geq 0 \tag{10}$$

$$z_{k,t} \in \mathbb{Z}^+ \tag{11}$$

$$\delta_{k,t} \in \{0,1\}. \tag{12}$$

3 Case Study: Fertilizers Company

3.1 Fertilizer Production Process

Fertilizers are chemical substances, intended to supply essential nutrients to crops, contributing to its growth, fertility and therefore, increasing its productivity. There are different types of chemical fertilizers, some of them are produced from synthesizing ammonia: the NPK fertilizers [2] and calcium nitrate [1]. Figure 2 shows a basic overview of the fertilizer production process.

The production process of NPK fertilizers requires Nitrogen (N), Phosphorus (P), and Potassium (K). Some necessary substances include Nitric acid, Phosphoric rock, and Ammonia. The fertilizer is measured and prepared according to the percentage of N, P, and K. For example: a 10–30–30 grade fertilizer, has 10% Nitrogen, 30% phosphorus as P_2O_5, and 10% potassium as K_2O. The remaining percentage is conformed by fillers (clay), moisture, and a portion of free salts resulting from the production process. NPK fertilizers are produced and packaged according to the volumes, and customer's needs.

On the other hand, the production process of calcium nitrate involves two raw materials: calcium carbonate ($CaCO_3$) and nitric acid (HNO_3). Gaseous NH_3 is used to adjust the pH of the solution and ammonium nitrate in solution [20]

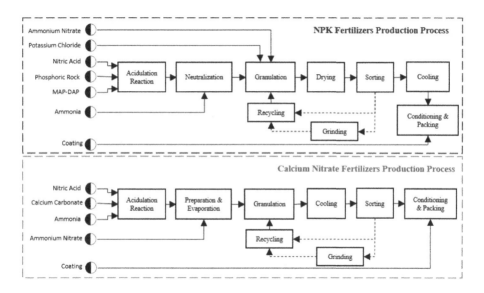

Fig. 2. Production process of fertilizers.

4 Results

4.1 Parameters

The information of products, costs, demand, times, resources and capacities are shown in the following tables. Table 2 shows the codes and descriptions for the 16 products or SKUs. Additionally, the basic costs like holding, production and purchase costs are listed.

Figure 3 shows an example of a bill of materials for a final product. This representation is a result from the information shown in Table 3 and Table 4.

The Stroke Inputs Matrix allows identifying the necessary raw materials for every Stroke. For example, Stroke 2 (K2) needs 0.71t of Ammonia and 0.18t of Nitric Acid. Those Strokes without inputs are considered as Purchase Strokes and are related to purchased substances like Phosphoric Rock, and Clay. On the other hand, The Stroke Outputs Matrix allows identifying the resulting products from every Stroke.

A complete data set is available at GitHub and includes all necessary parameters for GMOP model: Demand, Setup Cost, Resources Capacities, etc.

Table 2. Product description and basic costs in $COP

Code	Description	Holding Cost $h_{i,t}$	Production Cost $CO_{k,t}$	Purchase Cost $CB_{i,t}$
AMO	Ammonia	$ 4.100,0	$ 790.000,0	$ -
ACN	Nitric Acid	$ 2.650,0	$ 420.000,0	$ -
NAM	Ammonium Nitrate	$ 4.856,0	$ 443.000,0	$ -
MPKCl	Potassium Chloride	$ 8.900,0	$ -	$ 720.060,0
MPRF	Phosphoric Rock	$ 9.870,0	$ -	$ 366.300,0
MPMAP	Mono-ammonium Phosphate	$13.200,0	$ -	$ 985.380,0
MPDAP	Di-ammonium Phosphate	$ 13.200,0	$ -	$ 1.158.000,0
MPARC	Clay	$ 2.050,0	$ -	$ 57.920,0
MPCAL	Calcium Carbonate	$ 3.650,0	$ -	$138.740,0
NPK1	NPK 10-20-20	$14.760,0	$ 896.460,0	$ -
NPK2	NPK 10-30-10	$15.980,0	$ 950.600,0	$ -
NPK3	NPK 15-15-15	$12.750,0	$ 910.000,0	$ -
NPK4	NPK 12-24-12	$ 14.500,0	$ 897.300,0	$ -
NPK5	NPK 20-5-20	$ 17.320,0	$ 913.000,0	$ -
NPK6	NPK 10-15-25	$ 24.600,0	$ 1.187.300,0	$ -
NICA	Calcium Nitrate	$17.300,0	$ 680.000,0	$-

4.2 Computational Results

The GMOP mathematical model was programmed using Gurobi (version 6.5). The model contains 7697 equations and 7452 continuous variables. Hardware specifications included a personal computer with Intel® CoreTM i5-4300U 2.5 GHz processor and 8 GB of RAM. An integration with Java was necessary to connect Gurobi with Microsoft Excel. This interface allowed to import and export data sets and solutions. Computational times were acceptable, an optimal solution was obtained in approximately 1.6 s. The results from the mathematical model included production quantities, purchase quantities, and inventory levels for every product in all periods of the planning horizon. The analysis of results will be carried out considering the inventory turnover and the reduction of production, purchase, and holding costs. The inventory turnover was calculated for every product, according to the Eqs. 1, 2, 3, and 4.

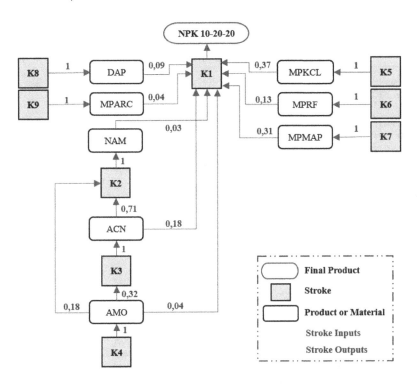

Fig. 3. Bill of materials for 10-20-20 NPK fertilizer.

Table 3. Stroke Inputs Matrix. In Tonnes (t)

Code	Stroke In (SI_{ik})															
	K1	K2	K3	K4	K5	K6	K7	K8	K9	K10	K11	K12	K13	K14	K15	K16
AMO	0,04	0,71	0,32	0	0	0	0	0	0	0,05	0,04	0,07	0,02	0,06	0,01	0
ACN	0,18	0,18	0	0	0	0	0	0	0	0,13	0,07	0,26	0,08	0,23	0,69	0
NAM	0,03	0	0	0	0	0	0	0	0	0,08	0,38	0,12	0,41	0,07	0,14	0
MPKCl	0,37	0	0	0	0	0	0	0	0	0,19	0,29	0,24	0,38	0,44	0	0
MPRF	0,13	0	0	0	0	0	0	0	0	0,14	0,15	0,14	0,06	0,13	0	0
MPMAP	0,31	0	0	0	0	0	0	0	0	0,49	0,28	0,45	0,12	0,31	0	0
MPDAP	0,09	0	0	0	0	0	0	0	0	0,11	0,13	0,08	0,14	0,07	0	0
MPARC	0,04	0	0	0	0	0	0	0	0	0,06	0,04	0,08	0,05	0,04	0	0
MPCAL	0	0	0	0	0	0	0	0	0	0	0	0	0	0	0,54	0
NPK1	0	0	0	0	0	0	0	0	0	0	0	0	0	0	0	0
NPK2	0	0	0	0	0	0	0	0	0	0	0	0	0	0	0	0
NPK3	0	0	0	0	0	0	0	0	0	0	0	0	0	0	0	0
NPK4	0	0	0	0	0	0	0	0	0	0	0	0	0	0	0	0
NPK5	0	0	0	0	0	0	0	0	0	0	0	0	0	0	0	0
NPK6	0	0	0	0	0	0	0	0	0	0	0	0	0	0	0	0
NICA	0	0	0	0	0	0	0	0	0	0	0	0	0	0	0	0

Table 4. Stroke outputs matrix. In Tonnes (t)

Code	Stroke Out ($SO_i k$)															
	K1	K2	K3	K4	K5	K6	K7	K8	K9	K10	K11	K12	K13	K14	K15	K16
AMO	0	0	0	1	0	0	0	0	0	0	0	0	0	0	0	0
ACN	0	0	1	0	0	0	0	0	0	0	0	0	0	0	0	0
NAM	0	1	0	0	0	0	0	0	0	0	0	0	0	0	0	0
MPKCl	0	0	0	0	1	0	0	0	0	0	0	0	0	0	0	0
MPRF	0	0	0	0	0	1	0	0	0	0	0	0	0	0	0	0
MPMAP	0	0	0	0	0	0	1	0	0	0	0	0	0	0	0	0
MPDAP	0	0	0	0	0	0	0	1	0	0	0	0	0	0	0	0
MPARC	0	0	0	0	0	0	0	0	1	0	0	0	0	0	0	0
MPCAL	0	0	0	0	0	0	0	0	0	0	0	0	0	0	0	1
NPK1	1	0	0	0	0	0	0	0	0	0	0	0	0	0	0	0
NPK2	0	0	0	0	0	0	0	0	0	1	0	0	0	0	0	0
NPK3	0	0	0	0	0	0	0	0	0	0	1	0	0	0	0	0
NPK4	0	0	0	0	0	0	0	0	0	0	0	1	0	0	0	0
NPK5	0	0	0	0	0	0	0	0	0	0	0	0	1	0	0	0
NPK6	0	0	0	0	0	0	0	0	0	0	0	0	0	1	0	0
NICA	0	0	0	0	0	0	0	0	0	0	0	0	0	0	1	0

Table 5 shows the results of the inventory turnover calculation for the empirical model and the GMOP model. Table 6 shows a comparison of the total costs obtained with both methods. Overall, the GMOP model demonstrated its

Table 5. Inventory turnover results

Code	Description	Empirical model	GMOP model	Variation
AMO	Ammonia	25	55,1	120,40%
ACN	Nitric Acid	69,8	172,1	146,56%
NAM	Ammonium Nitrate	113,2	146	28,98%
MPKCl	Potassium Chloride	7,6	18,4	142,11%
MPRF	Phosphoric Rock	4,6	10,4	126,09%
MPMAP	Mono-ammonium Phosphate	8,1	20,3	150,62%
MPDAP	Di-ammonium Phosphate	3,3	5,9	78,79%
MPCAL	Calcium Carbonate	7,4	26,2	254,05%
MPARC	Clay	4,9	11,4	132,65%
NPK1	NPK 10-20-20	28,9	26,2	−9,34%
NPK2	NPK 10-30-10	20,4	30,3	48,53%
NPK3	NPK 15-15-15	21,4	27,5	28,50%
NPK4	NPK 12-24-12	20,8	33,3	60,10%
NPK5	NPK 20-5-20	23,2	31,4	35,34%
NPK6	NPK 10-15-25	20,5	29,3	42,93%
NICA	Calcium Nitrate	40,4	50,9	25,99%

effectiveness optimizing the inventory turnover for the majority of final products and raw materials. For NPK fertilizers, turnover improvements were between 28.5% and 60,1%. On the other hand, Calcium Nitrate Fertilizer had an improvement close to 26%. Particularly, the biggest variations in turnover were related to raw materials, especially for solid substances like Calcium Carbonate (254,05%), and Monoammonium Phosphate (150,62%).

In terms of costs, holding and purchase costs had the highest reductions (Up to 29% and 21% respectively). On the contrary, Production costs had a slight variation, close to 2.84%. In a similar way to turnover results, raw materials obtained the highest reductions in holding costs. Table 6, shows reductions from 55,5% in the case of Nitrogenous raw materials (liquid) and 63.5% for solid raw materials.

As can be seen in Table 3, raw materials are the most used SKUs by the strokes. The GMOP model allowed obtaining a solution that optimizes the inventory levels and ensures the availability of these substances. This is a reasonable explanation for the higher reductions in costs and inventory turnover. In other words, inventory levels are intended to be the minimum and to be consumed as soon as they are needed. Also, purchase costs were minimized due to the reduction in purchase orders: only orders when is necessary.

Table 6. Costs variations (US dollars)

Cost	Empirical model	GMOP model	Variation
Production costs	$ 388.333,11	$ 377.307,07	−2,84%
Purchase costs	$ 280.897,60	$ 221.804,18	−21%
Holding costs for final products	$ 3.038,51	$ 2.158,06	−28,98%
Holding costs for solid raw materials	$ 7.705,54	$ 2.807,61	−63,56%
Holding costs for Nitrogenous raw materials	$ 707,82	$ 314,74	−55,53%

5   Conclusions

An implementation of the Generic Materials and Operational Planning Model (GMOP) was proposed for optimizing the inventory turnover in a fertilizer company. The model allowed considering complex constraints usually present in large-scale chemical processes, obtaining outstanding results when compared with an empirical production planning methodology.

Inventory turnover improvements were significantly high for raw materials, especially for the most common SKUs in the bills of materials. Holding and purchase costs were significantly minimized, due to the reduction of inventory levels and purchase orders. The GMOP model demonstrated its efficiency in dealing with a real-scaled problem with a complex set of constraints and product structures, optimizing an indirect variable like inventory turnover.

Further research approaches will consider extending the current model, including all packaging and transportation constraints. Also, is necessary to extend the analysis to broader cases and comparisons. This could lead to important contributions, combining approaches related to modular plants and sustainable chemical supply chain concepts.

References

1. Calcium Nitrate - an overview—ScienceDirect Topics. https://www.sciencedirect.com/topics/chemistry/calcium-nitrate
2. NPK Fertilizers - an overview—ScienceDirect Topics. https://www.sciencedirect.com/topics/agricultural-and-biological-sciences/npk-fertilizers
3. Allman, A., Palys, M.J., Daoutidis, P.: Scheduling-informed optimal design of systems with time-varying operation: a wind-powered ammonia case study. AIChE J. **65**(7) (2019). https://doi.org/10.1002/aic.16434
4. Amaran, S., et al.: Long-term turnaround planning for integrated chemical sites. Comput. Chem. Eng. **72**, 145–158 (2015). https://doi.org/10.1016/j.compchemeng.2014.08.003
5. Burawat, P.: Guidelines for improving productivity, inventory, turnover rate, and level of defects in manufacturing industry. Int. J. Econ. Perspect. **10**(4), 88–95 (2016)
6. Castillo, P.C., Castro, P.M., Mahalec, V.: Multiperiod inventory pinch algorithm for integrated planning and scheduling of oil refineries. In: Computing and Systems Technology Division 2016 - Core Programming Area at the 2016 AIChE Annual Meeting, pp. 402–404 (2016)
7. Cunha, A.L., Santos, M.O.: Mathematical modelling and solution approaches for production planning in a chemical industry. Pesquisa Operacional **37**(2), 311–331 (2017). https://doi.org/10.1590/0101-7438.2017.037.02.0311
8. Dziurzanski, P., Zhao, S., Swan, J., Indrusiak, L.S., Scholze, S., Krone, K.: Solving the multi-objective flexible job-shop scheduling problem with alternative recipes for a chemical production process. In: Kaufmann, P., Castillo, P.A. (eds.) EvoApplications 2019. LNCS, vol. 11454, pp. 33–48. Springer, Cham (2019). https://doi.org/10.1007/978-3-030-16692-2_3
9. Garcia-Sabater, J.P., Maheut, J., Marin-Garcia, J.A.: A new formulation technique to model materials and operations planning: the generic materials and operations planning (GMOP) problem. Eur. J. Ind. Eng. **7**(2), 119–147 (2013). https://doi.org/10.1504/EJIE.2013.052572
10. Kwak, J.K.: Analysis of inventory turnover as a performance measure in manufacturing industry. Processes **7**(10) (2019). https://doi.org/10.3390/pr7100760
11. Li, D., Zhang, X.: How time horizons and arbitrage cost influence the turnover premium? Appl. Econ. **51**(44), 4833–4848 (2019). https://doi.org/10.1080/00036846.2019.1602713
12. Maheut, J., Garcia-Sabater, J.P.: Algorithm for complete enumeration based on a stroke graph to solve the supply network configuration and operations scheduling problem. J. Ind. Eng. Manage. **6**(3 SPL.ISS), 779–795 (2013). https://doi.org/10.3926/jiem.550
13. Maheut, J., Garcia-Sabater, J.P., Mula, J.: The generic materials and operations planning (GMOP) problem solved iteratively: a case study in multi-site context. In: Frick, J., Laugen, B.T. (eds.) APMS 2011. IAICT, vol. 384, pp. 66–73. Springer, Heidelberg (2012). https://doi.org/10.1007/978-3-642-33980-6_8

14. Maheut, J., Garcia-Sabater, J.P.: A parallelizable heuristic for solving the generic materials and operations planning in a supply chain network: a case study from the automotive industry. IFIP Adv. Inf. Commun. Technol. **397**, 151–157 (2013). https://doi.org/10.1007/978-3-642-40352-1_20

15. Mostafaei, H., Harjunkoski, I.: Continuous-time scheduling formulation for multipurpose batch plants. AIChE J. **66**(2) (2020). https://doi.org/10.1002/aic.16804

16. Nugroho, Y.K., Zhu, L.: An integration of algal biofuel production planning, scheduling, and order-based inventory distribution control systems. Biofuels, Bioprod. Biorefin. **13**(4), 920–935 (2019). https://doi.org/10.1002/bbb.1982

17. Odongo, I., Nag, B.: Achieving quality by rapid inventory turnover in the supply chain. Int. J. Prod. Qual. Manage. **19**(2), 209–241 (2016). https://doi.org/10.1504/IJPQM.2016.078888

18. Otashu, J.I., Baldea, M.: Scheduling chemical processes for frequency regulation. Appl. Energy **260** (2020). https://doi.org/10.1016/j.apenergy.2019.114125

19. Pacheco Velásquez, E.A.: Un modelo para la optimización de políticas de inventario conjuntas en cadenas de suministro. INGE CUC **9**(1), 11–23 (2013). http://revistascientificas.cuc.edu.co/index.php/ingecuc/article/view/105

20. Reetz, H.F.: Fertilizantes e seu Uso Eficiente, vol. 2 (2016). www.anda.org.br

21. Romero-Conrado, A.R., Coronado-Hernandez, J.R., Rius-Sorolla, G., García-Sabater, J.P.: A Tabu list-based algorithm for capacitated multilevel IoT-sizing with alternate bills of materials and co-production environments. Appl. Sci. (Switz.) **9**(7), 1464 (2019). https://doi.org/10.3390/app9071464

22. Sabah, B., Nikolay, T., Sylverin, K.T.: Production planning under demand uncertainty using Monte Carlo simulation approach: a case study in fertilizer industry. In: Proceedings of the 2019 International Conference on Industrial Engineering and Systems Management, IESM 2019, pp. 1–6. Institute of Electrical and Electronics Engineers (IEEE), January 2019). https://doi.org/10.1109/IESM45758.2019.8948112

23. Talay, I., Özdemir-Akyıldırım, Ö.: Optimal procurement and production planning for multi-product multi-stage production under yield uncertainty. Eur. J. Oper. Res. **275**(2), 536–551 (2019). https://doi.org/10.1016/j.ejor.2018.11.069

24. Wiemer, P.: Production planning and scheduling in chemical and pharmaceutical industry. In: European Control Conference, ECC 1999 - Conference Proceedings, pp. 4836–4841 (2015). https://doi.org/10.23919/ecc.1999.7100102

25. Yang, L.: Design of production management system in ERP of coal chemical industry. Chem. Eng. Trans. **65**, 475–480 (2018). https://doi.org/10.3303/CET1865080

26. Zheng, H.: Chemical enterprise production management system based on ERP. Chem. Eng. Trans. **62**, 763–768 (2017). https://doi.org/10.3303/CET1762128

Evolutionary Adaptation of (*r*, *Q*) Inventory Management Policy in Complex Distribution Systems

Przemysław Ignaciuk[iD] and Łukasz Wieczorek[(✉)][iD]

Lodz University of Technology, 215 Wólczańska Street, 90-924 Łódź, Poland
przemyslaw.ignaciuk@p.lodz.pl,
lukasz.wieczorek.1@edu.p.lodz.pl

Abstract. The paper addresses the inventory control problem in logistic networks with complex, mesh-type interconnection structure. Contrary to the majority of previously analyzed models, the considered topology does not assume any simplifications nor restrictions in the way the nodes are linked with each other. The system encompasses two types of actors – retailers and suppliers – connected via unidirectional links with non-negligible transshipment delay. The uncertain external demand may be imposed on any retailer and backordering is not allowed. The resource distribution is governed using the classical (*r*, *Q*) inventory management policy implemented in a distributed way. In this work, the continuous genetic algorithm is applied for automatic selection of reorder point *r* and shipment quantity *Q*. The optimization process aims to provide a trade-off between the economic costs and customer satisfaction. Numerous simulations are performed to evaluate the effectiveness of genetic algorithm performance in the considered class of problems.

Keywords: Inventory control · (*r*, *Q*) policy · Periodic-review systems · Computational intelligence · Continuous genetic algorithms

1 Introduction

The last decades have led to the significant development of world-wide supply chains (SCs). The globalization and international commerce influence various sectors of the industry. As a consequence, both the regional markets and international trade routes have been enlarged and upgraded. Besides the growth of existing SCs, new ones are created. However, the foreign, as well as the domestic trade partners need to face a dynamically changing market situation, set according to political conditions [1], innovation deployment [2], and environmental factors [3]. Moreover, the future forecasts indicate even bigger unpredictability, e.g., in the context of the One Belt One Road Chinese (OBOR) initiative that aims at establishing long-term cooperation among various countries throughout Eurasia [4, 5]. The creation of OBOR would influence the current trade routes and thus change the conditions of multiple economic sectors in different areas and commerce zones.

© Springer Nature Switzerland AG 2020
K. Saeed and J. Dvorský (Eds.): CISIM 2020, LNCS 12133, pp. 146–157, 2020.
https://doi.org/10.1007/978-3-030-47679-3_13

A considerable amount of research is being devoted to the optimization of real-life control systems. In the context of logistics, new inventory management strategies and heuristics are proposed and examined in diverse environments, especially with respect to the uncertainties observed in the current network architectures [6, 7]. However, the majority of those works assume significant simplifications of the system structure, i.e., single-stage connections [8, 9], serial systems [6, 10], or arborescent linkage (with separate paths of goods flow) [11, 12]. In addition to the tuning difficulties, adoption of such narrow perspective could lead to serious side-effects, e.g., excessive costs, or unsatisfied demands, implying the loss of customer trust [13].

The effective optimization of models closely resembling the actual systems is a challenging task due to system nonlinearities and uncertainties. It is difficult to obtain the desired solution in an analytical way. On the other hand, recently, the intelligent computing based on evolutionary phenomena has found a wide range of applications in complex optimization tasks. As indicated in [14], despite a large number of new soft computing techniques proposed, genetic algorithms (GAs) remain a powerful tool to face the complexity of sophisticated numerical problems. They are widely used in the simulation-based optimization in many branches of science and industry, including logistics and transport [15, 16].

In this paper, logistic systems having non-trivial interconnection structure are investigated. The system nodes are linked via unidirectional channels affected by lead-time delays. The uncertain customer demand may be imposed on any controlled node in the distribution network. The novelty of this work is twofold. First, the application of the classical reorder point/order quantity policy – the (r, Q) policy – to govern the resource flow in periodic-review logistic systems with mesh topology is investigated in the analytical terms. Secondly, a continuous genetic algorithm (CGA) is proposed to adjust the policy parameters to the requirements of a given distribution environment. Those requirements balance reduction of economic costs against maintaining high customer satisfaction. The outcome of the optimization procedure provides a high-quality candidate solution that meets the objectives defined by the system managers in the decision-making process.

2 System Model

2.1 Network Topology

The model of the considered class of logistic systems assumes that the distribution of resources (single product) is realized between two types of entities:

- external sources (suppliers) that contribute goods to the network but do not handle the customer demand directly,
- controlled nodes (retailers) that are influenced by the customer demand and replenishment requests from the neighbors in the network structure (the nodes directly linked).

No structural simplifications with respect to the interconnection topology are assumed. The suppliers have an infinite capacity. The retailer can be replenished from any number of nodes (both external sources and other controlled nodes) and serve as a goods provider for multiple nodes. No isolated nodes (without input connections) are permitted. Also, no controlled node can be a source for itself. The uncertain external demand may be imposed on any controlled node during the entire planning horizon. The demand requests are satisfied immediately, without any delivery delays, if the amount of resources accumulated at the corresponding node allows it. In the case of insufficient on-hand stock the imposed demand is declined, i.e., backordering is not accepted. The node interconnection is described by a pair of attributes:

- supply factor (SF) that determines a part of the required resources that should be acquired from a neighbor or an external source,
- lead-time delay (LT) that determines the time – the number of periods – that passes from issuing a replenishment order to the goods delivery.

In order to establish a baseline, the nominal case of resource distribution process is considered. Hence, no perturbations in the transport channels that could lead to unplanned delays are expected. Moreover, neither damage, nor loss of resources occurs.

2.2   Node Operations

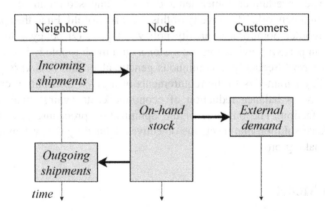

Fig. 1. Node operational diagram.

The sequence of operations that are performed at a controlled node in each time period is illustrated in Fig. 1. First, the resources from the incoming shipments are registered in the on-hand stock. Then, based on the current on-hand stock level, the node strives to satisfy the external demand imposed by the customers. Afterwards, the replenishment orders from the neighbors are fulfilled, depending on the availability of accumulated resources. Finally, the node generates the replenishment requests for its suppliers based on the inventory position comprising the on-hand stock level plus the goods in-transit.

2.3   Resource Distribution Metrics

Two metrics are considered to evaluate the effectiveness of resource distribution:

- holding cost (HC) – the cost of storing resources at the controlled nodes,
- customer satisfaction (CS) that determines the quality of services from the customer point of view, quantified here through the percentage of satisfied demand.

3   Mathematical Description

3.1   Node Interaction

The network under consideration consists of P nodes – M external sources (suppliers) and N controlled nodes (retailers) – with finite stock capacities. The nodes are connected using directed links parametrized by two attributes. The first attribute, α_{ij}, is the nominal SF between nodes i and j, $\alpha_{ij} \in [0, 1]$. Owing to the unidirectionality assumption, if $\alpha_{ij} > 0$, then $\alpha_{ji} = 0$. In addition, the SFs at node i sum to one, i.e., $\sum_{j=1}^{P} \alpha_{ji} = 1$. In the case of insufficient stock, i.e., when a node cannot fulfill all the replenishment requests from its neighbors, the SF is lowered as $0 \leq \alpha_{ij}(t) \leq \alpha_{ij}$. The second attribute – β_{ij} – denotes the LT between nodes i and j, $\beta_{ij} \in [0, B]$, where B is the longest LT at any link. The LT encompasses the time of all the activities from issuing an order to the goods delivery.

The on-hand stock level at controlled node i – $x_i(t)$ – evolves according to

$$x_i(t+1) = x_i(t) + u_i(t) - h_i(t) - o_i(t), \tag{1}$$

where:

- $t = 0, 1, \ldots$ – the independent variable marking subsequent instants of stock review,
- $u_i(t)$ – the quantity of resources from the incoming shipments at node i in period t,
- $h_i(t)$ – the demand satisfied by node i in period t,
- $o_i(t)$ – the replenishment quantity sent by node i to its neighbors in period t.

The model accepts the situation when the external demand is imposed on any controlled node. The demand requests are not known *a priori*. In order to obtain a high service level, they should be satisfied immediately. In the considered model, backordering is not allowed, which implies nonlinear dynamics with respect to the demand [17]. Denoting the quantity from the customer requests placed at node i in period t (external demand) by $d_i(t)$, the satisfied demand

$$h_i(t) = \min(d_i(t), x_i(t) + u_i(t)), \tag{2}$$

i.e., the customer demand is served up to the level of resources available in stock and currently realized orders $x_i(t) + u_i(t)$.

Let $q_i(t)$ denote the replenishment order generated by node i for all its neighbors and suppliers in period t. With α_{ji} being the lot part requested from node j and β_{ji} the delay in order procurement between nodes j and i, the goods quantity from the incoming shipments received by node i may be determined from

$$u_i(t) = \sum_{j=1}^{P} \alpha_{ji}(t - \beta_{ji})q_i(t - \beta_{ji}).$$ (3)

Finally, the quantity of resources sent by node i in the outgoing shipments to its neighbors is described as

$$o_i(t) = \sum_{j=1}^{N} \alpha_{ij}(t)q_j(t).$$ (4)

Note that if nodes i and j are not connected, then $\alpha_{ij}(t) = \alpha_{ji}(t) = 0$.

3.2   State-Space Representation

In order to implement the considered model on a computing machine, it is convenient to formulate the mathematical relationships describing the distribution system dynamics in a matrix-vector form. First, the elements from (1) and (3) are grouped into

- $\mathbf{x}(t)$ – the vector of node on-hand stock levels in period t,
- $\mathbf{q}(t)$ – the vector of replenishment signals issued in period t,
- $\mathbf{h}(t)$ – the vector of satisfied demands in period t.

Let us introduce a matrix that will hold the information about the network topology and goods in-transit to be delivered with delay of k periods, $k \in [0, B]$, as

$$\mathbf{A}_k(t) = \begin{bmatrix} \sum_{i:\beta_{i1}=k} \alpha_{i1}(t) & \gamma_{12}(t) & \cdots & \gamma_{1N}(t) \\ \gamma_{21}(t) & \sum_{i:\beta_{i2}=k} \alpha_{i2}(t) & \cdots & \gamma_{2N}(t) \\ \vdots & \vdots & \ddots & \vdots \\ \gamma_{N1}(t) & \gamma_{N2}(t) & \cdots & \sum_{i:\beta_{iN}=k} \alpha_{iN}(t) \end{bmatrix}.$$ (5)

The entries on the main diagonal represent the goods from the replenishment orders generated k periods before t, whereas the off-diagonal ones

$$\gamma_{ij} = \begin{cases} -\alpha_{ij}, & \text{if } \beta_{ij} = k, \\ 0, & \text{otherwise.} \end{cases}$$ (6)

Then, the state-space equation describing the system dynamics may be formulated as

$$\mathbf{x}(t+1) = \mathbf{x}(t) + \sum_{k=1}^{B} \mathbf{A}_k(t-k)\mathbf{q}(t-k) - \mathbf{h}(t).$$ (7)

3.3 Reorder Point/Order Quantity Policy

In order to manage the flow of goods in the logistic system under consideration the (r, Q) policy is applied. The policy is implemented in a distributed way – at each controlled node independently. In this approach, the replenishment order is generated as soon as the inventory position $y_i(t)$ (the sum of on-hand stock and in-transit orders) at node i falls below r_i – the reorder point (RP). The order is realized in batches of a certain size Q_i – the order quantity (OQ). In the steady state, at the end of each period, $r_i \leq y_i(t) < r_i + Q_i$. For the policy deployment, the pair of parameters, r_i and Q_i, needs to be determined for each controlled node. The quantity of goods to be ordered by node i in period t is established according to

$$q_i(t) = \begin{cases} Q_i, & \text{if } y_i(t) < r_i, \\ 0, & \text{otherwise.} \end{cases} \tag{8}$$

Let the RPs and OQs be grouped into vectors $\mathbf{r}$ and $\mathbf{Q}$, respectively.

3.4 Search Space Boundaries

It is necessary to find such vectors $\mathbf{r}$ and $\mathbf{Q}$ that enable one to perform efficient resource distribution. For this purpose, the CGA-based optimization procedure will be conducted. First, in order to limit the space of problem solutions, the baseline solution for the optimization should be established. Let $\mathbf{r}^{\text{init}}$ and $\mathbf{Q}^{\text{init}}$ denote the vectors of the initial RPs and OQs, respectively. The initial solution should provide full CS, possibly with excessive on-hand stock at the controlled nodes [18]. Let us consider the resource distribution scenario in which the fixed external demand with maximum intensity is imposed on all the retailers in each period. Denoting the vector containing the highest value of external demand imposed on the controlled nodes by $\mathbf{d}_{\max}$, the initial RPs may be determined as

$$\mathbf{r}^{\text{init}} = \left(\mathbf{I_N} + \sum_{k=1}^{B} k\mathbf{A}_k \right) \mathbf{A}^{-1}\mathbf{d}_{\max}, \text{ where } \mathbf{A} = \sum_{k=1}^{B} \mathbf{A}_k. \tag{9}$$

Moreover, according to [19], the initial batch size should satisfy the inequality (with component-wise ">" comparison)

$$\mathbf{Q}^{\text{init}} \cdot \mathbf{I} > \mathbf{A}^{-1} \cdot \mathbf{d}_{\max}. \tag{10}$$

4 GA-Based Optimization

In this work the optimization aims to balance two criteria – reducing the HC and maintaining CS. Due to the model nonlinearities, it is difficult to determine the input vectors $\mathbf{r}$ and $\mathbf{Q}$ in an analytical way for arbitrary demand distribution. Hence, the CGA is applied to adjust the parameters of the (r, Q) policy to given requirements of network managers. This adaptation is realized via simulation-based optimization.

4.1 Continuous Genetic Algorithm

This paper investigates the application of CGAs to adjust both RPs and OQs for the optimal resource reflow in logistic networks with complex topology. First, the fitness function that combines both optimization criteria is formulated. As opposed to the classical multi-objective approach, the CGA requires a single fitness function. The one proposed in this work allows one to smoothly balance the economic and customer-related objectives.

Since the domain of model variables is continuous, CGA is applied instead of the classical binary form of GAs [20]. Let the pair of vectors $\mathbf{r}$ and $\mathbf{Q}$ constitute a candidate solution. Then, the value of the node reorder point and batch size will be a gene in the chromosome. Contrary to the classical approaches to the (r, Q) policy tuning, that first establish the vector of RPs and afterwards OQs [21], the algorithm considered in this paper modifies both of them simultaneously.

The initial population is randomly generated using the baseline candidate solution set through (9) and (10). This solution provides a constraint of the search space. Hence, considering the baseline $\mathbf{r}^{init}$ and $\mathbf{Q}^{init}$, all the candidate solutions in the initial population are determined as

$$\mathbf{r} = [r_1, \ldots, r_N]^T, \text{ where } \forall_i r_i \in (0, r_i^{init}] \tag{11}$$

and

$$\mathbf{Q} = [Q_1, \ldots, Q_N]^T, \text{ where } \forall_i Q_i \in (0, Q_i^{init}]. \tag{12}$$

The simulation of resource distribution process is performed for each candidate solutions in the population. Afterwards, the values of the fitness function are calculated based on the performance metrics – HC and CS – described in Sect. 2. The selection, crossover, and mutation operations are performed sequentially. The evolution of the consecutive populations continues until one of the stop conditions: the generations limit or the number of generations without improvement of the highest fitness value; is satisfied.

4.2 Genetic Operations

The considered CGA has been tested using the classical genetic operations discussed in the current literature [20]. Conducting numerous simulations allow one to choose the most suited one for the problem at hand.

First, the tournament selection has been applied. This method takes into account the fitness values of the candidate solutions; hence, it allows one to privilege the "better" candidate solutions over "weaker" ones. Moreover, the computational cost is smaller than in other classical selection methods, e.g., roulette-wheel selection, or stochastic universal sampling [20].

Second, the recombination mechanism is realized according to the multi-point crossover that translates the pair of individuals (parents) into two candidate solutions (children). The considered algorithm assumes the two-point crossover that generates two arbitrary different divider points, in the range [1, N–1], for each pair of parents.

Finally, the uniform mutation operation is performed with the probability of 15%, which is a suggested value for CGAs in order to avoid local minima.

4.3   Fitness Function

The purpose of the discussed optimization process is to find the best trade-off between two conflicting objectives. The first objective aims at minimizing the economic cost that arises from the excess stock gathered at the controlled nodes. The overall HC may be calculated as

$$f_{HC} = \sum_{t=1}^{T} \mathbf{x}(t), \tag{13}$$

where T denotes the planning horizon.

The second objective is to maximize the CS that is expressed as the percentage of the imposed demand actually satisfied. It is quantified by

$$f_{CS} = \frac{\sum_{t=1}^{T} \mathbf{h}(t)}{\sum_{t=1}^{T} \mathbf{d}(t)}. \tag{14}$$

Let f_{HC}^{init} denote the initial HC that is the metric of the baseline candidate solution ((9) and (10)). Then, the fitness function that combines all the metrics and the managers' preferences may be formulated as

$$f_{fitness}(f_{HC}, f_{CS}) = \left(1 - \frac{f_{HC}}{f_{HC}^{init}}\right)^{\varphi} (f_{CS})^{\omega}, \tag{15}$$

where φ and ω allow for balancing the optimization objectives priority.

5   Numerical Study

In order to evaluate the effectiveness of CGA in solving the optimization problem under consideration, over 10^4 numerical studies were performed. The experiments assumed various scenarios in terms of network topology, customer demand, and planning horizon. Moreover, different objective priorities have been taken into account to observe the tendencies of the evolution process.

Two optimization scenarios have been chosen for closer examination. They are discussed in the framework a twenty-node network illustrated in Fig. 2. The interconnection structure encompasses five external sources ($M = 5$) and fifteen controlled nodes ($N = 15$). The links between the nodes are affected by LTs that are assigned randomly in the range [1, 5]. The planning horizon assumes 50 periods. The stochastic external demand is imposed on all the controlled nodes. The demand requests are generated using Gamma distribution with the shape and scale parameters set as 5 and 10, respectively. The initial on-hand stock levels at the controlled nodes are adjusted equal to the reorder point, i.e., $\mathbf{x}(0) = \mathbf{r}$.

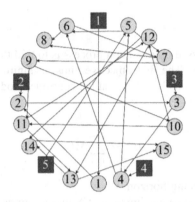

Fig. 2. Interconnection structure.

Two cases of priority setting have been examined. The first one – the satisfaction-oriented scenario – focuses on the maximization of CS while allowing excess stock. The second scenario aims for maximum reduction of the HC, accepting a loss of CS. Table 1 provides the applied priority coefficients.

Table 1. Optimization priorities

Scenario	φ (HC)	ω (CS)
Satisfaction-oriented	1	2
Cost-oriented	10	1

In both cases, the CGA-based optimization assumes 50 individuals in the population and two stop criteria: the maximum number of 200 generations, and the limit of 50 successive generations without an improvement of the best fitness function value.

The initial vector of RPs calculated according to (9), that limits the search space of the candidate solutions:

$$\mathbf{r}^{\text{init}} = \begin{bmatrix} 354, 2056, 1892, 2547, 1854, 3938, 2928, 1161, \\ 3163, 2723, 803, 632, 875, 431, 456 \end{bmatrix}. \tag{16}$$

Similarly, the initial OQs established from (10):

$$\mathbf{Q}^{\text{init}} = \begin{bmatrix} 178, 624, 592, 690, 928, 939, 814, 324, \\ 621, 515, 335, 227, 220, 136, 92 \end{bmatrix}. \tag{17}$$

By examining the baseline candidate, a full-search algorithm would need to perform almost $4.57 \cdot 10^{85}$ simulations to explore all the possibilities. The common computing machines do not permit solving such a problem in reasonable time. That is why the use of CGA is explored.

The initial HC, based on (16) and (17), equals $9.16 \cdot 10^5$ units (full CS is guaranteed with this setting). In the satisfaction-oriented scenario one may satisfy the imposed demand in full whereas reducing the overall HC to $7.3 \cdot 10^4$ units. On the other hand, in the cost-oriented one, the HC are decreased to $1.34 \cdot 10^4$ units, yet 16% of the demand requests need to be rejected.

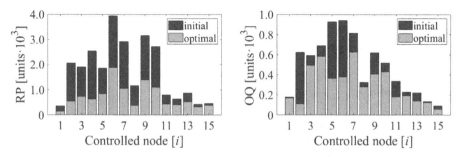

Fig. 3. (r, Q) policy parameters for satisfaction-oriented scenario.

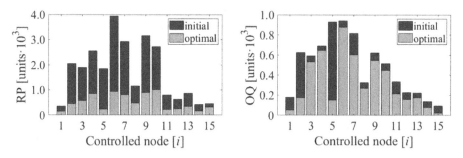

Fig. 4. (r, Q) policy parameters for cost-oriented scenario.

Figures 3 and 4 present the optimal solutions. The reduction of both RPs and OQs is clearly perceptible in each case. Moreover, the result of cost-oriented scenario indicates lower RPs than in the satisfaction-oriented one, albeit bigger values of OQs. It is caused by the fact that the transportation costs are not taken into account. Hence, from the cost perspective, it is more profitable to have in-transit orders than storing resources in the on-hand stock.

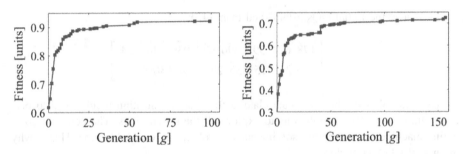

Fig. 5. Fitness improvements for (a) the satisfied-oriented and (b) the cost-oriented scenario.

Figure 5 depicts the fitness value improvement in either scenario. The observed tendencies indicate fast rise of the fitness values at the beginning of the optimization process. After the initial increase, the improvement speed lowers, especially near the optimal state. Before reaching the stop criteria, the satisfaction-oriented optimization performed 149 generations, i.e., $7.45 \cdot 10^3$ simulations. In turn, in the cost-oriented scenario, all the possible 10^4 simulations, i.e., 200 generations have been evaluated, since the optimal solution was found in the 157th generation.

6   Conclusions

The paper addresses the optimization of periodic-review resource distribution systems with complex topology. The distributed (r, Q) inventory management policy is applied to control the reflow of goods. The investigated CGA-based procedure allows one to efficiently adjust the policy parameters – reorder point and order quantity – to given requirements of system managers with respect to holding costs and customer satisfaction. The system performance has been verified through numerical studies assuming various goods distribution scenarios. As a further work, in addition to demand uncertainty, other perturbations, e.g., imperfect transportation channels, will be considered.

References

1. Dhingra, S., Ottaviano, G., Sampson, T., Van Reenen, J.: The consequences of Brexit for UK trade and living standards. Centre for Economic Performance, London School of Economics and Political Science (2016)
2. Fan, T., Tao, F., Deng, S., Li, S.: Impact of RFID technology on supply chain decisions with inventory inaccuracies. Int. J. Prod. Econ. **159**, 117–125 (2015)
3. Mamalis, A.G., Spentzas, K.N., Mamali, A.A.: The impact of automotive industry and its supply chain to climate change: somme techno-economic aspects. Eur. Transp. Res. Rev. **5** (1), 1–10 (2013)
4. Sarker, M., Hossin, M., Yin, X., Sarkar, M.: One Belt One Road Initiative of China: implication for future of global development. Mod. Econ. **9**, 623–638 (2018)

5. Liu, X., Zhanga, K., Chen, B., Zhou, J., Miao, L.: Analysis of logistics service supply chain for the One Belt and One Road initiative of China. Transp. Res. Part E Logist. Transp. Rev. **117**, 23–39 (2018)
6. Ignaciuk, P.: Discrete inventory control in systems with perishable goods – a time-delay system perspective. IET Control Theory Appl. **8**(1), 11–21 (2014)
7. Wieczorek, Ł., Ignaciuk, P.: Robust tuning of order-up-to policy in goods distribution networks with lead-time perturbations. In: 8th International Conference on Digital Information and Communication Technology and its Applications, Poland, pp. 22–27 (2018)
8. Ignaciuk, P., Bartoszewicz, A.: Linear-quadratic optimal control of periodic-review perishable inventory systems. IEEE Trans. Control Syst. Technol. **20**(5), 1400–1407 (2012)
9. Yang, C.T., Ouyang, L.Y., Wu, K.S., Yen, H.F.: Optimal ordering policy in response to a temporary sale price when retailer's warehouse capacity is limited. Eur. J. Ind. Eng. **6**(1), 26–49 (2012)
10. Arts, J., Kiesmüller, G.P.: Analysis of a two-echelon inventory system with two supply modes. Eur. J. Oper. Res. **225**(2), 263–272 (2013)
11. Ignaciuk, P.: Discrete-time control of production-inventory systems with deteriorating stock and unreliable supplies. IEEE Trans. Syst. Man Cybern. Syst. **45**(2), 338–348 (2015)
12. Ignaciuk, P.: Nonlinear inventory control with discrete sliding modes in systems with uncertain delay. IEEE Trans. Ind. Inform. **10**(1), 559–568 (2014)
13. Cattani, K.D., Jacobs, F.R., Schoenfelder, J.: Common inventory modeling assumptions that fall short: arborescent networks, Poisson demand, and single-echelon approximations. J. Oper. Manag. **29**(5), 488–499 (2011)
14. Jauhar, S.K., Pant, M.: Genetic algorithms in supply chain management: a critical analysis of the literature. Sadhana **41**, 993–1017 (2016)
15. Amaran, S., Sahinidis, N.V., Sharda, B., Bury, S.J.: Simulation optimization: a review of algorithms and applications. Ann. Oper. Res. **240**, 351–380 (2016)
16. Lee, C.K.H.: A review of applications of genetic algorithms in operations management. Eng. Appl. Artif. Intell. **76**, 1–12 (2018)
17. Axsäter, S.: Inventory Control. Springer, New York (2015). https://doi.org/10.1007/978-3-319-15729-0
18. Ignaciuk, P., Wieczorek, Ł.: Networked base-stock inventory control in complex distribution systems. Math. Probl. Eng. **2019**(3754367), 1–14 (2019)
19. Ignaciuk, P.: DSM relay control of logistic networks under delayed replenishments and uncertain demand. In: 24th Mediterranean Conference on Control and Automation, pp. 250–255, Greece, (2016)
20. Simon, D.: Evolutionary Optimization Algorithms. Wiley, New York (2013)
21. Federgruen, A., Zheng, Y.S.: An efficient algorithm for computing an optimal (*r*, *Q*) policy in continuous review stochastic inventory systems. Oper. Res. **40**(4), 633–825 (1992)

Design of a Decision Support System for Multiobjective Activity Planning and Programming Using Global Bacteria Optimization

Miguel Angel Jimenez-Barros[1], Diana Gineth Ramirez Rios[2],
Carlos Julio Ardila Hernandez[3], Lauren Julieth Castro Bolaño[1],
and Dionicio Neira Rodado[1($\boxtimes$)]

[1] Universidad de la Costa, Barranquilla, Colombia
{mjimenez47,ljcastro24,dneiral}@cuc.edu.co
[2] Fundación Centro de Investigación en modelación Empresarial del Caribe,
Barranquilla, Colombia
dramirez@fcimec.org
[3] Fundación Universidad del Norte, Barranquilla, Colombia
cardila@uninorte.edu.co

Abstract. The success of any project lies in a great manner on keeping costs in the estimated values, as well as meeting customer required due date. Therefore, there is a current need of developing an information system that facilitates the creation and managing of projects and their processes, including costing schemes, as well as monitoring an optimizing project's makespan. In order to address this situation a user-friendly information system (IS) was developed. This IS includes an optimization module that reduces the project's execution time, thus, minimizing costs and ultimately providing the manager with the right tools for the correct development of the project. Therefore, a better planning of activities in a reduced time is accomplished. In this way, the project manager is equipped with a decision support system (DSS) that allows a better decision making and, thanks to this performance optimization, a cost-effective solution can be delivered to the company. The optimization module is the main innovative component in this IS, considering that addresses the problem as a multiobjective one, considering at the same time makespan and cost. This module is based on global bacteria optimization (GBO). This becomes the most relevant improvement when compared to other ISs in the market.

Keywords: Information systems · DSS · GBO · Multiobjective optimization · Engineering projects · Project management

1 Introduction

Whether this duty is performed by many people or just one, scheduling of operations is hard work; especially if the assignment is a sequence or schedule of tasks that lead to the conclusion of a project. This is a common problem for any company delivering services through the execution of projects. In addition to this inconvenient, there is the

K. Saeed and J. Dvorský (Eds.): CISIM 2020, LNCS 12133, pp. 158–171, 2020.
https://doi.org/10.1007/978-3-030-47679-3_14

haste to send any quote where a project needs to be programmed. In this sense, a suitable enterprise resource planning –ERP- with an easy access is a requirement for an efficient allocation of resources and an adequate ending in a project.

The problem addressed in this research is oriented towards the way in which current computing platforms perform project programming. In general terms, they are rigid systems lacking a web environment, flexibility and options for the user to achieve alternative schedules, and most importantly, lacking integrated optimization algorithms. One of the most well-known computer-based project tools is Microsoft Project. This tool allows tracking and managing projects, but a multi-scenario assessment option for a project is absent. As it can be inferred, this instrument lacks an optimization module; some tools are used to give priority to certain tasks or activities, but it cannot show an optimal scenario by itself.

Total planning, programming and monitoring of projects grants optimal resource use, leading to a better quality, cost-effective process and with a reduced execution time. Acknowledging the need for competitiveness and promotion of Colombia's economic improvement, this research for a decision support system was put forward. For this, the development of an information system that allows the operational planning of projects and costing optimization of tasks and resources in service-oriented companies.

The research and development of the project was based on the need companies have for a project creation and management system with optimization methods for decision-making support. The advantage of this sort of system is to free project managers from the exhausting reprogramming and resource allocation tasks. This type of computer-based tools is vital for companies so as to automate projects. Today, available tools offer separate services, so a project's overall assessment cannot be carried through. Besides, projects always have unforeseen events arising, thus, as existing instruments provide rigid planning systems, these unexpected circumstances cannot be considered for decision making.

The solution to project programming in the IS is delivered using Resource Constrained Project Scheduling Problem –RCPSP-. This type of problem comprises activities or tasks, resources, precedence relationships, and assessment options. Projects, especially resources, have limitations. Among the applications of this kind of problems are: construction projects (e.g. skyscrapers, bridges or highways), on order and small lot production planning problems, development and research projects, and logistics for massive meetings (international political conferences or world sports events).

2   State of the Art

2.1   Costing Systems Definition

For a better definition of a *costing system*, the context of a company is necessary, going from the general to the particular. Katz and Kahn [1], under the approach of the theory of systems, conceive organizations as open systems in permanent interaction with their environments that receive supplies in order to transform them into products or services.

Kast and Rosenzweing [2] pose organizations as a set of interconnected subsystems: the goals and values subsystem, the technical subsystem, the managerial subsystem, the psychosocial subsystem and the structural subsystem.

The managerial subsystem is the group of interconnected functions and activities used to plan, lead and control the strategy and operation of the company in relation to the objectives established so as to permit the interaction of the company with its environment. For this purpose, the managerial subsystem uses other subsystems like the planning, information and management subsystems.

An information system is a permanent process of collecting, processing, storing, distributing and using information to support decision making and redefine the objectives and resources for the planning, coordination and monitoring process in a company [3]. Within the information subsystem, the accounting system constitutes the backbone, which is defined by Hansen and Mowen [4] as a system "that consists of interrelated manual and computer parts and uses processes such as collecting, recording, summarizing, analyzing (using decision making models), and managing data to provide information to users". According to Horngren et al. [5], accounting information serves managers to administer and coordinate the activities or functional areas they oversee. A costing system can be implemented in any organization that produces goods or provides services.

2.2 Resource Constrained Project Scheduling Problem–RCPSP

In project programming, the main objective is the reduction of *makespan*; this is, the total time it takes to start and finish a task schedule [6]. Other objectives include the reduction of task delays [7] and the maximization of the net present value of projects [8].

2.3 Algorithms Used to Solve the RCPSP

Ballestin [9] explains that with the development of heuristic and exact procedures for the RCPSP, the need to create evaluation and comparison instances arose. For several years, most researchers performed their own tests [10–12], making difficult to compare methods since, on the one hand, algorithms were not tried on the same problem (except for few authors who managed to put together more than one algorithm), and on the other hand, problems were frequently generated from a very restricted set of the project's features. [12] collected 110 problems from different authors to which an optimal solution was estimated. The Patterson problems, as this set of instances was called, were used for many years to measure algorithm efficiency [13–16].

2.4   Schedule Generation Schemes

Schedule Generation Schemes or SGS are a fundamental part in many RCPSP procedures. SGS build a feasible schedule by a stepwise extension of a partial schedule that at first assigns start 0 to task 1. In a partial schedule, only a subset of n activities has been scheduled. There are two types of SGS: serial and parallel. While the former uses activity incrementation in the stepwise procedure, the parallel performs time incrementation [9].

2.5   Information Systems for Project Programming

Information and Communication Technologies have a relevant role in every single organization and in society as they are used in every sector [17]. Commercial software tools that automate many project management procedures facilitate their overall management. Generally, project management software has tools to define, order and assign resources to the different tasks, establish initial end ending dates, track progress, and facilitate modifications to tasks and resources. Many of them automate the creation of Gantt and PERT charts.

Some of these tools are hefty and sophisticated software used to manage large projects, scattered work groups and corporate functions. Besides complex relations, these high-end tools can manage vast amounts of tasks and activities. Microsoft Office Project® has become the most widely used project management software today. Its PC-based software features has tools to produce PERT and Gantt charts and to support critical path analysis, as well as resource allotting, project tracking, and status reports. Microsoft Project also tracks how changes in a project's feature will affect the rest. Products such as EasyProjects ® and Vertabase Project Management Software ® are also useful for organizations that wish to implement web-based project management tools.

In the Colombian Market there can be found information systems for project management, but they tend to be oriented to the working area; there is not an information system product that can work oriented to a general project management. [18] concludes that Colombia's software industry have problems due the lack of knowledge of the software industry systemic dynamic, which implies that there's no general developments but specific ones based on specific business needs or custom software by contract.

2.6   Global Bacteria Optimization

There are a set of tools that allow decision makers to have a proper approach to an optimal solution of a problem. Relevant contributions have been proposed by different researchers regarding the MCDM. These approaches are more mainly descriptive, defining possible solutions, including the attributes and evaluation of the criteria, but most importantly, there is a utility function where the criteria is incorporated. This utility function has to be maximized during this process and that is how optimal solutions are reached.

However, there are some relevant topics that must be taken into consideration in MCDM [3]: 1) The purpose is the maximizitation of an utility function, that might be previously expressed by the decision maker; 2) There is always a feasible optimal solution for every situation; 3) Decision maker must always choose or sort between a pair of decisions, 4) Decision maker's preferences can depend upon two binary relations: preference (P) and indifference (I).

However, MCDM has also some limitations because problems are considered not to be realistic, making the theory not as useful as it should be. Additionally, Zeleny [19], considers that MCDM is not useful when there are time constraints, when the problem is more completely defined, when there is a strict hierarchical decision system, when there is a changing environment, when there is limited or partial knowledge of the problem and when there is collective decision making in businesses; all this because it reduces the number of criteria being considered, leaving behind other possible alternatives [4].

On the other hand, Carlsson [20], states that the traditional assumption used in MCDM, in which the criteria are taken as independent, is very limited and ideal to complex real life situations [5]. In the same line, Reeves and Franz [21] introduced a multicriteria linear programming problem, in which they presumed that the decision maker has more than an intuitive understanding of the trade-offs involved in the problem in order to determine his preferences in terms of the objectives [6]. In this sense, it is assumed, that the decision maker is considered to be a rational thinker. It is also important to point out that the decision maker has a complete understanding of the whole situation in which his preferences have some basis with the use of a utility function.

Considering that researchers agree there is no valid optimal solution for any multi-objective problem, different approaches have been proposed in order to tackle MCDM problems. For example, Delgado et al. [22] used fuzzy sets and possibility theory in their approach, and also, multiobjective programming. Similarly, Felix [23] worked with fuzzy relations among criteria. In order to achieve this, he used his awn approach for multiple attribute decision making [8]. Finally, Carlsson found the best compromise solution to MCDM problems with interdependent criteria using a fuzzy Pareto optimal set of non-dominated alternatives.

Zeleny [19] also pointed out that it is a mistake to use weights for criteria in MCDM problems independent from criterion performance [4], considering that there are objectives that might support others. In traditional MCDM it has been found that the criteria should be independent. However, there are some methods that deal with conflictive objectives without recognizing other interdependencies that can be present, making the problem more unrealistic.

A. Multiobjective Optimization Problems
When problems have more than one objective, they are known as multicriteria-based or multiobjective. It is important to understand the theory that they have considered to

solve these types of problems. The multicriteria optimization theory takes basically a set of priorities established by the decision maker and provides the best solution under their preferences. Kindt and Billaut [24] showed a mathematical definition of the multicriteria optimization problems expressing them as a special case of vector optimization problems where the solution space is S and the criteria space, Z(S), are vectorial euclidian spaces of finite dimension [3].

For single criterion problems it is not possible to compare between two solutions and therefore the optimum is given right away. On the other, in the case of multiple objectives, this is no longer the case because there will be a set of alternative solutions that minimize different criteria and they need to be compared. To approach it, Pareto Optima, a general definition of optimality, is used.

B. MCDM Theory to Solve Multi-objective Problems

In MCDM problems, the decision maker has to look for the "best trade-off" solutions between conflicting criteria, and it is assumed to be done by optimizing a utility function. When searching for the solution, the decision maker must choose for an algorithm or heuristic that can determine the whole Pareto optima set. Then, the decision maker, ranks the objectives depending of the importance they have in his decision-making process. This ranking process is done, through the assignation of weights to the different objectives. There are many ways in literature, that have been proposed to determine Pareto optima. The selection of the appropriate method depends on the quality of the calculable solutions and the ease of the application [3].

There are many methods available in MCDM theory to generate Pareto Optima solutions, such as Convex Combination of Criteria, Parametric Analysis, means of the ϵ-constraint approach, Tchebycheff Metric, Goal-Attainment Approach and Use of Lexicographical Order. This paper references the method that uses Convex Combination of Criteria and solutions generated were compared with the proposed Global Bacteria Optimization (GBO) metaheuristic.

T'Kindt and Billaut [24] introduced how graphical representations of the different optimization problems can be done by using level curves. For minimizing the convex combination of criteria, problem (Pα) can be represented by defining first the set of level curves in the decision space, using the conditions for this specific approach:

C. Global Bacteria Optimization (GBO) Algorithm

GBO is a population-based metaheuristic that combines both diversification and intensification concepts that characterize today's metaheuristics. It was developed as a result of a graduate thesis [9] which was directly applied to a scheduling problem whose results improve MOEA algorithm.

After observing the behavior of Bacteria phototaxis and the different processes that bacteria incur naturally, this metaheuristic was developed based on these processes, which was converted into a mathematical function that works as a process within the algorithm. The algorithm is described as follows:

```
START
Generate Valid Bacteria Colony: C
Assign Size of population: Tam
Assign Bacterial Loop Size: A
DO WHILE (A >= 1){
  NewTam = 0;
  REPEAT integer i=1:Tam times
      IF C[i].energy > rotation-and-race-wear THEN
      DO Bacteria Rotation for C[i]
      Save directions sets in D[i]
      Race to Light C[i] Random Select of the Best
      Light Direction in D[i]
      C[i].energy=C[i].energy – rotation-and-race-wear
      END IF
  END REPEAT
  REPEAT integer i=1:Tam times
    IF C[i].energy > binary-fission-wear THEN
    Create CT set with the Bacteria separated;
    CT[NewTam] = C[i]
    C[i].energy = C[i].energy – binary-fission-wear NewTam = NewTam + 1;
    END IF
  END REPEAT
  REPEAT i=1:Tam times
    IF C[i].energy > Spontaneous-Mutation-wear THEN
      IF random > Spontaneous-Mutation-Probability THEN
      Save C*[i] = C[i]
      Mutate C[i] to feasible solution
      C[i].energy = C[i].energy – Spontaneous-Mutation-wear
      END IF
    END IF
    IF random > Reverse-Mutation-Probability THEN {
      IF C[i].energy > Reverse-Mutation-wear THEN {
        IF C[i] was not better than C*[i] THEN {
        Apply reverse mutation C[i] = C*[i]
        C[i].energy = C[i].energy – Reverse-Mutation-wear
        END IF
      END If
      IF C[i].energy > 0 THEN
      Add energy to C[i] with GLS
      CT[NewTam] = C[i]; NewTam = NewTam + 1;
      END IF
    END IF
  END REPEAT
  A = A - 1
END WHILE
Select NO-Dominated
END
```

The main functions that are described in this algorithm are the following:

a. Generation of Initial Bacteria Colonies: Initial bacteria colonies are generated as feasible solutions to the problem.

b. Bacteria Rotation refers to a function (GLS) that is created to measure the amount of energy the bacteria is able to release in a rotation, in which the search directions are based on by the program.

c. Race to light: a function that is known for each bacteria colony in order to move from one location to another where the light is more intense. The intensity of the light is determined randomly among the four mayor intensities in order to get a more diverse search space.

d. Binary Fission: a process that undergoes bacteria when it is duplicated, generating a new bacterium with the initial intensity and position as the mother bacteria.

e. Spontaneous mutation: a function where bacteria change its structure and its position with respect to the light, which can improve a solution or make it worse. This mutation is only done to some bacteria chosen randomly, but subject to the energy they have to divide.

f. Mutation by reversion: this process is done on a percentage of mutated bacteria, given that some bacteria were made worse during mutation process.

g. Bacteria selected for death: Some bacteria do not have sufficient energy to rotate, so they will no longer continue in the colony. These bacteria are selected for death, while the rest undergo binary fission.

h. Photosynthesis: bacteria are fed by energy, according to the natural process of photosynthesis, expressed in ATP. This function assigns an ATP to each bacterium.

3   Methodology

Based on the theory implemented for the application of the cost theory, the organization of activities and the optimization of target functions, the design is created through UML diagrams as this language provides standard tools to perform the documentation analysis and the information system (IS) design [28]. Additionally, interviews were conducted to experts and knowledgeable people in project management who validated the theories and showed activities carried out in a real context.

For the IS design, a process diagram for the project management, the use cases, and a data model were defined. Once the IS designs were validated, a development platform and its execution environment were chosen. Based on the information gathered, we defined 4 phases for the software developing.

3.1   Phase 1: Process Diagrams and Use Cases

Process diagrams, also known as activity diagrams, illustrate the order in which activities are executed in one or various processes [28]. In order to understand the problem to be solved, an activity diagram was generated based on the information collected through the interviews conducted in the different organizations: a vital step to create the IS design.

Use cases show the options that the system provides to users [28]. Each use case is a view of the IS and it can be delivered to one or more users, for example, read-only and execute permission users and the system administrators.

3.2 Phase 2: Data Modelling (Database Design)

Data generated by a business process must be stored in a database. This database permits storing, securing, and managing data [28]. Data modeling is the graphic representation of data structures. For this IS, the database was designed using a relational model, this is, a model that organizes the data structure using tables (relations) [29].

3.3 Phase 3: Development Platform

The development platform consists of the software used to develop a computer-based solution, either through the generation of an automated source code or through people with the necessary language and programming knowledge to write the source code [30]. To develop the IS the Microsoft.NET platform was chosen to use C# language and perform data management with persistence in the Microsoft SQL Server database. Some features like its interoperability with other applications, its execution engine, its broad class library, and its simplified deployment model were considered to select this platform [31].

3.4 Phase 4: Execution Environment

There are two ways of presenting the IS' execution environment: as server and as client. For the server, an IS deployment (Internet Information Services - IIS) is necessary, as well as the implementation of a Microsoft Windows database environment. The client needs a Web explorer to access the application and Microsoft Project to see some reports. It is important to mention that the server (as a hardware) is not a mandatory issue and the information system can be installed in a PC.

4 Description of the Solution

The need to create an information system for project management originates in a research project that generated a decision-making IS for civil engineering projects. Modeling and development of this IS was based on the conceptual and practical analysis of this type of companies. From this point, a broader case for application was noticed: changing this information system into a project management tool (in general terms) in which some components are included, for example, staff, materials, equipment, time, and money. Considering these components, the integration with the application's conceptual model is carried out, thus, leading to this research.

The suggested architecture's design for the IS has three stages:

- Configuration
- Projects and activities
- Optimization

At the configuration stage, the basic starting data is managed for the IS functioning.

Data like providers, supplies, parameters, and unit prices are essential for information management at the moment of creating and managing projects and activities.

One advantage is the option for the end user to partially or totally reuse the information provided for the creation and management of future projects. Just as any information system, an adequate setting of parameters is crucial for project management as they are the basis for decision making.

During the creation and management of projects and activities, and using the data input from the configuration, projects begin to take shape. For every project, the activities to be performed, their schedule (precedence relationships), and the generation of quotes (internal or external depending on the company) are created. The most relevant part of this stage is the creation of each of the activities to be performed in the project. This task is accomplished by the project's director and team and it is fundamental for the project's success in every aspect: on time achievement of goals and within the budget for associated costs. Precedence relationships are directly linked to the latter as they are vital for the performance of activities, one after the other and in the correct order so as to avoid any delays due to the dependent nature of this process.

Finally, the objective of the optimization stage is not only to minimize time and costs, but also personnel, material, and equipment usage (if such case occurs), primarily based on the activities described in the project and its precedence relationships. Hence, the optimization process is in charge of programming activities and taking advantage of resources available at a lower cost and in the shortest time possible. In order to accomplish these objectives, the optimization process estimates the makespan considering the maximum time value in hours of composite unit prices included in the project. Once this is finished, personnel are included, and the activity length is estimated to enter a new personnel resource. Optionally, an estimation of the activity length is performed after entering equipment to determine its length and cost. As a result, solutions to the project are generated with Microsoft Project and, in this way, the activities and costs will be monitored by the project's director. In addition, any modification can also be done to the project.

The information system design with the cost and optimization modules was modeled using UML. In this design, use cases, data model and the system's flow processes are included. Based on the literature and the experience in project management, certain basic functions in the IS and the suggested modules were considered. However, some functions were added, making this a robust system capable of understanding more variables and allowing the user more flexibility to generate very basic or complex projects.

4.1 Entity Module

These are general system configurations associated to a project's supplies. Options in this module are of general use in the company and this non-specific nature permits to use it in one or more projects. In this module, the following options are included: Categories, Subcategories, Groups, Cities, Brands, Measurements, Suppliers, Assets, Staff, Staff Groups, and Supplies.

4.2   General Configuration Module

In this general configuration, all the costs and expenses related to the elements used in the project are managed. This module has the options of: Cost and Expenses Classification, Parameters, and Parameter Classification.

4.3   Unit Price Configuration Module

In this configuration, the description of unit prices is defined. These are associated to personnel, supplies, and services. The following options are included in this module: Capitulations, Sub-capitulations, Basic Unit Prices and Composite Unit Prices.

4.4   Project and Activity Module

In the projects and activities, the projects executed by the organization are managed. This embraces the generation of the necessary activities for the process and the consideration of personnel, products and services demanded to execute such project. The options included here are: Projects, Activities, Quotes, and Precedence Relationships.

4.5   Optimization Module

This module countenances cost and time optimization in a project. After the selection of the project and the number of loops to be performed, results are delivered through Microsoft Project. Thus, control on the suggested optimal execution of the project is granted to the project manager by allowing a tracking option and being able to make all the necessary modifications to improve the process or, on the contrary, state certain constraints when needed. The optimization module allows the reduction of execution costs in a project. For example, in an application corresponding to an office remodeling (construction industry), in which all the required resources were considered and containing more than 10 activities, it is not possible to find an exact solution. Therefore, the best way to tackle the problem is with the help of a metaheuristic. In this GBO is used. Once possible solutions are generated two Pareto points can be identified and the associated curve can be identified as well. The Pareto front can be observed on Fig. 1. In this figure x axis represents cost and y axis represents makespan.

Results were compared and can be found on Fig. 2. As it can be observed on Fig. 2, there were important improvements in both the cost and the makespan when tackling the problem with GBO. Now, depending on the weight assigned to each of the objectives (cost and makespan) the optimal solution can be point 1 or point 2. No matter this, the relevant fact is that the software is successfully looking for near optimal solutions, that could not be achieved when working manually.

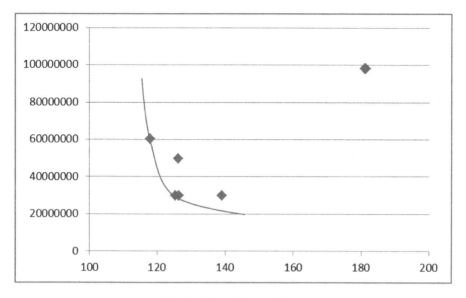

Fig. 1. Pareto front solutions

	COST	MAKESPAN
(1) GBO - Point 1	$ 60.375.250	118
(2) GBO - Point 2	$ 29.925.144	125,3
(3) ORIGINAL	$ 98.088.627	181
% DESV (1) vs (3)	-38%	-35%
% DESV (2) vs (3)	-69%	-31%

Fig. 2. GBO results

5   Conclusions

A user-friendly IS was designed and developed for service-oriented companies in order to manage projects. Through the implementation of optimization algorithms, the reduction of costs and execution times was accomplished in a flexible manner.

This IS accounts for a web-based interface that enables an easy management of information in the creation of projects. Besides, a costing module based on the cost per activity system is implemented, hence, giving way to an optimal definition of the project's budget. In addition, an activity programming module was also integrated to optimize project management.

Finally, an optimization algorithm grounded in an adequate activity programming was used to reduce the costs and execution times estimated in the project's budget.

References

1. Katz, D., Kahn, R.L.: The Social Psychology of Organizations. Wiley, New York (1966). 838 p.
2. Kast, F.E., Rosenzweig, J.E.: Administración de las organizaciones, Madrid (1987)
3. Laudon, K., Laudon, J.: Sistemas de información gerencial, 12th edn. Pearson, Mexico (2011). 640 p.
4. Hansen, D., Mowen, M.: Administración de Costos: Contabilidad y Control, p. 31. Internacional Thomson Editores, Mexico (2003)
5. Horngren, C.T., Foster, G., Datar, S.M.: Contabilidad de Costos, un enfoque gerencial, p. 501. Prentice Hall, México (2007)
6. Alvarez-Valdés, R., Tamarit, J.M.: Heuristic algorithms for resource-constrained project scheduling: a review and empirical analysis. In: Slowinski, R., Weglarz, J. (eds.) Advances in Project Scheduling, pp. 113–134. Elsevier, Amsterdam (1989)
7. Bell, C.G., Han, J.: A new heuristic solution method in resource-constrained project scheduling. Naval Res. Logist. **38**, 315–331 (1991)
8. Blazewicz, J., Lenstra, J.K., Rinnooy Kan, A.H.G.: Scheduling projects to resource constraints: classification and complexity. Discrete Appl. Math. **5**, 11–24 (1983)
9. Dumond, J., Mabert, V.A.: Evaluating project scheduling and due date assignment procedures: an experimental analysis. Manage. Sci. **34**, 101–118 (1988)
10. Elmaghraby, S.E., Herroelen, W.S.: The scheduling of activities to maximize the net present value of projects. Eur. J. Oper. Res. **49**(1), 35–49 (1990)
11. Ballestín, F.G.: Nuevos metodos de resolucion del problema de secuenciacion de proyectos con recursos limitados. Doctorate dissertation, Universitat de Valencia, Servei Publicaciones, Valencia (2002)
12. Christofides, N., Alvarez-Valdés, R., Tamarit, J.M.: Project Scheduling with resource constraints: a branch and bound approach. Eur. J. Oper. Res. **29**, 262–273 (1987)
13. Boctor, F.F.: Some efficient multi heuristic procedures for resource constrained project scheduling. Eur. J. Oper. Res. **49**, 3–13 (1990)
14. Patterson, J.H.: A comparison of exact approaches for solving the multiple constrained resource. Manage. Sci. **30**, 854–867 (1984)
15. Davis, E.W., Patterson, J.H.: A comparison heuristic and optimal solution in resource-constrained project scheduling. Manage. Sci. **21**, 944–955 (1975)
16. Demeulemeester, E.: Minimizing resource availability cost in time-limited project networks. Manage. Sci. **41**, 1590–1598 (1995)
17. Kolisch, R.: Efficient priority rules for the resource-constrained project scheduling problem. J. Oper. Manage. **14**, 179–192 (1996)
18. Lee, J.K., Kim, Y.D.: Search heuristics for resource-constrained project scheduling. J. Oper. Res. Soc. **47**, 678–689 (1996)
19. Leon, V.J., Ramamoorthy, B.: Strength and adaptability of problem space based neighbourhoods for resource-constrained scheduling. OR Spectr. **17**, 173–182 (1995)
20. Patterson, J.H.: Project Scheduling: the effects of problem structure of heuristics. Naval Res. Logist. Q. **23**, 95–123 (1976)
21. Patterson, J.H., Roth, G.W.: Scheduling a project under multiple resource constraints: a zero-one programming approach. AIIE Trans. **8**, 449–455 (1976)
22. Sampsom, S.E., Weiss, E.N.: Local searches techniques for the generalized resource constrained project scheduling problem. Naval Res. Logist. **40**, 665–675 (1993)

23. Mercado, D.R., Sepúlveda, J.A., Pedraza, L.E., Hernández, H.: Modelo de implementación de TIC en el sector transporte de la ciudad de Barranquilla utilizando dinámica de sistemas. Revista Dimensión Empresarial **12**(1), 3645 (2014)
24. Microsoft, Project Software. https://products.office.com/es/Project/project-standard-desktop-software. Accessed on: May. 04, 2016
25. Logic Software, Easyprojects Software. https://www.easyprojects.net. Accessed 04 May 2016
26. Vertabase, Project Management Software. http://www.vertabase.com. Accessed 04 May 2016
27. Martínez Marín, S.J., Arango Aramburo, S., Robledo Velásquez, J.: El crecimiento de la industria del software en Colombia: Un análisis sistémico, Revista EIA, Edición 23, pp. 95–106 (2015)
28. Kendall, K.E., Kendall, J.E.: Análisis y diseño de sistemas, 6th edn. Pearson, Mexico (2005). 752 p.
29. Coronell, C., Morris, S.: Database Systems: Design, Implementation, and Management, 11th edn. Cengage Learning, Stamford (2015). 751 p.
30. Amaya Amaya, J.: istemas de información Gerenciales, 2nd edn. Ecoe Ediciones, Bogota (2010). 207 p.
31. Troelsen, A.: Pro C# 5.0 and the.NET 4.5 Framework, 6th edn. Apress, New York (2012)

Quality Improvement in Ammonium Nitrate Production Using Six Sigma Methodology

Olmedo Ochoa-González[1], Jairo R. Coronado-Hernández[2]([⊠])(ID),
Mayra A. Macías-Jiménez[2](ID), and Alfonso R. Romero-Conrado[2](ID)

[1] Universidad Tecnológica de Bolívar, Cartagena 130001, Colombia
[2] Universidad de la Costa, Barranquilla 080001, Colombia
{jcoronad18,mmacias3,aromero17}@cuc.edu.co

Abstract. Six sigma has been used in different industries to reach operational excellence. However, in the chemical industry, the application of this methodology is limited. This research presents an implementation of the six sigma method for ammonium nitrate (AN) content optimization in condensate production for a fertilizer company in Colombia. The paper aims to determine the levels for input variables in the process, to meet desirable standards for condensate quality in terms of ammonium nitrate content. Based on the DMAIC steps implementation, it was possible to establish the main variables affecting the condensate quality and their optimal levels to reach an ammonium nitrate content below 15,000 ppm. These results demonstrate the impact that a six sigma project may have on operational effectiveness and quality improvement for meeting the customer requirements.

Keywords: Six sigma · Chemical industry · Ammonium nitrate · Condensate production · Fertilizer industry · Quality improvement

1 Introduction

Variability in a process could generate defects and nonconformity levels in products and services [1]. Therefore, keeping the process within tolerance limits is critical for consumer satisfaction, and to survive in a competitive market.

A business strategy for reducing process variation is six sigma [2]. Lately, this approach has been adopted by many companies to achieve operational excellence in terms of meeting quality standards and minimizing related costs [3].

A successful six sigma implementation implies the application of a methodology called DMAIC (define, measure, analyze, improve, and control), which mainly is a five steps problem-solving approach based on continuous improvement [4].

The industries in which the six sigma method had been applied vary from manufacturing, health, education, chemical, electronics, among others [5]. However, the use of this strategy in the chemical industry is limited [6]. Even more in the ammonium nitrate (AN) production industry, which has not been previously discussed in the literature.

We aim to address this gap by using six sigma for quality improvement in condensates production, which is a by-product of the AN production process. Specifically,

© Springer Nature Switzerland AG 2020
K. Saeed and J. Dvorský (Eds.): CISIM 2020, LNCS 12133, pp. 172–183, 2020.
https://doi.org/10.1007/978-3-030-47679-3_15

in this paper, we determined the optimal levels for the input variables in the AN production to obtain a by-product meeting a desirable standard.

This application is of fundamental importance as regards the AN is a growing industry around the world, considering its use as an oxidizing agent or for fertilizer purposes [7].

The remainder of the paper is organized as follows. Section 2 presents a brief literature review related to six sigma applications in the chemical production process and the fertilizer industry. The research methodology is described in Sect. 3. While the implementation of DMAIC steps is discussed in Sect. 4. Finally, Sect. 5 presents the main conclusions and further research directions.

2 Brief Literature Review

According to literature, six sigma methodology was developed by Motorola in 1987 [5], and due to its success, it was spread into other fields such as manufacturing, food, healthcare, service, etc. [6, 8, 9].

Six Sigma means that the distance between specification limits and the average of the current process output is six standard deviations [10]. Therefore, this methodology implies a disciplined improvement philosophy, where a low number of defects (less than 3.4 parts per million) are allowed [11].

In the chemical industry, six sigma has been used for achieving manufacturing excellence [12]. Such is the case of a research in Dow Chemical in which six sigma along with discrete event simulation was used to create a proposal to meet a variable demand and get savings. This research proves two main points. First, the use of six sigma in the chemical industry, and second, the use of simulation techniques to test some operational rules formulated as a result of the project [13].

Simulation is not the only complementary technique combined with six sigma for a successful implementation. Also, the lean-approach has been used in this kind of projects. Lean six sigma was used to reduce the delays in delivering reports in a chemical lab dedicated to the analysis of mineral concentrates. The results demonstrated that the average delays presented were due to retests which could be reduced by 5% for getting savings of at least 200 USD for every 40 trials [14]. In a Turkish fertilizer company, lean six sigma has been applied for optimizing logistics operations in terms of reducing inventory time (from 82 to 51 days) and the average inventory amount in 36% [15].

Also, this combined methodology was used for reducing the turnaround time in a biochemistry lab dedicated to testing fertilizers [16]. Another study in the fertilizer industry using the six sigma method was conducted in Indonesia, where this technique was used to analyze and formulate an improvement proposal to meet national quality standards. This plan consists of improving the raw material handling conditions and the storage facilities [17].

Therefore, six sigma implementation has demonstrated many benefits for the companies that have decided to use it for quality improvement. However, despite the reported use of six sigma in the fertilizer industry, no one of the above research presented a study where it was used to improve an ammonium nitrate by-product quality. Then, we aim to fill this gap through this research.

3   Methodology

A case study approach was used to perform the research. For conducting the six sigma project, the DMAIC steps were followed.

First, it was established the problem statement and the dependent and independent variables for the case. After, we collected and analyzed data from two years in the company.

Then, a control chart is proposed to analyze the process capability regarding the tolerance limits. Also, a factorial design of experiments (DOE) was used in the improvement phase, to determine the influence that the input variables have over the condensate quality, as the response variable, and to determine the best levels for input variables to get the desirable standard for the products. Finally, some actions to keep the improvements are proposed. Table 1 summarizes the DMAIC phases.

Table 1. DMAIC steps

Phase	Purpose
Define	Description of the process or the selected problem, project objective, critical to quality (CTQ), customer requirements, among others
Measure	Selecting one or more critical characteristics, measuring the process and data related
Analyze	In this phase, the problem diagnosis is issued based on the measuring step
Improve	Determine factors to control, planning and implementation of improvement proposals for optimal performance
Control	Ensure the sustainability of the improvement. Process monitoring and prevention of future failures

4   Improvement of Condensate Quality Through DMAIC Steps Implementation

4.1   Define

In this case study, the define step implies the description of the production process and the overall current performance for ammonium nitrate production.

The objective is minimizing the AN content in the condensates generated from the production process in the selected company. The desirable level for AN content in condensates is below 15,000 ppm (parts per million), which could represent an average of 350,000 USD per year in profits.

Production Process. The production process starts with the ammonia (NH_3) and nitric acid (HNO_3) supply to the production plant. In this facility, AN is produced with an 83% concentration and it is stored in a tank. After, this product is transported to a filling plant for distribution and commercialization purposes.

Condensates are by-products of the described production process. These are sent to nitric acid plants for being used during the absorption stage. Figure 1 presents the production process graphically.

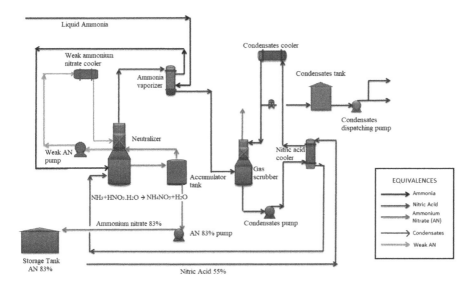

Fig. 1. Ammonium Nitrate (AN) production process

Problem Statement. The production rate depends on the demand of AN which is not stable and has variations over time. It causes operational adjustments to ensure that the product quality fits specifications around desirable concentrations and pH levels.

Considering the above, when an operational condition changes, the quality of the obtained condensates is affected. This situation is critical because if the concentration levels of AN are above 15,000 ppm and its pH reaches 1.5 or more, it would not be used at the nitric acid plant. After all, the parameters are out of specifications. This condition affects the efficiency of the plant and the opportunity costs.

4.2 Measuring

In this phase, the AN content in condensates and other independent variables were pre-selected based on a process analysis. However, the final selection of the independent variables affecting the response variable depends on the results of the project.

For analyzing the performance of the variables, historical data were collected in a bi-monthly based for two years. Statgraphics Centurion XVIII [18] was used to conduct statistical analysis.

The independent variables (Xn) of the process were:

- Reaction temperature (X_1)
- Reaction pH (X_2)
- Weak Ammonium nitrate (NH_4NO_3) temperature (X_3)
- Production concentration (X_4)
- Recirculation pressure with pump P-5301 (X_5)
- Weak ammonium nitrate pH (X_6)
- Diluted ammonium nitrate pH (X_7)
- Weak ammonium nitrate concentration (X_8)
- Diluted ammonium nitrate concentration (X_9)

Figure 2 indicates that most of the time the concentration levels for ammonium nitrate in condensates are surrounded by 21,000 ppm, instead of the average value of 18,500 ppm.

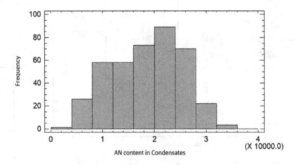

Fig. 2. Histogram of frequency for AN content in condensates.

Also, we measured the tons of condensates that could be generated per a production cycle in the AN plant (Table 2).

Table 2. Generated condensate per production rate.

Production rate	Condensates generated (t per day)
100	37
150	55
200	73
250	92
300	110

4.3 Analyze

According to the sampling results for the AN condensates, the performance of the main operational variables was analyzed for identifying their principal effects over the by-product quality. For this purpose, an analysis of variance (ANOVA) was used employing general linear models.

ANOVA results indicated that there is a statistically significant association of condensates with process variables at a 95% level of confidence (Table 3).

Table 3. General ANOVA for variables correlation

Source	Sum of squares	DF	Mean square	F-Ratio	p-value
Model	3.8673E8	9	4.297E7	3.19	**0.0128**

Also, Table 4 showed that the principal variables affecting the condensate quality were X_3, X_5, and X_8, which have the highest F-Ratios in the analysis (Table 4).

Table 4. ANOVA correlation among variables

Source	Squared sum	LD	Average squared	F-Ratio	p-value
X_1	7.6671E6	1	7.6671E6	0.57	0.4589
X_2	3.82816E7	1	3.82816E7	2.84	0.1062
X_3	1.73428E8	1	1.73428E8	**12.86**	**0.0016**
X_4	8.42935E6	1	8.42935E6	0.62	0.4376
X_5	8.14156E7	1	8.14156E7	**6.04**	**0.0224**
X_6	2.78549E7	1	2.78549E7	2.07	0.1648
X_7	1.80211E7	1	1.80211E7	1.34	0.2601
X_8	2.13988E8	1	2.13988E8	**15.87**	**0.0006**
X_9	2.66698E7	1	2.66698E7	1.98	0.1736

Besides, Figs. 3, 4 and 5 suggest that the condensate concentration is proportional to the weak ammonium nitrate temperature and concentration, and inversely proportional to the recirculation flow rate (recirculation pressure in the p-5301 pump).

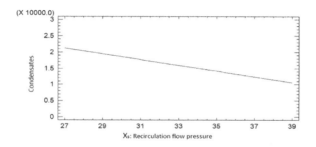

Fig. 3. Relationship recirculation flow pressure vs. condensates

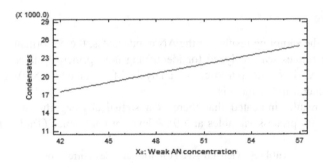

Fig. 4. Relationship weak AN concentration vs. condensates

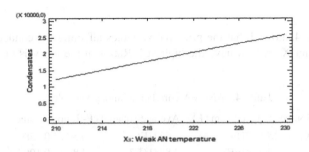

Fig. 5. Relationship weak AN temperature vs. condensates

Also, the results indicate that the weak ammonium nitrate concentration variable (X_8) depends on the recirculation flow in the P-5301 pump (X_5). Then, this variable was removed from the analysis, and a deeper study of the condensate performance was carried out. For this purpose, only X_3 and X_5 were considered.

Control Limits. It was determined three times the standard deviation (3σ) over and below the average value as UCL (upper control limit) and lower control limit (LCL), respectively. This procedure was applied for each variable under study (Figs. 6 and 7).

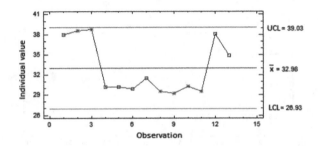

Fig. 6. Control chart for X_5 (Recirculation flow pressure)

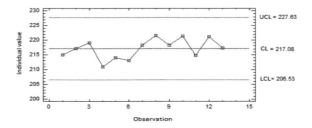

Fig. 7. Control chart for X_3 (Weak AN temperature)

After studying the control charts, a design of experiment (DOE) was conducted to define the appropriate levels of the input variables to reduce the ammonium nitrate concentration in the condensates under 15,000 ppm.

4.4   Improvement

Based on the established control limits, a factorial DOE [19] was carried out to identify the association that the input variables have over the response variable (AN concentration in the condensates).

DOE Configuration. It is a 3^2 factorial design (two factors with three levels each one). The response variable was the AN concentration in the condensates. The factors used and their levels are presented in Table 5.

Table 5. Factors and levels for factorial DOE.

Factor	Levels	Value
A: Recirculation pressure of weak AN in the p-5301 pump (psig)	High	39
	Medium	33
	Low	27
B: Weak AN temperature (°F)	High	228
	Medium	217
	Low	206

DOE Results. The p-values in bold in Table 6 indicate that the recirculation pressure in P-5301 pump (A) and the weak ammonium nitrate temperature (B) have statistically significant effects over the response variable at 95% of confidence (p-values less than 0.05).

In the interaction context, it shows that the effects among factors are not statistically significant. Due to this, the statistical model could be summarized as follows:

$$Y_{ij} = \mu + \alpha_i + \beta_j + \varepsilon_{ij} \tag{1}$$

Table 6. Analysis of variance for DOE.

Source	Sum of squares	DF	Mean square	F-Ratio	p-value
A: X_5	1.09198E7	1	1.09198E7	14.70	**0.0011**
B: X_3	1.87249E7	1	1.87249E7	25.21	**0.0001**
AA	397857.	1	397857.	0.54	0.4731
AB	47169.2	1	47169.2	0.06	0.8037
BB	193071.	1	193071.	0.26	0.6160
Blocks	182536.	2	91268.0	0.12	0.8851

Where Y_{ij} is the ammonium nitrate content in the condensates with an i-th level of pressure and j-th level of temperature. While μ is the average ammonium nitrate content in the condensates. α_i is the additional effect due to the i-th level of weak recirculation AN pressure. β_j is the additional effect due to the j-th level of weak recirculation AN temperature. And ε_{ij} the random error related to Y_{ij} measurement.

Therefore, the equation for the adjusted model to condensates estimation is:

$$Y = 30,789.3 - 1091.62X_5 + 90.13X_3 + 989.65 \tag{2}$$

Figure 8 shows that the effect with more influence over the response variable is in the first place the temperature, and secondly the pressure. While Fig. 9 indicates that to get a condensate with concentration levels below 15,000 ppm, the plant should operate with a recirculation temperature for weak nitrate ammonium as low as possible around 205°F and the highest pressure about 40 psig.

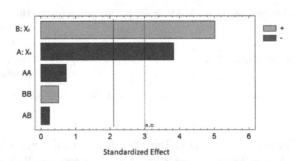

Fig. 8. Pareto chart of the standardized effects.

Figure 10 confirms that the two selected factors do not interact, because the lines are approximately parallel. This means that the effect of one factor will not depend upon the level of the other.

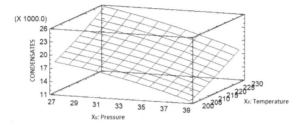

Fig. 9. Response surface plot.

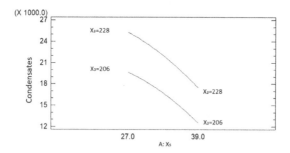

Fig. 10. Interaction effects plot for condensates.

4.5   Control Stage

One of the most challenging stage, when six sigma is applied, is the control phase [20]. This step pursues to keep the optimal results during the regular performance of the process. Therefore, the following actions were proposed to reach this goal:

- Three values for production rate were fixed: 100, 200 and 289 t per day. This decision was made for avoiding changes in this parameter each time the demand fluctuates.
- A strategy for public communication consisting of control charts for continuous monitoring of AN concentration in the condensate with a frequency of eight times per work shift.
- Use of poka-yoke devices whose purpose was set alarms for values of recirculation temperature and pressure out of permissible standards.
- A corrective action plan was put in place, in the event of nonconformity in the condensates due to operational or quality problems.

5   Conclusion

This work demonstrates the application of six sigma tools in a kind of industry and process until now unexplored. In this research, we have determined the optimal levels for input variables to improve the quality of condensates in an ammonium nitrate production process using six sigma methodology. Also, we identified the main effects caused by the variables over the AN content in the analyzed by-product.

The DMAIC steps were followed for a successful implementation of the six sigma project and statistical tools such as ANOVA and DOE were used to fulfill the research objective.

Therefore, it was possible to determine that the main variables affecting the condensate quality were the temperature of the weak ammonium nitrate recirculation and the flow pressure. Also, it was determined that the best quality in the condensates would be obtained when the variables mentioned before have values surrounded 205°F and 40 psig, respectively.

However, to optimize the ammonium nitrate production the company must implement the suggested actions in the control phase. With this proposal, it could improve its efficiency, costs minimization and desirable levels of ammonium nitrate in the by-products under study.

Also, these actions must be accompanied for a production rate control, because if the business would not implement this monitoring, the condensates can be out of control. Therefore, one of the main proposed changes was the standardization of production rate in three fixed charges: 100, 200 and 280 t per day.

Further studies within the same process, should consider using an improved method for the recirculation flow pressure measuring more directly. Because in the paper the discharge pressure from the p-5301 pump was used for estimating the variable. In future studies of the process, a flow measuring device for a direct and more accurate estimation could be used instead.

References

1. Baýa, A.P.: Achieving customer specifications through process improvement using six sigma: case study of NutriSoil – Portugal. Qual. Manag. J. **22**, 48–60 (2015). https://doi.org/10.1080/10686967.2015.11918428
2. Bunce, M.M., Wang, L., Bidanda, B.: Leveraging six sigma with industrial engineering tools in crateless retort production. Int. J. Prod. Res. **46**, 6701–6719 (2008). https://doi.org/10.1080/00207540802230520
3. Ortiz Barrios, M., Felizzola Jiménez, H.: Reduction of average lead time in outpatient service of obstetrics through six sigma methodology. In: Bravo, J., Hervás, R., Villarreal, V. (eds.) AmIHEALTH 2015. LNCS, vol. 9456, pp. 293–302. Springer, Cham (2015). https://doi.org/10.1007/978-3-319-26508-7_29
4. Kim, Y., Han, S.H.: Implementing lean six sigma: a case study in concrete panel production. In: 20th Annual Conference of the International Group for Lean Construction (2012)
5. Patel, M., Desai, D.: Critical review and analysis of measuring the success of Six Sigma implementation in manufacturing sector. Int. J. Qual. Reliab. Manag. **35**(8), 1519–1545 (2018). https://doi.org/10.1108/IJQRM-04-2017-0081
6. Sreedharan, V.R., Raju, R.: A systematic literature review of lean six sigma in different industries. Int. J. Lean Six Sigma. **7**, 430–466 (2016). https://doi.org/10.1108/IJLSS-12-2015-0050

7. Abbasfard, H., Rafsanjani, H.H., Ghader, S., Ghanbari, M.: Mathematical modeling and simulation of an industrial rotary dryer: a case study of ammonium nitrate plant. Powder Technol. **239**, 499–505 (2013). https://doi.org/10.1016/j.powtec.2013.02.037
8. Johannsen, F., Leist, S., Zellner, G.: Implementing six sigma for improving business processes at an automotive bank. In: vom Brocke, J., Rosemann, M. (eds.) Handbook on Business Process Management 1. IHIS, pp. 393–416. Springer, Heidelberg (2015). https://doi.org/10.1007/978-3-642-45100-3_17
9. Nabhani, F., Shokri, A.: Reducing the delivery lead time in a food distribution SME through the implementation of six sigma methodology. J. Manuf. **20**, 957–974 (2010). https://doi.org/10.1108/17410380910984221
10. Boltić, Z., Jovanović, M., Petrović, S., Božanić, V., Mihajlović, M.: Continuous improvement concepts as a link between quality assurance and implementation of cleaner production - case study in the generic pharmaceutical industry. Chem. Ind. Chem. Eng. Q. **22**, 55–64 (2016). https://doi.org/10.2298/CICEQ150430019B
11. Garland, R.W.: Six sigma project to improve a management of change process. AIChE (2010). https://doi.org/10.1002/prs.10402
12. Harold, M.P., Ogunnaike, B.A.: Process engineering in the evolving chemical industry. AIChE J. **46**, 2123–2127 (2000). https://doi.org/10.1002/aic.690461105
13. Buss, P., Ivery, N.: Dow Chemical design for six sigma rail delivery project. In: Proceedings of the 2001 Winter Simulation Conference, pp. 226–232 (2001)
14. Arone, A., Pariona, M., Hurtado, Á., Chichizola, V., Alvarez, J.C.: Improvement of chemical processes for the analysis of mineral concentrates using lean six sigma. In: Iano, Y., Arthur, R., Saotome, O., Vieira Estrela, V., Loschi, H.J. (eds.) BTSym 2018. SIST, vol. 140, pp. 533–540. Springer, Cham (2019). https://doi.org/10.1007/978-3-030-16053-1_52
15. Dönmez, C.Ç., Yakar, B.: A systematic perspective on supply chain improvement by using lean six sigma and an implementation at a fertilizer company. Electron. J. Soc. Sci. **18**, 1377–1396 (2019). https://doi.org/10.17755/esosder.513530
16. Wegner, K.A.: Process improvement for regulatory analyses of custom-blend fertilizers. J. AOAC Int. **97**, 759–763 (2014). https://doi.org/10.5740/jaoacint.13-401
17. Desyane, H.K., Fantinus, A.W.: Proposed quality improvement of liquid organic fertilitizers "Herbafarm" to meet national standards in Indonesia. Indones. J. Bus. Adm. **1**, 343–352 (2012)
18. Statgraphics Technologies Inc.: Statgraphics Centurion XVIII (2019). https://www.statgraphics.com/
19. Romero-Conrado, A.R., Suárez-Agudelo, E.A., Macías-Jiménez, M.A., Gómez-Charris, Y., Lozano-Ayarza, L.P.: Diseño experimental para la obtención de compost apto para uso agrícola a partir de lodo papelero Kraft. Rev. Espac. **38**, 1–14 (2017)
20. Kwak, Y.H., Anbari, F.T.: Benefits, obstacles, and future of six sigma approach. Technovation **26**, 708–715 (2006). https://doi.org/10.1016/j.technovation.2004.10.003

Multicriteria Strategic Approach for the Selection of Concrete Suppliers in a Construction Company in Colombia

Jorge E. Restrepo, Dionicio Neira Rodado[✉],
and Amelec Viloria Silva

Universidad de la Costa, Barranquilla, Colombia
york_5200@hotmail.com, {dneiral,aviloria7}@cuc.edu.co

Abstract. Within companies of the construction sector, the evaluation and selection of concrete suppliers, is considered a fundamental topic for the adequate construction of buildings, due to the impact it has, not only in the stability of the building but also in the productivity and the profitability. For an adequate selection, decision-makers must consider a series of criteria of different kind. However, as in different fields, due to the numerous criteria and alternatives that should be considered in the construction industry, the choice of an appropriate multicriteria decision-making approach has become a critical step in the selection of suppliers. Therefore, the objective of this research is to define the most adequate supplier of suitable concrete through the integration of powerful multicriteria decision-making methods. For this purpose, a fuzzy analytical hierarchy process (FAHP) is initially applied to define initial factor weights under uncertainty, followed, using the decision-making test and the evaluation laboratory (DEMATEL) to evaluate the interrelationships between elements of the hierarchy. Then, after combining FAHP and DEMATEL to calculate the final contributions of the factors including the interdependence, finally TOPSIS technique is used for the order of preference for the similarity with the ideal solution to evaluate and determine the best supplier.

Keywords: Construction industry · DEMATEL · AHP · Selection of suppliers

1 Introduction

In recent years the construction sector has had an important boom in Colombia, this largely due to social housing programs, in addition to this, according to [1, 2], construction sector in Colombia currently generates annual investments by US 25 billion representing 7% of GDP.

In Colombia there are currently 6859 companies in the construction sector, generating a fierce competition. This competition forces the companies to provide a quality service to the final customer in order to remain competitive. One of the most used and important strategies to reduce costs and keep competitive is the subcontracting of suppliers, although it is possible to reduce operating costs also generates a dependence on suppliers in processes that use the figure of subcontracting. In this sense it is

© Springer Nature Switzerland AG 2020
K. Saeed and J. Dvorský (Eds.): CISIM 2020, LNCS 12133, pp. 184–194, 2020.
https://doi.org/10.1007/978-3-030-47679-3_16

important to perform a constant assessment of the performance of suppliers keep the control of the operation and make decisions according to the performances of the suppliers.

Among the materials used in the construction is the concrete, which within its main characteristics are:

- Ability to withstand a variety of extreme exposure conditions during its lifetime.
- Aesthetic properties allowing architectural innovations and flexibility in their design.
- High availability that can be manufactured anywhere in the world, which helps optimize costs and reduce the carbon footprint.

Ceballos [1] indicates that the importance of concrete in infrastructure projects lies in its versatility, development of technologies that have taken it to unsuspected limits in its performance, uses and applications. When analyzing the concrete as a critical input for construction companies, it is important to carry out a multidisciplinary study to determine the most critical factors in the evaluation of suppliers, seeking to evaluate multiple factors such as timely delivery and the fulfillment of the expectations of the interested parties. Supply managers must consider all these factors to make an adequate selection and evaluation of suppliers. The value of this process lies on the fact that it has a direct impact on the performance of companies [3].

One of the critical aspects in the construction sector lies on the number of items or item types that are managed its operation, which leads to a high number of suppliers and makes it more complex to measure and control the quality of suppliers, impacting its costs, the duration of the projects, reprocessing, among others. The use of optimization techniques is proposed to evaluate the suppliers, however, as indicate Meysam, [4] many mathematical models do not address all the perspectives of the decision maker and become ambiguous. Therefore these models are not considered as reliable for decisions processes of this kind [5]. These methods usually handle only quantitative variables which causes the ambiguity of their solutions [6]. Bruno [6] mentions some advantages of the AHP method with respect to the previously analyzed, among which it can be mentioned the hierarchical representation of the model, the management of tangible and intangible attributes, since it can be combined with other strategies such as TOPSIS, DEMATEL and others, such as the multidisciplinary characteristic that gives weights to each factor analyzed according to the objective of the decision maker [6].

According to the latest "Hydraulic Cement: World Production, by Country" report provided by Index Mundi, the production of cement amounts 2310 million tons. This value not only show the relevance of the concrete industry for trade and the global economy, but also imply the need to find ways to improve the performance of companies and their value chain.

Relevant sectors (for example, construction and consumption) represent 28% of world production [7]. In all these sectors, cement is one of the most critical raw materials. In this regard, concrete suppliers must be carefully selected to guarantee meaningful information focused on the aspects. Therefore, this document aims to develop a fuzzy analytical hierarchy process (FAHP) for the selection of suppliers. The study was conducted considering concrete as the most critical raw material for a company in the construction industry.

2 Methodology

The proposed methodology is composed of six phases and aims for the selection of the best concrete supplier. In the first phase, a decision-making group is formed to design the hierarchy and make pairwise comparisons between factors for the FAHP and DEMATEL methods. A hierarchical structure is organized in phase two. In order to achieve this, the evaluation of experts in the area is used as well as the literature compiled, then the criteria based on comparisons of importance (FAHP) are evaluated. to obtain their global weights. Subsequently, DEMATEL is applied to evaluate the interrelationships between criteria. Finally, TOPSIS method is used in order to rank the different company concrete suppliers.

2.1 Fuzzy Analytic Hierarchy Process

Considering that AHP does not include the vagueness of human judgments, the theory of fuzzy logic was introduced thanks to its ability to represent inaccurate data. In FAHP, paired comparisons are represented by triangular numbers [8], as described below (see Table 1). Considering the findings of the literature review, decision-makers have adopted a reduced AHP scale when making comparisons. Next, the description of the FAHP algorithm is shown:

- Paired comparisons are made between criteria using the linguistic terms and the corresponding fuzzy triangular numbers established in Table 1. With these data, a fuzzy evaluation matrix is obtained as described below in Eq. 1:

$$\tilde{A}^K = \begin{bmatrix} \tilde{d}_{11}^k & \tilde{d}_{12}^k & \cdots & \tilde{d}_{1n}^k \\ \tilde{d}_{21}^k & \tilde{d}_{22}^k & \cdots & \tilde{d}_{2n}^k \\ \cdots & \cdots & \cdots & \cdots \\ \tilde{d}_{n1}^k & \tilde{d}_{n2}^k & \cdots & \tilde{d}_{nn}^k \end{bmatrix}, \tag{1}$$

indicates the preference of the (k) expert of (i) criterion on (j) criterion through fuzzy triangular numbers.

- In the case of a focus group, the judgments are averaged according to Eq. 2, where K represents the number of experts involved in the decision-making process. Then, the fuzzy evaluation matrix is updated as shown in Eq. 3.
- The geometric mean of the diffuse judgment values of each factor is calculated using Eq. 4. Here, $\tilde{r}_i$ denotes triangular numbers.

$$\tilde{d}_{ij} = \frac{\sum_{k=1}^{K} \tilde{d}_{ij}^k}{K}, \tag{2}$$

$$\tilde{A} = \begin{bmatrix} \tilde{d}_{11} & \cdots & \tilde{d}_{1n} \\ \vdots & \ddots & \vdots \\ \tilde{d}_{n1} & \cdots & \tilde{d}_{nn} \end{bmatrix}. \tag{3}$$

Table 1. Linguistic terms and their fuzzy triangular numbers

Reduced AHP scale	Definition	Fuzzy triangular number
1	Equally important	[1, 1, 1]
3	More important	[2, 3, 4]
5	Much more important	[4, 5, 6]
1/3	Less important	[1/4, 1/3, 1/2]
1/5	Much less important	[1/6, 1/5, 1/4]

$$\tilde{r}_j = \left(\prod_{j=1}^{n} \tilde{d}_{ij} \right)^{1/n}, \ i = 1, 2, \ldots, n. \tag{4}$$

- Determine the fuzzy weights of each factor (W_i) by applying Eq. 5.
- Defuzzify (W_i) through the use of Eq. 6, denoted by M_i. Then, M_i is normalized by applying Eq. 7.

$$\tilde{w}_i = \tilde{r}_i \otimes (\tilde{r}_1 \otimes \tilde{r}_2 \otimes \ldots \otimes \tilde{r}_n)^{-1} = (lw_i, mw_i, uw_i). \tag{5}$$

$$M_i = \frac{lw_i + mw_i + uw_i}{3}, \tag{6}$$

$$N_i = \frac{M_i}{\sum_{i=1}^{n} M_i}. \tag{7}$$

2.2   Decision Making Trial and Evaluation Laboratory

DEMATEL is an MCDM technique that effectively identifies causal relationships in complex decision-making hierarchies [9, 10]. The final product of DEMATEL is a visual representation that classifies the factors into two groups: receptors and dispatchers [8]. Dispatchers are the criteria that heavily influence other criteria, while the affected factors are called receivers. To do this, DEMATEL converts the relationships between the causes and effects of the criteria into a structural mapping model [11]. In addition, this method indicates the degree of influence of each element, so that they can be identified significant interdependencies [12].

The steps of the DEMATEL method are explained below:

- The matrix of direct relationship is found: to analyze interdependence, a committee of experts is asked to make paired comparisons between criteria according to their personal experience. Each decision maker specifies how the criterion i influences the criterion j by applying a comparison scale from 0 to 4: no influence (0), low influence (1), medium influence (2), high influence (3), and very high influence (4). With these judgments, an average matrix called the direct relation matrix Z is obtained (Eq. 8). Each z_{ij} value denotes the average degree in which criterion i affects criterion j. The value in any element on the diagonal is 0.

$$Z = \begin{bmatrix} 0 & z_{12} & \cdots & z_{1n} \\ z_{21} & 0 & \cdots & z_{2n} \\ \cdots & \cdots & \ddots & \cdots \\ z_{n1} & z_{n2} & \cdots & 0 \end{bmatrix}. \tag{8}$$

- The normalized direct relation matrix is calculated: using Eqs. 9 and 10, the normalized matrix can be derived from the direct relation matrix Z:

$$X = s \cdot Z, \tag{9}$$

$$s = \min \left(\frac{1}{\max\limits_{1 \le i \le n} \sum_{j=1}^{n} |z_{ij}|}, \frac{1}{\max\limits_{1 \le j \le n} \sum_{i=1}^{n} |z_{ij}|} \right), (i,j) \in \{1, 2, \ldots n\} \tag{10}$$

- Calculate the total relation matrix: after calculating the matrix X of normalized direct relation, the matrix T of total relation is obtained by applying Eq. 11, where I represents the identity matrix:

$$T = X + X^2 + X^3 + \ldots = \sum_{i=1}^{\infty} X^i = (1 - X)^{-1}. \tag{11}$$

- Dispatchers and receivers are identified: using D-R values, where R_i is the sum of the columns of matrix T (Eqs. 12–13) for each factor i and D_j is the sum of rows (Eqs. 12 and 14) for each factor j, and make $D_i = D_j$. Cause and effect groups can be determined. In this respect, factors with a negative value of D − R are classified as receivers, while positive values indicate elements of the dispatcher. On the other hand, the D + R values represent the force of influence between the elements of the system; however, significant interdependencies are identified using the Len method as described in the next step.

$$T = [t_{ij}]_{nxn}, (i, j) \in \{1, 2, \ldots, n\}, \tag{12}$$

$$R = \sum_{j=1}^{n} t_{ij}, \tag{13}$$

$$D = \sum_{i=1}^{n} t_{ij}. \tag{14}$$

- The threshold value is defined and identifies significant influences.

2.3   Technique for Order of Preference by Similarity to Ideal Solution

TOPSIS is a decision-making technique that involves selecting the alternative with the shortest distance from the positive ideal solution (PIS) and the farthest distance from the negative ideal solution NIS [13]. PIS is composed of all the best achievable attribute values, while NIS considers the worst attribute measures [13]. However, the selected alternative that has the minimum Euclidean distance of PIS can also have a short distance of NIS. In addition, a simple assumption is that each criterion is characterized by a monotonously increasing or decreasing utility. Therefore, TOPSIS tries to find alternatives that are simultaneously close to PIS and far from NIS by using the relative proximity coefficient [13]. This The relative proximity coefficient (R_i) is obtained by applying the following equation. If $R_i = 1$, the provider performs according to d_i^+; therefore, higher values of R_i represent satisfactory overall performance.

$$R_i = \frac{d_i^+}{d_i^+ + d_i^-}, \quad 0 \leq R_i \leq 1, \quad i = 1, 2, \ldots, m.$$

- Concrete suppliers are classified according to the order of preference of R_i.

3 Study Case

The case study is illustrated in a medium-sized construction company located in La Guajira region in Colombia. The company is responsible for the construction of different types of infrastructures in the department, which are built mostly of concrete, for better durability and resistance. In this sense, the company focuses on continuously satisfying customer requirements (for example, delivery date, standard compliance is quality, durability, zero leaks, etc.) to improve the performance of the company and, subsequently, address the growing number of competitors in the construction industry. To support these strategies, the construction company has determined the need to adequately select their concrete suppliers and, therefore, it is necessary to design a decision-making model that classifies potential suppliers according to a set of criteria and criteria defined.

Thanks to several meetings with some experts in the organization, eight factors were identified to select the best concrete supplier. In this case, five concrete suppliers were evaluated (P1, P2, P3, P4 and P5). All the criteria were determined considering the experience of the experts, the measures of the industry and the relevant scientific literature.

The description of each factor is detailed below:

- CONCRETE TEMPERATURE (C1): The concrete after mixing stiffens with time, a phenomenon that should not be confused with cement setting. What happens is that the mixing water is lost, because the aggregates absorb part of it, it evaporates, especially if the concrete is exposed to the sun and the wind and another part is eliminated by the initial chemical reactions.

  Concrete temperature is the most important, since it controls the chemical reactions that occur in the mixture and therefore modifies the properties of the concrete in a fresh and hardened state. Standard NTC 3357 sets the temperature limits for fresh concrete. The measurement of the temperature is made when the concrete is received in the work, while it is placed, with glass thermometers or with armor, which must have an accuracy of 1 °C and must be introduced into the representative sample for at least two minutes or until the reading stabilizes. It is also possible to determine the temperature by means of electronic temperature meters with precision digital displays.

- WORKABILITY OR HANDLING (C2): It is the capacity of concrete that allows it to be placed and compacted properly without any segregation. The workability is represented by the degree of compatibility, cohesiveness, plasticity and consistency.
- SEGREGATION (C3): An important aspect of workability and that is generally considered as another property, is the tendency to segregation, which is defined as the tendency of separation of coarse particles from the mortar phase of concrete and the collection of those deficient particles of mortar in the perimeter of the placed concrete, due to lack of cohesiveness, in such a way that its distribution and behavior ceases to be uniform and homogeneous. This leads to the fact that non-segregation is an implicit condition of concrete to maintain adequate workability.

  On the other hand, the main causes of segregation in concrete are the difference in density between its components, the size and shape of the particles and the particle size distribution, as well as other factors such as bad mixing, an inadequate transport system, poor placement and excessive vibration compaction.
- EXUDATION (C4): It is a form of segregation or sedimentation, in which part of the mixing water tends to rise to the surface of a freshly placed concrete mixture. This is because the solid constituents of the mixture cannot retain all the water when they settle during the setting process. The exudation of the concrete is influenced by the proportions of the mixture and the properties of the materials, the air content, shape and texture of the aggregates, quality of the cement and the use of the additives.
- UNIT MASS AND VOLUMETRIC PERFORMANCE (C5): Taking into account that the concrete is dosed by weight and is supplied by volume, therefore, it is important to determine the unit mass of the concrete to calculate the volume or volumetric yield produced by the known weights of each one of the materials that constitute it and to determine the content of cement per cubic meter of concrete. The test procedure is described in the NTC 1926 standard.
- CONCRETE FORGING (C6): Concrete forging corresponds to the hardening process of the concrete mixture, where a transition from a plastic state to a hardened state is experienced under certain conditions of time and temperature. The setting time is an arbitrary value that has been taken during the concrete hardening process, and the NTC 890 standard describes the procedure for its calculation.
- AIR CONTENT (C7): This element is present in all types of concrete, located in the non-saturable pores of the aggregates and forming bubbles between the concrete components, either because it is trapped during the concrete mixing or when it is incorporated using air-entraining agents, such air-entraining additives. The air content of a concrete if inclusion agents is usually between 1% and 3% of the volume of the mixture, while a concrete with air inclusions can obtain air content between 4% and 8%.

- BENDING RESISTANCE OF CONCRETE BEAMS (C8): It is the bending of a beam when having a weight on top.

In order to select the best concrete supplier by the application of the proposed approach the first step is the design of the AHP and DEMATEL surveys. Then, using Eqs. 1–7, weights of the criteria against themselves were determined. The participants responded the AHP survey using the scale described on Table 1. This process was repeated until all the trials were completed. In particular, the design of this instrument contributed to minimize discrepancies and lack of understanding. In addition, it excluded intransitive comparisons during the decision-making process.

Likewise, a similar survey was designed for DEMATEL, in order to analyze the interdependence between criteria. Then, when applying Eqs. 8–14, the normalized influence matrix was obtained. Respondents were asked to determine; how much influence does each element have on other element. The experts answered using a 5-point scale going from 0 (no influence) to 4 (high influence).

Through the integration of the FAHP and DEMATEL techniques, it was determined that the contributions of the criteria consider the environments of linear dependence, interrelations and uncertainty. To do this, the fuzzy judgment matrices were initially calculated based on the pairwise comparisons made by the decision makers. An example of this matrix is shown in Table 2.

Table 2. Matrix of fuzzy judgments for the criteria.

	(C1)	(C2)	(C3)	(C4)	(C5	(C6)	(C7)	(C8)
(C1)	[1, 1, 1]	[0.25, 0.33, 0.5]	[0.25, 0.33, 0.5]	[0.25, 0.33, 0.5]	[0.17, 0.2, 0.25]	[0.17, 0.2, 0.25]	[0.25, 0.33, 0.5]	[0.17, 0.2, 0.25]
(C2)	[2, 3, 4]	[1, 1, 1]	[0.25, 0.33, 0.5]	[0.25, 0.33, 0.5]	[0.25, 0.33, 0.5]	[0.25, 0.33, 0.5]	[0.17, 0.2, 0.25]	[0.17, 0.2, 0.25]
(C3)	[2, 3, 4]	[2, 3, 4]	[1, 1, 1]	[1, 1, 1]	[0.25, 0.33, 0.5]	[0.25, 0.33, 0.5]	[0.25, 0.33, 0.5]	[0.17, 0.2, 0.25]
(C4)	[2, 3, 4]	[2, 3, 4	[1, 1, 1]	[1, 1, 1]	[1, 1, 1]	[0.25, 0.33, 0.5]	[0.25, 0.33, 0.5]	[0.25, 0.33, 0.5]
(C5)	[4, 5, 6]	[2, 3, 4]	[2, 3, 4]	[1, 1, 1]	[1, 1, 1]	[1, 1, 1]	[0.25, 0.33, 0.5]	[0.25, 0.33, 0.5]
(C6)	[4, 5, 6]	[2, 3, 4]	[2, 3, 4]	[2, 3, 4]	[1, 1, 1]	[1, 1, 1]	[1, 1, 1]	[0.25, 0.33, 0.5]
(C7)	[2, 3, 4]	[4, 5, 6	[2, 3, 4]	[4, 3, 4]	[2, 3, 4]	[1, 1, 1]	[1, 1, 1]	[1, 1, 1]
(C8)	[4, 5, 6]	[4, 5, 6]	[4, 5, 6]	[2, 3, 4]	[2, 3, 4]	[2, 3, 4]	[1, 1, 1]	[1, 1, 1]

Then, applying Eq. 4, the geometric means of the fuzzy comparisons were calculated. In addition, when using Eqs. 5–7, the normalized weights of the criteria factors were obtained (see Table 3).

Table 3. Normalized weights for the criteria

	lw_i	mw_i	uw_i	Nonfuzzy weights	Normalized weights (NF_i)
CONCRETE TEMPERATURE (C1)	0.02	0.02	0.04	0.03	0.03
WORKABILITY OR HANDLING (C2)	0.02	0.04	0.07	0.04	0.04
SEGREGATION (C3)	0.04	0.06	0.11	0.07	0.07
EXUDATION (C4)	0.05	0.08	0.13	0.09	0.08
UNIT MASS AND VOLUMETRIC PERFORMANCE (C5)	0.07	0.12	0.19	0.13	0.12
CONCRETE FORGING (C6)	0.1	0.16	0.25	0.17	0.16
AIR CONTENT (C7)	0.15	0.22	0.34	0.24	0.22
BENDING RESISTANCE OF CONCRETE BEAMS (C8)	0.18	0.3	0.46	0.31	0.29
Total				1.07	

The inconsistency values (consistency ratio) was calculated and it was 9.3% indicating that the comparisons are reliable, considering that this index is not greater than 10%. Then the weights obtained with the AHP process can be accepted.

Therefore, the process of data collection can be considered satisfactory and, subsequently, the decision-making process with highly reliable results. Then, to estimate the weights of the criteria on the basis of interdependence (*WFc*), the weights obtained from the application of FAHP are multiplied with the matrix of normalized direct relations X (Table 4), obtained after the application of DEMATEL, as indicated in the following equation, in order to obtained *WFc* shown on Table 5:

$$WFc = \begin{bmatrix} x_{11} & x_{12} & \cdots & x_{1n} \\ x_{21} & x_{22} & \cdots & x_{2n} \\ \cdots & \cdots & \cdots & \cdots \\ x_{n1} & x_{n1} & \cdots & x_{nn} \end{bmatrix} \times \begin{bmatrix} NF_1 \\ NF_1 \\ \cdots \\ NF_1 \end{bmatrix}$$

Table 4. Normalized direct relations matrix (X)

	(C1)	(C2)	(C3)	(C4)	(C5)	(C6)	(C7)	(C8)
(C1)	0.000	0.333	0.500	0.286	0.000	0.167	0.000	0.000
(C2)	0.000	0.000	0.250	0.143	0.125	0.167	0.500	0.000
(C3)	0.500	0.333	0.000	0.571	0.375	0.250	0.167	0.100
(C4)	0.500	0.333	0.250	0.000	0.125	0.083	0.167	0.100
(C5)	0.000	0.000	0.000	0.000	0.000	0.167	0.000	0.300
(C6)	0.000	0.000	0.000	0.000	0.000	0.000	0.167	0.300
(C7)	0.000	0.000	0.000	0.000	0.125	0.000	0.000	0.200
(C8)	0.000	0.000	0.000	0.000	0.250	0.167	0.000	0.000

Table 5. Weights of the criteria on the basis of interdependence

(C1)	(C2)	(C3)	(C4)	(C5)	(C6)	(C7)	(C8)
0.095	0.179	0.222	0.136	0.114	0.124	0.073	0.056

Considering the importance (weights) assigned to each criteria, it is necessary which one is the best supplier considering the its performance in each of the criteria, in order to determine the most appropriate. A score from 1 to 5 was assigned to each supplier in each criterion according the evaluation rules the company has (Table 6).

Table 6. Supplier comparison matrix according to each criterion.

	C1	C2	C3	C4	C5	C6	C7	C8
P1	2	1	2	1	5	1	1	2
P2	5	4	4	2	2	4	3	3
P3	3	2	1	3	4	5	4	1
P4	1	3	5	4	1	2	2	5
P5	4	5	3	5	3	3	5	4

Taking in consideration this evaluation matrix and the weights on Table 5 TOPSIS method was performed obtaining the values shown on Table 7 for the distance to PIS, NIS as well as the relative proximity coefficient for each supplier. These results allows the company to identify which one is the best supplier considering the importance given to each factor and the performance of the different suppliers in each criterion.

Table 7. TOPSIS relative proximity coefficient for each supplier.

	d_i^+	d_i^-	R_i
P1	0.024	0.009	0.285
P2	0.012	0.019	0.617
P3	0.02	0.014	0.4
P4	0.015	0.02	0.564
P5	0.01	0.021	0.668

4 Conclusions

The selection of suppliers is an important process for the management of the supply chain of construction companies. However, studies that concentrate directly on the selection of suppliers with the use of MCDM techniques in construction companies are largely limited. Therefore, this research proposed a hybrid method of multiple criteria decision making to properly select concrete suppliers based on FAH P and DEMATEL methods. This approach is useful for managers to increase competitiveness and sustainability performance of their organizations.

In a decision-making process, experts cannot express their judgments exactly in numerical values due to human vagueness. To address this problem, FAHP was implemented to represent inaccurate data and determine linear dependencies in the hierarchy. This method was combined with DEMATEL to also evaluate the interdependencies between factors to determine possible improvement strategies. That is, this document provides an efficient and accurate approach that can also be used to address other managerial decision-making problems that contain many criteria with vague interrelations. Thus, it is scalable and adaptable in any context.

According to the results obtained in the study, it was possible to determine that the best supplier is number 5 considering that it has the highest value of R_i. An alternative supplier could be number 2 considering that its R_i score is close to the one obtained by supplier 5. The company could share the total order between these two suppliers, in order to increase resilience in case of a disruption of the supply chain.

References

1. Ceballos Arana, M.A.: EL CONCRETO, MATERIAL FUNDAMENTAL PARA LA INFRAESTRUCTURA, México (2016)
2. Revista Dinero: Aporte del sector de la construcción a la economía colombiana
3. Wetzstein, A., Hartmann, E., Benton Jr., W.C., Hohenstein, N.-O.: A systematic assessment of supplier selection literature – state-of-the-art and future scope. Int. J. Prod. Econ. **182**, 304–323 (2016)
4. Shaverdi, M., Akbari, M., Fallah Tafti, S.: Combining fuzzy MCDM with BSC approach in performance evaluation of iranian private banking sector. Adv. Fuzzy Syst. **2011**, 1–12 (2011)
5. Shaverdi, M., Heshmati, M.R., Ramezani, I.: Application of fuzzy AHP approach for financial performance evaluation of iranian petrochemical sector. Procedia Comput. Sci. **31**, 995–1004 (2014)
6. Bruno, G., Esposito, E., Genovese, A., Passaro, R.: AHP-based approaches for supplier evaluation: problems and perspectives. j. Purch. Supply Manage. **18**(3), 159–172 (2012)
7. Paloma, M., Ortiz, C.: Presente Futuro de la Industria del Plástico en México (2012)
8. Ortiz-Barrios, M.A., Kucukaltan, B., Carvajal-Tinoco, D., Neira-Rodado, D., Jiménez, G.: Strategic hybrid approach for selecting suppliers of high-density polyethylene. J. Multicriteria Decis. Anal. **24**(5–6), 296–316 (2017)
9. Liou, J.J.H., Tzeng, G.-H., Chang, H.-C.: Airline safety measurement using a hybrid model. J. Air Transp. Manage. **13**(4), 243–249 (2007)
10. Wu, W.-W., Lee, Y.-T.: Developing global managers' competencies using the fuzzy DEMATEL method. Expert Syst. Appl. **32**(2), 499–507 (2007)
11. Wang, C.H.: An integrated fuzzy multi-criteria decision making approach for realizing the practice of quality function deployment. In: IEEM2010 - IEEE International Conference on Industrial Engineering and Engineering Management, pp. 13–17 (2010)
12. Tzeng, G.-H., Chiang, C.-H., Li, C.-W.: Evaluating intertwined effects in e-learning programs: a novel hybrid MCDM model based on factor analysis and DEMATEL. Expert Syst. Appl. **32**(4), 1028–1044 (2007)
13. Ortíz-Barrios, M., Neira-Rodado, D., Jiménez-Delgado, G., Hernández-Palma, H.: Using FAHP-VIKOR for operation selection in the flexible job-shop scheduling problem: a case study in textile industry. In: Tan, Y., Shi, Y., Tang, Q. (eds.) ICSI 2018. LNCS, vol. 10942, pp. 189–201. Springer, Cham (2018). https://doi.org/10.1007/978-3-319-93818-9_18

Machine Learning and High
Performance Computing

Representation Learning for Diagnostic Data

Karol Antczak[(⊠)] [iD]

Military University of Technology, Sylwestra Kaliskiego 2, Warsaw, Poland
karol.antczak@wat.edu.pl

Abstract. Representation learning algorithms have recently led to a significant progress in knowledge extraction from network structures. In this paper, a representation learning framework for the medical diagnosis domain is proposed. It is based on a heterogeneous network-based model of diagnostic data combined with an algorithm for learning latent node representation. Furthermore, a modification of metapath2vec algorithm is proposed for representation learning of heterogeneous networks. The proposed algorithm is compared with other representation learning approaches in two practical case studies: symptom/disease classification and disease prediction. A significant performance boost can be observed for these tasks, resulting from learning representations of domain data in a form of a heterogeneous network. It is also shown that in certain situations the modified algorithm improves the quality of learned embeddings compared to reference methods.

Keywords: Representation learning · Feature learning · Network embedding · Heterogeneous networks · Medical diagnosis

1 Introduction

Representation learning is a group of machine learning methods that aims to find useful representations of the data. The "usefulness" is typically understood in terms of extraction of features that are meaningful from the point of view of the target objective. For neural networks, such representation is defined as a mapping f of input representations to d – dimensional vector space: $f : V \rightarrow \mathbb{R}^d$. The development of representation learning is motivated by numerous experimental results showing that extracting the features of the data improves the performance of the network compared to the "naive" data encoding schemes such as binary or one-hot encoding. This is further encouraged by the observations that many deep learning architectures seem to naturally learn the layer-wise representation of the features during the training – a phenomenon which some researchers point out as an important factor contributing to great performance of DL methods. Not without the significance is also the fact that the such internal representations can be, at least in some cases, interpreted by humans, which is a step toward improving the explainability of deep neural models.

Until recently, machine learning applications for medical diagnosis did not utilize representation learning, relying on either non-neural feature extraction methods or naive encoding schemes. On the other hand, other deep learning models are used extensively. The primary motivation of our research was to introduce RL into to diagnosis area, encouraged by performance boost observed in other fields. However,

© Springer Nature Switzerland AG 2020
K. Saeed and J. Dvorský (Eds.): CISIM 2020, LNCS 12133, pp. 197–207, 2020.
https://doi.org/10.1007/978-3-030-47679-3_17

this requires to define a data model that can be utilized by representation learning frameworks. Thus, we propose a formal model of diagnostic data, by means of a heterogeneous network, allowing us to use modern representation learning algorithms for graph-like structures, such as node2vec and metapath2vec. We also develop an extension of metapath2vec that further improves the quality of learned representations.

The structure of this paper is as follows. In Sect. 2 we briefly present current research status regarding representation learning for networks. Section 3 contains a proposed model of diagnostic data and representation learning framework for such data. In Sect. 4 we present experimental results. Obtained results are discussed in Sect. 5. Implementations of all algorithms used for this paper are available on the author's GitHub repository [1].

2 Related Research

A major milestone in the representation learning field was the development of the word2vec algorithm which finds efficient representations of words in text corpora [2]. The basic idea of word2vec was soon applied to other types of data, including networks. This resulted in the development of algorithms such as DeepWalk [3] and its generalization, node2vec [4]. Both of these algorithms are designed to learn representations of nodes V given a graph $G = (V, E)$. If input nodes are encoded by one-hot vectors, then the target mapping is a linear transformation and can be represented by the transformation matrix A. The learning process is then an optimization task that finds the matrix which, given feature representation of another node, maximizes the log-probability of observing a certain "neighbor" node. In other words, we aim for nodes with similar neighbors to have similar feature representation. Since it is ineffective to compute the target function for each possible pair of nodes, they are sampled from the network using random walks instead. The general algorithm of node representation learning is given below.

Algorithm 1. General node representation learning algorithm.

Input: Network $G=(V, E)$, Dimensions d, Walks per node r, Walk length l, Context size k
Output: node embedding matrix $A \in \mathbb{R}^{|V| \times d}$

Initialize *walks* to empty
for $i = 1 \rightarrow r$ **do**
  for $v \in V$ **do**
    walk = RandomWalk(G, v, l)
    Append *walk* to *walks*
  end
end
A = StochasticGradientDescent(k, d, *walks*)
return A

The exact algorithm of random walks differs between the algorithms. Authors of node2vec note that there are two distinct kinds of node similarities: homophily (occurring in nodes that are close to each other) and structural equivalence (occurring in nodes that have similar structural roles in the network but are not necessarily closely interconnected). The random walk procedure used in word2vec is characterized by two hyperparameters that incorporate both notions of similarity. The unnormalized transition probability from node v^i to node v^{i+1} given previous node v^{i-1} is:

$$\alpha\left(v^{i+1}|v^i\right) = \begin{cases} \frac{1}{p} & \text{if } d(v^{i-1}, v^{i+1}) = 0 \\ 1 & \text{if } d(v^{i-1}, v^{i+1}) = 1 \\ \frac{1}{q} & \text{if } d(v^{i-1}, v^{i+1}) = 2 \end{cases} \tag{1}$$

The $d(v^{i-1}, v^{i+1})$ denotes the shortest path between previous and next node. The return parameter p controls the likelihood of returning to already visited node, while the in-out parameter q controls the tendency to explore outward nodes. Thanks to this, the random walk can result in different pairs of neighbors, depending on which similarity seems more suitable to the target task. Specifically, the sampling strategy used in DeepWalk is the one where $p = 1$ and $q = 1$, meaning that each node has the same probability of being visited.

Metapath2vec is a modification of node2vec for heterogeneous networks. A heterogeneous network is defined as a graph $G = (V, E, T)$ in which each node and each edge is associated with respective mapping functions $\phi(v) : V \to T_v$ and $\phi(e) : V \to T_e$. Instead of using random walks scheme with explicit p and q parameters, metapath2vec utilizes additional information about node types to provide an alternative method, called meta-path-based random walks. The flow of the walk is determined by the so-called meta-path defined as a scheme $P : V_1 \xrightarrow{R_1} V_2 \xrightarrow{R_2} \ldots V_t \xrightarrow{R_t} V_{t+1} \ldots V_l$ wherein $V_1 \ldots V_l$ are respective node types and $R_1 \ldots R_{l-1}$ are relations between them. The transition probability for a node at step i is given by:

$$p\left(v^{i+1}|v^i, P\right) = \begin{cases} \frac{1}{|N_{t+1}(v_t^i)|} & \left(v^{i+1}, v_t^i\right) \in E, \phi(v^{i+1}) = t+1 \\ 0 & \left(v^{i+1}, v_t^i\right) \in E, \phi(v^{i+1}) \neq t+1 \\ 0 & \left(v^{i+1}, v_t^i\right) \notin E \end{cases}$$

Each of the above algorithms can also be used for learning edge representations. They are obtained by combining node representations of adjacent nodes using binary operators, for example average or Hadamard product. This allows to use these algorithm for edge-related tasks, such as link prediction (predicting whether two nodes should be connected or not) or edge classification.

Current applications of network-based representation learning in the medical domain include diverse areas such as genetics [5, 6], biomedical pathways discovery [7], or learning relationships between doctors, patients and medical services from electronic health records [8]. All of them use, or are based on, node2vec algorithm and therefore work on homogeneous networks, making no distinction between node or edge types. A novel algorithm, called edge2vec, was proposed for edge-based representation learning with potential applications for broad range of biomedical data [9]. Similarly to metapath2vec, it works on heterogeneous networks, however, it utilizes knowledge about edge types rather than nodes. As such, it is not directly comparable to our method. To the best of our knowledge, meta-path-based approach was not used for the medical data yet. Additionally, we have not found any research regarding representation learning used specifically for the symptomatic diagnosis task.

3   Proposed Framework

3.1   Domain Model

The model of diagnostic data proposed here is inspired by observations in [10]. Let $G = (V, E, T)$ be a heterogeneous undirected network, with nodes V, edges E, and node/edge types $T = T_v \cup T_e$. Let $\phi(v)$ and $\phi(e)$ be appropriate type mapping functions for nodes and edges. We define four types of nodes: $T_v = \{D, S, N, W\}$. Each node type represents a category of entities from diagnostic domain. They are:

- D - diseases
- S - symptom occurrences
- N - symptom names
- W - symptom values

We also define three types of edges: $T_e = \{SD, SN, SW\}$. SD represents a relationship between disease and symptom occurrences including both non-specific, specific and pathognomonic ones. It is interpreted as "symptom s can occur in disease d". SN is a relationship between symptom and symptom name. Its semantics means "symptom s has name n". Each symptom occurrence is associated with exactly one name. Relationship type SW occurs between symptom and symptom value and can be interpreted as "symptom s has value w". Each symptom occurrence is associated with at least one value. An example of such network is shown in Fig. 1.

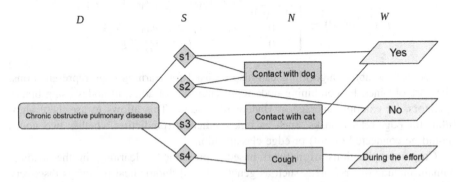

Fig. 1. Example of diagnostic data network.

The proposed model has several important practical features. First, it represents a heterogeneous network, allowing us not only to use homogeneous network-based representation learning frameworks but also utilize the knowledge of edge type. The second feature is ease to gather the actual data. It can be gathered in a form of triplets $\langle d, n, w \rangle \in D \times N \times W$. The symptom node s is then created as an intermediate node between the last two. Another practical assumption was the compliance of diseases and symptom names with International Statistical Classification of Diseases and Related Health Problems (ICD-10) to minimize the risk of data inconsistency.

3.2   Representation Learning Algorithm

The original metapath2vec algorithm uses only a single meta-path to generate walks. This may be an issue if relationships that we want to be included do not form a path. An example can be seen in Fig. 2.

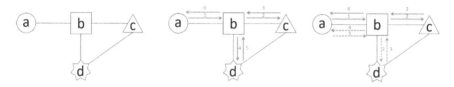

Fig. 2. Illustration of multiple relationships problem. Left: Sample meta-graph consisting of 4 types of nodes. Middle: Meta-path for classical metapath2vec algorithm. Right: meta-paths for proposed algorithm.

The sample graph contains four types of nodes. Suppose we would like to generate embedding vectors that will incorporate two kind of relationships: between nodes a and c and between nodes a and d. Due to the requirement of recursion, the simplest meta-path to include both of them must have 7 steps, for example $a \rightarrow b \rightarrow c \rightarrow b \rightarrow d \rightarrow b \rightarrow a$. This may result in generation of redundant skip-grams as well as lead to learning undesired features like, for example, relationships between nodes c and d. As a consequence, embeddings will require more iterations to train. We propose a simple modification of the metapath2vec algorithm, by allowing to use multiple shorter meta-paths instead of a single one. This way, the desired relationships can be learned by using two meta-paths: $a \rightarrow b \rightarrow c \rightarrow b \rightarrow a$ and $a \rightarrow b \rightarrow d \rightarrow b \rightarrow a$. The full modified algorithm, which we call multi-metapath2vec, is listed below.

Algorithm 2. Multi-metapath2vec algorithm.

Input: Heterogeneous network $G=(V,E,T)$, Dimensions d, Walks per node r, Walk length l, Context size k, Meta-paths M
Output: node embedding matrix $A \in R^{|V| \times d}$

Initialize *walks* to empty
$\rho = \lfloor r/|M| \rfloor$
for $i=1 \rightarrow \rho$ **do**
  for $v \in V$ **do**
    for $m \in M$ **do**
    walk = MetaPathRandomWalk(G,v,l,m)
    Append *walk* to *walks*
    end
  end
end
 A = StochasticGradientDescent(k,d, *walks*)
return

MetaPathRandomWalk(G,v,l,m)
 $MP[1]=v$
 for $i=2 \rightarrow l$ **do**
   u = draw node from V according to equation (2)
   $MP[i]=u$
 end
 return MP

---

4 Experiments

4.1 Experimental Setup

The goals of the experiments are twofold. First, we want to study the effect of using representation learning for diagnostic data by comparing neural networks trained in a supervised way with networks of the same architecture, but trained in a semi-supervised way, with first layers containing embeddings pretrained with representation learning. Second, we want to compare the proposed algorithm with other representation learning algorithms. We study two practical tasks involving medical diagnostic data: grouping of symptoms and diseases according to ICD-10 taxonomy and disease prediction.

The dataset used for experiments consists of 14086 triplets $\langle d,n,w \rangle$ which results in an undirected graph consisting of 91 D nodes, 91 W nodes, 728 N nodes, and 1327 S nodes. The data was harvested by a team of physicians as a part of the POIG.02.03.03-00-013/08 project [11] and covers dermatology and pulmonology areas.

Four methods were used to create embedding layers. Each embedding layer was represented by the matrix of the size $d \times k = 2238 \times 100$ with elements sampled uniformly from $[-0.05, 0.05]$ range. Following methods were used:

- No pretraining. The embedding layer was only initialized with random values.
- node2vec: an original implementation, with parameter values $p = 1$, $q = 1$, $d = 2238$, $k = 100$, $r = 10$, $l = 80$.
- metapath2vec: a custom implementation based on the original code provided by the authors, with parameters values the same as above (excluding p and q) and with a single meta-path $d \rightarrow s \rightarrow n \rightarrow s \rightarrow w \rightarrow s \rightarrow d$.
- multi-metapath2vec: a custom implementation, with parameters values the same as above and with two meta-paths: $d \rightarrow s \rightarrow n \rightarrow s \rightarrow d$ and $d \rightarrow s \rightarrow w \rightarrow s \rightarrow d$.

Embedding layers were pretrained with 1 million skip-gram pairs generated using above methods. A single epoch of RMSProp algorithm was used for the training, with default hyper-parameter values and binary cross-entropy as a loss function.

4.2   Case Study: ICD-10 Node Classification

The task here is to classify disease or symptom name nodes according to the subgroup in ICD-10 classification. For example, disease 'Other atopic dermatitis' (ICD code L20.8) is assigned to the subgroup 'L20-L30 Dermatitis and eczema'. The input vector consists of a single node whereas the output vector is one-hot-encoded subgroup label. The full dataset contains a total of 43 classes – 33 classes of symptom names and 10 classes of diseases. To make the task non-trivial, a certain percentage of the training data is not used. The network should, therefore, rely on the knowledge about neighbor nodes, in a form of embedding layer, in order to make correct classification. We analyze two ranges of such incompleteness: from 0% to 90% (with step 10%) and from 90% to 99% (with step 1%) of missing data.

The neural networks used in experiments are two-layer feed-forward networks. The first layer is a pretrained embedding layer. The second layer contains 43 neurons with sigmoid activations. Networks are trained with 10 epochs of RMSProp algorithm and binary cross-entropy loss. Cross-validation is used for model evaluation, with $k = 10$ and two metrics: F1 micro- and macro- averaged.

The results for respective ranges and metrics are presented in Fig. 3. Pretrained embeddings give a significant performance boost compared to the network without pretrained embeddings. Of the representation learning methods, node2vec obtained the worst values, being outperformed by both metapath2vec and multi-metapath2vec for most of the ranges. However, for the 97–99% of missing data, one can observe a rapid performance decrease of metapath2vec. On the other hand, multi-metapath2vec is steadily the best method in terms of both metrics. The performance gap between multi-metapath2vec and other methods, while relatively small for the 90–99% range, becomes particularly visible for the missing data range 90–98%. On the average, multi-metapath2vec obtains 8% higher F1 micro score than metapath2vec (25% for F1 macro score) and 220% higher F1 micro score compared to non-pretrained network (471% for F1 macro score).

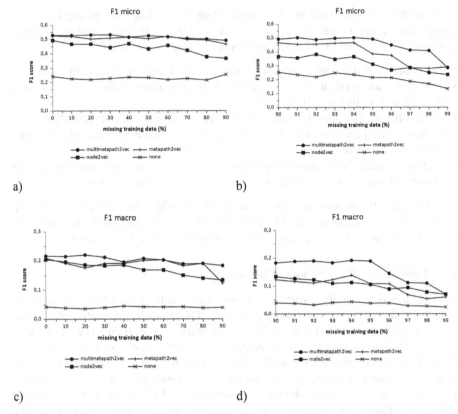

Fig. 3. F1 scores for node classification task. a) F1 micro score for 0%–90% range. b) F1 micro score for 90%–99% range. c) F1 macro score for 0%–90% range. d) F1 macro score for 90%–99% range.

4.3 Case Study: Disease Prediction

In this case, we aim to predict whether a set of symptoms is caused by a specific disease. While it represents a real-life task of diagnosis, we generate artificial samples using the proposed network model. Each sample is modeled as a pair of vectors of symptoms and a single disease. The symptoms are selected from associated symptom nodes. Additionally, we use a parameter that incorporates data incompleteness by removing randomly $\alpha\%$ of nodes (along with associated edges) from the graph. The full algorithm is given below.

Algorithm 3. Algorithm for patient case generation for disease prediction.

Input: Heterogeneous network $G=(V,E,T)$, Node mapping function $\phi(v)$, Cases per disease n, Maximum symptoms per case h, Missing data percentage α

Output: patient cases C

Initialize C to empty
$V'=V$ with $\alpha\%$ nodes removed
$V_d=\{v\in V':\phi(v)=D\}$
$V_s=\{v\in V':\phi(v)=S\}$

for $v\in V_d$ **do**
  for $i=1\rightarrow n$ **do**
    $c=$ GenerateCase(V_s,E,v,h,α)
    Append c to C
  end
end
return C

GenerateCase(V_s,E,v,h,α)
  $S=\{s\in V_s:\langle v,s\rangle\in E\}$
  $c=$ select at most h elements from S
  return c

We generate 10 cases per each disease, using two ranges of α parameter: 0–90% (with 10% step) and 90–99% (with 1% step). Each case contains at most $h=10$ symptoms. Neural networks used for tests are two-layer feed-forward networks with one embedding layer and an output layer with sigmoid activation for one-hot-encoded diseases. Each network is trained with 10 epochs of RMSProp algorithm, with mean squared error loss. A separate dataset is generated for validation, using the whole graph, with 10 cases per disease. The training and validation are repeated 10 times with different datasets and the performance metrics are averaged.

Test results are presented in Fig. 4. Pretrained networks achieve better performance compared to non-pretrained ones, which is particularly visible in cases where relatively few data (0–90%) is missing. For values of $\alpha>90\%$, a benefit of representation learning becomes less significant. Similarly, the advantage of metapath2vec and multi-metapath2vec over node2vec is visible when the training graph is mostly untrimmed (0–20% missing nodes). In this case, we have not observed a significant advantage of multi-metapath2vec over non-modified metapath2vec approach.

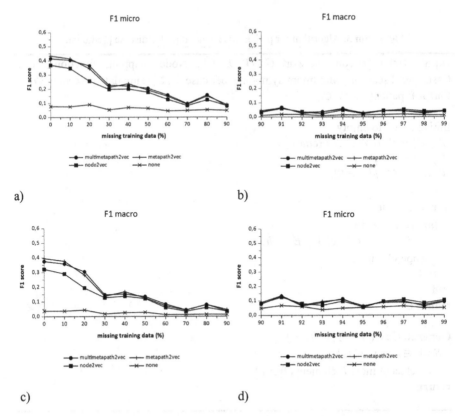

Fig. 4. F1 scores for diagnosis task. a) F1 micro score for 0%–90% range. b) F1 micro score for 90%–99% range. c) F1 macro score for 0%–90% range. d) F1 macro score for 90%–99% range.

5 Conclusions

As indicated by test results, using representation learning improves the performance of neural networks for medical diagnosis-related tasks such as disease/symptom classification and disease prediction. It follows that the knowledge incorporated in the heterogeneous network model can be efficiently learned and utilized in order to improve machine learning methods used in the diagnostic domain to date.

We have shown that the heterogeneous network-based metatpath2vec algorithm improves the final performance of the neural network compared to node2vec. Node2vec requires additional parameters to be specified, that control the influence of structural equivalence and homophily. In most practical cases, relationships between nodes are not exclusively one of them but rather some mixture of both. However, it is difficult to determine the exact proportions of them before the training. On the other hand, in meta-path-based approach, while they are not specified explicitly, the complex relationships can still be incorporated in a form of meta-paths, which are more natural to specify and interpret. Moreover, by allowing to use multiple meta-paths we can avoid traversing unimportant edges that are introduced by the recursion constraint. For certain applications, this results in the better performance compared to unmodified metapath2vec, which is especially visible in case of training data shortage.

References

1. multimetapath2vec. https://github.com/KarolAntczak/multimetapath2vec. Accessed 04 Feb 2020
2. Mikolov, T., Chen, K., Corrado, G., Dean, J.: Efficient estimation of word representations in vector space. arXiv:13013781 Cs (2013)
3. Perozzi, B., Al-Rfou, R., Skiena, S.: DeepWalk: online learning of social representations. In: Proceedings of the 20th ACM SIGKDD International Conference on Knowledge Discovery and Data Mining, KDD 2014, pp. 701–710. ACM Press (2014). https://doi.org/10.1145/2623330.2623732
4. Grover, A., Leskovec, J.: node2vec: scalable feature learning for networks. arXiv:160700653 Cs Stat (2016)
5. Peng, J., Guan, J., Shang, X.: Predicting Parkinson's disease genes based on node2vec and autoencoder. Front. Genet. **10** (2019). https://doi.org/10.3389/fgene.2019.00226
6. Shen, F., et al.: Constructing node embeddings for human phenotype ontology to assist phenotypic similarity measurement. In: 2018 IEEE International Conference on Healthcare Informatics Workshop (ICHI-W), pp. 29–33 (2018). https://doi.org/10.1109/ichi-w.2018.00011
7. Kim, M., Baek, S.H., Song, M.: Relation extraction for biological pathway construction using node2vec. BMC Bioinform. **19**, 206 (2018)
8. Wu, T., et al.: Representation learning of EHR data via graph-based medical entity embedding. arXiv:191002574 Cs Stat (2019)
9. Gao, Z., et al.: edge2vec: representation learning using edge semantics for biomedical knowledge discovery. arXiv:180902269 Cs (2019)
10. Walczak, A., Paczkowski, M.: Medical data preprocessing for increased selectivity of diagnosis. Bio-algorithms Med.-Syst. **12**, 39–43 (2016)
11. Budowa nowoczesnej aplikacji ICT do wsparcia badań naukowych w dziedzinie innowacyjnych metod diagnostyki i leczenia chorób cywilizacyjnych. https://isi.wat.edu.pl/sites/default/files/isi_ver8/proj_POIG.html. Accessed 04 Feb 2020

A Machine Learning Approach for Severe Maternal Morbidity Prediction at Rafael Calvo Clinic in Cartagena-Colombia

Eugenia Arrieta Rodríguez[1]([✉]) [ID], Fernando López-Martínez[2] [ID],
and Juan Carlos Martínez Santos[3] [ID]

[1] Universidad del Sinú Cartagena Elias Bechara Zainum, Cartagena, Colombia
investigacionsistemas@unisinucartagena.edu.co
[2] Department of Computer Science, Oviedo University, Oviedo, Spain
felmco@gmail.com
[3] Universidad Tecnológica de Bolívar, Cartagena, Colombia
jcmartinezs@utb.edu.co
http://www.unisinucartagena.edu.co, http://www.uniovi.es/,
http://www.utb.edu.co

Abstract. There is a huge problem in public health around the world called severe maternal morbidity (SMM). It occurs during pregnancy, delivery, or puerperium. This condition establishes risk for babies and women lives since it's earlier detection isn't easy [8]. In order to respond to such a situation, the current study suggests the use of logistic regression, and supports vector machine to construct a predicting model of risk level of maternal morbidity during pregnancy. Patients for the current study was the pregnant women who received prenatal care at Rafael Calvo Clinic in Cartagena, Colombia and final attention in the same clinic. This study presents the results of two machine learning algorithms, logistic regression and support vector machine. We validated the datasets from the first, second and third quarter of pregnancy with both techniques. The study shows that logistic regression achieves the best results with the prenatal control dataset from the first and second quarter and the support vector machine algorithm achieves the best prediction results with the data set from the third quarter. We generated two datasets using the information of medical records on pregnancy patients at Maternidad Rafael Calvo Clinic. The first dataset contains the six initial months of pregnancy data and the second dataset contains the last quarter of pregnancy data. We trained the first model with logistic regression and the datasets corresponding to the first semester of pregnancy. We obtained a classification of 97% sensibility, 51.8% positive predictive value and F1 score of 67.7%. The support vector machine model was implemented with the datasets obtained from the third quarter of pregnancy. We obtained a classifier with 100% of sensibility, 27.0% of precision.

Keywords: Severe maternal morbidity · Machine learning · Logistic regression · Support Vector Machine

© Springer Nature Switzerland AG 2020
K. Saeed and J. Dvorský (Eds.): CISIM 2020, LNCS 12133, pp. 208–219, 2020.
https://doi.org/10.1007/978-3-030-47679-3_18

1   Introduction

Nowadays more people are using artificial intelligence techniques in different industry areas, especially in the medical field. The most common uses are prediction and classification of diseases, as in the case of this work, which attempts to apply supervised learning techniques to reduce morbidity rates in maternal complication scenarios, more commonly called Severe Maternal Morbidity (SMM) [3,8,11,15,17].

Maternal morbidity summarizes a set of complications that can have severe adverse effects on a woman's health, and that occurs during pregnancy, childbirth, or puerperium; putting at risk the lives of the women and their babies. When it appears, it is necessary to provide immediate attention to the patient to avoid death [14].

Despite advances in maternal health, complications related to pregnancy remain a significant public health problem in the world. Approximately 500,000 women die every year during pregnancy, delivery, or puerperium [6]. There are about 50 million problems in maternal health annually, and approximately 300 million women suffer in the short and long-term from diseases and injuries related to pregnancy, childbirth, and puerperium [14,19].

This condition is complicated to detect at an early stage. In response to the above, this article proposes the use of one of the techniques of machine learning that are considered more relevant in biomedical studies, such as Logistic Regression [7,9,13]. This technique performs a training and learning process to then predicts the level of risk for severe maternal morbidity in patients during pregnancy. The studied population corresponds to pregnant women who received prenatal care and final care at Clínica Rafael Calvo (CMRC) in Cartagena, Colombia.

Several studies worldwide for classification of diseases have shown excellent results when implementing logistic regression and support vector machine [2,10], being the logistic regression the most used and accepted regarding sensitivity and specificity. For this reason, in this study, logistic regression and support vector machine were implemented.

In a previous study called "Early Prediction of Severe Maternal Morbidity Using Machine Learning Techniques" [18], was presented the feature selection by classical statistical techniques. Fairly results with these techniques were obtained. Thus, for this study, we decided to test with a more significant sample data set along with machine learning feature selection techniques [5].

In the previous study, we applied an ANOVA analysis for features elimination. In this study, we used features selection of machine learning to reduce the number of features. Although the selected variables were not the same, the results obtained showed no improvement. We concluded that the problem was not the filtering techniques used, but the limited data set used for the training of the model.

For this work, logistic regression and support vector machine were considered the machine learning techniques [20], implemented with sklearn libraries (Python) [16].

2 Dataset Construction

For the dataset construction, a population of about 1300 patients who fulfilled the criteria of inclusion was selected, and it is shown in Table 1.

Table 1. Inclusion criteria

Criteria	Description
Age	13 to 45 years-old
Prenatal visits	Yes
Controls number	2 or more

Dataset construction was performed with the same inclusions criteria of the previous study as well as the variables collected by each patient.

A sample population of 479 patients was obtained, a retrospective analysis was performed to data contained in the clinical histories of pregnant patients, where between 6,000 to 8,000 cesarean sections are attended per year.

This list of variables collected for the study was provided by the area of surveillance in Public Health and Obstetrics and Gynecology of the research center at CMRC.

Two data sets were created with the collected information. The first data set corresponds to the prenatal controls of the first and second quarter. The second data set corresponds to the prenatal controls of the third quarter. It was necessary to include filtering techniques to the variables due to the enormous amount of cases. This type of technique is being used to reduce the number of variables in the data set to improve the outcome and performance of the classifier. The function RFECV (Recursive Features Elimination with Cross Validation) of the python sklearn library was used for variable elimination *Sklearn* [16].

3 Logistic Regression

Logistic regression notwithstanding its name is a linear classification model instead of a regression model [12].

Logistic regression is a widely linear classification algorithm used in medicine where a sigmoidal function is coupled with a linear regression model [22].

In this work, the logistic regression was applied with cross validation and a regularization component was introduced to control the complexity of the model. In order to evaluate the results of the statistical analysis, it uses the cross validation technique to ensure that the results of each layer or segment are independent between the test data and the training data, consists of analyzing the data in a subset called "training set" and validating the analysis in the other subset called "test set".

In order to solve this optimization problem, the cost function is minimized and an adjustment factor is added that allows controlling the complexity of the model. The regularization consists in reducing the importance of the θ parameters being modified the function of costs by the addition of the sum of all the θ parameters with a factor called parameter of regularization, λ. Getting Eq. 1 as a result [1].

$$min \ -\frac{1}{N} \sum_{i=1}^{N} [y_n \log h_\theta(x) + (1 - y_n) \log(1 - h_\theta(x))] + \lambda ||\theta||^2 \qquad (1)$$

Knowing that N is the number of variables, θ are the parameters of each variable, y is the vector of response (only manages binary values (0,1)), and λ is the parameter of regularization.

3.1 Variable Filtering

we used a recursive feature elimination and cross-validated selection to select best number of features, using 80% of data to train and 20%. Dataset was corresponding to the first and second trimester data of pregnancy were used, called on SUBSET1. The outcome by using this method was that 11 of 60 variables were recommended as predictors or parameters for the classifier. The output of the algorithm is a matrix with false and true values that indicates the 11 selected variables. Table 2 shows the selected variables.

Table 2. Features filtered of 1st y 2nd trimester

Type	Feature
Personal information	Age
Ethnicity	Palequero
Health care regulation	Contributory Regime (CR)
Antecedent	Preeclampsia
Antecedent	Eclampsia
Antecedent	Diabetes
Antecedent	Urinary tract infection
Diagnosis	E20–E35: Disorders of other endocrine glands
Diagnosis	O20–O29: Respiratory tract infection
Diagnosis	N30–N39: Diseases that can affect the fetus
Diagnosis	Z30–Z39: Medical care for reproduction

The same algorithm was applied to the another dataset (third trimester), and this will be called SUBSET2. In this case, the algorithm selected 5 variables as predictors of 59. Which can be seen in Table 3.

Table 3. Features 3rd trimester

Type	Feature
Personal information	Multiparity
Antecedent	Urinary tract infection
Diagnosis	O10–O16: Edema proteinuria and hypertension
Diagnosis	O30–O48: Complications of pregnancy that require attention to the mother
Diagnosis	O60–O75: Complications of pregnancy and delivery

3.2 Training and Validation

An analysis was necessary to be performed to validate the effectiveness of the created data set and verify if this dataset provides a good predictor to predict severe maternal morbidity in its early stages. A cross-validation analysis of the behavior of SUBSET1 and SUBSET2 was performed with logistic regression, and both showed 63% of accuracy. Based in this result we decided to use only SUBSET1 and the third trimester data will not longer be used, only will be use the 223 examples of data set of first and second trimester. In this work was implemented 5-fold cross-validation like see in Fig. 1.

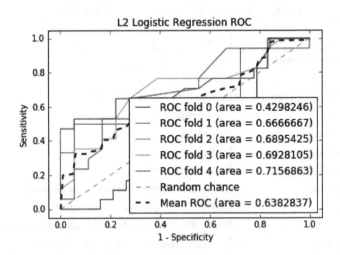

Fig. 1. ROC curve

The dotted line shows the average of the AUC for all folds. The prediction model should detect a mayor number of patients with SMM risk, and for this reason, the selected metrics were sensitivity and precision. DATASET1 contain 223 examples and data analysis was performed only with the 80% of this data with a balanced distribution of the positive and negative class, and 11 variables were used as predictor variables for the classifier (View Table 3).

The result of the prediction can be seen in the confusion matrix shown in Fig. 2. When total of $TP = 85$ (True Positive), this means that 85 patients were classified as SMM and is true that they have SMM. $TN = 12$ (True Negatives), which indicates that 12 patients were classified as no with SMM condition and they were not with SMM condition. $FN = 2$ (False Negatives), this indicates classification errors of patients with SMM condition. Finally, $FP = 79$ (False Positive), patients that were classified with SMM but did not have the condition.

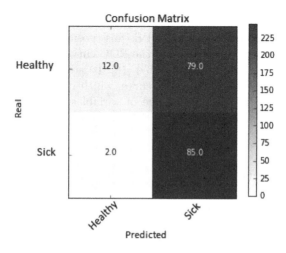

Fig. 2. Confusion matrix 1st y 2nd quarter

The confusion matrix results allow calculate the Precision, Recall and F1 Score. As seen in the Table 4.

Table 4. Results

Metrics	Result
Precision	51.8%
Recall	97.7%
F1 score	67.7%

Precision is defined as the proportion of positive predictions made correctly by the model and denoted by Eq. 2.

$$Precision = \frac{TP}{TP + FP} \tag{2}$$

Recall or Sensitivity that is the ability of the model to predict positive cases that are really sick, like see in Eq. 3.

$$Recall = \frac{TP}{TP + FN} \tag{3}$$

F1 score is a harmonic measure between recall and precision. It is a weighted average between these two measures as indicated in Eq. 4.

$$F1 = 2 * \frac{Precision * Recall}{Precision + Recall} \tag{4}$$

Another way to evaluate the classifier is by calculating the ROC curve as shown in Fig. 1 [4,21]. This is a graphical representation of the sensitivity vs. the specificity for a binary classifier. This can also be expressed as the reason between TPR (true positive rate) and FPR (false positive rate) which is equal to 1 − specificity. Comparing the result of the ROC curve of this work that is 63% and the previous one corresponds to 66% [18]. It is evident that there was no improvement in the terms of the ROC curve, which indicates that the problem is not related to the selection technique of variables. The reason is the small number of prenatal exams or examples that are counted in the first and second trimesters of pregnancy.

4   Support Vector Machine Algorithm

Support Vector Machines (SVM) are one of the methods of supervised learning for two types of classification problems. SVMs work with linearly non-separable problems, and seek to separate the data with a large gap or hyperplane. For the case of problems that are not linearly separable, it is recommended the inclusion of core functions or Kernel which has the effect of mapping the inputs to a high-dimensional space, where the data will be linearly separable. The core function objective is to separate the support vectors from the rest of the training data, this is a quadratic programming problem (QP). See Eqs. 5 and 6.

$$minw, \xi \frac{1}{2}||w||^2 + C \sum_{i=1}^{n} \xi_i \tag{5}$$

subject to

$$y(\boldsymbol{x_i} \cdot w + b) \geq 1 - \xi \tag{6}$$

In this study the core function RBF (Radial Base Function) is used. It is also known as the "exponential" core. The RBF core functions take the form $\exp(-\gamma|x - x'|^2) \cdot \gamma$. Where γ eis a constant of proportionality whose range of useful values must be estimated for each particular application.

4.1   Variable Filtering

We used the same algorithm RFECV, but with linear support vector machine as classifier and the second dataset SUBSET1. As result was eliminated 45 features and finally it is obtain 15 features, as shown Table 5.

Table 5. Features for third and second quarter con technique SVM

Classification	Feature
Personal date	Age
Ethnicity	Raizal
Socio economic	Strata
Marital status	Consensual union couples
Marital status	Widowhood
Gynecological data	Parity
Multiple pregnancy	Multiple
History of	Preeclampsia
History of	Eclampsia
History of	Urinary Tract Infection
ICD-10	I11–I15: Hypertensive diseases
ICD-10	J00–J06: Acute upper respiratory infections
ICD-10	N30–N39: Other diseases of urinary system
ICD-10	O10–O16: Oedema, proteinuria and hypertensive disorders in pregnancy, childbirth and the puerperium
ICD-10	O20–O29: Other maternal disorders predominantly related to pregnancy

Followed by this the feature, the selection algorithm was applied again, but taking 80% of the data set of the SUBSET2. For a total of 59 variables, the algorithm eliminates 45 variables and only 13 are recommended, as shown in Table 6.

4.2   Training and Validation

The result of the SVM classifier is shown in Fig. 3 which is the confusion matrix. In this one, we can see a total of $VP = 95$ (True Positives), this means that 95 patients who really had it were classified with risk for SMM. $VN = 8\,8$ (True Negatives), which indicates that 8 patients were classified as healthy or without risk for SMM and that in reality they did not have SMM. On the other hand, $FN = 0$ (False Negatives) which means that no patient who ended up in SMM was classified without risk for SMM. Finally, false positives $FP = 245$ False Positive), this indicates that 245 patients without risk were classified with risk for SMM.

Table 6. SVM third quarter predictor

Classification	Feature
Ethnicity	Black Race
Insurance	Commercial
Insurance	Self-pay
Insurance	Government
Location	Urban
Location	Rural
Marital status	Preeclampsia
History of	Eclampsia
History of	Urinary Tract Infection
ICD-10	E00–E07: Disorders of Thyroid Gland
ICD-10	O10–O16: Edema, proteinuria and hypertensive disorders in pregnancy, childbirth and the puerperium
ICD-10	O60–O75: Complication of labor and delivery, unspecified
ICD-10	Z30–Z39: Persons encountering health services in circumstances related to reproduction

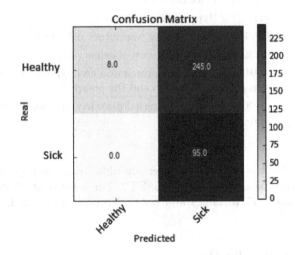

Fig. 3. SVM third quarter confusion matrix

Based on the confusion matrix, the measures that are of interest to evaluate the performance of the SMM classifier, which are the Precision, Recall and measure F1, are calculated. Table 7 shows the results of each one.

The result of 100% sensitivity or recall indicates that all patients with potential SMM risk will be detected by the classifier. A precision of 27% which indicates that there will be many patients who may not develop SMM but the

Table 7. Performance metrics of support vector machine

Metric	Result
Precision	27.9%
Recall	100%
F1	43.6%

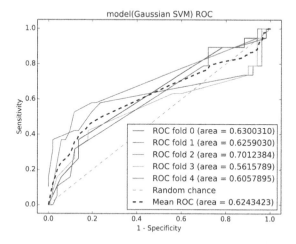

Fig. 4. ROC graph using SVM for the third quarter

predictor will tend to classify them as sick. In the same way as the ROC curve is used in the logistic regression to evaluate the performance, it is also used for support vector machine, the ROC graphic represents sensitivity vs specificity, as shown in the Fig. 4.

5   Conclusion and Future Work

In this research work, we present the use of features selection, logistic regression, and support vector machine applied to the data set obtained from the prenatal controls of the CMRC.

Cross validation and test error were applied to determine the supervised learning technique best suited to the early prediction problem of severe maternal morbidity. The logistics regression technique was selected for the first and second quarter dataset. The result was a classifier with 97% recall, 51.8% precision and 67.7% measurement F. For the third quarter dataset, support vector machine were selected with results of 100% recall and 27% precision.

In the previous study we made an analysis of variance (ANOVA) while in the current one we use recursive features elimination with cross-validation (RFECV) technique. The results in the first work are similar than second, 65% and 63% in ROC graphic respectively. Whereby it is concluded that filtering technique not improvement the performance of the model, but this may be being affected by the no quality of data, then, required apply other techniques of datamining and data cleaning. I Nevertheless, in current study we show logistic regression technique presents more realistic results than the SVM technique in terms of precision and recall.

Recommendations for future work is aimed at improving the collection of information by CMRC, guiding clinical histories and the prevention of variables recommended by the bibliography.

As a result of observation of the data, it is identified that there are few data in the first and second quarters of pregnancy, which shows a culture of not attending prenatal check-ups, so it is recommended that health institutions implement the campaigns to obtain information about the early stages of pregnancy that facilitate the use of artificial intelligence techniques for the early detection of SMM. Finally, this work builds a baseline for future projects in SMM prediction using machine learning techniques.

References

1. Caicedo-Torres, W., Paternina, Á., Pinzón, H.: Machine learning models for early dengue severity prediction. In: Montes-y-Gómez, M., Escalante, H.J., Segura, A., Murillo, J.D. (eds.) IBERAMIA 2016. LNCS (LNAI), vol. 10022, pp. 247–258. Springer, Cham (2016). https://doi.org/10.1007/978-3-319-47955-2_21
2. Calvert, J.S., et al.: A computational approach to early sepsis detection. Comput. Biol. Med. **74**, 69–73 (2016)
3. Farran, B., Channanath, A.M., Behbehani, K., Thanaraj, T.A.: Predictive models to assess risk of type 2 diabetes, hypertension and comorbidity: machine-learning algorithms and validation using national health data from Kuwait—a cohort study. BMJ Open **3**(5), e002457 (2013)
4. Fawcett, T.: ROC graphs: notes and practical considerations for researchers. Mach. Learn. **31**(1), 1–38 (2004)
5. Feizi-Derakhshi, M.R., Ghaemi, M.: Classifying different feature selection algorithms based on the search strategies. In: International Conference on Machine Learning, Electrical and Mechanical Engineering (2014)
6. Haaga, J.G., Wasserheit, J.N., Tsui, A.O., et al.: Reproductive Health in Developing Countries: Expanding Dimensions, Building Solutions. National Academies Press, Washington, D.C. (1997)
7. Huang, G.B., Zhu, Q.Y., Siew, C.K.: Extreme learning machine: theory and applications. Neurocomputing **70**(1–3), 489–501 (2006)
8. Jahan, S., Begum, K., Shaheen, N., Khandokar, M.: Near-miss/severe acute maternal morbidity (SAMM): a new concept in maternal care. J. Bangladesh Coll. Phys. Surg. **24**(1), 29–33 (2006)
9. Lorduy Gómez, J., Carrillo González, S., Muñoz Baldiris, R.E., Díaz-Pérez, A., Perez, I.: Prognostic factors of early neonatal sepsis in the city of Cartagena Colombia (2018)
10. Mani, S., et al.: Medical decision support using machine learning for early detection of late-onset neonatal sepsis. J. Am. Med. Inform. Assoc. **21**(2), 326–336 (2014)
11. Nanda, S., Savvidou, M., Syngelaki, A., Akolekar, R., Nicolaides, K.H.: Prediction of gestational diabetes mellitus by maternal factors and biomarkers at 11 to 13 weeks. Prenat. Diagn. **31**(2), 135–141 (2011)

12. Ng, A.: Machine learning: Stanford machine learning course materials
13. Nilashi, M., bin Ibrahim, O., Ahmadi, H., Shahmoradi, L.: An analytical method for diseases prediction using machine learning techniques. Comput. Chem. Eng. **106**, 212–223 (2017)
14. World Health Organization, UNICEF: Revised 1990 estimates of maternal mortality: a new approach. World Health Organization (1996)
15. Park, F.J., Leung, C.H., Poon, L.C., Williams, P.F., Rothwell, S.J., Hyett, J.A.: Clinical evaluation of a first trimester algorithm predicting the risk of hypertensive disease of pregnancy. Aust. N. Z. J. Obstet. Gynaecol. **53**(6), 532–539 (2013)
16. Pedregosa, F., et al.: Scikit-learn: machine learning in python. J. Mach. Learn. Res. **12**, 2825–2830 (2011)
17. Poon, L.C., Kametas, N.A., Maiz, N., Akolekar, R., Nicolaides, K.H.: First-trimester prediction of hypertensive disorders in pregnancy. Hypertension **53**(5), 812–818 (2009)
18. Rodríguez, E.A., Estrada, F.E., Torres, W.C., Santos, J.C.M.: Early prediction of severe maternal morbidity using machine learning techniques. In: Montes-y-Gómez, M., Escalante, H.J., Segura, A., Murillo, J.D. (eds.) IBERAMIA 2016. LNCS (LNAI), vol. 10022, pp. 259–270. Springer, Cham (2016). https://doi.org/10.1007/978-3-319-47955-2_22
19. Tsui, A.O., Wasserheit, J.N., Haaga, J.G., et al.: Healthy pregnancy and childbearing (1997)
20. Witten, I.H., Frank, E., Hall, M.A., Pal, C.J.: Data Mining: Practical Machine Learning Tools and Techniques. Morgan Kaufmann, Burlington (2016)
21. Yang, Z., Zhang, T., Lu, J., Zhang, D., Kalui, D.: Optimizing area under the ROC curve via extreme learning machines. Knowl.-Based Syst. **130**, 74–89 (2017)
22. Zheng, Z., Li, Y., Cai, Y.: The logistic regression analysis on risk factors of hypertension among peasants in east china & its results validating. Int. J. Comput. Sci. Issues (IJCSI) **10**(2 Part 1), 416 (2013)

Collaborative Data Acquisition and Learning Support

Tomasz Boiński$^{(\boxtimes)}$ and Julian Szymański

Faculty of ETI, Gdańsk University of Technology, Gdańsk, Poland
{tomasz.boinski,julian.szymanski}@eti.pg.edu.pl

Abstract. With the constant development of neural networks, traditional algorithms relying on data structures lose their significance as more and more solutions are using AI rather than traditional algorithms. This in turn requires a lot of correctly annotated and informative data samples. In this paper, we propose a crowdsourcing based approach for data acquisition and tagging with support for Active Learning where the system acts as an oracle and repository of training samples. The paper presents the CenHive system implementing the proposed approach. Three different usage scenarios are presented that were used to verify the proposed approach.

Keywords: Data annotation · Annotation verification · Active Learning

1 Introduction

With the constant development of neural networks, traditional algorithms relying on data structures lose their significance. More and more problems, even those solved using traditional algorithms, due to the sheer size of the problem space, rely on deep neural networks. Teaching neural networks in turn rely heavily on data acquisition, tagging and the ability to increase the network quality rather than pure algorithms creation and optimization.

The difficulty of data acquisition itself varies depending on the domain and usually cannot be easily automated. Furthermore without annotation the data is usually useless. The process of data annotation or tagging can be time-consuming and requires human effort. For example we can quite easily record bees entering and exiting a hive. Tagging the images correctly marking all the bees in the pictures is however a very difficult task. This can be done correctly either by a very refined neural network (which had to be prepared beforehand and required an annotated set of training for the learning process) or a team of humans. In some cases the humans involved have to be experts in the field.

In this paper, we propose a crowdsourcing based approach for data acquisition and tagging with support for Active Learning [8,11] where the system acts as an oracle and repository of training samples. The training samples, as they can be

K. Saeed and J. Dvorský (Eds.): CISIM 2020, LNCS 12133, pp. 220–229, 2020.
https://doi.org/10.1007/978-3-030-47679-3_19

the neural network output verified by humans, can serve as correctional samples that were previously wrongly classified by the network.

The structure of this paper is as follows. In Sect. 2 available approaches to data acquisition are presented. Next, in Sect. 3 architecture of the proposed system for data acquisition and verification is described. Later on in Sect. 4 usage of the proposed system is presented in terms of data acquisition and verification. Section 5 discusses system usage for Active Learning support and finally a summary is given.

2 Data Acquisition and Annotation Approaches

Crowdsourcing-based data acquisition and annotation are well established. Companies, like VoiceLab[1], often rely on crowdsourcing for gathering real-life data (e.g. voice recordings of an average human rather than a professional speaker). For such solutions, dedicated systems are created. Another example of such approach is Snapshot Serengeti[2] project [12], where users can browse pictures of the park's wildlife captured by automated cameras and describe the contents of the images in terms of the number and type of animals visible. The project was created to help track the migration habits of the park animals. This solution is based purely on the will of the community to help in solving the problem, the only reward is the possibility of watching the wildlife. Other solutions like Duolingo[3] [5] aim at providing commercial data transformation. In this case the users, while learning the foreign language of their choice, perform translation of texts that are further used by news agencies.

One of the first and the most general examples of crowdsourcing data acquisition is, however, Amazon Mechanical Turk[4] [3]. This platform allows the creation of tasks that can vary from simple data annotation (e.g. tags creation or verification) to data creation (e.g. review writing, providing voice recordings etc.). The platform allows giving the users some small financial gratitude for every task solved. The money is paid by the problem supplier. Other solutions, like Google reCaptcha [16] or Foldit [2] are proof of the viability of this approach, especially that all aforementioned systems proved to be successful.

Currently, available solutions are usually very specific. They are used to solve a one, well-defined problem, provide one, dedicated interface or are commercial solutions, usually designed for certain companies.

During our works we designed and implemented a universal service called CenHive[5], that aims at providing a general solution allowing acquisition and annotation of different data types, like any combination of text, images, audio and video. Furthermore the data can be easily distributed to the users for verification through multiple clients. The solution allows also usage of the data acquired for active learning through simple REST interfaces.

---

[1] https://www.voicelab.ai/.
[2] https://www.snapshotserengeti.org/.
[3] https://en.duolingo.com/.
[4] https://www.mturk.com/mturk/welcome.
[5] https://kask.eti.pg.gda.pl/cenhive/.

3   CenHive Architecture

CenHive is a web-based solution. Originally it was designed as a service dedicated to Wikipedia-WordNet mappings verification [1, 7, 13] using crowdsourcing approach via Games with a Purpose [15]. It supported only one client and only text questions with a fixed amount of answers.

Currently the system has been extended considerably. The current version allows multiple clients and data types. It supports not only text data but audio, video, images and any combination of aforementioned input types both in questions and answers. The amount of answers sent to the client is also configurable and the client, instead of an answer to a question presented, can create a new set of data for verification.

The main part of the system is a database containing the data aggregated into so-called *contents*. It can be a form of mappings (e.g. Wikipedia – WordNet mappings), tags (e.g. pictures with tags describing the content or images with yes/no tags answering *contents* questions, like "Is this a human face?" or "Is this a bee?") or plain data that can be used for detailed data acquisition (like crowd or beehive pictures, camera footage, etc.). The tags or descriptions associated with given *content* are called *phrase*.

The *contents* elements are distributed to CenHive clients with client-defined number of randomly selected *phrases* associated with given *content* using RESTful [10] API calls.

The system supports multiple clients that can use any technology. Currently, the system has 4 official clients:

- TGame[6] – Android platform game originally designed for Wikipedia – WordNet mappings verification, currently supporting audio, video, text and images as elements within the *contents* and *phrases* parts. The player has to answer provided questions to activate checkpoints in the game;
- Truth or Bunk[7] – web-based client designed for Wikipedia – WordNet mappings – it takes a form of a quiz game where players have to answer as many questions as possible;
- 2048 clone[8] – an extension of the original 2048 game[9]. The player, during the game, donates some of his or her computer computation power to create new data (in this case finding human faces on photos provided from CenHive) in volunteer computing mode and can also verify the detected faces using crowdsourcing model when he or she wants to undo a move;
- captcha element – used for securing user login and registration page, can provide a set of questions based on *content* elements with tags stored as *phrases* in the system.

There are 2 other clients in the works that will support general image tagging and will help to produce data for teaching neural networks using an active

---

[6]   https://play.google.com/store/apps/details?id=pl.gda.eti.kask.tgame.
[7]   https://kask.eti.pg.gda.pl/cenhive/tob/.
[8]   https://kask.eti.pg.gda.pl/cenhive/2048/.
[9]   http://git.io/2048.

learning approach. Both aim at allowing users verification of bee detection done by a neural network and the ability to tag bees on random images.

4   Data Acquisition and Verification

The main usage of the system is data verification. CenHive was created as crowd-sourcing based solution for Wikipedia – WordNet mappings verification and was later extended for general relations verification. In this case a mapping can be treated as a relation, where a Wikipedia article is a description for a Word-Net synset. Relations can thus take any form – keyword-based tags, mappings, yes/no answers, image to text mappings, etc. Currently we are aiming at automatic usage of gathered data for deep neural network learning. In all cases the data can be in any form supported by CenHive.

4.1   Wikipedia-WordNet Mappings Verification

The system was used for assessing and verification of mappings between Word-Net synsets and Wikipedia articles. The task was done using a 2D platform game called TGame [1,13] ("Tagger Game"). It followed the output-agreement model [15]. We aimed at introducing higher replayability by providing a trophy system based on the player in-game performance; both in the form of collectibles gathering, like coins or hearts, and in the number of tasks solved.

In TGame activating a checkpoint requires reaching it and answering the question provided by CenHive, which in turn is a mapping verification. The checkpoint is activated when the user answer (the mapping) is similar to the one in CenHive. All the answers are logged so if the players will mark other answers the administrator can correct the answer. The players also have a chance of reporting the questions thus strongly indicating wrong mapping. We asked for mappings using two approaches (Fig. 1):

- we extended the automatically generated mappings with other, false mappings based on the Wikipedia search functionality, the users should choose the correct mapping;
- we presented only the mapping from the database, the user should choose whether it is correct or not.

During two months-long test period, players gave 3731 answers in total to 338 questions. The total number of answers for different mapping types is shown in Table 1.

The tests proved that the proposed solution is useful in mappings verification. During the evaluation, however, we had to deal with malicious users. As in the crowdsourcing we cannot verify the user we decided that only questions that had multiple answers will be taken into consideration. Furthermore to eliminate blind shots we decided to take into account only questions that the user had displayed for at least 5 s. Such time proved to be enough for the user to actually read the question and the answers presented. Fortunately the number of answers that could be considered blind shots is very low, in worst case it did not go above 3% and across all tests the ratio of blind shots was no greater than 1.56%.

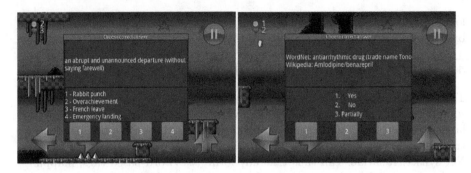

Fig. 1. TGame user interface (left: extended mappings, right: yes/no question).

Table 1. Number of answers for different type of questions

Question type	Questions	Answers	Answers per question	Reports	Blind shots	Blind shots %
Extended mappings	239	3,308	13.84	625	16	0.48
Yes/No	99	423	4.27	10	12	2.84
Total	338	3731	11.04	685	62	1.34

4.2 Face Recognition and Verification

The first major extension introduced into CenHive was the ability to not only verify already gathered data but to create the data itself. This can be done twofold – the users can either generate data manually, e.g. by mapping the images with a dedicated client, or provide the computation power in the volunteer computing model. During this stage we decided to use the second approach. For that we created a dedicated client in the form of a 2048 game clone. It is available as a web application, where, during the gameplay, the photos are downloaded from the CenHive server. The photos are then analyzed using Viola-Jones Feature Detection Algorithm using Haar Cascades [14] algorithm (more specifically using HAAR.js implementation[10]). Detected faces are then sent back to CenHive and stored as tags where the original image is tagged with coordinates of the rectangle containing the detected face (Fig. 2).

During the test period, where tests were done with the help of the research team and students, for 64 multi-person photos the game players generated 628 face-detects (giving 303 distinct face detects). In this case multiple face detects are not necessary as they are done arithmetically using the player's device processing power. The detected faces were further verified – for 92 images we got 181 verification responses and 7 reports.

---

[10] https://github.com/foo123/HAAR.js.

Fig. 2. Detected faces.

5   Active Learning Support

Traditionally when performing a deep network learning for detection problems a big learning set is needed. The data for teaching is randomly selected, which can lead to longer teaching times as newly selected samples can contain partial or even no new information. This leads to a small or even lack of increase in network quality.

This problem can be mitigated with the Active Learning approach [4,8,11]. In this approach the network should choose whether the sample is informative enough for the sample to be a part of the teaching set.

In our approach CenHive is treated as an oracle that should be able to answer the question of whether the given sample is informative or not. In our test case (bee detection on video streams) we consider a sample informative when the network wrongly detects bees on the given image.

For each step of Active Learning we select samples based on the following criteria:

- For each sample, detection is done using the network from the previous iteration, only tags with the detection certainty over given value were taken into account,
- Tagging done by the network is compared to the ones from the oracle
  - If the number of objects detected by the network differs from the number of objects returned by the oracle the sample is added to the training set,
  - If the number of detected objects is the same, for each object we calculate Intersection over Union (IoU) value, meaning how well the area of detected objects covers the area of objects returned by the oracle. If it is below the predefined threshold we add the sample to the training set.

In this case the oracle works as described in the previous chapter. The CenHive system distributes neural network tagged images to the users for verification. The implemented clients also allow manual image tagging allowing thus the creation of new training data. Using friends and family approach we managed to tag 2500 photos, what after splitting the images to the size appropriate for the network gave 60423 samples.

To test the viability of the proposed solution we performed Active Learning on Faster R-CNN [9] network implemented using Detectron[11] [6] system. The network was first trained using 12000 random pictures during 13 epochs. During the tests we considered two parameters: IoU value over 70% and over 80% and minimal certainty score over 30% and over 60%. Three models were thus trained:

- min IoU 70%, min score 30% – the image was considered correctly detected by the network when IoU had value at least 70% and certainty score was over 30%,
- min IoU 80%, min score 30% – the image was considered correctly detected by the network when IoU had value at least 80% and certainty score was over 30%,
- min IoU 80%, min score 60% – the image was considered correctly detected by the network when IoU had value at least 80% and certainty score was over 60%.

The images tagged by the network were then verified using CenHive and the network was trained again for 12 iterations with the extended training set. For each iteration up to 1000 new images verified using CenHive were added to the training set. During each iteration The aforementioned approach showed constant network quality upgrade, however with each new data set from CenHive the training set grew and thus the training time increased. We then tested the approach with limited usage of the old training set. Altogether we used three approaches to introduce new samples:

- Full training set – the training set for given iteration consisted of new samples and all previously used samples,
- Only new data – the training set for given iteration consisted of only the new images verified through CenHive,
- New data with memory – the training set contained new images and 10000 randomly selected images from previous learning sessions.

The results are presented in Fig. 3. Using only new samples did not provide good results. However when the size of the training set was limited to new samples and 10000 randomly selected previously used the network quality was nearly as good as when the training was done with all the samples. The training time, however, was much smaller, as can be seen in Fig. 4. The data acquired proves that the proposed approach is viable to serve as an oracle in the Active Learning approach.

---

[11] https://github.com/facebookresearch/Detectron.

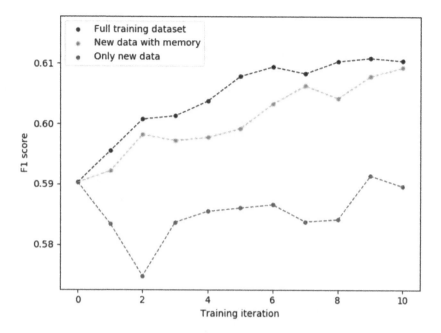

Fig. 3. Network quality.

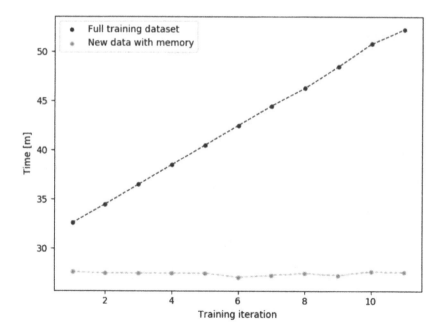

Fig. 4. Training times.

6   Summary and Future Work

We believe that with each day we are more justified to state, that modern solutions are less algorithms and more annotated data. More and more modern problems cannot be easily solved using traditional approaches and require the usage of deep neural networks which in turn require a high amount of data for a precise learning process.

The paper presents a crowdsourcing based approach to data gathering, annotation and verification. The approach also allows automatic usage of gathered data for Active Learning where the implemented system acts as an oracle.

We tested the approach for different data types and tasks. The approach proved to be viable in data gathering, annotation and verification. We also showed that it can be used effectively in the Active Learning as an oracle that determines whether the given sample is informative and should be used in the learning process or not.

The drawback of the solution is that it relies on heavy community involvement. This can be however achieved by embedding the problems in so-called Games with a Purpose. We implemented a few such clients and performed preliminary tests proving such approach viable.

In the future we plan on extending the solution with more and better-suited clients. We also plan on using captcha element in real-life systems that will guarantee a steady stream of solutions provided by the users of such systems. This will allow us optimization of the whole verification process generating more training data for deep learning.

Acknowledgements. We would like to thank Agata Krauzewicz and Łukasz Lepek who implemented part of the presented solution during their studies.

References

1. Boiński, T.: Game with a purpose for mappings verification. In: 2016 Federated Conference on Computer Science and Information Systems (FedCSIS), pp. 405–409. IEEE (2016)
2. Curtis, V.: Motivation to participate in an online citizen science game: a study of foldit. Sci. Commun. **37**(6), 723–746 (2015)
3. Fort, K., Adda, G., Cohen, K.B.: Amazon mechanical turk: gold mine or coal mine? Comput. Linguist. **37**(2), 413–420 (2011)
4. Gal, Y., Islam, R., Ghahramani, Z.: Deep bayesian active learning with image data. In: Proceedings of the 34th International Conference on Machine Learning-Volume 70, pp. 1183–1192. JMLR. org (2017)
5. Garcia, I.: Learning a language for free while translating the web. Does Duolingo work? Int. J. Engl. Linguist. **3**(1), 19 (2013)
6. Girshick, R., Radosavovic, I., Gkioxari, G., Dollár, P., He, K.: Detectron (2018). https://github.com/facebookresearch/detectron
7. Jagoda, J., Boiński, T.: Assessing word difficulty for quiz-like game. In: Szymański, J., Velegrakis, Y. (eds.) IKC 2017. LNCS, vol. 10546, pp. 70–79. Springer, Cham (2018). https://doi.org/10.1007/978-3-319-74497-1_7

8. Kellenberger, B., Marcos, D., Lobry, S., Tuia, D.: Half a percent of labels is enough: efficient animal detection in uav imagery using deep cnns and active learning. IEEE Trans. Geosci. Remote Sens. **57**(12), 9524–9533 (2019)

9. Ren, S., He, K., Girshick, R., Sun, J.: Faster R-CNN: towards real-time object detection with region proposal networks. In: Advances in Neural Information Processing Systems, pp. 91–99 (2015)

10. Richardson, L., Ruby, S.: RESTful Web Services. O'Reilly Media, Inc., Newton (2008)

11. Settles, B.: Active learning literature survey. University of Wisconsin-Madison, Department of Computer Sciences, Technical report (2009)

12. Swanson, A., Kosmala, M., Lintott, C., Simpson, R., Smith, A., Packer, C.: Snapshot serengeti, high-frequency annotated camera trap images of 40 mammalian species in an african savanna. Sci. Data **2**, 150026 (2015)

13. Szymański, J., Boiński, T.: Crowdsourcing-based evaluation of automatic references between wordnet and wikipedia. Int. J. Softw. Eng. Knowl. Eng. **29**(03), 317–344 (2019). https://doi.org/10.1142/s0218194019500141

14. Viola, P., Jones, M.: Rapid object detection using a boosted cascade of simple features. In: Proceedings of the 2001 IEEE Computer Society Conference on Computer Vision and Pattern Recognition, CVPR 2001, vol. 1, p. I. IEEE (2001)

15. Von Ahn, L.: Games with a purpose. Computer **39**(6), 92–94 (2006)

16. Von Ahn, L., Maurer, B., McMillen, C., Abraham, D., Blum, M.: reCAPTCHA: Human-based character recognition via web security measures. Science **321**(5895), 1465–1468 (2008)

Benchmarking Deep Neural Network Training Using Multi- and Many-Core Processors

Klaudia Jabłońska and Paweł Czarnul[(✉)](iD)

Faculty of Electronics, Telecommunications and Informatics,
Gdańsk University of Technology, Narutowicza 11/12, 80-233 Gdańsk, Poland
klajablo@gmail.com, pczarnul@eti.pg.edu.pl

Abstract. In the paper we provide thorough benchmarking of deep neural network (DNN) training on modern multi- and many-core Intel processors in order to assess performance differences for various deep learning as well as parallel computing parameters. We present performance of DNN training for Alexnet, Googlenet, Googlenet_v2 as well as Resnet_50 for various engines used by the deep learning framework, for various batch sizes. Furthermore, we measured results for various numbers of threads with ranges depending on a given processor(s) as well as compact and scatter affinities. Based on results we formulate conclusions with respect to optimal parameters and relative performances which can serve as hints for researchers training similar networks using modern processors.

Keywords: Deep neural network training · Benchmarking · Parallel computations · Caffe · MKL

1 Introduction

Deep neural networks have become an important method for modeling various hard to describe phenomena and subsequent usage of the models for making out for other data sets. As an example, such networks allow for detection of object classes, prediction of trends, modeling complex functions etc. While trained networks can be used for inference using less powerful devices, training, using typically data sets of very large sizes, is computationally very demanding and requires high performance computing systems. Parallelization of deep neural network training is a key challenge nowadays. Consequently, finding powerful and cost-effective compute devices is of paramount importance. In this paper, we investigate performance of various many- and multi-core CPUs for assessment of deep neural network training. This research is complementary to the typically analyzed training using GPUs. One of the goals of the paper is to investigate relative performance of multi- and many-core CPUs for deep learning frameworks. Similar research has already been performed for other applications, for example: parallel chess search [1], parallel computation of similarity measures between large vectors [2], Fast Fourier Transform (FFT), numerical integration and heat distribution [3].

© Springer Nature Switzerland AG 2020
K. Saeed and J. Dvorský (Eds.): CISIM 2020, LNCS 12133, pp. 230–242, 2020.
https://doi.org/10.1007/978-3-030-47679-3_20

2   Related Work

There are several works available that aim at comparative studies of various deep learning frameworks run on various hardware configurations. As an example, paper [4] includes a comparison of Caffe, the Microsoft Cognitive Toolkit (CNTK), MXNet, TensorFlow, and Torch. Preliminary tests were performed for both synthetic as well as real data sets for networks such as: feed-forward neural networks (FCNs), convolutional neural networks (CNNs) AlexNet and ResNet-50 as well as FCN, AlexNet, ResNet-56 and recurrent neural networks (RNNs) long short-term memory (LSTM) respectively. Successive tests were performed for various combinations of the aforementioned networks, CPUs and GPUs. Multi-GPU tests (2, 4 GPUs) were performed for FCN, AlexNet and ResNet-56 networks. CPU thread scaling showed benefits, albeit up to 4 or 8 threads on the i7 and up to 16 or 32 threads on the E5s, depending on the network. Scaling up to 4 GPUs gave best results in practically all cases. For tests performed on E5, TensorFlow is best for AlexNet, CNTK for FCN real, Torch for FCN synthetic, Torch for ResNet and CNTK for LSTM. In this context, the contribution of our paper is similar but focuses more on various CPU types, not only desktop (i7-3820) and server Xeon (E5-2630x2) benchmarked in the aforementioned paper. We also investigate multi- vs many-core CPUs for deep learning.

For CPUs, paper [5] shows the importance of proper thread affinity and data mapping in machine learning algorithms, tested on Intel® Xeon® X7550 (Nehalem) as well as Intel Xeon Phi 7250 Knights Landing (KNL). The authors tested machine learning applications from the Rodinia Benchmark Suite: backprop, kmeans, knn and sc. They have concluded that especially the scatter affinity and interleaved data mapping can reduce execution times by up to 25.2% and 18.5% for Xeon and Xeon Phi, respectively, albeit depending to a degree on the application. Specifically, largest savings were obtained for backprop for Xeon and knn for Xeon Phi.

As for actual applications, some works investigate performance of running deep learning models on actual users' devices. For instance, paper [6] discusses RSTensorFlow extending the TensorFlow framework and investigates performance of various CPU and GPU models available on phones running Android. The authors investigate key operations for the inference run using TensorFlow identifying two key operations: convolution and matrix multiplication. The modified implementation of TensorFlow can run these operations in parallel through RenderScript. The authors have demonstrated improvements in running matrix multiplication using a GPU on Nexus 5X but generally worse than the original TensorFlow results on Nexus 6. Forward pass of the Inception model run with the proposed solution for matrix multiplication brings improvements for the two phones. The same applies for LSTM tests for batch size equal 64.

From the perspective of this paper i.e. focus on CPU-based DNN training, paper [7] provides excellent and very interesting background. That work takes a look at the whole multilayered software stack i.e. from top to bottom deep learning applications, framework, Basic Linear Algebra Subprograms (BLAS) libraries that, next to ATLAS and OpenBLAS, involve Intel® Math Kernel Library (MKL) and cuDNN/cuBLAS and finally hardware like GPUs/CPUs. The paper presents performance results for training AlexNet and ResNet-50 using various stacks and compute devices including NVIDIA P100, K40, K80, as well as CPUs of generations such as KNL, Broadwell, Haswell. The paper concludes that the optimized version using KNL (using MKL 2017) can even match P100 (using cuDNN/cuBLAS) for training ResNet-50 while being slightly slower for AlexNet. Various Caffe variants were used. This means that CPU-based stacks, when configured properly, can be interesting. Performance scaling vs the number of nodes is presented, better for OSU-Caffe (GPU) compared to Intel-Caffe (CPU).

In line with other works proposing sets of representative benchmarks concerning AI using deep neural networks, paper [8] describes DNNMark, a highly configurable GPU benchmark package that includes a set of deep neural network primitives, covering a rich set of GPU computing patterns such as: convolution, pooling, local response normalization, activation, fully connected layer.

Following this trend and addressing portability, paper [9] proposes a DNN benchmark framework that is able to run on any platform supporting OpenCL and CUDA. Tests within the benchmark include: Convolutional Neural Networks (CNN): CifarNet, AlexNet, ResNet, SqueezeNet, VGGNet, Recurrent Neural Networks (RNN): Long Short Time Memory (LSTM), Gated Recurrent Unit (GRU).

Paper [10] demonstrates that selected FPGA solutions can be very interesting not only in terms of performance/Watt but also pure performance against even GPUs. As an example, the paper shows better performance and performance-Watt estimates of Stratix 10 FPGA 750 MHz compared to NVIDIA Titan X for ResNet-50 results and ImageNet problem size. Comparative results of FPGA performance and performance/Watt against base results using Intel Xeon E3 CPU and NVIDIA GTX 650 Ti can be found in [11]. Performance/Watt values are always better compared to the CPU, mostly better compared to the GPU. Performance is worse using the Zynq board compared to both CPU and GPU, better compared to CPU for Stratix and worse or better depending on a neural network model compared to the GPU and mostly better using Arria.

Work [12] focuses on performance investigation of inference using deep neural networks using CPUs and GPUs in mobile devices. It demonstrates that acceleration of inference using GPU can provide benefits over using mobile CPUs, in terms of both performance and power consumption, assuming proper support from the software layer used. Such support needs to be extended in the existing deep neural network frameworks. Performance of smartphones for performing AI related tasks using deep neural networks can be assessed using AI Benchmark[1]. Study of several devices using this benchmark is available in [13].

Paper [14] deals with performance of processing using deep neural networks using CPUs. It demonstrates that using optimizations such those concerning

---

[1] http://ai-benchmark.com/.

data layout, SSE instructions including fixed point instructions it is possible to obtain a speed-up of 3x compared to original floating point version.

A recent paper [15] presents a cross platform comparison of TPUs, GPUs and CPUs for fully connected (FC), convolutional neural networks (CNN), and recurrent neural networks (RNN). Several insights have been provided based on experiments. For instance, the authors have concluded that large FC models benefit from CPUs due to memory constraints, for large batch sizes TPUs and for small batch sizes GPUs are preferred, TPUs benefit more and more over GPUs for larger CNNs.

3   Benchmarking Methodology

A general idea behind this research was to find optimal configurations for training deep neural network using a selected framework. An optimal configuration is thought of as a recommended set of parameters and hardware. For meaningful and comparable results all tests were executed on selected framework and one simple application. Additionally, all tests were run on several processors, with a set of parameters that was changed in search of an optimal solution.

3.1   Framework

The framework used for the needs of this benchmarking was Intel Distribution of Caffe – based on the open source Caffe framework for deep learning. It was extended by Intel for better performance results when running computations on CPU, especially Intel Xeon processors and Intel Xeon Phi processors. The name CAFFE of the original framework comes from Convolutional Architecture for Fast Feature Embedding. It is especially dedicated to image classification.

The Intel Distribution of Caffe framework was chosen for several reasons. First of all, it was optimized for the Intel Xeon and Intel Xeon Phi processors by the Intel itself. It means that it can use dedicated libraries for low level mathematical operations on these hardware, like the Intel MKL library. Therefore, we can omit the influence of low-level mathematical operations and assume that the performance depends solely on the deep learning framework, deep learning parameters and parallel computation parameters. Secondly, this framework allows for defining most of the parameters through json configuration files without interfering the network structure for each parameter.

3.2   Parameters

We can distinguish two types of parameters used in this benchmark.

Deep learning parameters include:

– network topology – each network topology has a different level of complexity; usually networks that are more complex take more time but provide results that are more accurate,

– data type – randomly generated or real objects,
– engine – indicates which underlying library for math operations is used,
– batch size.

Parallel computation parameters include: number of 'cpus', number of threads - OMP_NUM_THREADS', affinity of threads - KMP_AFFINITY.

4   Experiments and Results

4.1   Testbed Environments and Libraries

All tests were executed using platforms from the Intel vLab Machine Learning cluster. The Intel vLab Machine Learning cluster provides an environment to support and advance academic research in Machine Learning and Deep Learning. Tests were run on a single node, used exclusively. Table 1 presents a list of all types of processors used in the benchmark containing name and basic parameters like number of sockets, cores or threads. For tests, we used Intel Math Kernel Library (Intel MKL) 2017 Beta update 1 and mkl-dnn v0.11, (mklml_lnx_2018.0.1.20171007).

Table 1. Tested processors

Parameter/processor name	Knights Landing (KNL)	Knights Mill (KNM)	Skylake Xeon (SKL)
Logical processors	272	288	112
Thread(s) per core	4	4	2
Socket(s)	1	1	2
Core(s) per socket	68	72	28
Model name	Intel(R) Xeon Phi(TM) CPU 7250 @ 1.4 GHz	Intel(R) Xeon Phi(TM) CPU 7295 @ 1.50 GHz	Intel(R) Xeon(R) Platinum 8180 CPU @ 2.50 GHz
L1d cache	32K	32K	32K
L1i cache	32K	32K	32K
L2 cache	1024K	1024K	1024K
TDP	215W	320W	205W

4.2   Results

Default Configuration on all Processors. The following default configuration test will act as a basic test and a reference point. The default configuration is a set of parameters values either set in the script (when a user does not specify a value themselves) or commonly used values. Table 2 shows the default configuration of parameters. Some of these are hardware-dependent and thus they will differ for each of the processors used.

Table 2. Default configuration parameters

Number of 'cpus'	Maximum possible
OMP_NUM_THREADS	Equal to number of all cores
KMP_AFFINITY	granularity = fine, compact, 1, 0;
Data	Random, <batch_size> × 3 × 227 × 227
Batch size	256
Engine	MKLDNN

Tests with this configuration were run on all network topologies and all processors used in this benchmark. Results of this default testing can be found in Table 3.

Table 3. Default configuration results – images per second

	SKL	KNL	KNM
Alexnet	1046	458	978
Googlenet	334	143	236
Googlenet_v2	261	107	160
Resnet_50	127	50	57

In terms of network topologies, we can see that Alexnet is always the quickest network. Then comes Googlenet, Googlenet_v2 and Resnet_50, always in the same order. This is a proper behavior if we take into account complexities of those models. This order of networks corresponds to the complexity of network, which considers the number of hidden layers and total number of operations. That is why Alexnet, the simplest network, is always the fastest. However, it might not satisfy us with its accuracy. On the other hand, the Resnet_50 network is a 50 layers Residual Network, making it a more sophisticated model.

Comparing results between processors, we can observe that for this configuration, Skylake Xeon (SKL) is generally the fastest and Knights Landing (KNL) is the slowest. While proportions of speed between topologies are very similar for those two, the difference in their speed is almost double. For Knights Mill (KNM), however, it looks slightly different. For a simple model – Alexnet – it was almost as fast as the Skylake processor. However, with each following topology, the efficiency of this processor drops down.

Engine Comparison for Real Data. Tables 4, 5 and 6 show results for the tested processors, for real life images. All other parameters besides 'engine' stayed the same as in default configuration. ImageNet (ILSVRC2012) was used as input dataset. The performance strongly depends on the engine, with MKLDNN achieving best results.

Experiments with Various Batch Sizes. We have performed experiments with various batch sizes for all four networks – Alexnet, Googlenet, Googlenet_v2 and Resnet_50, for which results are presented in Tables 7, 8 and 9 for Skylake, Knights Landing and Knights Mill processors respectively.

Table 4. Real data on Skylake results – images per second

	Alexnet	Googlenet	Googlenet_v2	Resnet_50
CAFFE	389	143	49	19
MKL2017	938	267	195	61
MKLDNN	1019	332	260	93

Table 5. Real data on Knights Landing results – images per second

	Alexnet	Googlenet	Googlenet_v2	Resnet_50
CAFFE	113	46	19	7
MKL2017	444	161	120	25
MKLDNN	530	178	133	54

Table 6. Real data on Knights Mill results – images per second

	Alexnet	Googlenet	Googlenet_v2	Resnet_50
CAFFE	153	54	20	7
MKL2017	696	191	147	28
MKLDNN	852	219	167	63

Table 7. Performance for various batch sizes on Skylake Xeon – images per second

Batch size	Alexnet	Googlenet	Googlenet_v2	Resnet_50
64	786	294	223	94
128	897	318	245	93
256	1002	311	257	94
512	1024	338	260	95
1024	1049	337	270	94

Table 8. Performance for various batch sizes on Knights Landing – images per second

Batch size	Alexnet	Googlenet	Googlenet_v2	Resnet_50
64	423	152	123	70
128	473	186	143	51
256	521	131	131	54
512	540	162	124	48
1024	507	144	119	47

Table 9. Performance for various batch sizes on Knights Mill – images per second

Batch size	Alexnet	Googlenet	Googlenet_v2	Resnet_50
64	494	174	136	81
128	580	192	157	66
256	629	177	138	57
512	642	171	132	53
1024	771	157	127	49

Thread Number Tests for Compact Affinity. This and the following sub-sections focus on parallel computing parameters. In this scenario, the default application was tested with various thread numbers and thread affinity set to compact.

Results of this test are presented in Figs. 1, 2 and 3.

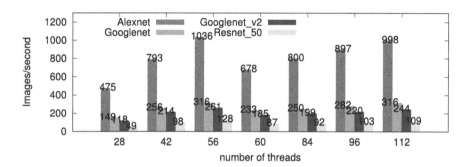

Fig. 1. Skylake Xeon results for various number of threads and compact affinity

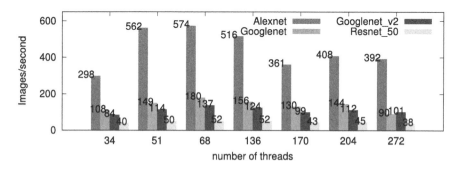

Fig. 2. Knights Landing results for various number of threads and compact affinity

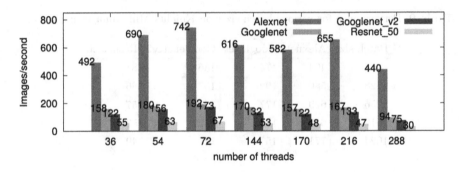

Fig. 3. Knights Mill results for various number of threads and compact affinity

For the Skylake Xeon system we can observe the best performance for the thread number equal to the number of physical cores (56) and a trend of growing performance from above 56 threads to 112 threads which is the number of all logical cores.

In the case of Xeon Phi co-processors the best performance is also for the thread number equal to the number of physical cores. The difference is in the trend of the other results. Performance drops down for too small or too big numbers of threads, especially when equal to number of all logical cores. However, there are local peaks for numbers that can be divided by four.

Additional Thread Number Tests for Compact Affinity and Core Granularity. In addition to the previous tests for various thread numbers with the compact affinity and fine granularity, a few tests for granularity set to core were run to check if there will be influence on performance. The tested configuration includes the Skylake processor with 112 threads and Knights Landing with 272 threads. The results are as shown in Tables 10 and 11 respectively.

Table 10. Core granularity results on Skylake Xeon

	Alexnet	Googlenet	Googlenet_v2	Resnet_50
112 threads	1001	317	245	116

Table 11. Core granularity results on Knights Landing

	Alexnet	Googlenet	Googlenet_v2	Resnet_50
272 threads	402	137	102	39

Results of this test scenario were almost the same as for granularity fine. We can assume that granularity has no significant influence on the performance.

Thread Number Tests for Scatter Affinity. This section refers to tests performed for various number of threads for thread affinity set to scatter. Figures 4, 5 and 6 present results for Skylake, Knights Landing and Knights Mill processors respectively.

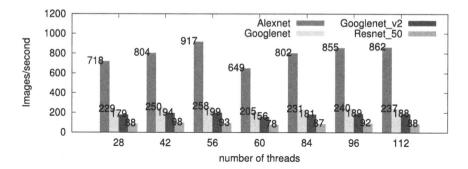

Fig. 4. Skylake Xeon results for various number of threads and scatter affinity

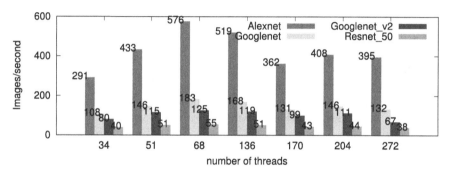

Fig. 5. Knights Landing results for various number of threads and scatter affinity

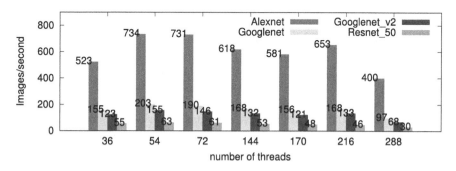

Fig. 6. Knights Mill results for various number of threads and scatter affinity

4.3 Discussion

From the collected results, we can draw conclusions with respect to both relative performances of the used CPUs as well as recommended configurations as well as parameters for deep neural training.

We can formulate the following conclusions, based on the performed tests and obtained results:

1. In terms of relative performance, from best the obtained order is as follows: Skylake Xeon, Knights Mill and Knights Landing which suggests that the architecture with fewer but more powerful cores has an edge for this type of applications.
2. Relative order of engines in terms of performance is MKL2017 and MKLDNN being much better that CAFFE and largest differences between MKL2017 and MKLDNN observed for Resnet_50, for all devices.
3. Tests with various batch sizes, between 64 and 1024 revealed that best throughput depends on the network tested:

   Skylake – 1024 for Alexnet and Googlenet_v2, 512 for Googlenet and Resnet_50,

   Knights Landing – 512 for Alexnet, 128 for Googlenet and Googlenet_v2, 64 for Resnet_50,

   Knights Mill – 1024 for Alexnet, 128 for Googlenet and Googlenet_v2, 64 for Resnet_50.

   Configurations are apparently similar for Knights Mill and Knights Landing.
4. Generally, the best numbers of threads appeared to be the numbers of physical cores with the exception of Knights Mill and scatter affinity for which best results were obtained for 54 threads. In terms of thread affinity, the compact mode proved visibly better for Skylake. For Knights Mill, Googlenet obtained better results for scatter while Googlenet_v2 and Resnet_50 for compact. For Knights Landings, Googlenet and Resnet_50 obtained better results for scatter while Googlenet_v2 for compact.

5 Summary and Future Work

Within the paper, we performed thorough benchmarking of deep neural network training using both multi- and many-core processors, including Intel Skylake, Knights Landing and Knights Mill. Various engines were used for testing including CAFFE, MKL2017 and MKLDNN. Tests have been performed for 4 different networks including Alexnet, Googlenet, Googlenet_v2 and Resnet_50, for various numbers of threads, batch sizes as well as thread affinities. Results can serve as guidelines which architectures appear to be better, performance-wise as well as which configurations on multi- and many-core CPUs are preferred for training particular network types. Our research has shown that these differ based on the CPU architectures, network types and structures as well as sizes of benchmarks and thus are of interest to the community performing research in this area.

Future work includes incorporation of energy consumption [16] in benchmarking, along performance, specifically with the use of software supporting finding preferred performance-energy consumption configurations [3].

Acknowledgments. This work is partially supported by Intel and the Intel Labs Academic Compute Environment. Work partially performed within statutory activities of Dept. of Computer Architecture, Faculty of ETI, Gdańsk University of Technology.

References

1. Czarnul, P.: Benchmarking parallel chess search in Stockfish on Intel Xeon and Intel Xeon Phi processors. In: Shi, Y., et al. (eds.) ICCS 2018. LNCS, vol. 10862, pp. 457–464. Springer, Cham (2018). https://doi.org/10.1007/978-3-319-93713-7_40
2. Czarnul, P.: Benchmarking performance of a hybrid Intel Xeon/Xeon Phi system for parallel computation of similarity measures between large vectors. Int. J. Parallel Program. **45**, 1091–1107 (2017). https://doi.org/10.1007/s10766-016-0455-0
3. Krzywaniak, A., Proficz, J., Czarnul, P.: Analyzing energy/performance trade-offs with power capping for parallel applications on modern multi and many core processors. In: FedCSIS, pp. 339–346 (2018)
4. Shi, S., Wang, Q., Xu, P., Chu, X.: Benchmarking state-of-the-art deep learning software tools. In: 2016 7th International Conference on Cloud Computing and Big Data (CCBD), pp. 99–104 (2016)
5. Serpa, M.S., Krause, A.M., Cruz, E.H.M., Navaux, P.O.A., Pasin, M., Felber, P.: Optimizing machine learning algorithms on multi-core and many-core architectures using thread and data mapping. In: 2018 26th Euromicro International Conference on Parallel, Distributed and Network-Based Processing (PDP), pp. 329–333 (2018)
6. Alzantot, M., Wang, Y., Ren, Z., Srivastava, M.B.: RSTensorFlow: GPU enabled TensorFlow for deep learning on commodity Android devices. In: MobiSys, pp. 7–12 (2017). https://doi.org/10.1145/3089801.3089805
7. Awan, A.A., Subramoni, H., Panda, D.K.: An in-depth performance characterization of CPU- and GPU-based DNN training on modern architectures. In: Proceedings of the Machine Learning on HPC Environments, MLHPC 2017, pp. 8:1–8:8. ACM, New York (2017)
8. Dong, S., Kaeli, D.: DNNMark: a deep neural network benchmark suite for GPUs. In: Proceedings of the General Purpose GPUs, GPGPU 2010, pp. 63–72. ACM, New York (2017)
9. Karki, A., Keshava, C.P., Shivakumar, S.M., Skow, J., Hegde, G.M., Jeon, H.: Tango: a deep neural network benchmark suite for various accelerators (2019)
10. Barney, L.: Can FPGAs beat GPUs in accelerating next-generation deep learning? (2017). The Next Platform. https://www.nextplatform.com/2017/03/21/can-fpgas-beat-gpus-accelerating-next-generation-deep-learning/
11. Sharma, H., et al.: From high-level deep neural models to FPGAs. In: 2016 49th Annual IEEE/ACM International Symposium on Microarchitecture (MICRO), pp. 1–12 (2016)
12. Seppälä, S.: Performance of neural network image classification on mobile CPU and GPU. Master's thesis, Aalto University (2018)
13. Ignatov, A., et al.: AI benchmark: running deep neural networks on Android smartphones. CoRR abs/1810.01109 (2018)

14. Vanhoucke, V., Senior, A., Mao, M.Z.: Improving the speed of neural networks on CPUs. In: Deep Learning and Unsupervised Feature Learning Workshop, NIPS 2011 (2011)
15. Wang, Y., Wei, G., Brooks, D.: Benchmarking TPU, GPU, and CPU platforms for deep learning. CoRR abs/1907.10701 (2019)
16. Czarnul, P., Proficz, J., Krzywaniak, A.: Energy-aware high-performance computing: survey of state-of-the-art tools, techniques, and environments. Sci. Program. **2019** (2019). Article ID. 8348791. https://doi.org/10.1155/2019/8348791

Binary Classification of Cognitive Workload Levels with Oculography Features

Monika Kaczorowska(ID), Martyna Wawrzyk,
and Małgorzata Plechawska-Wójcik(✉)(ID)

Lublin University of Technology, Nadbystrzycka 36B, 20-618 Lublin, Poland
{m.kaczorowska, m.plechawska}@pollub.pl,
martyna.wawrzyk@pollub.edu.pl

Abstract. Assessment of cognitive workload level is important to understand human mental fatigue, especially in the case of performing intellectual tasks. The paper presents a case study on binary classification of cognitive workload levels. The dataset was received from two versions of the digit symbol substitution test (DSST), conducted on 26 healthy volunteers. A screen-based eye tracker was applied during an examination gathering oculographic data. DSST test results such as total number of matches and error ratio were also applied. Classification was performed with several different machine learning models. The best accuracy (97%) was achieved with linear SVM classifier. The final dataset for classification was based on nine features selected with the Fisher score feature selection method.

Keywords: Cognitive workload · Binary classification · SVM · Eye-tracking signal

1 Introduction

According to the literature, the term "cognitive workload" is a quantitative measure of the amount of mental effort necessary to perform a task [1]. Estimation of cognitive workload is of great importance in understanding human mental fatigue related to performing tasks of various complexity requiring different concentration level. Moreover, assessment of mental effort might be useful in the process of modeling information processing capabilities.

The Digit Symbol Substitution Test (DSST) [2, 3], known from over a century ago, was introduced as a tool to understand human associative learning. Currently it is one of the most commonly used tests in clinical neuropsychology to measure cognitive dysfunction. Its popularity is related to its brevity and high discriminant validity [4]. The DSST enables to check the processing speed, memory and executive functioning of the patient. It is prevalent in cognitive and neuropsychological test batteries [5, 6]. Originally, the DSST was designed as a paper-and-pencil cognitive test presented on a single sheet of paper.

In the present study, user performance in a computerised version of the DSST test is analysed. The DSST was performed on a homogeneous group of participants,

© Springer Nature Switzerland AG 2020
K. Saeed and J. Dvorský (Eds.): CISIM 2020, LNCS 12133, pp. 243–254, 2020.
https://doi.org/10.1007/978-3-030-47679-3_21

composed of twenty six healthy students aged 20–24. The data analysed in the study originate from two boards of the DSST differing in their difficulty.

The literature proves that eye-tracking features might be applied in prediction of cognitive states. Benfatto et al. [7] used eye-tracking combined with machine learning in detecting psychological disorders. In [8] and in [9] eye-movement features were applied in the classification of visual tasks. Eye-tracking was applied in order to assess the workload and performance of skill acquisition [10]. Other studies examined cognitive workload using eye-tracking features among such groups as surgeons [11] or pilots [12].

In statistical and correlation analysis the parameters such as pupil dilation or pupil diameter size are the most often used to distinguish the state of cognitive workload [13, 14]. Additionally, such features as fixation rate and duration, saccade duration and amplitude or the number of blinks can be used in statistical analysis in the context of cognitive workload [15, 16].

The aim of the study is to verify whether features based on eye tracking might be used to classify cognitive workload level in the DSST test. The evaluation is based on eye-tracking features (fixations and saccades, blinks and pupil size) and test results (total number of matches and error ratio). The novelty of the paper is focused on the classification rate of cognitive workload level based on eye-tracking features.

The rest of the paper is structured as follows. The research procedure covering the computer application, equipment and experiment details is discussed in Sect. 2. Section 3 presents the methods applied in data processing, classification and statistical analysis procedures. The results are discussed in Sect. 4, whereas conclusions are presented in Sect. 5.

2   The Research Procedure

2.1   The Computer Application

The Digit Symbol Substitution Test (DSST) applied in the study was a computerised version of the DSST developed on the basis of the original paper-and-pencil cognitive test [15, 16]. The test requires a subject to match symbols to numbers according to a key located at the bottom of the screen. A symbol is assigned by clicking it on the key. A currently active letter is marked with a graphical frame. After assigning a symbol to a letter, the frame is moving to the next letter. The subject matches symbols to subsequent letters within specified time. Subsequent letters were generated randomly, with repetitions and continuously within a defined period of time.

The number of symbols and the time is defined in the application settings. In the case study two DSST parts were applied:

– 4 different symbols to assign; the test lasted 90 s.
– 9 different symbols to assign; the test lasted 180 s.

The application was developed in Java and is operated using a computer mouse. The interface of the computerised version is presented in Fig. 1.

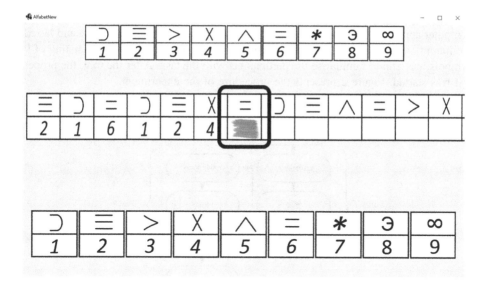

Fig. 1. The interface of the application.

2.2 Setup and Equipment

The experiment was conducted in a laboratory, in a testing room illuminated with standard fluorescent light. Eye activity was recorded using screen-based eye tracker Tobii Pro TX300 (Tobii AB, Sweden). The Tobii Pro TX300 uses video-oculography based on the dark pupil and corneal reflection method. It collects binocular gaze data with the frequency of 300 Hz.

The experiment was designed in Tobii Studio 3.2, the software compatible with the Tobii Pro TX300 eye tracker, dedicated for preparing and analysing eye-tracking experiments. Visual stimuli were presented on a separate monitor (23" TFT monitor at 60 Hz). During the experiment the participants were seated at a distance from the screen between 50 and 80 cm. The differences were insignificant for the results and they were depended on individual participant preferences (a comfortable position for working with a computer) and All participants were tested using the same software and hardware settings.

2.3 Experiment

The experiment was conducted in a dedicated laboratory with eye-tracker and computer. The 26 participants spanned the age range of 20 to 24 (mean = 20.77 years, std. dev. = 1.65). A single participant was examined for approximately 15 min. The experiment was divided into two parts, with calibration before each part.

At the beginning of each session a 9-point built-in calibration procedure was run on the eye-tracker. Then, the participants were provided with the instructions on the screen, in which they were asked to make as many matches as possible by assigning symbols to the appearing letters. The assigning was to be done by clicking a key with a particular symbol. Next, the participants completed two parts of the DSST using the computer application. Each part had a different number of symbols to assign and lasted a different amount of time. At the beginning of each part, a short trial, consisting of 9 symbols, was run to familiarise the participants with the task. After the trial, the proper test was started. Figure 2 presents the procedure of the experiment.

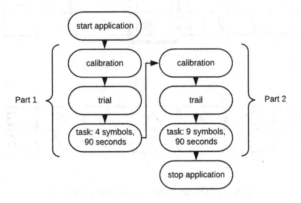

Fig. 2. The procedure of the experiment.

2.4   The Data Set

The dataset consists of 52 files generated from the eye-tracker and 52 files from the application. The total number of files generated per participant is (equal to) 4:2 files from the eye-tracker and 2 from the application. The data from each task were saved in a separate file

3   Applied Methods

3.1   Data Processing

Figure 3 presents the procedure of data processing. The procedure of data processing is divided in several main steps: data acquisition, data pre-processing, feature extraction, feature selection and the classification process.

The pre-processing step consisted of data synchronisation. Four files were generated for one participant. Since two files were generated for each part: one file from the computer application and one from the eye-tracker, the synchronisation procedure was necessary to prepare the dataset. The data from the two files were synchronised on the basis of the time stamps contained in the files. One file per part for each participant was created as the result of the synchronisation. After pre-processing, the feature selection step was performed, during which twenty features were extracted. These features are: the mean, standard deviation, maximum value, minimum value, duration of fixations and saccades and the maximum amplitude of saccades. The mean and standard deviation were calculated for the left and right pupil. The number of blinks and duration of

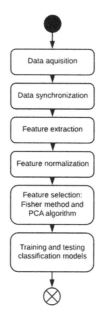

Fig. 3. The procedure of data processing.

a blink were also extracted. Additionally to eye-tracking features, a DSST-based set of features were obtained: number of responses, number of error responses and mean time of responses. The next step in the procedure of data processing was feature normalisation, which was necessary to ensure a uniform scale of the features.

In order to reduce the input dimension number and ensure higher classification accuracy, feature selection was performed. Two methods of feature selection were applied in the analysis:

- Fisher score feature selection method [17] to select the most valuable features, and
- Principal Component Analysis (PCA) [18] to find principal components with high variance.

The following classification models were applied: The following features appeared at the top of the ranking: standard deviation of the right pupil, the mean of the left pupil, standard deviation of the saccades, the mean blink duration and the number of blinks.

After the selection feature step was completed, the final step – training and classifying was started. The following classification models were applied

- Support Vector Machine (SVM) with linear kernel
- Support Vector Machine (SVM) with polynomial (poly) kernel
- Support Vector Machine (SVM) with radial based function (rbf) kernel
- K nearest neighbours (kNN)
- Random forest

The dataset was shuffled in random order and divided into train and test datasets. The test part was 20% of the entire dataset. After the learning process, the correctness of the classifier was tested using the test dataset.

3.2 Statistical Analysis

The Kolmogrov-Smirnov (K-S test) test was performed for all 20 features to determine whether the variables have a normal distribution. In order to compare the mean values from two DSST parts, the independent-samples t-Test was used. Furthermore, the Pearson correlation coefficients between each features was calculated. The analysis was performed in the MATLAB software using the Statistical and Machine Learning Toolbox.

4 Results

4.1 Classification Results

A two class classification was conducted. Observations with a low level of cognitive workload were labelled as class 1, whereas high level observations were grouped in class 2. Two approaches are presented below: the first one is based on the feature selection method and the second resorts to the application of the PCA algorithm. The following classifiers were chosen: SVM with linear kernel, SVM with poly kernel, SVM with RBF kernel, KNN and random forest. Each learning process was repeated 200 times. The number of repetitions was established on the basis of simulation of the results presented in Fig. 4. It can be observed that 200 repetitions is enough for the partial standard deviation of the partial mean (1) of classification accuracy to reach the value of 0.01. The partial mean of classification accuracy is defined as:

$$mean(acc)_i = \frac{1}{i}\sum_{j=1}^{i} acc_j, i \in \mathbb{N}, \ i \leq n \tag{1}$$

The partial standard deviation of $std(mean(acc))_i$ is defined as the standard deviation of first i partial means.

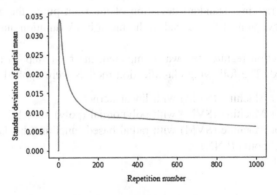

Fig. 4. Selection of repetition number – standard deviation of mean(acc).

Table 1 presents the results obtained for selected classifiers for 9 features selected by the Fisher score based feature selection method. The accuracy was calculated for each classifier. The best accuracy score was obtained for the SVM classifier with linear kernel – 0.94 and for random forest – 0.93. The worst result was obtained for the SVM classifier with poly kernel – 0.79. Tables 2 and 3 present the mean confusion matrix for the SVM with linear kernel and for random forest. Table 2 shows that on average 5.33 observations coming from the first class were classified in the proper way and only 0.285 observations from the first class were classified as second class observations. On average 5.055 observations from the second class were classified properly and 0.33 observations from the second class were classified as a first class. A similar situation occurred with the random forest confusion matrix.

Table 1. Selected classifier accuracies for 9 features selected by Fisher score based feature selection method.

Classifier	Type	Accuracy
SVM	Linear	0.94
	Poly	0.79
	Rbf	0.89
KNN		0.84
Random forest		0.93

Table 2. Confusion matrix for SVM classifier with linear kernel.

	Class 1	Class 2
Class 1	5.33	0.285
Class 2	0.33	5.055

Table 3. Confusion matrix for random forest classifier with linear kernel.

	Class 1	Class 2
Class 1	5.01	0.385
Class 2	0.375	5.23

Table 4 presents the results obtained for selected classifiers for 2 principal components. Accuracy was calculated for each classifier. Application of the PCA algorithm instead of feature selection methods ensured a high accuracy score obtained using feature selection methods and allowed to obtain even better results. The best accuracy was calculated for the SVM classifier with linear kernel. However, all the results obtained are acceptable.

Table 4. Selected classifiers accuracies for 2 principal components.

Classifier	Type	Accuracy
SVM	Linear	0.97
	Poly	0.93
	Rbf	0.93
KNN		0.95
Random forest		0.95

Figure 5 presents an example of the scatter plot for two first principal components. It can be observed that class 1 and class 2 are separable both for the train and test set observations.

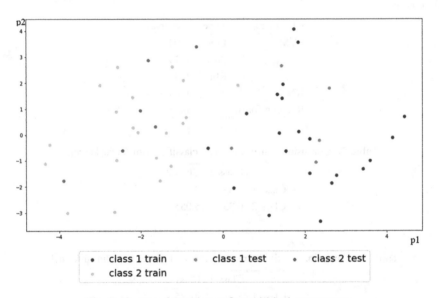

Fig. 5. Scatter plot with two first principal components.

4.2 Statistical Analysis

The K-S test found four features with non-normal distribution (Min Fix, Mi Saccade, Blinks No and Error No). The number of responses was not included in statistical analysis. The independent-samples t-Test revealed statistically significant differences for some features. Table 5 presents features revealed with the t-Test for p-value 0.05.

Table 5. The results of independent-samples t-Test

Features	P_{value}	Features	P_{value}
Mean Response Time	<0.001	Std Saccade	0.034
Fix No	<0.001	Max Saccade	<0.001
Max Fix	0.005	Mean Saccade Amplitude	<0.001
Std Fix	0.028	Std Pupil Left	0.006
Saccade No	<0.001	Std Pupil Right	0.015

Tables 6 and 7 present the statistically significant (p-value 0.05) correlation coefficients for first and second part of the DSST examination. In the case of the first DSST part, 11 pairs of correlated features were observed. The strongest correlation was observed for the Mean Pupil Right – Mean Pupil Left pair. The Blinks No – Saccade No pair also presents a high value of correlation coefficient.

The second part of the examinations revealed 11 pairs of correlated features. As in the first part of the DSST examination, the highest correlation has been found for the Mean Pupil Left – Mean Pupil Right pair. The most frequent feature to appear in the first and second examination is Blinks No, which is correlated with Mean Saccade Amplitude, Saccade No, Std Saccade No and Std Pupil Left.

Table 6. The values of correlation coefficient for first part of examination

Features	Correlation coefficient	Features	Correlation coefficient
Mean Fix Duration-Saccade No	−0.5814	Mean Saccade Duration-Std Pupil Left	−0.5002
Max Saccade-Std Fix	−0.5163	Blinks No-Mean Saccade Amplitude	−0.6109
Max Saccade-Max Fix	−0.3989	Blinks No-Saccade No	0.7049
Mean Pupil Right-Mean Pupil Left	0.9127	Blinks No-Std Saccade	0.7144
Mean Pupil Right-Std Pupil Left	0.5637	Blinks No- Std Pupil Right	0.4728
Std Pupil Left-Std Pupil Right	0.7538		

Table 7. The values of correlation coefficient for second part of examination

Features	Correlation coefficient	Features	Correlation coefficient
Mean Fix Duration-Saccade No	−0.5079	Mean Pupil Right-Std Pupil Left	0.4971
Saccade No-Max Fix	−0.5717	Std Pupil Left-Std Pupil Right	0.5776
Max Saccade-Std Pupil Left	0.4145	Std Pupil Right-Saccade No	0.4355
Max Saccade-Std Pupil Right	0.4073	Blinks No-Std Saccade	0.5874
Mean Blinks Duration-Max Fix	0.4000	Blinks No-Mean Saccade Amplitude	0.4933
Mean Pupil Right-Mean Pupil Left	0.9014		

5 Discussion and Conclusions

The aim of the paper was to verify whether eye tracking-based features might be used to classify cognitive workload level. Mental fatigue was measured during two sessions of the DSST test run in different conditions.

The binary classification was performed with different machine learning models based on such algorithms as the SVM with kernels: linear, poly and radial basis function, KNN and random forest. The evaluation was based on eye-tracking features (mean, standard deviation, maximum and minimum value, duration of fixations and saccades, maximum amplitude of saccades, the number and duration of blinks, and pupil size) and test results (the number of responses and error responses and the mean time of responses). The Fisher score feature selection method was applied to select nine of the most informative features used to build models. The learning process for each model was repeated 200 times.

The results show that the highest accuracy was achieved for the linear SVM model (94%), although the random forest algorithm (93%) occurred to be efficient as well. Confusion matrices for these models, where type I and type II errors are at relatively low levels, proved the stability of the models. Data analysis with the PCA algorithm for the first two principal components showed linear separability of classes, which corresponds to the fact that the linear model occurred to be the most efficient. The worst classification results was reached for the SVM with polynomial kernel.

Statistical analysis revealed the most significant features, which are: mean response time, standard deviation of saccades, standard deviation of fixation, fixation number, maximum saccade, maximum fixation, mean saccade amplitude, standard deviation of the right and left pupil. Most of these features were found by the Fisher score feature selection method. Statistical analysis did not reveal very strong significant correlations between features.

References

1. Gevins, A., Smith, M.E., McEvoy, L., Yu, D.: High-resolution EEG mapping of cortical activation related to working memory: effects of task difficulty, type of processing, and practice. Cereb. Cortex **7**, 374–385 (1997)
2. Boake, C.: From the Binet-Simon to the Wechsler-Bellevue: tracing the history of intelligence testing. J. Clin. Exp. Neuropsychol. **24**, 383–405 (2002)
3. Wechsler, D.: The Measurement of Adult Intelligence. The Williams & Wilkins Company, Baltimore (1939)
4. Jaeger, J.: Digit symbol substitution test: the case for sensitivity over specificity in neuropsychological testing. J. Clin. Psychopharmacol. **38**(5), 513 (2018)
5. Sicard, V., Moore, R.D., Ellemberg, D.: Sensitivity of the Cogstate Test Battery for detecting prolonged cognitive alterations stemming from sport-related concussions. Clin. J. Sport Med. **29**(1), 62–68 (2017)
6. Cook, N.A., et al.: A pilot evaluation of a computer-based psychometric test battery designed to detect impairment in patients with cirrhosis. Int. J. Gen. Med. **10**, 281–289 (2017)
7. Benfatto, M.N., Seimyr, G.Ö., Ygge, J., Pansell, T., Rydberg, A., Jacobson, C.: Screening for dyslexia using eye tracking during reading. PLoS One **11**(12) (2016)
8. Coco, M.I., Keller, F.: Classification of visual and linguistic tasks using eye-movement features. J. Vis. **14**(3), 11 (2014)
9. Henderson, J.M., Shinkareva, S.V., Wang, J., Luke, S.G., Olejarczyk, J.: Predicting cognitive state from eye movements. PLoS ONE **8**(5), 1–6 (2013)
10. Mark, J., et al.: Eye tracking-based workload and performance assessment for skill acquisition. In: Ayaz, H. (ed.) AHFE 2019. AISC, vol. 953, pp. 129–141. Springer, Cham (2020). https://doi.org/10.1007/978-3-030-20473-0_14
11. Ortega-Morán, J.F., Pagador, J.B., Luis-del-Campo, V., Gómez-Blanco, J.C., Sánchez-Margallo, F.M.: Using eye tracking to analyze surgeons' cognitive workload during an advanced laparoscopic procedure. In: Henriques, J., Neves, N., de Carvalho, P. (eds.) MEDICON 2019. IP, vol. 76, pp. 3–12. Springer, Cham (2020). https://doi.org/10.1007/978-3-030-31635-8_1
12. Van Acker, B.B., et al.: Mobile pupillometry in manual assembly: a pilot study exploring the wearability and external validity of a renowned mental workload lab measure. Int. J. Ind. Ergon. **75** (2020). https://doi.org/10.1016/j.ergon.2019.102891
13. Marshall, S.P., Pleydell-Pearce, C.W., Dickson, B.T.: Integrating psychophysiological measures of cognitive workload and eye movements to detect strategy shifts. In: Proceedings of the 36th Annual Hawaii International Conference on System Sciences, Big Island, HI, USA, p. 6 (2003)
14. Marshall, S.P.: The index of cognitive activity: measuring cognitive workload. In: Proceedings of the IEEE 7th Conference on Human Factors and Power Plants, Scottsdale, AZ, USA, p. 7 (2002)
15. Chen, S., Epps, J., Ruiz, N., Chen, F.: Eye activity as a measure of human mental effort in HCI. In: Proceedings of the 16th International Conference on Intelligent User Interfaces, Palo Alto, CA, USA, pp. 315–318 (2011)
16. Tokuda, S., Obinata, G., Palmer, E., Chaparro, A.: Estimation of mental workload using saccadic eye movements in a free-viewing task. In: 2011 Annual International Conference of the IEEE Engineering in Medicine and Biology Society, pp. 4523–4529. IEEE Engineering in Medicine and Biology Society (2011)

17. Gu, Q., Li, Z., Han, J.: Generalized fisher score for feature selection. arXiv preprint arXiv: 1202.3725 (2012)
18. Pechenizkiy, M., Tsymbal, A., Puuronen, S.: PCA-based feature transformation for classification: issues in medical diagnostics. In: Proceedings of the 17th IEEE Symposium on Computer-Based Medical Systems, pp. 535–540. IEEE (2004)

Machine Learning Approach Applied to the Prevalence Analysis of ADHD Symptoms in Young Adults of Barranquilla, Colombia

Alexandra Leon-Jacobus[1], Paola Patricia Ariza-Colpas[2(✉)],
Ernesto Barcelo-Martínez[2,3], Marlon Alberto Piñeres-Melo[4],
Roberto Cesar Morales-Ortega[2],
and David Alfredo Ovallos-Gazabon[5]

[1] Universidad Metropolitana, Barranquilla, Colombia
`aleonj@unimetro.edu.co`
[2] Universidad de la Costa, CUC, Barranquilla, Colombia
`{parizal,ebarcelo,rmoralesl}@cuc.edu.co`
[3] Instituto Colombiano de Neuropedagogia, Barranquilla, Colombia
[4] Universidad del Norte, Barranquilla, Colombia
`pineresm@uninorte.edu.co`
[5] Universidad Simón Bolivar, Barranquilla, Colombia
`david.ovallos@unisimonbolivar.edu.co`

Abstract. Disorder Attention Deficit/Hyperactivity Disorder, or ADHD, is recognized as one of the pathologies of high prevalence in children and adolescents from the global environment population; this disorder generates visible symptoms usually diminish with the passage of time in adulthood, however they remain concealed by demonstrations damnifican personal stability and human development apt. This article shows the results of the research aimed at determining the prevalence of symptoms of attention deficit hyperactivity disorder in Young Adults University of Barranquilla and its Metropolitan Area. The sample of 1600 young adults between 18 and 25 years, which has been estimated at 95% confidence level and a margin of error of 2.44%. The information was acquired through the application of exploratory instruments symptoms of attention deficit hyperactivity disorder. With the application of the algorithm different machine learning algorithms such as: Bagging, MultiBoostAB, DecisionStump, LogitBoost, FT, J48Graft, a high performance in the Bagging algorithm could be identified with the following results in quality metrics: Accuracy 91.67%, Precision 94.12%, Recall 88.89% and F-measure 91.43%.

Keywords: ADHD disorder · Prevalence of symptoms · Pathology · Hyperactivity · Impulsivity · Classification techniques

1 Introduction

Attention Deficit Hyperactivity Disorder, known by its Spanish acronym as ADHD, is a topic of interest for professionals in psychology, psychiatry and neurosciences in general, due to the negative impact it has on social functioning, personal, work,

© Springer Nature Switzerland AG 2020
K. Saeed and J. Dvorský (Eds.): CISIM 2020, LNCS 12133, pp. 255–265, 2020.
https://doi.org/10.1007/978-3-030-47679-3_22

academic and family of those who suffer from it. Recently, the American Psychiatric Association defines Attention Deficit Hyperactivity Disorder in DSM V as a persistent pattern of inattention and/or hyperactivity-impulsivity that interferes with functioning or development, which is characterized by the presence of symptoms of inattention and/or hyperactivity and impulsivity that are present before the age of 12, and affect the different contexts in which an individual develops.

Attention Deficit Hyperactivity Disorder has been considered one of the most prevalent diseases in children and adolescents worldwide [1]; and although their externally visible symptoms tend to decrease over the years in adulthood, they tend to remain hidden, behind manifestations that affect the personal stability and proper development of the human being [2]. Studies show; that of the patients diagnosed with ADHD in childhood, from 30% to 70% continue to present symptoms that generate difficulties during adolescence and adulthood, in addition; at the age of 19, 38% still fully meet the diagnostic criteria of the pathology (without remission); and 72% have at least one third of the symptoms required for diagnosis (with partial remission) [3, 4].

On the other hand, according to the Diagnostic and Statistical Manual of DSMV IV Mental Disorders, subjects diagnosed with ADHD can reach academic levels lower than those presented by their peers, and their functionality may also be compromised in the work, family, social environment, low self-esteem, disorder, poor planning capacity, lack of concentration, inadequate time management, among others. However, this research aims to determine the prevalence of symptoms of Attention Deficit Hyperactivity Disorder in Young Adults of the Universities of Barranquilla and its Metropolitan Area. The results of the study will also allow obtaining useful information at the level of higher education, since the symptoms may be having an impact on academics, leading to poor performance or even dropout associated with academic factors [5].

2 Methods

The present investigation is framed in the Empirical-Analytical paradigm because this study handles a theoretical approach of analytical cut. Regarding its approach, this research is considered quantitative; of descriptive scope. Finally, it is revealed that the temporality of the research responds to a Transactional study. The population under study corresponds to young university adults, men and women, residing in the city of Barranquilla and its metropolitan area, aged between 18 and 25 who voluntarily express their interest in participating in this research.

For the sample calculation, the reports of the ANDA National Data Archive, arranged by the National Administrative Department of Statistics DANE5, were taken into account, according to which the projection of the population to 2014 by sex and age group between 18 and 25 years in Barranquilla, Atlántico is the following (Table 1):

The final sample was found from a calculation of sample error for finite populations (taking into account a total population of 164,676). The result of the size is 1600 young adults between the ages of 18 and 25, which has been estimated with 95% confidence and a margin of error of 2.44%. The sampling implemented was non-probabilistic, intentional and the selection technique is of the Expert type [6] based on the following inclusion and exclusion criteria:

Inclusion Criteria. Schooling: University, age: 18–25 years, place of residence: Barranquilla and Metropolitan area, absence of significant clinical history.

Exclusion Criteria

- Individuals who are not in school or have a lower level of education than the University.
- Individuals with ages that do not meet the age range between 18 and 25 years of age.
- Individuals whose fixed place of residence is not Barranquilla and/or municipalities of the metropolitan area.
- Individuals with significant clinical/neuropsychiatric history, for which the presence of the symptoms to be evaluated could be explained.

Table 1. Population projection by sex and age

Gender	2014	Frequency
Men	83.950	51%
Women	80.726	49%
Total	*164.676*	100%

It is necessary to rescue, that in the present study the instruments were applied to 1,674 subjects, of which 74 were excluded from the analyzes for not meeting the inclusion criteria, or for not having answered the questionnaire in its entirety.

3   Materials and Methods

3.1   Procedure

To carry out the present investigation, a procedure described in 6 phases was established:

- Phase 1: Theoretical review and state of the art in ADHD Adult.
- Phase 2: Selection and delimitation of the sample.
- Phase 3: Administration of the instruments.
- Phase 4: Analysis of results.
- Phase 5: Preparation of the report, conclusions and recommendation.

3.2   Dataset Description

For this study, the dataset selected was generate for the experimentation during the applications of neuropsychological tests to 1674 patients and 184 features, its description in the following.

- Sex: Nominal values Male or Female.
- Age: values of the age of a person in years.
- Utah Rating Scale (WURS): Wender Utah Rating Scale for the Attention Deficit and Hyperactivity Disorder, is a battery created to measure the severity of the symptoms presented by adults with ADHD. This test that applies in 10 to 15 min, measure symptoms in seven categories, see Table 2:

Table 2. WURS scale

Number	Description
1	Attention difficulties
2	Hyperactivity/Agitation
3	Humor
4	Affective lability
5	Emotional hyperreactivity
6	Disorganization
7	Impulsivity

Manages a scale of individual item types: 1 to 5, where 1 corresponds to (not at all), 2 (a little), 3 (moderately), 4 (quite a lot) and 5 (a lot). Its heading refers to the symptoms presented in childhood through sentences as: "As a child I was (or had) (or was) [7]. This test can be used to assess status mood and emotional changes that occur.

- ADHD: It is a self-applied checklist designed for ongoing research, the time of which application ranges from 10 to 15 min. Saying instrument is made up of 18 items, its objective is to measure the prevalence of ADHD symptoms in young adults through the quantification of related symptoms with this, taking into account the different factors that affect the configuration of these symptoms and that compromise the feasibility of the evaluation diagnostic and neuropsychological of this disorder [8]. The Exploratory Inventory of ADHD Symptoms (IES-ADHD), is made up of two subscales, which evaluate two criteria such as inattention and hyperactivity/impulsivity respectively. The first of them covers items 1 to 9 and is aimed at evaluating the symptoms of inattentive type in which questions arise on the presence of errors in activities labor, deficiency in the maintenance of attention in certain activities, the difficulty in following instructions etc. The second subscale includes items from 10 to 18, and evaluates the characteristic symptoms of the hyperactive type/impulsive, denoting questions like tinkling of the hands, feeling restless or constant anxiety, impatience, etc. The list of individual item types comprises 0 (never), 1 (rarely), 2 (sometimes), 3 (with frequency) and 4 (very frequently). The objective of this checklist is to establish with both subscales, the prevalence in the symptomatology of the ADHD in young adults whose childhood they presented the disorder and how these affect the functionality of the individual in the different aspects of your life.
- CIE-10 Test: This checklist is self-applicable and is developed to be answered in a period of 10 to 15 min. This instrument is consisting of 18 items, which are divided

according to the evaluation criteria ranging from inattention to hyperactivity to impulsiveness [9]. The first measures the difficulty itself of inattentive predominance that goes from items 1 to 9, the second of hyperactive predominance that goes from items 10 to 14 and the third of impulsive predominance which ranges from 15 to 18. It comprises a scale of individual element types from 0 to 4, where 0 corresponds to (never), 1 (rarely), 2 (sometimes), 3 (with frequency) and 4 (very frequently). The objective of this instrument is to assess membership and prevalence of ADHD symptoms in adults based on the evaluation system criteria OMS ICD-10 in 1992

- Class: In case 0 is a control patients, 1 is the patients has the ADHD disease.

3.3 Algorithms Description

3.3.1 Bagging

Bootstrap aggregation, also known as Bagging, is really a meta-algorithm designed to get model combinations from an initial family, causing a decrease in variance and avoiding overfitting. Although the most common is to apply it with the methods based on decision trees, it can be used with any family. Bagging has been shown to tend to produce improvements in cases of unstable individual models (such as neural networks or decision trees), but it can produce mediocre results or even worsen results with other methods, such as the K closest neighbors [10].

3.3.2 Algorithms Based on Boosting

Unlike bagging, boosting does not create versions of the training set, but always works with the complete input set, and manipulates the weights of the data to generate different models. The idea is that in each iteration the weight of the objects misclassified by the predictor is increased in that iteration, so in the construction of the next predictor these objects will be more important and will be more likely to classify them well [11].

3.3.3 DecisionStump

DecisionStump is an algorithm implemented in weka, where the tree decision will have three branches: one of them will be in case the attribute is unknown, and the other two will be in the case that the value of the attribute of the test example is equal to a specific value of the attribute or different from said value, in case of symbolic attributes, or that the value of the test example is greater or less than a certain value in the case of numerical attributes [12].

3.3.4 J48

Algorithm C4.5 builds decision trees of a training data system of the same form that the ID3 algorithm, which uses the concept of information entropy. Training data they are a system $S = s_1, s_2, \ldots$ of samples already classified. Each example $s_i = \{x_1, x_2, \ldots\}$ is a vector where $x_1, x_2, \ldots$ represent the attributes or characteristics of the example. Training data they are augmented with a vector $C = \{c_1, c_2, \ldots\}$ where c_1, c_2,... represent the class to which it belongs each sample. C4.5 is an extension of the ID3 algorithm previously developed by Quinlan. The trees decision generator by C4.5 can be used for classification, and for this reason, C4.5 is almost always referred to as a statistical classifier [13].

4   Experimentation

For this experimentation the confusion matrix was used. This is a fundamental tool when evaluating the performance of a classification algorithm, since it will give a better idea of how the algorithm is being classified, based on a count of the successes and errors of each of the classes in the classification. This way you can check if the algorithm is classifying the classes poorly and to what extent.

Table 3. Confusion matrix.

		Classification	
		Negative	Positive
True Values	Negative	a	b
	Positive	c	d

The meaning of the result of confusion matrix are explained next:

- a, is the number of correct predictions that a case is negative, called *True Negative*.
- b, is the number of incorrect predictions that a case is positive, that is, the prediction is positive when the value should really be negative, called *False Positive*.
- c, is the number of incorrect predictions that a case is negative, that is, the prediction is negative when the value should really be positive, called *False Negative*.
- d, is the number of correct predictions that a case is positive, called *True Positive*.

From the results obtained in the confusion matrix, other quality metrics can be analyzed, which are explained below:

4.1. *Accuracy (AC):* refers to the dispersion of the set of values obtained from repeated measurements of a magnitude [14]. The smaller the dispersion, the greater the accuracy. It is represented by the ratio between the number of correct predictions (both positive and negative) and the total predictions, and is calculated using the equation.

$$AC = \frac{a+d}{a+b+c+d} \tag{1}$$

4.2. *Precision (P):* It refers to how close the result of a true value measurement is. In statistical terms, accuracy is related to the bias of an estimate [15]. It is also known as True Positive (or "True positive rate"). It is represented by the proportion between the real positives predicted by the algorithm and all positive cases, and is calculated using the equation.

$$P = \frac{d}{b+d} \tag{2}$$

4.3. Recall (TP): Sensitivity, is also known as the True Positive Rate (TP). It is the proportion of positive cases that were correctly identified by the algorithm [16]. It is calculated according to the equation:

$$TP = \frac{d}{c+d} \tag{3}$$

4.4. Specificity (TN): Also known as the True Negative Rate (TN). These are the negative cases that the algorithm has correctly classified [17]. It is calculated according to the equation:

$$TP = \frac{a}{a+b} \tag{4}$$

4.5. False Positive Rate (FP): Is the proportion of negative cases that were mistakenly classified as positive by the algorithm [18]. It is calculated according to the equation:

$$FP = \frac{b}{a+b} \tag{5}$$

4.6. False Negative Rate (FN): Is the proportion of positive cases incorrectly classified [118]. It is calculated according to the equation:

$$FN = \frac{c}{c+d} \tag{6}$$

4.7. F-measure: With the concepts of precision and recall it is possible to define another type of metric called "F-Measure" [19]. This occurs depending on the two metric already seen and can be interpreted as the harmonic mean of both. In particular measure F handles a parameter "α" of as follows:

$$F_\alpha = \frac{1+a}{\frac{1}{Precision} + \frac{\alpha}{Recall}} \tag{7}$$

5 Results

After carrying out the process of cleaning and preprocessing the data, different algorithms of automatic learning support were used to identify patients with ADHD problems. These results are shown in Table 3 and Figs. 1, 2, 3 and 4.

Table 4. Algorithms results

Algorithm	Accuracy	Precision	Recall	F-measure
Bagging	91,67%	94,12%	88,89%	91,43%
MultiBoostAB	90,74%	94,00%	87,04%	90,38%
DecisionStump	89,81%	93,88%	85,19%	89,32%
LogitBoost	88,89%	90,38%	87,04%	88,68%
FT	87,96%	90,20%	85,19%	87,62%
J48Graft	87,04%	88,46%	85,19%	86,79%

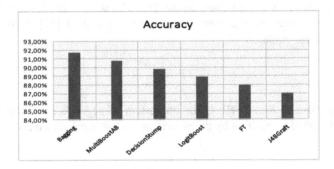

Fig. 1. Algorithms accuracy results. Source: Created by author.

In Fig. 1 the results of the algorithms accuracy can be found: Bagging 91,67%, MultiBoostAB 90,74%, DecisionStump 89,81%, LogitBoost 88,89%, FT 87,96% and J48Graft 87,04% (Table 4).

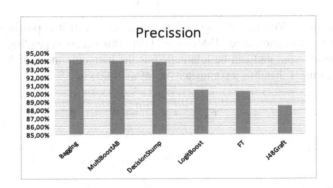

Fig. 2. Algorithms precission results. Source: Created by author.

In Fig. 2 the results of the algorithms precission can be found: Bagging 94,12%, MultiBoostAB 94,00%, DecisionStump 93,88%, LogitBoost 90,38%, FT 90,20% and J48Graft 88,46%.

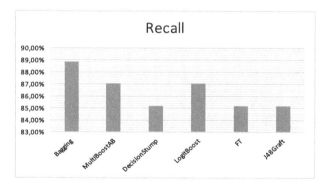

Fig. 3. Algorithms recall results. Source: Created by author.

In Fig. 3 the results of the algorithms recall can be found: Bagging 88,89%, MultiBoostAB 87,04%, DecisionStump 85,19%, LogitBoost 87,04%, FT 85,19% and J48Graft 85,19%.

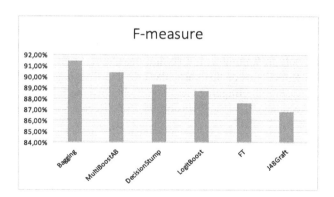

Fig. 4. Algorithms f-measure results. Source: Created by author.

In Fig. 4 the results of the algorithms F-measure can be found: Bagging 91,43%, MultiBoostAB 90,38%, DecisionStump 89,32%, LogitBoost 88,68%, FT 87,62% and J48Graft 86,79%.

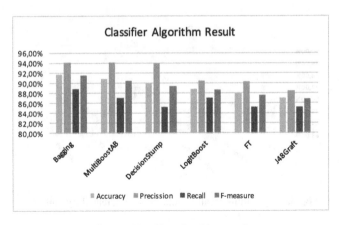

Fig. 5. Classifier algorithm result

6 Conclusions and Discussion

As a result of this experimentation, it can be concluded that machine learning techniques are an effective means or form for the analysis process of neurosychological variables related to attention deficit disorders that currently affect a large population conglomerate not only in Colombia, but around the world. From the results shown in Fig. 5, it can be defined that for this particular case the Bagging technique.

In the specific case of the Bagging classifier, it obtained the following results in the quality metrics. Accuracy 91,67%, Precision 94,12%, Recall 88,89% and F-measure 91,43%

References

1. Velázquez, J., García, M.: Trastorno por déficit de la atención e hiperactividad de la infancia a la vida adulta. Red de Revistas Científicas de América Latina, el Caribe, España y Portugal **9**(4), 176–181 (2007)
2. Ramos-Quiroga, J., Chalita, P., Vidal, R., Bosch, R., Palomar, G., et al.: Diagnóstico y tratamiento del trastorno por déficit de atención/hiperactividad en adultos. Rev. Neurol. **54** (1), 105–115 (2000)
3. Cabanyes, J., García, D.: Trastorno por déficit de atención e hiperactividad en el adulto: perspectivas actuales. Psiquiatría Biol. **13**(3), 86–94 (2006)
4. Faraone, S.V., Biederman, J., Spencer, T., Wilens, T., Seidman, L.J., et al.: Attention-deficit/hyperactivity disorder in adults: an overview. Biol. Psychiatry **48**(1), 9–20 (2000)
5. DANE: Archivo Nacional de Datos ANDA (2014). http://formularios.dane.gov.co/Anda_4_1/index.php/home. Citado 20 Marzo 2016
6. Pimienta-Lastra, R.: Encuestas probabilísticas vs. no probabilísticas. Polít. Cult. **13**, 263–276 (2000)
7. León-Jacobus, A., Valle-Cordoba, S., Florez-Niño, Y.: Diseño y validación piloto del inventario exploratorio de síntomas de TDAH (IES-TDAH) ajustado al DSM-V en jóvenes universitarios (Trabajo de Grado) (2007)

8. Adler, L., Kessler, R., Spencer, T.: Instrucciones para contestar la Escala de Auto-reporte de síntomas de TDAH en Adultos (ASRS-V1.1) (2003). http://www.neuropediatrica.com/descargas/tests/AUTOREPORTE%20TDA%20ADUL.pdf. Citado 15 Feb 2016
9. Barceló-Martínez, E., León-Jacobus, A., Cortes-Peña, O., Valle-Córdoba, S., Flórez-Niño, Y.: Validación del inventario exploratorio de síntomas de TDAH (IES-TDAH) ajustado al DSM-V. Rev. Mex. Neu. **17**(1), 1–113 (2016)
10. Breiman, L.: Bagging predictors. Mach. Learn. **24**(2), 123–140 (1996). https://doi.org/10.1007/BF00058655
11. Friedman, J.H.: Stochastic gradient boosting. Comput. Stat. Data Anal. **38**(4), 367–378 (2002)
12. Pang, J., Huang, Q., Jiang, S.: Multiple instance boost using graph embedding based decision stump for pedestrian detection. In: Forsyth, D., Torr, P., Zisserman, A. (eds.) ECCV 2008. LNCS, vol. 5305, pp. 541–552. Springer, Heidelberg (2008). https://doi.org/10.1007/978-3-540-88693-8_40
13. Bhargava, N., Sharma, G., Bhargava, R., Mathuria, M.: Decision tree analysis on J48 algorithm for data mining. Proc. Int. J. Adv. Res. Comput. Sci. Softw. Eng. **3**(6) (2013)
14. Ariza-Colpas, P., et al.: Enkephalon - technological platform to support the diagnosis of Alzheimer's disease through the analysis of resonance images using data mining techniques. In: Tan, Y., Shi, Y., Niu, B. (eds.) ICSI 2019. LNCS, vol. 11656, pp. 211–220. Springer, Cham (2019). https://doi.org/10.1007/978-3-030-26354-6_21
15. Davis, J., Goadrich, M.: The relationship between Precision-Recall and ROC curves. In: Proceedings of the 23rd International Conference on Machine Learning, pp. 233–240, June 2006
16. Powers, D.M.: Evaluation: from precision, recall and F-measure to ROC, informedness, markedness and correlation (2011)
17. Ye, K., Anton Feenstra, K., Heringa, J., IJzerman, A.P., Marchiori, E.: Multi-RELIEF: a method to recognize specificity determining residues from multiple sequence alignments using a Machine-Learning approach for feature weighting. Bioinformatics **24**(1), 18–25 (2008)
18. Yih, W.T., Goodman, J., Hulten, G.: Learning at low false positive rates. In: CEAS, July 2006
19. Lane, T., Brodley, C.E.: An application of machine learning to anomaly detection. In: Proceedings of the 20th National Information Systems Security Conference, Baltimore, USA, vol. 377, pp. 366–380, October 1997

Application of DenseNets
for Classification of Breast Cancer
Mammograms

Anita Rybiałek(ID) and Łukasz Jeleń(✉)(ID)

Department of Computer Engineering, Wrocław University of Science
and Technology, Wybrzeże Wyspiańskiego 27, 50-370 Wrocław, Poland
235133@student.pwr.edu.pl, lukasz.jelen@pwr.edu.pl

Abstract. In this study, we focus on the problem of a breast cancer
diagnosis using mammography images by classifying them as belonging
either to a *negative* or to a *malignant mass* class. We explore the potential
of densely connected convolutional neural network (DenseNet) architectures by comparing its three different variants that were trained to classify the abnormalities in breast tissue. The models have been tested in
a series of systematic experiments. With a limited dataset (2247 images
per class), it was necessary to perform tests to verify whether the amount
of data used in this work is sufficient to allow for the conclusion that the
experimental results are not dependent on the subset of the data. The
training was conducted using stratified 10-fold cross-validation to obtain
statistically reliable metrics estimates. DenseNet-201 was found to be the
best model achieving: 0.96 value for area under the curve (AUC), 0.92
for precision, 0.90 for recall, and 91% for accuracy.

Keywords: Deep Learning · DenseNet · Computer aided diagnosis ·
Breast cancer classification · Mammography

1 Introduction

Breast cancer is the most common type of cancer in the world. In 2018, it
was diagnosed in more than 2 million patients (out of whom 620 000 died) [4].
The prevalence of this disease depends on multiple factors related to women's
lifestyle, origin or even the economic development of the specific country. The
key to reducing mortality among women is the quick detection of the disease and
the appropriate treatment. Early enough detection of breast cancer significantly
increases the probability of patient's full recovery [19].

There are various methods that could yield its earlier detection: palpation
(self-examination of a breast), ultrasound, mammography, magnetic resonance
imaging, and biopsy. Unfortunately, the diagnosis itself is a highly demanding
process, sensitive to human error. A challenging part is mostly the correct identification of changes in the appearance of tissue structures that are frequently
undetectable to the human eye [3].

© Springer Nature Switzerland AG 2020
K. Saeed and J. Dvorský (Eds.): CISIM 2020, LNCS 12133, pp. 266–277, 2020.
https://doi.org/10.1007/978-3-030-47679-3_23

In this work, we focus specifically on mammography, the effectiveness of which is closely related to the experience and knowledge of radiologists responsible for its analysis. The problem is not only the aforementioned susceptibility to human error but also common inconsistencies originating from the doctors' disagreement about the condition of the particular patient [8]. In such a situation, a possible solution could be to use a dedicated software to support specialists in the detection of pathologies. It has been already shown that computer-aided diagnosis systems can help detect breast cancer based on mammograms [1]. Overall goal is to support the radiologists by drawing their attention to the suspicious parts of the tissue.

Deep Learning [15] was a breakthrough in image processing, leading to previously unachievable results for computer vision tasks such as classification [23], segmentation or detection [10]. Potentially, application of selected deep learning methods would not only increase the reliability of the diagnosis but also reduce the time spent by specialists on the analysis of images.

For many years now, machine learning community has been struggling with the problem of breast cancer classification. Multiple attempts have been made to mitigate this problem. In particular, approaches using classical machine learning methods [18,21] and deep neural networks [2,6,7,9,16,17,20,22] have been explored.

In [18], in order to address the problem of breast cancer classification based on the characteristic features of cells extracted from microscopic examinations. Authors proposed to combine *support vector machines (SVMs)* and *k-nearest neighbours (kNN)* into one. Another approach that has been presented in [21] is called an ensemble. Agarwal et al. proposed to use deep neural networks (VGG16, ResNet50, and InceptionV3, specifically) for automatic detection of masses in full–size digital mammograms [2]. Tsochatzidis et al. [20] presented an analysis of four common deep neural networks: AlexNet, VGG, GoogLeNet, and ResNet, and used them to classify the CBIS-DDSM dataset using transfer learning. The article [7] introduces the idea of application of several smaller ResNet networks to automatic mammogram classification as either benign or malignant.

The main contribution of our paper is the detailed comparison of the performance of various DenseNet architectures on a DDSM dataset.

2   Materials and Method

2.1   Dataset

The dataset used in this work was provided by Heath *et al.* [12] and is freely available from Kaggle website. This database contains mammograms from the Digital Database for Screening Mammography (DDSM) and Curated Breast Imaging Subset of DDSM (CBIS-DDSM). The data comes from mammographic examinations carried out in 1988–1999 [12].

Each image of a healthy tissue from the patient's breast (negative class) was divided into 598 × 598 pixel patches that were then resized to 299 × 299 pixels and converted to grayscale. The Regions of Interest (ROI) of size 598 × 598 pixels were also selected from pathologist-annotated images containing nodules/lesions (positive class). Such images were then resized to 299 × 299 pixels and also converted to grayscale. For the positive classes, the following sub-classes were selected: benign calcification, benign mass, malignant calcification, and malignant mass (as indicated in Table 1). All data was saved as TFRecords files (a binary data format designed to efficiently work with a TensorFlow library). The original, unprocessed dataset contains 71 249 patches, 14% of which are positive and 86% are negative.

Table 1. Number of labeled images per class. For the experiments, only negative and malignant mass classes were selected (as indicated in bold).

Label	Class	#Images per class
0	**Negative**	**61956**
1	Benign calcification	2661
2	Benign mass	2553
3	Malignant calcification	1832
4	**Malignant mass**	**2247**

The initial dataset preparation was to combine all the images available on the Kaggle platform. Since the TFRecords structure allowed to retrieve the data in either binary or multiclass classification format, the binary one was chosen for our study. To provide more details, the *negative* and *malignant mass* classes were selected for further analysis (see Fig. 1).

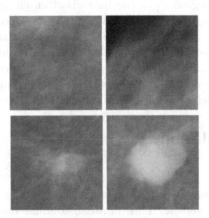

Fig. 1. Patches from mammograms without abnormalities (top) and malignant breast masses (bottom).

To normalize the data into a $[0.0, 1.0]$ range, each pixel of the mammogram was divided by 255. To balance the classes in the dataset, the *undersampling* method was used. Eventually, 2247 images per class were obtained. Final step was to randomly split the data into training (70%), validation (15%) and test (15%) subsets.

2.2   Densely Connected Convolutional Neural Network

In 2018, the *DenseNet* [13] architecture was introduced, often considered to be an improved (well-known in the field of computer vision) ResNet [11] model. "Dense" connections suggested by Huang et al. [13] mitigate the vanishing-gradient problem during the backpropagation procedure. To ensure maximum information flow, all layers in this network have been directly connected to one another so that each layer receives additional input from all previous layers (the l layer has l inputs, consisting of feature maps of all previous blocks) and passes its own feature maps to all subsequent layers $(L - l)$. Therefore, the number of direct connections in the DenseNet network consisting of L layers is $\frac{L(L+1)}{2}$ (see Fig. 2).

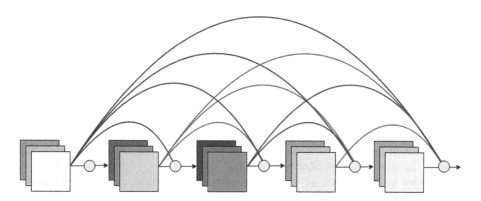

Fig. 2. Connectivity layers in the DenseNet network. Each layer (pink circles) takes all preceding feature maps as an input. (Color figure online)

Intuitively, it could be said that the more connections between layers in the network, the more parameters it has. Surprisingly though, the counter-intuitive result of this dense connectivity pattern in DenseNet is that it requires *less parameters* than similar traditional convolutional network [13].

Standard *feed-forward* architectures can be seen as the algorithms that states are passed from layer to layer. Each layer reads the state from the previous layer and writes its updated version to the subsequent layer. Such operations modify the current state in the layer, but also provide the piece of information that should be preserved. Combining feature maps of individual layers increases the input differences of subsequent layers, and also improves the performance. DenseNet is able to distinguish between information added to the network and information that is retained. Adding only a small set of feature maps to the network's "collective knowledge" (as well as maintaining the remaining feature maps unchanged) can result in a simplified final decision-making process for the final classifier based on all feature maps in the network [13]. DenseNet maximizes the full potential of its architectural design by intensive feature reuse that results in condensed models that are easier to train and highly efficient in terms of computational performance.

The improvement in the information and gradient flow has contributed to the simplified training. Each layer has direct access to the gradients from the loss function as well as the original input signal, which leads to indirect deep supervision that significantly affects the training of deeper network architectures (in a positive way). In the aforementioned article [13] the authors observed that dense connections have a regularizing effect that reduces *overfitting* for tasks that have a small amount of training data.

The DenseNet was chosen to be used in this work for its tractable computational and memory complexity. Additionally, high quality implementation of DenseNet is available in the Keras library [5], which simplifies the entry cost into the project.

2.3   Hyperparameters

The search for the best set of hyperparameters started with an attempt to determine the best value for the *learning rate*. Since learning rate is often said to have a great influence on the learning process, it was given the most attention from the very beginning of the search. The lower the value of the learning rate, the slower the learning process (it is even possible to stop it due to infinitesimal values of the gradients). However, if its value is too high, there is a risk of getting stuck in the unfavorable local minimum or even get a continuous increase in the value of the cost function (divergence).

In order to find the best learning rate (defined as 10^x), the methodology called a *coarse–to–fine random search over logarithmic space* was followed. Three search iterations were conducted, one for each variant of the DenseNet network (DenseNet-121, DenseNet-169, DenseNet-201) on the same dataset, that resulted in finding the best values of this parameter (within the searched ranges) and are as follows:

- 0.00019766 for DenseNet–121,
- 0.00048247 for DenseNet–169,
- 0.00055786 for DenseNet–201.

The next hyperparameter – batch size – was specified based on the amount of available GPU memory(in our case 12 GB). DenseNet models were trained using 5 different values of this parameter (batch size $\in 1, 2, 3, 4, 5$). For further experiments, the maximum batch size for the largest network was selected (the one that did not cause memory errors) and was used for all other DenseNet variants (to ensure a homogeneous experiment configuration). Taking into consideration the memory limitations mentioned above, a final batch size value of 4 was chosen. Other parameters were set as follows: binary cross-entropy as a loss function, Adam [14] as the optimization method and sigmoid as the final activation function.

2.4 Stratified 10–Fold Cross–Validation

In this work, the stratified 10-fold cross-validation methodology has been followed for better estimation of model prediction accuracy and for verification whether the number of available examples in the dataset is sufficient to conclude that the obtained results are independent of the way the data is split into training, validation and test subsets. The use of cross–validation allowed us to:

- use all of the available data,
- obtain statistically significant results,
- properly optimize for the hyperparameters values.

The dataset preparation involved all the data that has originally been split into training and validation subsets in order to create a raw dataset D_s. The original global test set remained unchanged and was marked as D_t. During each cross–validation iteration $i \in 0, 1, ..., 9$ the dataset D_s was randomly divided into 10 non–overlapping folds of equal size *Fold 0, Fold 1, ..., Fold 9*. Each fold was stratified to maintain the original proportions of classes. The model was trained on training data containing samples from 9 folds $(D_s \backslash D_i)$, then it was validated on the remaining i^{th} fold. Each of the 10 generated models were evaluated on the D_t global test set.

The aforementioned cross–validation methodology has been applied to all three variants of the DenseNet network (DenseNet-121, DenseNet-169, DenseNet-201). Importantly, all models were trained and evaluated on the identical data. This means that for the purpose of stratified 10–fold cross–validation, the folds were generated and saved to disk only once, and then each model was trained on such a statically created data directory.

When the cross-validation results (on the validation dataset) were obtained, the mean accuracy and mean standard deviation for each epoch were found, based on all 10 iterations. The epoch at which the model (on average) started to overfit was read from the resulting curve.

After performing a stratified 10–fold cross–validation and finding the key epochs, it was possible to combine data from the training and validation subset (within each iteration) to perform training on all the available data. The final model for each variant of the DenseNet network was trained on the full dataset D_s. This time, however, the training was stopped at the epoch when the overfitting was observed. Last but not least, the models were evaluated on the D_t test set, obtaining values of metrics such as accuracy, precision, recall, receiver operating characteristic (ROC) curve, and Area Under the ROC Curve (AUC).

3 Results and Discussion

In this section we present the results of individual stages of experiments with conclusions for all variants of tested DenseNet networks. To determine the performance for all the models we have adopted the AUC metrics. AUC provides a wider overview of the model's performance based on the classification thresholds and is also often used in the medical literature.

Stage 1 - manual, arbitrary splitting of the training and validation data.

Table 2. Selected metrics values obtained during model evaluation on test data from the DDSM dataset.

Model	#Parameters	Accuracy	Precision	Recall	AUC
DenseNet–121	7 032 257	0.92	0.94	0.89	0.96
DenseNet–169	12 638 273	0.92	0.93	0.91	0.98
DenseNet–201	18 317 633	0.90	0.88	0.93	0.97

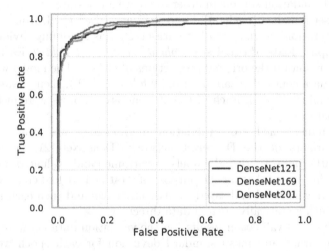

Fig. 3. The ROC curves for test data from the DDSM dataset.

Looking at the results in Table 2 and ROC curves presented in Fig. 3 for all variants of DenseNets we can draw the following conclusions:

- All analyzed models are performing well, with only subtle differences between one another.
- Table 2 contains a shining example of the relationship between precision and recall. The model that yields fewer false–negatives has higher recall (recall is sensitive to false negatives), yet precision can either increase or decrease because it is false negative invariant. In general, for models that perform better than a random classifier, precision and recall are inversely proportional (precision $\sim \frac{1}{\text{recall}}$).

- The accuracy of the system decreases with the increase of the number of network layers. This relationship suggest that for larger models the greater overfitting is observed.
- From analysis of ROC curves one can observe a slight advantage of the DenseNet-169 model in comparison to others and the relationship between the models' performance seems to be of the form:
  DenseNet–169 > DenseNet–201 > DenseNet–121.

Stage 2 - relaxation of a constraint imposed by an arbitrary division of data by stratified cross–validation.

Table 3. Values of selected metrics obtained during model evaluation on test data from the DDSM set using stratified 10–fold cross–validation.

Model	#Parameters	Accuracy	Precision	Recall	AUC
DenseNet–121	7 032 257	0.89 ± 0.01	0.87 ± 0.03	0.91 ± 0.02	0.95 ± 0.01
DenseNet–169	12 638 273	0.90 ± 0.02	0.89 ± 0.04	0.91 ± 0.02	0.96 ± 0.01
DenseNet–201	18 317 633	0.90 ± 0.01	0.89 ± 0.03	0.91 ± 0.02	0.96 ± 0.01

The results shown in Table 3 and Figs. 4, 5 and 6 allow us to make a following conclusions:

- The application of stratified 10–fold cross–validation allowed for a statistically significant conclusion that the larger the model, the better the performance.
- Trained models' performance is independent of the data splitting approach. The bias associated with division into training, validation and test subsets has been successfully eliminated.
- By comparing the results presented in Table 2 and Table 3 it was possible to observe that training the model on an arbitrarily split dataset leads to a misinterpretation of the obtained results and the seemingly satisfactory results were caused by pure coincidence (favorable random split of the data samples, to be precise).
- One can notice that the larger the model, the faster it gets to the local minimum (see Figs. 4, 5 and 6).

Figure 4(a) shows the mean validation accuracy value (with the standard deviation overlayed) for the DenseNet-121 model. A thick, solid red line indicates the accuracy value averaged over 10 different runs, while a "shadow" around it represents the standard deviation between those runs (similarly for the validation loss). The accuracy of the model gradually increases while the training proceeds. Figure 4(b) shows the average validation loss value (with the standard deviation overlayed). Analysis of the plot shows that with each epoch, the validation loss decreases, but only to a certain point. For the DenseNet-121 model, the loss value starts to increase at around 29th epoch - from that moment the overfitting starts to occur. Similar conclusions can be drawn from Fig. 5(a) and (b).

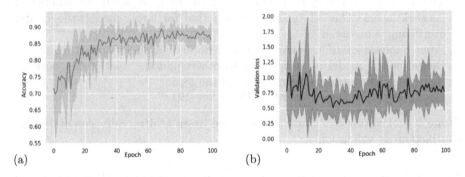

(a)   (b)

Fig. 4. Mean and standard deviation for: (a) the accuracy, (b) loss function on the validation set for the DenseNet-121 model.

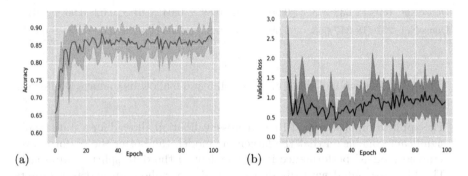

(a)   (b)

Fig. 5. Mean and standard deviation for: (a) the accuracy, (b) loss function on the validation set for the DenseNet-169 model.

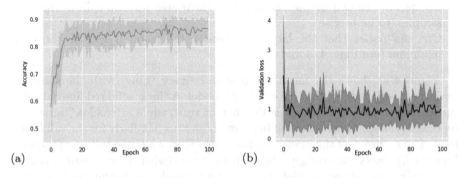

(a)   (b)

Fig. 6. Mean and standard deviation for: (a) the accuracy, (b) loss function on the validation set for the DenseNet-201 model.

The overfitting of the DenseNet-169 model starts to appear closer to the 30th epoch (see Fig. 5) and for DenseNet-201 approximately at the 73th epoch (see Fig. 6).

Stage 3 – choice of the final best–performing model.

Table 4. Results for selected metrics obtained during model evaluation.

Model	#Parameters	Accuracy	Precision	Recall	AUC
DenseNet–121	7 032 257	0.86	0.84	0.89	0.87
DenseNet–169	12 638 273	0.82	0.79	0.88	0.83
DenseNet–201	18 317 633	0.91	0.92	0.90	0.96

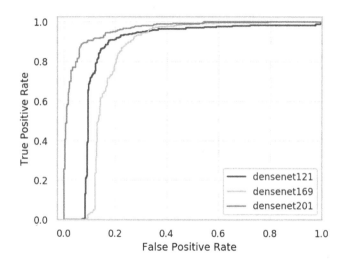

Fig. 7. ROC curves for the test data from the DDSM dataset.

From the results in Table 4 and ROC curves presented in Fig. 7 we can deduce that:

- The DenseNet-201 model is within the estimated error limits presented in Table 3. It also achieved relatively stable results across all the stages of the experiments.
- The upper right corner of the ROC curve illustrates that for low prediction thresholds, all DenseNet models make correct predictions for the positive class, with a large number of false positives.
- Analysis of the upper left corner of the ROC curve shows that for a certain prediction threshold, all three variants of the DenseNet model have a *TPR* value of approximately 0.9. However, the DenseNet–201 model achieves

a significantly lower *FPR* value. Slightly higher *FPR* value is generated by DenseNet–121 and the worst results are obtained with DenseNet–169. Such curves may suggest that DenseNet–201 probably has more contrasting predictions than for a negative class.

– The ROC curve and metric values shown in Table 4 for DenseNet–201 model indicate that this is the best–performing model to help address the problem of binary breast cancer classification from mammograms.

4   Conclusions and Final Remarks

The aim of this work was to compare the performance of various DenseNet architectures to be applied to a breast cancer classification problem when the mammography data is used. Evaluation of the three DenseNet variants (DenseNet-121, DenseNet-169, DenseNet-201) provided a very satisfactory outcomes. We can notice that DenseNet-201 was the best–performing model that achieved the highest values across all the metrics. Results described in this paper are both encouraging and very promising.

Therefore we can draw a final conclusion that DenseNets can be used to help diagnosticians in day–to–day work by limiting the number of slides they need to analyze. To further improve the classification results, a larger dataset should be used.

In the future we would like to increase the size of the existing dataset, as well as further explore other deep neural network architectures to eventually overcome the problem of breast cancer classification.

References

1. Abdel-Zaher, A.M., Eldeib, A.M.: Breast cancer classification using deep belief networks. Expert Syst. Appl. **46**, 139–144 (2016)
2. Agarwal, R., Diaz, O., Lladó, X., Yap, M.H., Martí, R.: Automatic mass detection in mammograms using deep convolutional neural networks. J. Med. Imaging **6**(3), 031409 (2019)
3. Astley, S.M., Gilbert, F.J.: Computer-aided detection in mammography. Clin. Radiol. **59**(5), 390–399 (2004)
4. Bray, F., Ferlay, J., Soerjomataram, I., Siegel, R.L., Torre, L.A., Jemal, A.: Global cancer statistics 2018: GLOBOCAN estimates of incidence and mortality worldwide for 36 cancers in 185 countries. CA Cancer J. Clin. **68**(6), 394–424 (2018)
5. Chollet, F., et al.: Keras. Software (2015). https://keras.io
6. Chougrad, H., Zouaki, H., Alheyane, O.: Deep convolutional neural networks for breast cancer screening. Comput. Methods Programs Biomed. **157**, 19–30 (2018)
7. Dhungel, N., Carneiro, G., Bradley, A.P.: Fully automated classification of mammograms using deep residual neural networks. In: 2017 IEEE 14th International Symposium on Biomedical Imaging (ISBI 2017), pp. 310–314. IEEE (2017)
8. Elmore, J.G., et al.: Variability in interpretive performance at screening mammography and radiologists' characteristics associated with accuracy. Radiology **253**(3), 641–651 (2009)

9. Geras, K.J., et al.: High-resolution breast cancer screening with multi-view deep convolutional neural networks. arXiv preprint arXiv:1703.07047 (2017)
10. He, K., Gkioxari, G., Dollár, P., Girshick, R.: Mask R-CNN. arxiv preprint arxiv: 170306870 (2017)
11. He, K., Zhang, X., Ren, S., Sun, J.: Deep residual learning for image recognition. In: Proceedings of the IEEE Conference on Computer Vision and Pattern Recognition, pp. 770–778 (2016)
12. Heath, M., Bowyer, K., Kopans, D., Moore,R., Kegelmeyer, W.P.: The digital database for screening mammography. In Proceedings of the 5th International Workshop on Digital Mammography, pp. 212–218. Medical Physics Publishing (2000)
13. Huang, G., Liu, Z., Van Der Maaten, L., Weinberger, K.Q.: Densely connected convolutional networks. In: Proceedings of the IEEE Conference on Computer Vision and Pattern Recognition, pp. 4700–4708 (2017)
14. Kingma, D.P., Ba, J.: Adam: a method for stochastic optimization. arXiv preprint arXiv:1412.6980 (2014)
15. LeCun, Y., Bengio, Y., Hinton, G.: Deep learning. Nature **521**(7553), 436 (2015)
16. McKinney, S.M., et al.: International evaluation of an AI system for breast cancer screening. Nature **577**(7788), 89–94 (2020)
17. Nawaz, M., Sewissy, A.A., Soliman, T.H.A.: Multi-class breast cancer classification using deep learning convolutional neural network. Int. J. Adv. Comput. Sci. Appl. **9**(6), 316–332 (2018)
18. Rong, L., Yuan, S.: Diagnosis of breast tumor using SVM-KNN classifier. In: 2010 Second WRI Global Congress on Intelligent Systems, vol. 3, pp. 95–97. IEEE (2010)
19. PDQ Screening and Prevention Editorial Board: Breast cancer screening (PDQ®). In: PDQ Cancer Information Summaries [Internet]. National Cancer Institute (US) (2019)
20. Tsochatzidis, L., Costaridou, L., Pratikakis, I.: Deep learning for breast cancer diagnosis from mammograms–a comparative study. J. Imaging **5**(3), 37 (2019)
21. Wang, H., Zheng, B., Yoon, S.W., Ko, H.S.: A support vector machine-based ensemble algorithm for breast cancer diagnosis. Eur. J. Oper. Res. **267**(2), 687–699 (2018)
22. Wu, N., et al. Deep neural networks improve radiologists' performance in breast cancer screening (2019)
23. Zoph, B., Vasudevan, V., Shlens, J., Le, Q.V.: Learning transferable architectures for scalable image recognition. In: Proceedings of the IEEE Conference on Computer Vision and Pattern Recognition, pp. 8697–8710 (2018)

Augmentation of Segmented Motion Capture Data for Improving Generalization of Deep Neural Networks

Aleksander Sawicki$^{(\boxtimes)}$ ⓘ and Sławomir K. Zieliński ⓘ

Faculty of Computer Science, Bialystok University of Technology,
Bialystok, Poland
{a.sawicki,s.zielinski}@pb.edu.pl

Abstract. This paper presents a method for augmenting the motion capture trajectories to improve generalization performance of recurrent long short-term memory (LSTM) neural networks. The presented algorithm is based on the interpolation of existing time series and can be applied only to segmented or easy-to-segment data due to the possibility of blending similar motion trajectories that are not significantly time-shifted. The paper shows the results of the classification efficiency with and without augmentation for two publicly available databases: Multimodal Kinect-IMU Dataset and National Chiao Tung University Multisensor Fitness Dataset. The former contains the data representing separate human computer interaction gestures, while the latter comprises the data of unsegmented series of body exercises. As a result of using the presented algorithm, the classification accuracy increased by approximately 11% points for the first dataset and 8% points for the second one.

Keywords: Augmentation · Classification · Deep learning · Motion capture

1 Introduction

Recently, the deep learning methods have been gaining in popularity in the area of gesture and motion recognition. In order to achieve a high classification efficiency, these methods typically have to be trained on big databases [1]. Gathering such datasets is usually time-consuming and expensive. In cases when data acquisition is too expensive or impractical, data augmentation could be applied. It constitutes a powerful tool for an artificial extension of the training dataset.

Data augmentation, in the form of affine transformations (e.g. scaling or rotation) is very popular in the area of image recognition. However, in the case of gesture or motion classification, this process is typically more challenging [1].

This paper presents the results of the augmentation method applied to the motion capture data representing a movement trajectory. The proposed method aims at averaging selected pairs of motion trajectories. The algorithm allows to obtain $N!(N-1)!/2$ additional time series from a data corpus of size N. The method can be used for segmented or easy-to-segmented motion trajectories with the same duration. The comparison of the

© Springer Nature Switzerland AG 2020
K. Saeed and J. Dvorský (Eds.): CISIM 2020, LNCS 12133, pp. 278–290, 2020.
https://doi.org/10.1007/978-3-030-47679-3_24

gestures (or movements) classification efficiency, obtained with and without data augmentation, is presented in the paper.

The experiments were undertaken using two publicly available databases: Multimodal Kinect-IMU Dataset [2] and National Chiao Tung University Multisensor Fitness Dataset [3]. The former one represents a set of 5 types of interaction gestures pertinent for human-computer interaction performed by one person. The latter one exemplifies selected fitness exercises, performed by 10 participants. In addition to the introduced augmentation method, the paper's added value consists in the original modifications of a simple but effective data segmentation method (motion activity detection [4]) as well as in the alterations of the advanced algorithm for a motion onset detection [5].

The paper is organized as follows. Relevant literate reports, in the area of data augmentation and motion capture, are overviewed in Sect. 2. The exploited databases are described in Sect. 3. Methodology is described in Sect. 4. The classification results are presented in Sect. 5. The last section presents the conclusions and gives a future work outlook.

2   Literature Review

According to the literature, motion data augmentation methods can be divided into three groups. The algorithms increasing the number of data samples captured with wearable sensors belong to the first one. The common feature of these algorithms is that the data transformation is directly related to inertial-measurement-units (IMU) signals processing. For example, to increase the number of accelerometer and gyroscope signals, researchers commonly use such transformations as time-warping, rotations, permutation, scaling, magnitude-warping, and jittering [1, 6, 7]. Since these methods are specific to wearable sensors data augmentation, they cannot be directly applied to data collected using depth cameras.

The second group of approaches includes the methods in which augmentation concerns the input data in the form of images. For example, Dawar et al. [8] proposed the methods simulating various orientations of the depth camera. For that purpose they processed the images using the rotation matrix. A more traditional approach was taken by Molchanov et al. [9]. They employed the popular methods of augmentation of video data by means of affine image transformations (rotation, scaling, mirroring). Additionally, they exploited the techniques of changing the sequence order of individual video frames (reverse ordering, mirroring, etc.).

In contrast to the methods described above, Huynh-The et al. [10] and Li et al. [11] adopted a slightly different approach to data augmentation. They proposed the methods which worked in two phases. Initially, time series of individual skeletal points (recorded by depth cameras) were converted to represent pseudo images, then they were subjected to image augmentation (standard affine transformations).

The last group of data augmentation methods are based on the distortion of existing motion trajectories. Cabrera and Wachs [12] as well as Guo et al. [13] proposed the methods in which an additional noise signal was added to the time series of joint positions. An interesting approach was introduced in the studies described in [14, 15].

Here, completely artificial motion trajectories were generated using a biomechanical operator. In this publication, the subject of synthetic data generation has not been discussed and is left for future work.

In addition to the three groups of motion data augmentation methods discussed above, the recent literature also includes some less popular methods, such as the one based on spatial-temporal perturbations [16]. However, these methods are not widely used and have not gained recognition in the scientific world so far.

The augmentation method described in the paper is similar to the algorithms presented in [12, 13]. It should be noted that the methods described in the above publications are based on a single or a limited number of motion trajectories, and the augmentation involves their distortion and noise addition. The algorithm proposed in this paper is also based on motion trajectory processing. However, unlike the above mentioned methods, it requires at least a few trajectories of movement, and the augmentation involves their mixing. To the best of the authors' knowledge, this solution has not been applied in the field of motion signal processing yet.

3   Datasets

The study uses two publicly available databases [2, 3] containing sets of human character movements recorded with depth cameras. The first one contains a set of right-handed gestures dedicated to human-computer interaction (HCI) applications. The second repository comprises the characters' motion trajectories recorded during a fitness session. The individual exercises represented the movement of both arms and legs. The devices used for data acquisition allowed to determine the trajectory of the movement of a character in a three-dimensional space, with a frequency of approximately 30 Hz. The skeletons used in the bases used were compared in Fig. 1. Due to different versions of depth cameras used for data acquisition, individual skeletons differ in the number of registered joints as well as the applied scale.

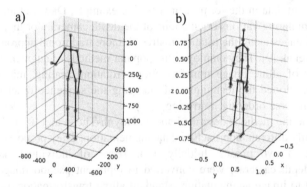

Fig. 1. The character skeleton used in: a) Multimodal Kinect-IMU Dataset [2], b) National Chiao Tung University Multisensor Fitness Dataset [3]

Multimodal Kinect-IMU Dataset
The authors of the publication [2] provided a multimodal data set containing signals from depth cameras and IMU devices. The original data corpus is divided into smaller sets, i.e. "Human Computer Interaction", "Background dataset" (Human Activity Recognition), etc. In this study, only the first one (HCI) was used. The data corpus contained a set of five gestures described as "circle", "infinity", "slider", "triangle" and "square". Each of the gestures was performed by one participant, who made approximately 50 repetitions (for the purpose of further experiment the data were balanced). The gestures were performed with the use of only the right limb, with unchanged orientation (azimuth) of the participant towards the depth camera lens. The Microsoft Kinect device, used in the above study, enabled the authors [2] to record the positions of 15 body joints (Fig. 1) with a frequency of approximately 30 Hz.

National Chiao Tung University Multisensor Fitness Dataset
In 2019, Soraya et al. [3] provided a multimodal repository of data representing selected fitness exercises. The publicly available collection contains the recordings of 10 participants performing 10 exercises. Data acquisition was carried out with the use of depth and vision cameras and a set of motion sensors, i.e. a gyroscope and an accelerometer (wearable sensors). The motion trajectories were tracked using the Kinect 2.0 sensor. This device allowed to estimate the position of 25 points (Fig. 1) representing selected joints of human body. The original data set includes exercises such as: "front raise", "jumping jacks", "lateral raise", "lunges", "mid-back turns", "running on the spot", "side lunges", "side reach", "skipping", and "squat". Due to the lack of segmentation possibilities, only part of the possible motion trajectories were used in the present experiment. The work involved activities such as "front raise", "jumping jacks", "lateral raise", "side reach", "squat".

4   Methodology

Independently of the selected database, the motion trajectories were subjected to the following processes: filtration, normalization, segmentation and interpolation. All of the above mentioned steps aimed at creating a coherent set of data, in which individual motions could be described by a fixed dimension matrix. The proposed processing steps are described in the following sections.

4.1   Preprocessing

For both databases [2, 3], position coordinates of individual joints exhibited a high degree of noise (jitter), which is typical for depth camera data. Noise reduction can be done using various methods, including Gaussian low-pass filter, discrete cosine transform (DCT), wavelet or Bandelet based methods [17–19]. The DCT method was used in this work since it was exploited for this purpose by such researchers as Li et al. [20] as well as by Kumar et al. [21].

For the first dataset [2], motion trajectories have been transformed into a frequency domain using DCT, where k = 50% of the high frequency components have been removed. Then the data were subsequently transformed back to time domain using the

Inverse Discrete Cosine Transformation. In the case of the second examined database [3] the magnitude of jitter in joints position signals was even higher. The signals of the lower limb joints were particularly noisy. This might have resulted from, among other factors, improperly chosen distance between the participants of the study and the camera, non-optimal orientation of the camera or disturbed lighting conditions. Due to the disturbances encountered, $k = 60\%$ of the highest frequencies were eliminated using Discrete Cosine Transformation.

Additional normalization has been applied due to the joint distortion. Its aim was to ensure a constant distance between the joints, interpreted as the length of the limbs. Using the time series for the first gesture ("circle") [2] and the first repetition, the average distances between all connected joints were determined. The calculated values were then taken as the pattern which was applied to all motion trajectories. In the case of the database [3], an analogous solution was used.

4.2 Segmentation and Interpolation

It is worth noting that the data provided by the authors of the publication [2] were already segmented. Nevertheless, the process was in many cases inaccurate. Within a given gesture there were both "correct" motion patterns as well as fragments of inactivity ("pause"). Additional segmentation to eliminate motionless data has been applied. Taking inspiration from the work of Lugade et al. [4], the division into movement and static segments was made with the use of the technique based on the signal magnitude area (SMA). For this purpose, a right hand position differential was used as input signal. A range of windows (13 ms - 4 frames) for SMA feature extraction was selected heuristically. When the value of the indicator exceeded 1.6, the window was counted as representing movement. The cleaned motion trajectories were interpolated to a fixed length of 128 samples.

The major challenge in analyzing the motion signals in the database [3] was the absence of their segmentation. Individual exercises were stored in blocks, differing in the number of repetitions. In order to apply the proposed augmentation algorithm it was necessary to separate individual repetitions. To this end, the modified version of the algorithm for the automatic data segmentation described in detail in [5] was utilized. Originally, the method was intended to detect gait cycles in a one-dimensional signal of the accelerometer magnitude.

The original segmentation algorithm [5] has two distinct features:

- The method does not allow to search for a defined pattern in the signal. The algorithm automatically finds the local minimum in the signal (the beginning of the walk cycle), and creates a pattern from the data in its boundary.
- The algorithm requires the *Range* parameter, indicating the maximum distance from the current walking cycle to the next supposed one.

By contrast, in the approach presented in this paper:

- The method requires a pattern to be searched, which is defined by the piecewise Bézier curve.
- The algorithm requires *Range* parameter which is modified inversely proportional to the main signal frequency *f*, defined as the exercise tempo.

The values of the *Range* parameter and the generated pattern were chosen experimentally. The patterns were generated using piecewise Bézier curves, according to manually set points. It should be emphasized that the determined curves do not have to strictly reflect the segmented signal, it is important that they keep an approximate trend and duration. Segmentation was carried out using a one-dimensional left forearm position signal (*OZ* or *OX*). The parameters necessary to reproduce the experiment are presented in Table 1.

Table 1. Specification of signal segmentation parameters

#	Exercise type	Signal #Joint, Axis	*Range*	Pattern							
0	Front raise	6, OZ	40	$\begin{bmatrix} 0 & 3 \\ 0.05 & 0.05 \end{bmatrix}$	$\begin{bmatrix} 3 & 10.5 & 10.5 & 18 \\ 0.05 & 0.05 & 0.7 & 0.7 \end{bmatrix}$
1	Jumping jacks	6, OZ	30	$\begin{bmatrix} 0 & 2 \\ 0.0 & 0.0 \end{bmatrix}$	$\begin{bmatrix} 2 & 9.5 & 9.5 & 17 \\ 0.0 & 0.0 & 1.5 & 1.5 \end{bmatrix}$
2	Lateral raise	6, OX	35	$\begin{bmatrix} 0 & 2 \\ 0.0 & 0.0 \end{bmatrix}$	$\begin{bmatrix} 2 & 9.5 & 9.5 & 17 \\ 0.0 & 0.0 & 0.7 & 0.7 \end{bmatrix}$
7	Side reach	6, OZ	34	$\begin{bmatrix} 0 & 2 \\ 0.0 & 0.0 \end{bmatrix}$	$\begin{bmatrix} 2 & 9.5 & 9.5 & 17 \\ 0 & 0 & 1 & 1 \end{bmatrix}$
9	Squat	6, OZ	45	$\begin{bmatrix} 0 & 2 \\ 0.0 & 0.0 \end{bmatrix}$	$\begin{bmatrix} 2 & 7 & 7 & 14 \\ 0.0 & 0.0 & 0.5 & 0.5 \end{bmatrix}$

In Fig. 2 the OZ position signal of the left wrist joint is shown. The signal represents one series of repetitions of the "squat" exercise. In the figure, the vertical lines mark the moments when the exercises start. Segmentation in this case was carried out correctly, despite the fact that the first repetition was significantly longer than the rest. It should be emphasized that eventually each of the gestures was interpolated to a fixed length of 128 samples.

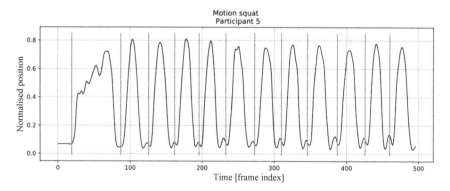

Fig. 2. "Squat" exercise segmentation

The condition of correct segmentation of gestures or exercises is necessary for correct operation of the data augmentation algorithm. In the case of a dataset [3], the segmentation was performed correctly only in 5 out of 10 available gestures. Moreover, the "side reach" exercise was performed alternately by the participants' left and right hands. As a result, the actual number of segmented exercise cycles (for which left hand movement is assumed) are twice as low as in the other exercises (5 repetitions compared to 9 or 10).

4.3   Augmentation

Individual motion trajectories were represented using matrices of dimensions of 45×128 for HCI gestures [2] and 75×128 for fitness exercises [3]. The first dimension was connected to the number of registered joints (*OX, OY, OZ* coordinates), while the second was related to time (time normalized to 128 time index). The proposed method of data augmentation consists in averaging two trajectories in order to generate an additional artificial gesture (Fig. 3).

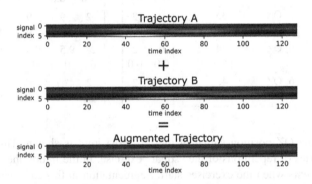

Fig. 3. Example of data augmentation

The illustration shows three matrices of a size of 6×128 each. The first two represent the trajectories of the "circle" gesture [2], whereas the third one is the result of their mixing. For the purpose of illustration only 6 signals were presented: right elbow (*OX, OY, OZ*) and right hand (*OX, OY, OZ*). The second dimension (128) corresponds to the duration of the gesture normalized to 128 samples for each cycle of a given gesture, all signals are interpolated to 128 samples along the time axis). The created matrix has a smaller numerical range (smaller color gradients), which is a consequence of averaging two patterns. The algorithm allows to generate $N!(N-1)!/2$ data samples from the set of N trajectories.

An example demonstrating a dataflow diagram is presented in Fig. 4. The illustration shows a procedure of augmenting each type of gesture. It should be noted that mixing is applied to two input trajectories. Additionally, both trajectories belong to the same gesture class.

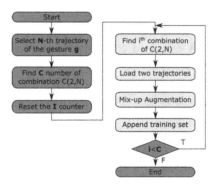

Fig. 4. Dataflow diagram

4.4 Data Division

The original Multimodal Kinect-IMU database contained a set of 5 gestures performed by one participant, each of them was repeated 50 times. In the further part of the work, a stratified 5-fold cross-validation was used. In Fig. 5 a detailed division of data is presented. The boxes show the number of repetitions for each of the five gestures.

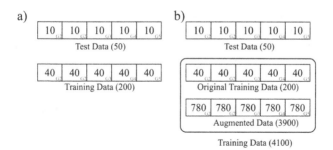

Fig. 5. Multimodal Kinect-IMU database division: a) without augmentation, b) with data augmentation

Without data augmentation, the corpus contained a relatively small number of 50 test and 200 training trajectories (Fig. 5a). After applying the proposed augmentation method, the number of trajectories in the training dataset increased from 200 to 4100. Hence, the training data have been increased more than 20 times.

The modified National Chiao Tung University Multisensor Fitness Dataset used in further experiments contained a set of 5 gestures performed by 10 participants. The number of repetitions of individual gestures was equal to 10 for "front raise", 9 for "jumping jacks", 9 for "lateral raise", 5 for "side reach and 9 for "squat". In this work, 10-times cross-validation in a leave-one-participant-out configuration was used. The number of individual motion trajectories for one of the validations is presented in Fig. 6. The boxes show the number of repetitions of each of the five gestures.

Without data augmentation, the corpus contained 42 and 378 trajectories intended for testing and training purposes, respectively (Fig. 6a). In the case of the proposed augmentation, the repository contained 15093 training trajectories. Thus, the training dataset has been increased approximately 40 times.

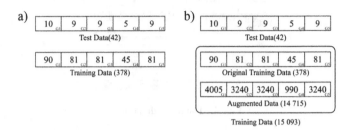

Fig. 6. National Chiao Tung University Multisensor Fitness Dataset Division: a) without augmentation, b) with data augmentation

5 Classification Results

For both datasets of motion trajectories [2, 3], the classification was made using LSTM recurrent networks in a "many to one" configuration. The studied networks included two LSTM cells, with 32 hidden neurons. The total number of network parameters (weights and bias) was equal to 16640. The rectified linear unit (ReLU) was used as an activation function in the input layer. In all the other network layers, a linear function was utilized. The network weights were optimized using the Adam algorithm (AdamOptimizer in the TensorFlow package), where cross-categorical entropy was used as a cost function. The network was trained with a learning rate of 0.0025 over a period of 300 epochs. The network input data were motion trajectories in the form of a $B \times 6 \times 128$ matrix, where B was a batch size. For the first Multimodal Kinect-IMU Dataset [2], a parameter B was equal to 200 (case without augmentation) and 4100 (case with augmentation). For the National Chiao Tung University Multisensor Fitness Dataset [3], the B parameter was set to 378 or 15 093 for the original and augmented condition, respectively.

For the Multimodal Kinect-IMU Dataset [2], the input data were right hand and right elbow position signals. In this work, a stratified 5-fold cross-validation was used. The cross-validation procedure was repeated 50 times [22]. The standard measures such as accuracy and F-score (Eq. 1) were used to assess the quality of classification.

$$Accuracy = \frac{TP + TN}{TP + TN + FP + FN}, \qquad Precision = \frac{TP}{TP + FP},$$

$$Recall = \frac{TP}{TP + FN}, \qquad Fscore = 2 \cdot \frac{Precision \cdot Recall}{Precision + Recall}, \qquad (1)$$

where:

TP – true predicted as positive (true positive);
TN – false predicted as negative (true negative);
FP – false predicted as positive (false positive);
FN – true predicted as negative (false negative)

The classification accuracy and F-score mean values, standard deviations and p-score coefficient (Student's t-test) are presented in Table 2. For all the considered cases, the augmentation process resulted in a statistically significant increase in classification accuracy ($p < 0.05$).

Table 2. Classification accuracy and F-score obtained for Multimodal Kinect-IMU Dataset

#Validation	Accuracy [%]		p-score	F-score		p-score
	No augmentation	Augmentation		No augmentation	Augmentation	
1	71.84 ± 10.70	84.44 ± 9.23	8.2×10^{-9}	0.71 ± 0.11	0.84 ± 0.09	6.4×10^{-09}
2	86.12 ± 9.12	97.68 ± 3.03	2.1×10^{-13}	0.86 ± 0.09	0.98 ± 0.03	4.2×10^{-13}
3	86.84 ± 6.60	98.12 ± 2.75	3.9×10^{-19}	0.87 ± 0.07	0.98 ± 0.03	9.7×10^{-19}
4	84.52 ± 12.74	96.4 ± 6.15	4.4×10^{-08}	0.83 ± 0.14	0.96 ± 0.06	1.0×10^{-07}
5	88.16 ± 6.93	98.12 ± 3.27	6.7×10^{-15}	0.87 ± 0.08	0.98 ± 0.03	3.3×10^{-14}
Mean	83.50 ± 11.13	94.95 ± 7.58	2.7×10^{-35}	0.83 ± 0.12	0.95 ± 0.08	4.5×10^{-35}

On the basis of all iterations carried out, an average confusion matrix was generated (see Fig. 7).

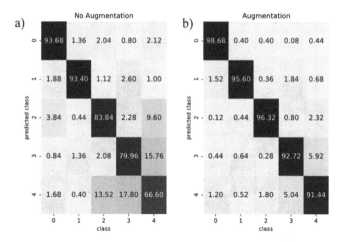

Fig. 7. Classification accuracy [%] obtained using the Multimodal Kinect-IMU Dataset: a) without augmentation, b) with data augmentation

In the case of the National Chiao Tung University Multisensor Fitness Dataset [3], the right and left wrist signals (6 signals) were used at the network input. A 10-fold cross-validation was utilized in the study, in a leave-one-participant-out configuration. For individual validations, 100 repetitions were performed. The number of repetitions was doubled compared to the previous experiment in order to increase the power of the statistical test [22]. The mean accuracy and f-score values, associated standard deviations and p-scores obtained from the t-test are presented in Table 3. In the 8-th validation the observed mean accuracies were not statistically different at a 0.05 statistical level. For the F-score parameter, a statistically significant difference is confirmed in all validation.

Table 3. Classification accuracy and F-score obtained for National Chiao Tung University Multisensor Fitness Dataset

Validation	Accuracy [%]		p-score	F-score		p-score
	No augmentation	Augmentation		No augmentation	Augmentation	
1	90.21 ± 10.22	98.31 ± 4.13	5.3×10^{-12}	0.63 ± 0.15	0.96 ± 0.10	4.7×10^{-23}
2	77.83 ± 9.54	80.17 ± 6.36	4.3×10^{-02}	0.67 ± 0.12	0.80 ± 0.10	4.8×10^{-08}
3	78.45 ± 11.86	90.05 ± 10.58	6.9×10^{-12}	0.68 ± 0.17	0.90 ± 0.11	4.2×10^{-11}
4	76.93 ± 7.15	83.50 ± 6.29	6.8×10^{-11}	0.67 ± 0.10	0.79 ± 0.09	1.1×10^{-08}
5	82.09 ± 7.37	98.26 ± 5.03	6.7×10^{-44}	0.69 ± 0.14	0.95 ± 0.10	4.9×10^{-18}
6	81.76 ± 11.34	89.12 ± 9.31	1.1×10^{-06}	0.61 ± 0.15	0.88 ± 0.14	1.2×10^{-14}
7	71.14 ± 13.83	89.21 ± 12.08	7.3×10^{-19}	0.49 ± 0.15	0.86 ± 0.14	7.5×10^{-22}
8	76.67 ± 5.48	77.40 ± 4.08	2.8×10^{-01}	0.68 ± 0.12	0.76 ± 0.04	5.9×10^{-05}
9	65.93 ± 10.76	72.60 ± 11.03	2.4×10^{-05}	0.51 ± 0.11	0.67 ± 0.11	4.1×10^{-10}
10	69.62 ± 15.91	77.14 ± 9.80	8.0×10^{-05}	0.63 ± 0.15	0.74 ± 0.11	2.8×10^{-07}
Mean	77.06 ± 12.60	85.58 ± 11.84	1.2×10^{-51}	0.62 ± 0.15	0.83 ± 0.14	1.2×10^{-89}

On the basis of all iterations carried out, an average confusion matrix was generated (see Fig. 8).

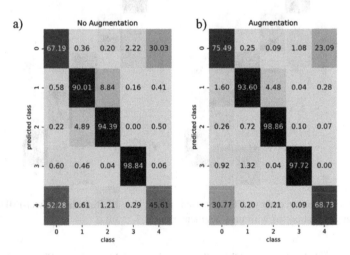

Fig. 8. Classification accuracy [%] obtained using the National Chiao Tung University Multisensor Fitness Dataset: (a) without augmentation, (b) with data augmentation

6 Conclusions and Future Work

The presented method of data augmentation allowed to significantly improve the average classification efficiency for both the Multimodal Kinect-IMU Dataset (Table 2) and the National Chiao Tung University Multisensor Fitness Dataset. For the former database [2], the average classification accuracy increased from 83.50% to 94.95% (Table 2) for the original and the augmented datasets, respectively. For the latter repository, the accuracy in the original version was equal to 77.06%, whereas for the augmented dataset the accuracy amounted to 85.58% (Table 3). The proposed method involves the averaging of pairs of motion trajectories. The algorithm allows to obtain $N!(N-1)!/2$ additional time series from the database of size N.

The presented method of data augmentation has some limitations. Namely, the augmented data must be easily segmented to individual, repeated gestures. This restriction makes it impossible to use the method for human activity recognition applications (HAR). Due to the necessity of having a few motion trajectories, the method cannot be used for a one-shot gesture recognition, either. Despite the above limitations, the presented methodology can be successfully applied for augmentation of gestures used in HCI applications, as demonstrated in the paper.

It should also be noted that the generated motion trajectories will inherently be "between" the original ones. Further work is planned to develop a hybrid algorithm that would generate trajectories with variance (as in [12]) and then interpolate them.

Acknowledgment. The work was supported by the grant from Bialystok University of Technology and funded with resources for research by the Ministry of Science and Higher Education in Poland and by grant WI/WI/1/2019 from Białystok University of Technology and funded with resources for research by the Ministry of Science and Higher Education in Poland.

References

1. Eyobu, O.S., Han, D.S.: Feature representation and data augmentation for human classification based on wearable IMU sensor data using a deep LSTM neural network. Sensors **18**, 2892 (2018)
2. Banos, O., et al.: Kinect=IMU? learning MIMO signal mappings to automatically translate activity recognition systems across sensor modalities. In: International Symposium on Wearable Computers, pp. 92–99 (2012)
3. Soraya, S.I, Chuang, S., Tseng, Y., İk, T., Ching, Y.: A comprehensive multisensor dataset employing RGBD camera, inertial Sensor and web camera. In: 20th Asia-Pacific Network Operations and Management Symposium (APNOMS), pp. 1–4 (2019). https://doi.org/10.23919/apnoms.2019.8892906
4. Lugade, V., Fortune, E., Morrow, M., Kaufman, K.: Validity of using tri-axial accelerometers to measure human movement Part I: posture and movement detection. Med. Eng. Phys. **36**(2), 169–176 (2013)
5. Gadaleta, M., Rossi, M.: Idnet: Smartphone-based gait recognition with convolutional neural networks (2016). arXiv:1606.03238

6. Um, T.T., et al.: Data augmentation of wearable sensor data for Parkinson's disease monitoring using convolutional neural networks. In: Proceedings of the 19th ACM International Conference on Multimodal Interaction, Glasgow, UK, pp. 13–17 (2017)

7. Ohashi, H., et al.: Augmenting wearable sensor data with physical constraint for DNN-based human-action recognition. In: Presented at ICML 2017 Times Series Workshop (2017)

8. Dawar, N., Ostadabbas, S., Kehtarnavaz, N.: Data augmentation in deep learning-based fusion of depth and inertial sensing for action recognition. IEEE Sensors Lett. 1 (2018). https://doi.org/10.1109/lsens.2018.2878572

9. Molchanov, P., Gupta, S., Kim, K., Kautz, J.: Hand gesture recognition with 3D convolutional neural networks, In: Proceedings of the IEEE Conference on Computer Vision and Pattern Recognition Workshops, pp. 1–7 (2015)

10. Huynh-The, T., Hua-Cam, H., Kim, D.: Encoding pose features to images with data augmentation for 3D action recognition. IEEE Trans. Ind. Inform. (2019). https://doi.org/10.1109/tii.2019.2910876

11. Li, B., Dai, Y., Cheng, X., Chen, H., Lin, Y., He, M.: Skeleton based action recognition using translation-scale invariant image mapping and multi-scale deep CNN, arXiv preprint arXiv:1704.05645 (2017)

12. Cabrera, M.E., Wachs, J.: A Human-centered approach to one-gesture learning. In: Frontiers in Robotics and AI, vol. 4 (2017)

13. Guo, D., Zhou, W., Wang, M., Li, H.: Sign language recognition based on adaptive HMMS with data augmentation. In: 2016 IEEE International Conference on Image Processing (ICIP), Phoenix, pp. 2876–2880 (2016). https://doi.org/10.1109/ICIP.2016.7532885

14. Liu, C., Liang, J., Li, T.S., Chang, K.: Motion imitation and augmentation system for a six degrees of freedom dual-arm robot. IEEE Access 7, 153986–153998 (2019). https://doi.org/10.1109/ACCESS.2019.2949019

15. Cabrera, M.E., Wachs, J.P.: Biomechanical-based approach to data augmentation for one-shot gesture recognition. In: 2018 13th IEEE International Conference on Automatic Face & Gesture Recognition (FG 2018), Xi'an, 2018, pp. 38–44 (2018). https://doi.org/10.1109/fg.2018.00016

16. Tu, J., Liu, H., Meng, F., Liu, M., Ding, R.: Spatial-temporal data augmentation based on LSTM autoencoder network for skeleton-based human action recognition. In: 2018 25th IEEE International Conference on Image Processing (ICIP), Athens, 2018, pp. 3478–3482 (2018). https://doi.org/10.1109/icip.2018.8451608

17. Xiao, J., Feng, Y., Ji, M., Yang, X., Zhang, J.J., Zhuang. Y.: Sparse motion bases selection for human motion denoising. Signal Process. 110, 108–122 (2015)

18. Wachowiak, M.P., Rash, G.S., Quesada, P.M., Desoky, A.H.: Wavelet-based noise removal for biomechanical signals: a comparative study. IEEE Trans. Biomed. Eng. 47(3), 360–368 (2000)

19. Song, B., Xu, L., Wenfang, S.: Image denoising using hybrid contourlet and bandelet transforms. In: Fourth International Conference on Image and Graphics (ICIG 2007), Sichuan, pp. 71–74 (2007). https://doi.org/10.1109/icig.2007.16

20. Li, Q., Wang, Y., Sharf, A., et al.: Classification of gait anomalies from Kinect. J. Vis. Comput. 34(2), 229–241 (2018)

21. Kumar, M., Babu, R.V.: Human gait recognition using depth camera: a covariance based approach. In: Proceedings of the 8th Indian Conference on Computer Vision, Graphics and Image Processing, pp. 20. ACM (2012)

22. Demsar, J.: Statistical comparisons of classifiers over multiple data sets. J. Mach. Learn. Res. 7, 1–30 (2006)

Improving Classification of Basic Spatial Audio Scenes in Binaural Recordings of Music by Deep Learning Approach

Sławomir K. Zieliński$^{(\boxtimes)}$ (iD)

Faculty of Computer Science, Białystok University of Technology,
Białystok, Poland
s.zielinski@pb.edu.pl

Abstract. The paper presents a deep learning algorithm for the automatic classification of basic spatial audio scenes in binaural music recordings. In the proposed method, the binaural audio recordings are initially converted to Mel-spectrograms, and subsequently classified using the convolutional neural network. The proposed method reached an accuracy of 87%, which constitutes a 10% improvement over the results reported in the literature. The method is capable of delivering moderate levels of classification accuracy even when single-channel spectrograms are directed to its input (e.g. solely from the left "ear"), highlighting the importance of monaural cues in spatial perception. The obtained results emphasize the significance of including multiple frequency bands in the convolution process. Visual inspection of the convolution filter activations reveals that the network performs a complex spectro-temporal sound decomposition, likely including a form of a foreground audio content separation from its background constituents.

Keywords: Spatial audio information retrieval · Binaural technology · Spatial scene classification · Deep learning

1 Introduction

Binaural technology is currently one of the most successful "tools" for delivering 3D audio content to the users of multimedia systems. It is commonly employed in virtual reality applications. Furthermore, it is gradually adopted by broadcasters [1]. In their pilot experiment, BBC employed binaural technology for broadcasting a series of prominent classical music concerts [2]. Moreover, binaural technology has recently been introduced to one of the most popular video sharing services (YouTube), allowing amateur film-makers to disseminate their video clips with accompanying 3D audio content [3]. For a comprehensive review of the fundamentals underlying binaural technology and its advancements, an interested reader is referred to [4].

Due to the increasing popularity of binaural technology, large repositories of 3D audio recordings (including music) will soon be created. This, in turn, may bring new challenges regarding semantic search or retrieval of such repositories in terms of their *spatial content*. Therefore, retrieval of *spatial information* from real-life binaural

© Springer Nature Switzerland AG 2020
K. Saeed and J. Dvorský (Eds.): CISIM 2020, LNCS 12133, pp. 291–303, 2020.
https://doi.org/10.1007/978-3-030-47679-3_25

recordings of music, such as those already available on the Internet, constitutes a new and challenging research topic. The existing state-of-the-art methods, allowing for spatial information extraction from binaural signals, have been developed almost solely for speech signals. Moreover, they were intended for very restricted use cases. Hence, due to the limitations of such methods, they cannot be directly applied to binaural real-life music signals. This points to the need for the development of new methods, tailored specifically for the retrieval of spatial information from binaural music signals.

In their recent study, Zieliński and Lee [5] proposed the method for the automatic classification of binaural recordings of music, according to the three basic spatial audio scenes, using a traditional hand-engineered feature extractor combined with the least absolute shrinkage and selection operator (LASSO). In their experiment, they exploited a corpus of several thousand synthetically generated but realistic binaural music recordings. The main purpose of this paper is to show that their results could be improved by as much as 10% using a deep learning approach.

The method introduced in this paper is based on the convolutional neural network (CNN), whose inputs are fed with the audio spectrograms. Converting audio signals into spectrograms and using them as "images" at the inputs of CNNs has recently become a successful technique in speech recognition [6], music genre recognition [7], audio event detection [8], and acoustic environment recognition [9, 10]. However, to the best of the authors' knowledge, nobody has applied such a technique for the classification of spatial scenes in binaural recordings of music yet.

The original contribution of the paper is as follows. Firstly, it demonstrates that the traditional approach to the automatic classification of the basic spatial audio scenes could be markedly improved by the incorporation of the deep learning classification method. Secondly, it gives an insight into how the performance of the method depends on the number of spectrum channels used. Furthermore, it investigates the effects of the size of the convolution kernels, highlighting the importance of using multiple spectral bands in the convolution process. Finally, it sheds some light on the way the neural network extracts spatial information from the audio recordings.

The proposed method could be used for the semantic search of binaural music recordings from the publicly available repositories. For example, a hypothetical user could search the Internet for music clips with musicians located in the front, behind, or surrounding the head (*height* dimension to be included in future developments). After its adaptation, the method might also be used for the segmentation of binaural recordings according to their spatial scenes. Ultimately, the method might be used as part of an artificial listener [11] as it could provide information conducive for auditory *spatial scene understanding* [12].

2   Related Work

Spatial hearing in humans is predominantly reliant on the binaural cues, namely, interaural time differences (ITD), interaural level differences (ILD), and interaural coherence (IC) [4]. Monaural spectral cues, accounting for pinnae, head and torso filtering effects, play an important role in reducing front-back localization errors and localization of sources in the median plane, allowing listeners for a limited perception of spatial impressions even with one ear [13]. It has to be stressed here that both the binaural and monaural cues can be *confounded* in the case of real-life music recordings due to two factors: (1) the mixtures of simultaneously sound-emitting audio sources in complex spatial audio scenes and (2) room reflections. Hence, the extraction of spatial information from *two-channel-only* binaural signals, accounting for acoustic waves reaching the listener's ears, constitutes not a trivial task.

The initial studies in the area of machine spatial listening were limited to the automatic localization of single audio sources in anechoic environments [14]. Later on, the researchers managed to develop models allowing to localize audio sources in reverberant environments, exploiting the so-called precedence effect [15]. More recent advancements include the attempts to localize the mixtures of simultaneous sound-emitting sources [16, 17]. Such authors as Ma, May, Brown, and Gonzalez [18–20], among others, showed that deep learning techniques could be successfully applied to localize several speakers (talkers) in complex binaural signals both under anechoic and reverberant conditions.

While the localization accuracy of the developed state-of-the-art models could be considered as satisfactory, they have several drawbacks preventing them from their direct application to the ecologically-valid binaural music recordings, such as those already available in the Internet. For example, they require *a priori* information about the number of audio sources and their spectral characteristics [18, 20]. Such information is normally unavailable in publicly available music repositories. Moreover, to reduce the number of front-back localization errors the state-of-the-art algorithms normally exploit an adaptive mechanism mimicking involuntary micro-head movements in humans [17, 18, 20]. While such an approach can be applied to binaural systems equipped with robotic dummy-head microphones, it cannot be utilized in the context of many ecologically-valid real-life recordings. Furthermore, most of the state-of-the-art models have been tested either with noise or speech signals, hence, there is no guarantee that such algorithms would perform correctly using music signals.

As already indicated above, the overall aim of the spatial information retrieval algorithms developed so far was predominantly limited to the localization of individual virtual audio sources. While the accurate localization of audio sources is important and potentially useful for communication applications (e.g. conducive for beamforming [21]), it might be of secondary importance in the case of music or multimedia applications. In the latter case, a *crude* localization of individual sources or *ensembles of sources* might be sufficient for aesthetical characterization of spatial scenes exhibited by binaural recordings, as exemplified by Rumsey's scene-based paradigm [22]. Based on such a simplified approach, Zieliński and Lee [5] developed an algorithm capable to automatically classify binaural recordings according to the basic spatial audio scenes. They demonstrated that the method could be applied both to the recordings acquired from the Internet [23] and to the synthesized but realistic recordings of music [5], yielding a satisfactory but moderate classification accuracy levels. This paper demonstrates that the performance of such a classification algorithm could be further improved by the incorporation of a deep learning technique.

3   Method

The aim of the proposed method is to classify binaural recordings of music according to one of the following three basic spatial audio scenes:

- *Foreground-Background (FB)* scene – foreground content (musicians) located in front of a listener whereas background content (room reflections) arriving from the back of a listener (this scene could be considered to be a conventional stage-audience scenario [24] and is often used in jazz and classical music recordings),
- *Background-Foreground (BF)* scene – background content in front of a listener with foreground content perceived from the back (reversed stage-audio scenario),
- *Foreground-Foreground (FF)* scene – a listener is surrounded by musicians in a horizontal plane (360° source scenario [24] – often used in pop music).

In the proposed method, the binaural audio recordings were initially converted into Mel-spectrograms and, subsequently, they were fed to the input of the neural network.

3.1   Binaural Music Recordings

For the sake of comparative consistency, the recordings formerly used by Zieliński and Lee [5] were also incorporated in this study. They were acquired from the publicly available repository [25]. The repository consisted of 5928 two-channel binaural music recordings (7 s each), out of which 4368 recordings were used for training and the remaining 1560 items were intended for testing purposes. They were all synthesized by convolving the multi-track music recordings with the *binaural-room-impulse-responses* (BRIRs). Thirteen sets of BRIRs were used in [5] to generate the recordings. They represented a variety of room acoustics (office, laboratory, recording studio, concert hall, etc.), with reverberation time ranging from 0.17 to 2.1 s. The repository was representative of a broad range of music categories including pop, orchestral, jazz, country, electronica, dance, rock, and heavy metal. The number of individual music sources present in each recording varied from 5 to 62. The recordings were labeled according to the aforementioned three spatial audio scene categories (*FB, BF, FF*).

3.2   Mel-Spectrograms

Each binaural audio recording was converted into four Mel-spectrograms, according to the procedure described below. The first two spectrograms were computed for the left (l) and right (r) audio channel signals, respectively. The remaining two spectrograms were obtained for the sum ($m = l + r$) and difference signals ($s = l - r$), respectively.

The spectral resolution of each spectrogram was determined by the number of Mel-frequency filters, which in this study was set to 150. The filter bank spanned a frequency range from 100 Hz to 20 kHz. The temporal resolution of the spectrograms was set to 40 ms with a 50% overlap. The above spectro-temporal resolution, adopted in this study, was similar to that used by other researchers in the area of machine listening [17–20]. The signals in each temporal frame were subject to a Hamming window. As a result of the above procedure, every recording was converted to four Mel-spectrograms, with the resolution of 150 × 349 pixels each (number of frequency-channels × number of time-frames). In line with the typical practice in machine listening [17–20], the spectrograms were standardized prior to their use by the convolutional network. The Mel-spectrograms were calculated in MATLAB.

3.3 Train and Test Datasets

According to the results of the pilot tests (omitted in this paper due to space limitations), the duration of each binaural recording could be reduced from 7 to 3.5 s. with only a marginal deterioration in the classification accuracy. Therefore, each spectrogram was split in half, effectively doubling the number of audio samples available for training and testing. Consequently, the resolution of each spectrogram was reduced from 150×349 to 150×174 pixels, while the number of audio samples available for *training* and *testing* increased to 8736 and 3120, respectively. The above procedure could be considered to constitute a form of data augmentation. In order to adjust the hyper-parameters of the neural network and monitor its performance during the training, 10% of the training data were set aside as a *validation dataset*.

3.4 Neural Network

The topology of the neural network used in this study was inspired by the architecture of the AlexNet [26]. The network consisted of 23 layers, out of which the first 14 layers constituted the convolutional part, whereas the remaining layers (15–23) formed the fully connected segment (see Fig. 1).

The spectrograms were combined into three-dimensional tensors ("images") which were then fed to the input layer. The size of each tensor was $150 \times 174 \times N_{ch}$, where N_{ch} represented the number of channels in a tensor. This number was equal to 4 for the model in which all four types of spectrograms were used. For some network models considered in this study, the number of channels N_{ch} was restricted to 3, 2 and even 1, e.g. for a model incorporating the spectrograms obtained from the left signals only.

Five convolutional layers were included in the network (layers 2, 5, 8, and 14). They all had the same the size of the kernel (3×3) but their number of filters progressively increased from layer to layer, which is typical for convolutional neural networks used for image classification [26] (the number of filters in the convolutional layers was equal to 64, 128, 256, 512, and 1024, respectively). In order to accelerate the learning procedure, batch normalization was employed after the first four convolutional layers (layers 3, 6, 9, and 12). Moreover, four max pooling layers were exploited to reduce the resolution of the processed spectrograms (layers 4, 7, 10, and 13). They all had a symmetric window size of 2×2.

The numbers placed in brackets Fig. 1 (positioned near the interconnections between the layers in) indicate the dimensions of the tensors. They show that the original resolution of the spectrograms fed to the input of the network was gradually reduced from 150×174 (at the network input) to 5×7 pixels (at the output of layer 14). The strategy of a progressive reduction of image resolution, combined with the increasing number of filters in each convolutional layer, is commonly employed in convolutional networks [26].

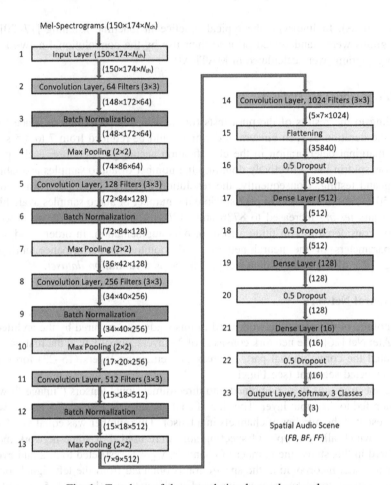

Fig. 1. Topology of the convolutional neural network.

The purpose of layer 15 was to convert the two-dimensional spectrograms into vectors, which were subsequently processed by the three fully connected dense layers consisting of 512, 128, and 16 nodes, respectively. In order to prevent the network from overfitting, four dropout layers were employed. The role of the last (output) layer was to provide the results of the classification in the form of one of the three classes, representing the basic spatial audio scenes (*FB, BF, FF*). A rectified linear unit (ReLU) was used in all the convolutional and dense layers, whereas a Softmax function was applied to the output layer.

In total, the network consisted of 24.7 million trainable parameters. The proposed structure of the network, along with its hyper-parameters, was selected heuristically, with the outcomes "monitored" using a hold-out validation technique (10% of the training data were used for validation purposes). Due to its superior properties, Adam [27] optimization technique was used to train the network with a categorical cross-entropy as a loss function. The learning rate was set to 2×10^{-4}, the size of a dropout rate was adjusted to 0.5, whereas the batch size was fixed to 128. The number of learning epochs was set to 100, with the weights reverted to those yielding the best validation accuracy.

The network was implemented in Python, using Keras, Tensorflow, scikit-learn, and numpy packages. The computations were performed on a graphical processing unit (NVIDIA RTX 2080Ti).

4   Results

The proposed method was evaluated on the test dataset. An accuracy metric, defined as the ratio of the number of correctly classified recordings to their total number, was used as a performance measure. For each experimental condition, the neural network was trained and tested 10 times. The mean values of the classification accuracy were used to assess the performance of the method.

4.1   Input Channels Effects

For each audio recording, four spectrograms were generated and stored in three-dimensional tensors, as explained in Sects. 3.2 and 3.4. These tensors could be fed to the network either in an unrestricted form ($N_{ch} = 4$) or with some channels being discarded. The boxplots, presented in Fig. 2, illustrate the results obtained for the selected combinations of input channels.

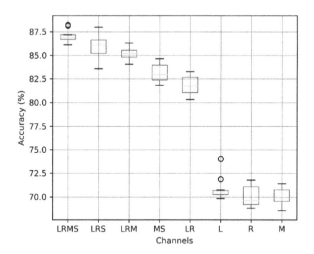

Fig. 2. Influence of the input channels on the classification accuracy.

The best mean classification accuracy, equal to 87%, was obtained when all four input channels were exploited by the network. This scenario is signified as *LRMS* in Fig. 2. It has to be emphasized here, that the reached accuracy level is 10% higher compared to the accuracy of 76.9% formerly reported by Zieliński and Lee [5], which demonstrates the superiority of a deep learning technique over the traditional classification method.

When the input tensors are limited to three channels (N_{ch} = 3), the accuracy drops by approximately 2%, which is illustrated in Fig. 2 by the *LRS* and *LRM* conditions, respectively. For the *LRM* condition the observed effect is statistically significant at p = 4.2×10^{-3} level according to the Tukey HSD test. Further deterioration of the accuracy, to a value of approximately 82.5%, is observed when the input tensors are limited to two channels (N_{ch} = 2), which is exemplified by the *MS* and *LR* scenarios. The above effect was statistically significant at $p < 10^{-3}$ level.

A substantial decline in the network performance is observed when the input tensors are limited to a single channel (N_{ch} = 1), as illustrated by the *L, R*, and *M* scenarios in Fig. 2. For these cases, the classification accuracy drops to approximately 70%. Interestingly, this outcome indicates the network is still capable to classify the spatial audio scenes at a moderate classification rate with only one "ear" (e.g. spectrograms obtained only for the left-ear signals) This phenomenon could be explained by the fact that, in addition to binaural cues, monaural cues are also pertinent to spatial perception [13].

Two examples of the confusion matrices are presented in Fig. 3. They were obtained for the best model (*LRMS* scenario) and the model trained only with the left-ear signal (*L* condition), respectively. It can be seen that in both cases the network is relatively good in identifying the *FB* and *BF* scenes, however, it "struggles" to recognize the *FF* scene.

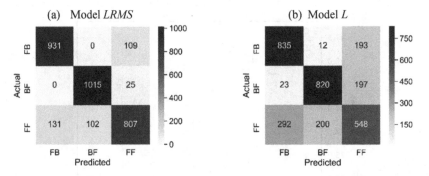

Fig. 3. Examples of the confusion matrices obtained for: (a) model incorporating all spectrogram types, (b) model using only spectrograms for the left-ear audio signals.

4.2 Kernel Size Effects

Figure 4 shows the results obtained for the model with four different kernels, varying in size and shape, used in the convolutional layers. It can be seen that the best results are obtained for the symmetric kernel of size 3 × 3. When the kernel size is increased to 5 × 5, the accuracy appears to drop. However, according to the Tukey HSD test performed at a 0.05 significance level, the observed effect is not statistically significant. Hence, it can be concluded that both kernels (3 × 3 and 5 × 5) yield similarly good results. A small but statistically significant deterioration in the classification accuracy occurs for asymmetric kernel of size 5×1.

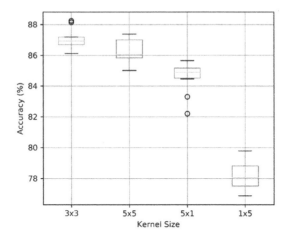

Fig. 4. Influence of the convolution kernel size on the classification accuracy.

The results obtained for the last kernel (1×5) are almost 10% worse than the ones achieved for the two best kernels (3×3 and 5×5). Following from this, it can be tentatively concluded that information conveyed by the first dimension (frequency) is more important, in terms of discrimination between the spatial audio scenes, than information represented by the second dimension (time). This observation highlights the importance of including multiple frequency bands in the convolution procedure.

4.3 Frequency Resolution Effects

In all the experiments presented so far, the spectrograms were generated with a fixed frequency resolution. This resolution was determined by the number of Mel-frequency filters which in this study was set to 150 (see Sect. 3.2). In order to investigate the influence of the frequency resolution on the classification accuracy, the experiments were repeated with the number of frequency filters ranging from 12 to 250. For some conditions, the adjustments to the network architecture had to be made to accommodate for the changes in the spectrogram size. For example, when the number of frequency filters was increased beyond 150, the size of the window in the first max pooling layer (layer 4) had to be enlarged from 2×2 to 3×2 pixels. Without this adjustment the size of the network could exceed 46 million trainable parameters. Hence, it could be impractical to train such a big network with a given dataset. On the other hand, for the spectrograms with a smaller frequency resolution (less than 100), some layers of the network had to be discarded. For example, when the resolution was set to 50, layers 13 and 14 were omitted. For the spectrograms with the resolution equal to 25, layers 11–14 were discarded. For the smallest frequency resolution examined in this work (resolution reduced to 12), the original network was simplified by removing layers 7–14.

According to the obtained results, presented in Fig. 5, the method yields a similar classification rate for the spectrograms with the frequency resolution of 150 or higher, reaching approximately 87.5%. The differences between the results obtained for the resolution equal to 150, 200, and 250 are statistically not significant. When the frequency resolution of the spectrograms is reduced to 100, the accuracy drops by approximately 2.5%. The above effect is statistically significant at $p < 0.001$ level. Further reduction of the resolution of the spectrograms results in additional deterioration of the classification accuracy.

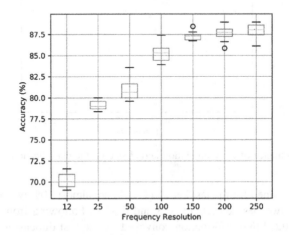

Fig. 5. Influence of the frequency resolution of the spectrograms.

According to the results discussed above, the classification accuracy tends to "saturate" for the frequency resolution values being equal or greater than 150. The observed tendency confirms the correctness of the original choice of the spectrogram frequency resolution (see Sect. 3.2), since the selected value of 150 appears to be a reasonable trade-off between the classification accuracy and the computational complexity of the proposed method.

4.4 Convolution Filter Activations

An interesting observation can be made by visual inspection of the convolution filter activations. Figure 6 shows the activations at the output of the first convolution layer for the selected audio recording fed to the network input. For clarity, only the results obtained for 4 (out 64) filters were shown (filter 1, 9, 45 and 61). It appears that filters 1 and 45 are responsible for foreground audio extraction, with clearly visible harmonic components (horizontal contours), whereas the role of filters 9 and 61 is to isolate predominantly background components. Hence, to an extent, these two pairs of filters complement each other. Visual inspection of the remaining filter activations (not presented in the paper) indicates that the network performs a very complex spectro-temporal sound decomposition, difficult to achieve using the traditional signal processing techniques [11, 12].

(a) Filter 1   (b) Filter 9

(c) Filter 45   (d) Filter 61

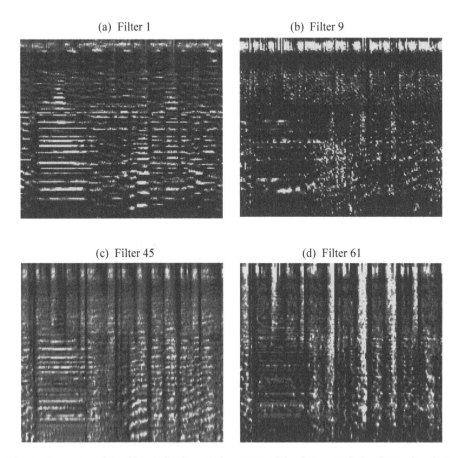

Fig. 6. Examples of the filter activations at the output of the first convolution layer (for clarity only 4 out 64 filter activations are shown).

5 Conclusions

This paper presents a deep learning algorithm intended for the automatic classification of basic spatial audio scenes in binaural music recordings. In the proposed method, the binaural audio recordings are initially converted to Mel-frequency spectrograms, and subsequently classified using the convolutional neural network. For the same repository of the binaural recordings of music, the proposed method reached the accuracy of 87%, which constitutes a 10% improvement over the results reported in the literature, obtained using the hand-engineered feature extractor combined with the LASSO classifier [5]. This outcome exemplifies the superiority of the proposed deep learning technique over the traditional one.

The best results were obtained when the convolutional network was fed with four types of spectrograms, obtained for the left, right, sum, and difference audio signals, respectively. Interestingly, the network was capable of delivering moderate levels of classification accuracy even when single-channel spectrograms were directed to its input, highlighting the importance of monaural cues in spatial machine perception.

The developed network was tested with different convolution kernels, varying in shape and size. The obtained results emphasize the importance of including multiple frequency bands in the convolution process. Moreover, visual inspection of the convolution filter activations revealed that the network performs a complex spectro-temporal sound decomposition, likely including a form of separation of foreground audio content from its background constituents.

Further work might involve extending the experimental context to additional spatial audio scenes, including a height dimension. Following the recent trends in audio signal processing [21], attempts to "bypass" a spectrogram extraction stage and develop the network with a "self-learning" frequency filter bank will also be considered in the future experimental work.

Acknowledgments. The work was supported by the grant from Białystok University of Technology and funded with resources for research by the Ministry of Science and Higher Education in Poland.

References

1. Binaural Sound. Immersive spatial audio for headphones. BBC R&D. https://www.bbc.co.uk/rd/projects/binaural-broadcasting. Accessed 07 Feb 2020
2. Firth, M.: Developing Tools for Live Binaural Production at the BBC Proms. BBC R&D. https://www.bbc.co.uk/rd/blog/2019-07-proms-binaural. Accessed 07 Feb 2020
3. Kelion, L.: YouTube live-streams in virtual reality and adds 3D sound. BBC News. https://www.bbc.com/news/technology-36073009. Accessed 07 Feb 2020
4. Blauert, J. (ed.): The Technology of Binaural Listening. MASP. Springer, Heidelberg (2013). https://doi.org/10.1007/978-3-642-37762-4
5. Zieliński, S.K., Lee, H.: Automatic spatial audio scene classification in binaural recordings of music. Appl. Sci. **9**(9), 1724 (2019). https://doi.org/10.3390/app9091724
6. Mao, Q., Dong, M., Huang, Z., Zhan, Y.: Learning salient features for speech emotion recognition using convolutional neural networks. IEEE Trans. Multimedia **16**(8), 2203–2213 (2014). https://doi.org/10.1109/TMM.2014.2360798
7. Oramas, S., Barbieri, F., Nieto, O., Serra, X.: Multimodal deep learning for music genre classification. Trans. Int. Soc. Music Inf. Retrieval **1**(1), 4–21 (2018). https://doi.org/10.5334/tismir.10
8. Stowell, D., Giannoulis, D., Benetos, E., Lagrange, M., Plumbley, M.D.: Detection and Classification of Acoustic Scenes and Events. IEEE Trans. Multimedia **17**(10), 1733–1746 (2015). https://doi.org/10.1109/tmm.2015.2428998
9. Han, Y., Park, J., Lee, K.: Convolutional neural networks with binaural representations and background subtraction for acoustic scene classification. In: Proceedings of the Conference on Detection and Classification of Acoustic Scenes and Events, Munich, Germany, 16 November 2017 (2017)
10. Barchiesi, D., Giannoulis, D., Stowell, D., Plumbley, M.D.: Acoustic scene classification: classifying environments from the sounds they produce. IEEE Signal Process. Mag. **32**(3), 16–34 (2015). https://doi.org/10.1109/MSP.2014.2326181

11. Brown, G.J., Cooke, M.: Computational auditory scene analysis. Comput. Speech Language. **8**(4), 297–336 (1994). https://doi.org/10.1006/csla.1994.1016
12. Szabó, B.T., Denham, S.L., Winkler, I.: Computational models of auditory scene analysis: a review. Front. Neurosci. **10**, 524 (2016). https://doi.org/10.3389/fnins.2016.00524
13. Blauert, J.: Spatial Hearing. The Psychology of Human Sound Localization. The MIT Press, London (1974)
14. Jeffress, L.A.: A place theory of sound localization. J. Comp. Physiol. Psychol. **41**, 35–39 (1948)
15. Hummersone, Ch., Mason, R., Brookes, T.: Dynamic precedence effect modeling for source separation in reverberant environments. IEEE Trans. Audio Speech Lang. Process. **18**(7), 1867–1871 (2010). https://doi.org/10.1109/TASL.2010.2051354
16. Benaroya, E.L., Obin, N., Liuni, M., Roebel, A., Raumel, W., Argentieri, S.: Binaural localization of multiple sound sources by non-negative tensor factorization. IEEE/ACM Trans. Audio Speech Lang. Process. **26**(6), 1072–1082 (2018). https://doi.org/10.1109/TASLP.2018.2806745
17. Ma, N., Brown, G.J.: Speech localisation in a multitalker mixture by humans and machines. In: Proceedings of INTERSPEECH. San Francisco, USA. https://doi.org/10.21437/interspeech.2016-1149
18. Ma, N., May, T., Brown, G.J.: Exploiting deep neural networks and head movements for robust binaural localization of multiple sources in reverberant environments. IEEE/ACM Trans. Audio Speech Lang. Process. **25**(12), 2444–2453 (2017). https://doi.org/10.1109/TASLP.2017.2750760
19. Ma, N., Gonzalez, J.A., Brown, G.J.: Robust binaural localization of a target sound source by combining spectral source models and deep neural networks. IEEE/ACM Trans. Audio Speech Lang Process. **26**(11), 2122–2131 (2018). https://doi.org/10.1109/TASLP.2018.2855960
20. May, T., Ma, N., Brown, G.J.: Robust localisation of multiple speakers exploiting head movements and multi-conditional training of binaural cues. In: Proceeding of IEEE International Conference on Acoustics, Speech and Signal Processing (ICASSP). Brisbane, QLD, Australia (2015). https://doi.org/10.1109/icassp.2015.7178457
21. Sainath, T., Weiss, R.J., Wilson, K.W., Li, B., Narayanan, A., Variani, E., et al.: Multichannel signal processing with deep neural networks for automatic speech recognition. IEEE/ACM Trans. Audio Speech Lang. Process. **25**(5), 965–979 (2017). https://doi.org/10.1109/TASLP.2017.2672401
22. Rumsey, F.: Spatial quality evaluation for reproduced sound: terminology, meaning, and a scene-based paradigm. J. Audio Eng. Soc. **50**(9), 651–666 (2002)
23. Zieliński, S.K.; Lee, H.: Feature extraction of binaural recordings for acoustic scene classification. In: Proceedings of the 2018 Federated Conference on Computer Science and Information Systems (FedCSIS), Poznań, Poland, 9–12 September 2018 (2018)
24. Lee, H., Millns, C.: Microphone Array Impulse Response (MAIR) library for spatial audio research. In: Proceedings of the 143rd Convention of the Audio Engineering Society. New York, NY, USA (2017)
25. Zieliński, S.K., Lee, H.: Database for automatic spatial audio scene classification in binaural recordings of music. Zenodo (2019). http://doi.org/10.5281/zenodo.2639058
26. Krizhevsky, A., Sutskever, I., Hinton, G.E.: ImageNet classification with deep convolutional neural networks. Commun. ACM **60**(6), 84–90 (2017). https://doi.org/10.1145/3065386
27. Kingma, D.P, Ba, J.: Adam: a method for stochastic optimization. (2014). arXiv:1412.6980

Modelling and Optimization

AutoNet: Meta-model for Seamless Integration of Timed Automata and Colored Petri Nets

Muhammad Waqas Ahmad[(✉)], Muhammad Waseem Anwar,
Farooque Azam, Yawar Rasheed, Usman Ghani, and Mukhtar Ahmad

Department of Computer and Software Engineering, College of Electrical and
Mechanical Engineering, National University of Sciences and Technology
Islamabad, Islamabad, Pakistan
{waqas,waseemanwar,farooq,
mukhtar.ahmad}@ceme.nust.edu.pk,
yawar.rasheed18@ce.ceme.edu.pk, ugghani@gmail.com

Abstract. Time dependent modeling paradigm has always remained the focus of study for embedded system designers. The reason behind this rationale is that the safety and reliability of real time systems mainly depend on how precisely and accurately the time domain is modeled. Several time-driven models have been proposed and attained the level of maturity through series of developments. The most adopted formal methods are 1) Timed Automata which extends the finite states with finite number of real valued clocks, and 2) Colored Petri net which extends finite set of directed graphs with finite number of tokens coupled with color. In this paper, we proposed a Meta model (named as AutoNet) aimed at integration of timed automata and colored petri net. The main purpose of AutoNet is the transformation i.e. a single design with basic classes and state transition diagrams can be transformed to both timed automata and colored petri net. We performed case study to show proof of our concept prototype at traffic light signal modeling. A single iteration through the AutoNet produced both the timed automata and colored petri net of our test case, which is the validation of our design. AutoNet will serve as an automated 'what you see is what you get' (WYSIWYG) tool for embedded system engineers.

Keywords: Timed automata · Colored Petri net · Meta model · Time domain

1 Introduction

Timed Automata (TA) are a predetermined set of automaton complimented with finite discrete clocks. During a particular execution of TA, clock value increases along the execution of the system. The clock value can be conceptualized as integer variable. The guards may be enabled or disabled at specific transitions by comparing the value of clock and by doing so possible behaviors of the execution may be constrained. A TA is kind of directed state/transition diagram, in which the nodes are represented as states (i.e. system status at particular point in time, depicted by rectangles/rounded rectangles) and the transitions are represented as arcs (i.e. trigger conditions, depicted by directed arrows).

© Springer Nature Switzerland AG 2020
K. Saeed and J. Dvorský (Eds.): CISIM 2020, LNCS 12133, pp. 307–319, 2020.
https://doi.org/10.1007/978-3-030-47679-3_26

Colored Petri nets (CPN) are a backward-compatible extension of the formal concept of Petri nets (PN). A Petri net is a directed graph (comprised of places, transitions, and arcs), in which states are represented as places (i.e. system status at certain point in time, represented by circles), transitions are represented as nodes (i.e. events that may occur, represented by bars) and the directed arcs (represented by arrows) describe which places are pre- and/or post-conditions for which transitions. Arcs (i.e. arrows) can run from a place (i.e. circle) to a transition (i.e. bar) or vice versa, it can never run in-between two places or in-between two transitions. The source places from which an arc goes to a transition are referred as the input places of the transition; and the destination places to which arcs go from a transition are referred as the output places of the transition. Graphically, there exists discrete no. of tokens at each place in a Petri net. A transition is first enabled to make it fire, i.e. there have produced defined count of tokens at each of its corresponding input places; the moment transition executes, it takes that required number of tokens and puts them at respective output place. CPN preserves essential characteristics of Petri nets along with extended formalism allowing differentiation among tokens. CPN tokens may have attached a data value, which is called color of the token.

Both TA and CPN have become the de facto standards for the representation of system evolution at occurrences of events over time. We have compiled various resemblances and distinctions between TA and CPN which were previously distributed across literature. Furthermore, despite tremendous applications of TA and CPN in embedded system modeling, there existed no common platform to define both formalisms at a single place i.e. designing either of TA or CPN and achieving the other through automated translation. We have fulfilled this research gap by developing an integrated meta-model for creation and inter-translation of timed automata and colored Petri nets at a single platform. The concept diagram of the proposed system is shown at Fig. 1.

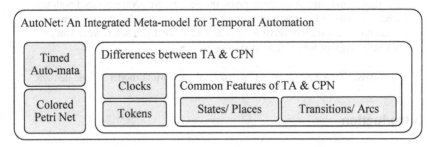

Fig. 1. AutoNet: concept diagram

We have identified similarities and differences between TA and CPN. Based on the common characteristics of both formal methods, an equivalent mapping has been drawn. Conflicting requirements have been modeled separately for each formalism. The proposed Meta model is capable to generate separate instance models of timed automata and colored Petri nets by single iteration thorough it. The Meta model is named as AutoNet, the first part word i.e. Auto is taken from Automata and Net is derived from Petri NET as second part of the acronym.

2 Literature Review

We started our literature review with an attempt to answer the very basic research question i.e. "What is the state-of-the-art in the field of Model Driven Engineering for Embedded Software?"

We found two of the very recent survey papers demonstrating up-to-date development in the applications of model driven approaches for embedded systems design. In [1], Akdur et al. carried out a study to find state-of-the-practices, advantages, challenges and ultimate consequences of using model driven approaches in the field of embedded software engineering. Furthermore, an empirical investigation on the use and the evaluation of model based engineering in embedded system domain was performed by Liebel et al. [2]. After learning state-of-the-art of subject under study, we further searched the literature and found a very current article which had cited both of the above listed papers. Authors [3] highlighted the identified strengths/weaknesses of various formal methods and proposed new opportunities by integrating formal methods. Furthermore, possible threats (along with probable mitigations) to the adoption of new opportunities have also been visualized. We set this paper as the baseline for our research work. According to Gleirscher, Foster and Woodcock [3], single formal method may be focused at considering very specific aspect or perspective of a system's behavior, which when used in isolation, may limit its effectiveness. Furthermore, authors suggested that various scholars recommend smart integration of formal methods to overcome this very inherent weakness.

From this point onward, we started searching for literature to hunt for integration opportunities among two of the well-known formal methods i.e. TA and CPN.

The concepts of timed automata have been taken from [4] and [5]. Dill [4] proposed TA targeted to model the time-behavior of real time applications, which proved to be a very simple and yet effective way to represent state/transition diagrams complemented with timing constraint by the integration of several clocks. TA was studied from the perspective of its application in formal language theory. Furthermore, Bengtsson and Yi [5] presented the abstract and concrete semantics of TA. The conceptual understandings of Petri nets have been developed from [6] and [7]. PN has been demonstrated as a graphical tool for the analysis/description of concurrent processes by Adam Petri and Reisig [6]. Furthermore, Pawlewski [7] briefly explained Petri nets as modeling tool for timed systems and their application in the perspective domain. The application of model-driven software engineering paradigm for embedded system domain and web based application development, has been drawn from the authors of [8] and [9].

Srba [10] suggested that the most studied formal methods among time-driven modeling techniques include TA and timed extensions of PN. Bouyer et al. [11] studied the probable relationship between timed PN and TA. Researchers [12] presented two translations: one from extended TA to CPN, another from a parallel combination of TA to timed PN. Byg et al. [13] proved the equivalence of bounded timed-arc PN coupled with read-arcs to network of TA and described a translation of extended timed-arc PN to network of TA. Cassez and Roux [14] proposed translation from timed PN to TA that is capable to preserve behavioral semantics of the timed PN. D'Aprile et al. [15] presented a method for the translation from timed PN to TA. Furthermore, Barkaoui et al. [16] established a method for mutual translations among timed-arc PN and networks of TA. Authors [17] highlighted that despite the fact that timed PN and TA models were developed independently, they still bear strong interconnection. Balaguer et al. [18] focused on translation of 1-bounded timed PN into a network of TA and considered a novel equivalence by accounting the distribution of actions. Xia C [19] surveyed various recent approaches of translating TA to timed PN. Bérard et al. [20] performed comparison of TA and CPN w.r.t weak time bisimilarity.

3   Proposed Solution

Timed automata and colored Petri net have been adopted by industry as effective formal methods. Both formalisms were developed as independent tool with respect to the each other. Despite the fact for sharing of many common features, yet there exist significant differences among them, as well.

We have performed comparative analysis of timed automata and colored Petri net from various perspectives. The similarities and differences between TA and CPN are compiled and summarized in Table 1. Aspect is used to represent the particular point of view for the system under study. Then implementation of each aspect by TA and CPN is described. Mapping decision is drawn based on the relationship between specific realizations of each aspect by both formal methods. Last but not the least, our implementation of each aspect for AutoNet has also been elaborated in the table.

3.1   Proposed Meta Model

The proposed Meta model is presented at Fig. 2. It is designed in Ecore using Eclipse Modeling Framework. The root element is defined as AutoNet which has composed all the other elements i.e. Node, Switch, Bar, Clock and Token. Node and switch are exactly similar concepts for TA and CPN, while both formalisms have different implementation for bar, clock and token.

The instance model for both formal methods i.e. timed automata and Colored Petri net can be defined by the instantiation of node, switch and bar. TA modeling will require an additional requirement of defining the clock instance and CPN modeling will be completed by defining the token. Clock and token may be interchangeably translated to each other with default values of color and nodeID for token. Hence timed automata and colored Petri net are seamlessly integrated at meta-model. Furthermore desired translation in-between TA and CPN is also achieved at instance model.

Table 1. Similarities and differences between TA & CPN

Aspect	Timed Automata (TA)	Colored Petri Net (CPN)
Essence	In automation theory, a Timed Automata is a finite set of states extended with a finite set of real-valued clocks	In mathematical modeling, a Colored Petri Net is a finite set of directed graphs extended with finite discrete tokens

As both TA and CPN represent time behavior of the system, there is an obvious opportunity to look for their probable integration.

We propose an abstract concept of temporal automation termed as AutoNet, to reflect both TA and CPN. It will contain an attribute (name) of string type and three operations i.e. run, pause and reset. Run() will be used to start particular execution of the system, pause() will halt the execution at certain point in time and reset() will be used to restore the system to initial default settings

Currency	States are used to represent current status of the system	Places are used to represent current status of the system

States and places are exactly similar concepts, hence ideal candidates for an equivalent mapping definition. Nodes are introduced as an equivalent concept to map states and places for TA and CPN respectively. Node will have an ID of int type, an attribute (name) of string type and three Boolean variables named as isInitial, isCurrent and isFinal to model instantaneous behaviors of system

Switching	Transition is used to switch from one state to other	Transition and Arcs are used to switch between places

Transition is the exactly matching concept. Transition is defined as an equivalent concept with an attribute (name) of string type, an operation named as fire() and two integer variables timeInterval and minTokens associated with TA and CPN respectively. Bar is additionally introduced with an attribute (name) of string type to define arc concept for CPN

Time Behavior	Clocks are used to reflect time behavior of the system. Transition is attached with time value of clocks	Tokens are used to reflect time behavior of the system. Transition is attached with count of tokens

Clocks and Tokens are dissimilar concepts, therefore can't be mapped to an equivalent definition. Clocks and Tokens are introduced as concrete classes to handle the time behavior of TA and CPN respectively. Both has an attribute name of string type, while token has further color (integer) and nodeID (Node) attributes. The difference is resolved with these two specific implementations

Sequencing	In TA, we have to define the entire set of states first and then the transition function is defined which controls the switching between states	In CPN, we have to define the transition function along with enabling mechanism first and states are then defined by the set of places along with tokens

State/Transition and Place/Transition are not exactly similar concepts but their underlying semantics are comparable. They are modeled with an equivalent custom mapping function

Active Node	TA may contain only a single current state	In CPN multiple locations can contain one or more tokens
Concurrency	TA is single threaded	CPN is concurrent
Triggers	In TA, the active state changes in response to an event or time interval	In CPN, transitions are executed as soon as all input places contain at least 1 token

(*continued*)

Table 1. (*continued*)

Aspect	Timed Automata (TA)	Colored Petri Net (CPN)
Visibility	In TA, state is global. Given two states, all we can say is "these states are different"	In CPN, place is a marking, which is to say "how many tokens are in each place"
Design	In TA, we have to look at each state individually and determine the possible transitions to other states	Each transition in CPN represents a whole group of transitions in the underlying reachability graph
Application	TA is recommended in situations when we have a problem that's small enough to be handled with finite states (preferably up to a few dozen)	CPN is preferred choice in situations when we need to create a model for distinguishable subsystems that interact with each other in real time

The above listings are the conflicting requirements. TA and CPN are specialized constructs with their own specific implementation for these aspects. Hence these concepts can't have equivalent mapping definitions. However, Meta Model is designed to be flexible enough for handling these contradictions

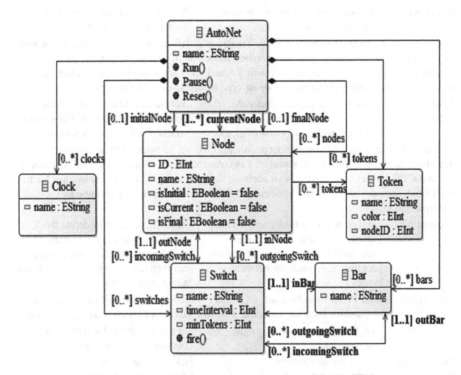

Fig. 2. Meta-model for seamless integration of TA & CPN

4 Case Study

We selected traffic light signal simulator as case study to validate proof of concept prototype for our Meta model. A simple graphical representation for traffic light signal simulator is shown at Fig. 3.

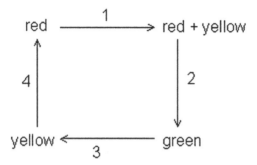

Fig. 3. Traffic light signal simulator: concept diagram

There are four states i.e. red, red + yellow, green, yellow and corresponding four transitions i.e. 1, 2, 3, 4. Transition 1 switches from state red to red + yellow, 2 transitions from red + yellow to green, 3 follows the path from green to yellow and 4 completes the loop from last state i.e. yellow to initial state i.e. red.

We created an EMF Project in Eclipse Modeling Framework to develop dynamic instance model for traffic light signal simulator. The developed XMI model is presented at Fig. 4.

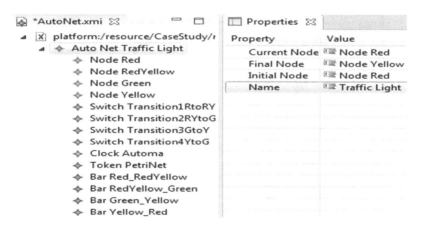

Fig. 4. XMI instance model - traffic light signal simulator's case study

The root element is defined as Traffic Light: AutoNet, which composes all other elements of the model. Initial and current node is set as Red: Node and final node is declared as Yellow: Node.

All the four nodes are presented at Fig. 5. These nodes represent states as well as places. The same concept i.e. node is being used to model both TA states and CPN places, hence first step towards seamless integration. Each node has one incoming and one outgoing switch i.e. transition.

Node Red		
Node RedYellow	Incoming Switch	Switch Transition4YtoG
Node Green	Is Current	true
Node Yellow	Is Final	false
Switch Transitior	Is Initial	true
Switch Transitior	Name	Red
Switch Transitior	Outgoing Switch	Switch Transition1RtoRY
Switch Transitior	Tokens	Token PetriNet

Node Red		
Node RedYellow	Incoming Switch	Switch Transition1RtoRY
Node Green	Is Current	false
Node Yellow	Is Final	false
Switch Transitior	Is Initial	false
Switch Transitior	Name	RedYellow
Switch Transitior	Outgoing Switch	Switch Transition2RYtoG
Switch Transitior	Tokens	

Node Red		
Node RedYellow	Incoming Switch	Switch Transition2RYtoG
Node Green	Is Current	false
Node Yellow	Is Final	false
Switch Transitior	Is Initial	false
Switch Transitior	Name	Green
Switch Transitior	Outgoing Switch	Switch Transition3GtoY
Switch Transitior	Tokens	

Node Red		
Node RedYellow	Incoming Switch	Switch Transition3GtoY
Node Green	Is Current	false
Node Yellow	Is Final	false
Switch Transitior	Is Initial	false
Switch Transitior	Name	Yellow
Switch Transitior	Outgoing Switch	Switch Transition4YtoG
Switch Transitior	Tokens	

Fig. 5. Nodes (TA states & CPN places)

✦ Switch Transition1RtoRY

Property	Value
In Bar	Bar Red_RedYellow
In Node	Node Red
Min Tokens	1
Name	Transition1RtoRY
Out Bar	Bar Red_RedYellow
Out Node	Node RedYellow
Time Interval	45

✦ Switch Transition2RYtoG

Property	Value
In Bar	Bar RedYellow_Green
In Node	Node RedYellow
Min Tokens	1
Name	Transition2RYtoG
Out Bar	Bar RedYellow_Green
Out Node	Node Green
Time Interval	5

✦ Switch Transition3GtoY

Property	Value
In Bar	Bar Green_Yellow
In Node	Node Green
Min Tokens	1
Name	Transition3GtoY
Out Bar	Bar Green_Yellow
Out Node	Node Yellow
Time Interval	60

✦ Switch Transition4YtoG

Property	Value
In Bar	Bar Green_Yellow
In Node	Node Green
Min Tokens	1
Name	Transition3GtoY
Out Bar	Bar Green_Yellow
Out Node	Node Yellow
Time Interval	60

Fig. 6. Switches (TA transitions & CPN arcs)

The four switches are described at Fig. 6. These switches represent transitions as well as arcs. The same concept i.e. switch is being used to model TA transition and CPN arcs, hence another move towards seamless integration. For TA, each switch has one incoming/outgoing pair of source/destination node, while for CPN each switch has two incoming/outgoing pairs i.e. a source node/destination bar and a source bar/destination node.

The four CPN bars are presented at Fig. 7. Each bar has one incoming and one outgoing switch. This specific concept has specialized definition for CPN transitions.

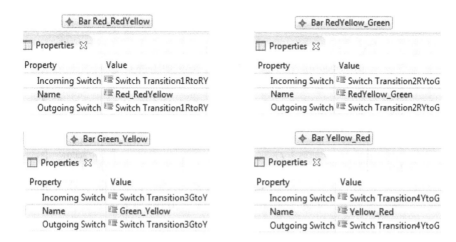

Fig. 7. Bars (CPN transitions)

We defined a clock for TA and a token for CPN (as shown at Fig. 8) to complete the instance model for both formal methods.

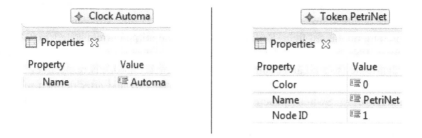

Fig. 8. TA clock and CPN token

Graphical representations of instance models for TA and CPN are shown at Figs. 9 and 10 respectively.

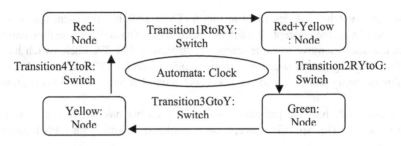

Fig. 9. AutoNet instance model for timed automata

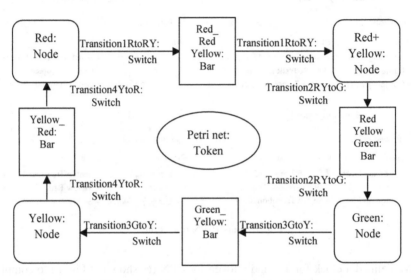

Fig. 10. AutoNet instance model for colored Petri net

The working instance model i.e. the model in execution is further elaborated to promote the practical understanding of AutoNet. Traffic Light Simulation (TLS) starts with switching on the red light. Red: Node is set as the initial/current state (for TA) and place (for CPN). Automata: Clock is declared/initialized and Petrinet: Token is defined/distributed to start and control the execution of TA and CPN respectively. As in TLS, first switching from red to red + yellow light is activated after defined interval, Transition1RtoRY: Switch is activated and TA is transitioned from red state to red + yellow state and CPN is arced from red place to red + yellow place through Red_RedYellow: Bar. Red + Yellow: Node becomes the current state (for TA) and active place (for CPN). The next switching of TLS follows the path from red + yellow to green light, AutoNet Transition2RYtoG: Switch is activated and TA undergoes transitioning from red + yellow state to green state and CPN undertakes arcing from red + yellow place to green place through RedYellow_Green: Bar. Green: Node turns into the current state (for TA) and active place (for CPN). Further in the series, TLS follows next switch from green to yellow light and Transition3GtoY: Switch is activated; TA undergoes through transition from green state to yellow state and CPN travels through arc from green place to yellow place via Green_Yellow: Bar. Yellow: Node develops into the current state (for TA) and active place (for CPN). Finally TLS executes last switching from yellow to red light, which triggers Transition4YtoR: Switch. TA completes its cycle by transitioning from yellow state to red state and CPN takes its closing arc from yellow place to red place through Yellow_Red: Bar. Red: Node builds up into the final state (for TA) and last place (for CPN). Hence one cycle of TA and CPN is completed with single iteration through AutoNet. The AutoNet may be set to loop through countable-infinite executions to reflect the real time behavior of traffic light signal simulation. It may be paused/resumed and reset/restarted at any point in time during its execution. Clock remains the switching agent for TA and Token plays the role of transitioning mechanism for CPN.

5   Discussion

Timed Automata and Colored Petri Net modeling paradigms have their own respective domain of users, and applications. Both formal methods have wide spread acceptance across industry. Despite the fact that both formalisms share common basic essence i.e. they describe evolution of system over time, yet there existed no common platform to develop both TA and CPN at one point in time through single scan of the system under study. AutoNet has fulfilled this research gap by integrating TA and CPN at meta-modeling level. The essential benefit of AutoNet is the provision of platform which automates the definition, description, development and execution of TA and CPN at one place. One particular understanding of the system can be decomposed into two types of visualizations. For instance, if one can envision the states and transition of any real time system, AutoNet has the capability to design its TA and CPN models. Another feature of AutoNet is the inter-translation mechanism from/to TA and CPN models, which is further value addition in modeling paradigm. TA and CPN model may be interchangeably translated to each other through AutoNet, with little customization. It is pertinent to mention that first edition of AutoNet is aimed at modeling basic versions of TA and CPN. It is limited at defining deterministic timed automata; we have planned to add provision of defining non-deterministic timed automata along with advanced features of TA and CPN modeling in its later versions.

6   Conclusion and Future Work

Behavioral aspect of embedded systems design is usually conceptualized by time dependent models. The precision and accuracy of time feature are the prime factors demonstrating safety and reliability of real time systems. Timed automata and colored Petri net stand among most widely adopted time-driven formal methods. We have proposed AutoNet, a Meta model by integrating the both formalisms. The seamless integration of TA and CPN is performed at Meta level. For the purpose, similarities have been worked out among TA and CPN to draw an equivalent mapping, and differences were identified to resolve the conflicts. The central aim of AutoNet is to transform each single design to both timed automata and colored petri net by going through one iteration along the system. We have validated our proposed design with the case study of traffic light signal modeling, which demonstrated proof of concept design. AutoNet is aimed at valued addition to time driven modeling paradigm for embedded systems designers. Furthermore, we have planned to work at designing Meta model for seamless integration of other candidate formal methods which share some common characteristics like B-method and Z-notation, in future.

References

1. Akdur, D., Garousi, V., Demirörs, O.: A survey on modeling and model-driven engineering practices in the embedded software industry. J. Syst. Architect. **91**, 62–82 (2018)
2. Liebel, G., Marko, N., Tichy, M., Leitner, A., Hansson, J.: Model-based engineering in the embedded systems domain: an industrial survey on the state-of-practice. Softw. Syst. Model. **17**(1), 91–113 (2018)
3. Gleirscher, M., Foster, S., Woodcock, J.: New opportunities for integrated formal methods. ACM Comput. Surv. (CSUR) **52**(6), 1–36 (2019)
4. Alur, R., Dill, D.L.: A theory of timed automata. Theoret. Comput. Sci. **126**(2), 183–235 (1994)
5. Bengtsson, J., Yi, W.: Timed automata: semantics, algorithms and tools. In: Desel, J., Reisig, W., Rozenberg, G. (eds.) ACPN 2003. LNCS, vol. 3098, pp. 87–124. Springer, Heidelberg (2004). https://doi.org/10.1007/978-3-540-27755-2_3
6. Petri, C.A., Reisig, W.: Petri net. Scholarpedia **3**(4), 6477 (2008)
7. Pawlewski, P. (ed.) Petri nets: manufacturing and computer science. BoD–Books on Demand (2012)
8. Anwar, M.W., Rashid, M., Azam, F., Kashif, M., Butt, W.H.: A model-driven framework for design and verification of embedded systems through System Verilog. Des. Autom. Embed. Syst. **23**(3–4), 179–223 (2019)
9. Rasheed, Y., Azam, F., Anwar, M.W., Tufail, H.: A model-driven approach for creating storyboards of web based user interfaces. In: Proceedings of the 2019 7th International Conference on Computer and Communications Management, pp. 169–173 (2019)
10. Srba, J.: Comparing the expressiveness of timed automata and timed extensions of Petri nets. In: Cassez, F., Jard, C. (eds.) FORMATS 2008. LNCS, vol. 5215, pp. 15–32. Springer, Heidelberg (2008). https://doi.org/10.1007/978-3-540-85778-5_3
11. Bouyer, P., Haddad, S., Reynier, P.A.: Timed Petri nets and timed automata: on the discriminating power of zeno sequences. Inf. Comput. **206**(1), 73–107 (2008)
12. Bouyer, P., Reynier, P.A., Haddad, S.: Extended timed automata and time Petri nets. In: Sixth International Conference on Application of Concurrency to System Design (ACSD 2006), pp. 91–100. IEEE, June 2006
13. Byg, J., Jørgensen, K.Y., Srba, J.: An efficient translation of timed-arc Petri nets to networks of timed automata. In: Breitman, K., Cavalcanti, A. (eds.) ICFEM 2009. LNCS, vol. 5885, pp. 698–716. Springer, Heidelberg (2009). https://doi.org/10.1007/978-3-642-10373-5_36
14. Cassez, F., Roux, O.H.: Structural translation from time Petri nets to timed automata. J. Syst. Softw. **79**(10), 1456–1468 (2006)
15. D'Aprile, D., Donatelli, S., Sangnier, A., Sproston, J.: From time Petri nets to timed automata: an untimed approach. In: Grumberg, O., Huth, M. (eds.) TACAS 2007. LNCS, vol. 4424, pp. 216–230. Springer, Heidelberg (2007). https://doi.org/10.1007/978-3-540-71209-1_18

16. Barkaoui, K., Couvreur, J.-M., Klai, K.: On the equivalence between liveness and deadlock-freeness in Petri nets. In: Ciardo, G., Darondeau, P. (eds.) ICATPN 2005. LNCS, vol. 3536, pp. 90–107. Springer, Heidelberg (2005). https://doi.org/10.1007/11494744_7
17. Bérard, B., Cassez, F., Haddad, S., Lime, D., Roux, Olivier H.: Comparison of the expressiveness of timed automata and time Petri nets. In: Pettersson, P., Yi, W. (eds.) FORMATS 2005. LNCS, vol. 3829, pp. 211–225. Springer, Heidelberg (2005). https://doi.org/10.1007/11603009_17
18. Balaguer, S., Chatain, T., Haar, S.: A concurrency-preserving translation from time Petri nets to networks of timed automata. Form. Method Syst. Des. 40(3), 330–355 (2012)
19. Xia, C.: 2009, December. Translation methods from timed automata to time Petri nets. In: 2009 International Conference on Computational Intelligence and Software Engineering, pp. 1–6. IEEE, December 2009
20. Bérard, B., Cassez, F., Haddad, S., Lime, D., Roux, O.H.: When are timed automata weakly timed bisimilar to time Petri nets? Theoret. Comput. Sci. 403(2–3), 202–220 (2008)

A Multi-purpose Model Driven Platform for Contingency Planning and Shaping Response Measures

Mukhtar Ahmad[(✉)], Farooque Azam, Yawar Rasheed,
Muhammad Waseem Anwar, and Muhammad Waqas Ahmad

Department of Computer and Software Engineering, College of E&ME, National
University of Sciences and Technology (NUST), H-12, Islamabad, Pakistan
{mukhtar.ahmad,farooq,waseemanwar,
waqas}@ceme.nust.edu.pk,
yawar.rasheed18@ce.ceme.edu.pk

Abstract. Effective emergency response requires situational awareness and preplanning for various contingencies inherent to the catastrophe; may it be a natural disaster or man-made crisis situation. Model Driven Software Engineering has contributed to the domain of contingency planning and response in a befitting manner by providing generic and scalable models, simulating emergency scenarios, in order to enhance the skills of responders. However, a thorough literature review identified that a comprehensive, intelligent and repository based model for planning of various contingencies inherent to the emergency situation is a mile stone to be achieved. Accepting the challenge, Interactive Contingency environment creation and Response Planning System (CRIPS) is proposed, which is a model driven platform/framework and facilitates a crisis manager to create a virtual emergency environment with its diverse ingredients and shaping an effective response actions to cater it. A multi perspective feedback mechanism and centrally administered picture board, which are the essential components of any response planning system, have also been incorporated. In addition, intelligent rater and repository concepts are introduced to rate the planning of crisis manager and save this whole contingency environment for analysis and self-learning of model, making it distinctive to the previous researches. The outcome of this research is a comprehensive meta-model, which can be further extended for model based development of an effective contingency and emergency response planning system. The validity of proposed meta-model is demonstrated through a real world case study of terrorists attack on Army Public School (APS) Peshawar/Pakistan. The results prove that proposed model is capable of modeling simple as well as complex scenarios and allows a crisis manager to effectively model a response in order to deter the catastrophic effects.

Keywords: Contingency · Meta-model · Model driven · Response · Feedback · Rating · Repository

© Springer Nature Switzerland AG 2020
K. Saeed and J. Dvorský (Eds.): CISIM 2020, LNCS 12133, pp. 320–331, 2020.
https://doi.org/10.1007/978-3-030-47679-3_27

1 Introduction

Communities and environments are facing climate change, such as floods, cyclones, earthquakes, fires, and tsunamis, and technological infrastructure failure (like computer system failure), added with human induced threat like terrorism and war. So considering the unpredictability of emergency, its impact, nature and scope, it is inevitable to be prepared for this type of impulsive scenarios. For better preparedness it is vital to train the crisis managers in such a way that they can visualize and create the hypothetical contingency environment with different intensities and know how to respond by utilizing all the actors and assets in unified way toward the end goal of minimizing emergency cost by sharing responsibilities with their best capabilities. Inherited nature of dynamicity and complexity of emergency give a great challenge to model and automate contingency creation and response environments.

Modern software development engineering approaches are helpful to make the task of simulating these environments like agent-oriented software engineering (AOSE) and model-driven development (MDD). In AOSE, different assets and responses are modeled by using the concept of agent. These agents are actually autonomous, reactive, and proactive software components [14]. In [22] provides the framework for developing multi-agent system (MAS) for emergency response environment (ERE) by using aforementioned concepts. It also provide a tool for developing model driven ERE system with code and an execution platform.

Despite enormous work done in modeling of contingency/emergency response planning, there is a big missing of a comprehensive, intelligent and repository based model to deal with inherent physical damages and life losses caused by emergency outbreaks. To address this missing, Interactive Contingency environment creation and Response Planning System (CRIPS) is proposed, a meta-model which facilitate the crisis manager to simulate hypothetical contingencies, its victims, responses and feedback mechanism. It also embodies centroid wholesome picture board, intelligent rater for planning and a repository to save strategic planning steps taken against these adverse situations. This will not only increase the skill and capability of crisis managers to face crisis in real scenario but also enrich the repository with diverse intellectual approaches to cater catastrophic endeavors. To grasp the concept easily a simple description model is shown in Fig. 1.

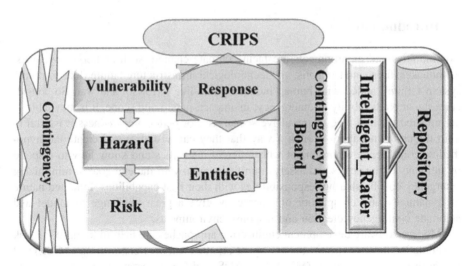

Fig. 1. Proposed meta-model of CRIPS

2 Literature Review

Emergency management is complete process to handle the emergency in different scenario and prospective. It is service sector to facilitate the individuals and societies to guide them and help them in the emergency environment. To respond the emergency situation, responders use lengthy pre-written steps and measures in the form of files and documents but these are inefficient and unproductive ways. A new mechanism consisting of software system to model, plan and respond is inevitable [1]. Local documented procedures and models with unformal symbols to represent entities in environment and flow are difficult to understand and execute. So formal entity notations, interaction with graphical flow such as Business Process Model and Notation (BPMN) [2], Unified Modeling Language (UML) [3], and User Requirements Notation (URN) [4], to represent processes are best alternative [5]. Extending these formal notation by using domain specific terms will make the representation of a system more comprehensive and effective [6]. Standard modeling languages aforementioned are common but to handle domain specific system modeling, they are to be extended with underlying meta-model [7].

Another method is to model emergency [8] in real world or simulate by software application (Emergency Response Application) [9]. Successive workshops [8] were conducted with realistic emergency setting as demo to understand the situation possibilities of happenings and responses. Garcia et al. [10] proposed a model driven application to simulate the ERES and interaction pattern between the victims and rescuers [11]. A modeling tool called IDK (INGENIAS Development Kit) was used to develop agents for the groups with goals and capabilities. It was also tested and optimized to better its performance of interaction but other aspects like team formation and task allocation was ignored [12]. Xie et al. [13] presented that social media is a rich source of real time and critical information about the disaster and emergency and can be used to collect, filter fake feeds and manage to respond the emergency and disaster in effective way.

Despite the applications and simulations for EREs, the dynamic nature makes the system difficult and complicated. To cater this challenge Model Driven Software Engineering and Model Driven Architecture provides a platform to develop generic and scalable models that have huge potential and power to simulate complex systems in an organized way. For example, authors of [14] developed a meta-model for creating story boards of web based user interfaces which is capable of generating simple as well as complex story boards. Anwar et al. [15] benefits the underneath power of model-driven engineering (MDE) and proposes a framework to simplify the design and verification process of embedded systems by using UML profile for SystemVerilog (UMLSV).

Mustafa et al. [16] proposed Conceptual Role Organizational Model (CROM) framework for modeling and simulating disasters using organizational methodological by following model driven architecture (MDA). Gascueña et al. [17] approach simplifies the model developing with interface code that can be completed to develop full-fledge system. In [18] propose a domain-specific software language (DSL) by using Use Case Map (UCM) meta-model, providing 3D environment for modeling disaster with location and social media type interaction. A flow model to detect information spread on Weibo, largest Chinese microblog site, indicating a rich-gets-richer phenomenon in the sense of relation network and authenticity features [19]. A design science research methodology [20] is proposed to develop a decision meta-model for response process and needs in Philippines context. Base line data for this methodology was used by a nongovernmental organization. A model-driven framework ERE-ML [21] is proposed by integrating multi-agent system (MAS) and domain-specific modeling language (DSML) named ERE-ML. This is a full-fledge system from language, developing tool, code generation mechanism and platform for execution However, ERE-ML does not facilitate interaction between agents and organization. Later on this discrepancy was enhanced by extending modeling framework, transformation code and platform by the researcher in ERE-ML 2.0 [22].

By exhaustive journey of proposed research till date related to model driven concept utilization for the EREs, few gaps like feedback mechanism, centroid wholesome picture, rating and database are identified and addressed in the proposed meta-model CRIPS which enables it to rate the planning of the crisis manager and save this whole environment creation and planning snapshot for analysis and self-learning of system itself. A case study is used to validate the proposed meta-model. The results shows that CRIPS is capable for contingency environment creation and response planning.

3 Proposed Meta-model

Emergency response management is real time task in case of emergency actually happens. The proposed meta-model (CRIPS) provides the basis for the crisis mangers to learn the environment creation and response management in planning phase to enhance their capability in actual task that is multi-perspective and multidimensional. Following Fig. 2 show the CRIPS meta-model developed in Papyrus, Eclipse.

Proposed meta-model has CPRIS as root element which embodies the most of the other element of the model. Root element can have one or more contingencies at a time encompassing vulnerabilities that may be available in real scenario and projecting hazards with different probabilities caused from theses vulnerabilities. Actual risks are created by these hazards with supposed intensities as per exposure of contingency. Every risk has victims, in our case, it threatens the entities (may be tangible or intangible or responders themselves), so after creating contingency, victims can be created as per type and exposure of mimicking emergency. Again considering quantity, worth and priority like parameters response by responsible organizations/teams is assigned and performed. Member and resource like elements are available to develop minor to huge level of response to the help and neutralize the risk threat. Timestamp of help call/allocation of resources with priority is maintained by responder organization to execute rescue operation in a logical and sensible flow. Although there are already more exhaustive simulations available in term of presenting a model (risks and victims) with its implementation but information flow and reporting mechanism is not present in those models which is one of the basic ingredient of the emergency response environment. This flow of information (feedback) leads toward vibrant and dynamic decision making and enriched planning. So modeling diverse feedback, diverse in the sense that coming from different entities with different perspective and situation at hand, in the model makes it more effective and useful tool for contingency planning. Current era is information based and importance of rightful information play important role with response action, even more than response for psychological reasons. So how to manage and process temporal and spatial inflow feedback and display (and communicate) in effective way to recipients. This concept with the name of CPB (Contingency Picture Board) is also added to make the crisis manager able to deal this scenario of response during planning. What about the metrics to assess the good or bad planning by the crisis manager? Proposed meta-model also provides this facility to the crisis manager to evaluate and consider its strategic decisions and steps during planning. This is named as intelligent Rater class, taking data as feature vector from CPB and repository, using regression algorithms to rate manager's contingency planning strategy. Repository is also there, bank of strategic planning processes, database for intelligence and rating. Snapshots of the whole contingency environment creation and response planning activity will be stored in repository for enriching the strategic data bank and making the CRIPS meta-model intelligent. It enhances the capability of model to evolve inwardly and evaluate crisis manager outwardly. Bounding crisis manager to follow a strict flow of actions against contingency event loses model robustness and variant rich knowledge for repository. Used concepts in the meta-model are illustrated in Table 1.

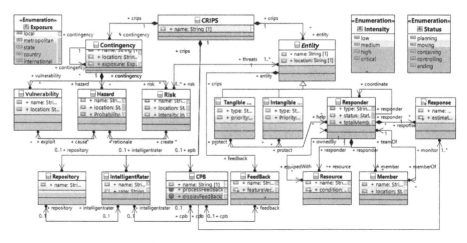

Fig. 2. Proposed meta-model of CRIPS

Table 1. Basic concepts (classes) definitions

Name	Definition
CRIPS	It stands for "Interactive Contingency Response Planning System". It is basic container class and depicts the actual theme of whole system
Contingency	Mimicking a sudden natural, manmade or technological phenomena which will embodies some properties like vulnerabilities, hazards and risks
Vulnerability	Any discrepancy or shortage or laps which is due to lack of management or resources
Hazard	An affect that can exploits the laps and vulnerabilities and results in risks
Risk	It is the probability of the hazard toward the assets and can range from low to high value (0–1)
Entity	It is the abstract and general class which has some specific class to be implemented with some common features in our model
Tangible	These are entities (assets) that have physical body and dimension like human, building
Intangible	These are the entities (assets) that have no physical body but has eminent worth and cost like reputation, morale, brand value and economy
Responder	This is modeled to represent the organization (teams) which will response against the risk and rescue the assets like security forces and fire brigade etc
Response	It is the task and role which every team or organization has to perform to mitigate the risk and save entities
Resource	It represents the tools and techs owned by the organizations used to neutralize the contingency

(continued)

Table 1. (*continued*)

Name	Definition
Member	It is the basic component of response system which develops teams and works collectively for the assigned response and task
Feedback	This is the collections of meaningful data of the environment under threat and reported to the CPB
CPB	CPB stands for "Contingency Picture Board" and it models the information desk center. It has responsibility to manipulate the input feedback and display information in meaningful formation
Intelligent_Rater	This class has the concept of rating. It takes input from the CPB and repository and analyze the recent task planning with the overall calculated average performance of the already completed task maps
Repository	This is the class to represent the database behind the system to save the snapshot of the whole contingency environment creation and planning environment. This maintains records of every planning assignment and can be accessed by the intelligent Rater to rate the current planning task

4   Case Study

A case study of APS Attack-2014 is modeled to appreciate the proposed CRIPS (Interactive Contingency environment creation and Response Planning System) meta-model. Even though model is robust and is scalable to cover more but considering the space limitation of this section, only few aspects are generated and modeled which serves the purpose.

4.1   APS Attack 2014 Peshawar, Pakistan

On 16 December 2014, six terrorist conducted a terrorist attack on the APS (Army Public School) situated in the renowned city Peshawar of Pakistan. This brutal attack caused 149 causalities including 132 school children. The exposure and intensity of emergency required many metropolitan and state level organizations to respond and coordinate mutually for neutralizing the situation. Pakistan Security forces, Fire Brigade, Rescue 1122, Hospitals (Military and Civil) and many more coped the attack and rescued 960 lives.

4.2   Modeling APS Attack-2014 with CRIPS

In this part of the section crisis manager intends to create the contingency of APS attack-2014 environment and streamline the response in the same context as shown in Fig. 3. The modeling of environment follow;

Emergency Environment Creation
Crisis Manager firstly create a contingency with the name of "APS (Army Public School) Attack" with hypothetical location. Its exposure is declared at "country" level and reason behind this emergency is terrorism, so type of this contingency is shown as

"man-made". As emergencies can create other emergencies, so "APS Attack", a terrorism act, contingency leads toward two sub-contingencies named as "Fire Explosion" and "Hospital Emergency" with exposure level as "Local" and "metropolitan" respectively.."Fire Explosion" is due to technological issue, so described as "tech" and "Hospital Emergency" is created by both other contingencies, so its type is both "tech" and "man-made".

Vulnerabilities, Hazards and Risks Encompassed
Essential ingredients of three contingencies, including vulnerabilities, hazards and risks are created. "APS Attack", "Fire Explosion" and "Hospital Emergency" contingencies have "Low boundary wall", "Electric wire naked" and "Low blood bank capacity" respectively. These vulnerabilities lead toward hazards "Intrusion", "Eclectic fire" and "Shortage of blood" with probabilities 0.8, 0.7 and 0.8 respectively. These hazards turns into risks of "Terrorists attack school", "Building on fire" and "Causalities of personals" with intensities of critical, high and high respectively.

Assets (Tangible or Intangible)
Risk created in previous step of contingencies ultimately threats entities, which may be tangible, "Students" and "Classrooms", and intangible like "Army Repute". First two are effected directly because these are physically in proximity of the event and third entity is indirectly effected. The threat level (intensity) faced by these entities is marked as "Critical", "Medium" and "High" respectively. These entities are located at diverse locations as mention in the Fig. 3 in their instance specification slots values.

Building Organizations with Response Assignment
During execution of the contingency, in this step, crisis managers responds, to the entities for their asked or perceived help request. In response, respective organizations, "Security Agencies", "Fire Brigade" and "CMH" are activated. These organization mobilize their teams named as "QRF", "Fire Squad" and "Medical staff" with assigned response of "Protecting students", "Extinguishing fire" and "Medical treatment" respectively. Teams are built with members and equipped with resources to deliver their responsibilities. "QRF" consist of three professional security members with names S1, S2 and S3 and rifles with good condition. "Fire Squad" consist of two professional fire fighters named as FF1 and FF2 having fire extinguishers with medium working state. "Medical staff" consist of M1, M2 members with good condition laboratories. These teams are given estimated time slab to complete these responses with values 120,240,420 min hypothetically.

Handling, Synthesizing Feedback and Managing CPB
Feedback can be initiated by any of the living victim, responder or indirect observer toward a control system for help, guide and show concern. In this context, student feeds a feedback with diverse, un-structured and semi-confirmed data as "Terrorist_6, location_room3, Casualities_50, Risk_high, Injured". Same way "QRF" reports a feedback to the CPB (Contingency Picture Board) as "Arrival_time_30Min, Expected_Completion_time_120Min, Terrorist_4, location_BackGate, Casualities_30, Risk_high, Controlling". Crises manager gets experience and handle this type of unstructured reported feedback to change strategic decisions. For example two different figures about terrorists and casualties from two different entities needs perspective,

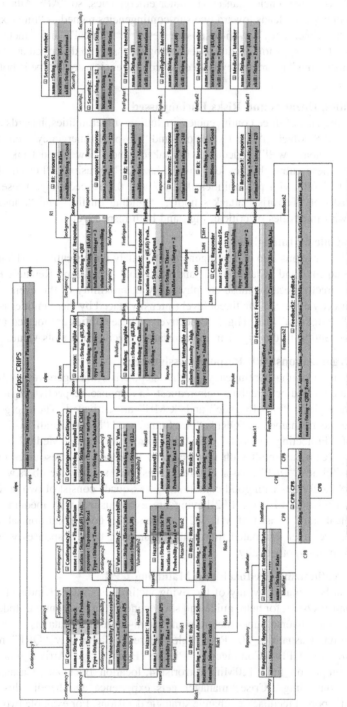

Fig. 3. CRIPS meta-model validation via Instance level diagram in UML

experience and authenticity measure to synthesize and enhance the response level and activation of other resources. Crisis manager go through whole process of managing (on CPB) diverse and overlapping reported data in peace time. It is also learn that how to synthesis this inflow data and only show relevant and well affirmed information as per perspective audience (on CPB). For example at initial stage, a non-confirmed causalities at huge number are reported and without sensing its concerns are propagated then parents of the students can suffer at large with anxiety and strain.

Intelligent Rating Mechanism in Action
By taking this synthesized information at the end of contingency planning process and comparing it with samples available in the repository, enables the "IntellegentRater" to rate the effort of crisis manager. It only takes those records which have matching type of emergency and exposure level. For example, in this contingency, type of emergency was man-made, a terrorism act, its exposure was country level, and victims was innocent student. Rater searches those previous executed contingencies management consisting of various stages which has same features, and then check, how many organizations were activated and, which type of resources was capitalized and how much time was taken to neutralize the contingency with rating. The whole process of rating is beyond the scope of meta-model and will be address in future work.

Saving Snapshot Planning Scenario with Rating
Synthesized information in a recorded format with rating is saved in the repository. This is the full fledge snapshot of the contingency environment creation and response planning system. This is not only used to rate the new crisis manager but also will be a rich knowledge bank consisting of strategic thinking templated for diverse and complex contingency scenarios. The structure and semantic of repository for this type of mixed data will be addressed in future work.

5   Discussion

Proposed CRIPS meta-model provides mechanism to crisis manager for contingency creation and response planning, making confident and effective him in actual disaster scenario when stress is prevailing around. By using this model, crisis manager is able to locate the vulnerabilities existing in the zone/organization, resulting hazards which will explode and associated risks created by these hazards. It gives him new vision and aspect to brainstorm all possibilities of happenings and their consequences by simulating real scenario of emergency environment and response requirement. Considering the type of disaster, man-made or natural, entities under threats are identified and priority of their worth is marked with intensity of threat. Understanding organization selection and activation with different capacities and skills are build and in true sense are executed, resulting to know every actor and stakeholder their responsibilities and assignments. Tangible entities bearing the disaster circumstances are able to feedback status of scene and sufferings to the responding organizations. In CRIPS, the crisis manager is able to explore these diverse feed from the direct and indirect effected ones. A multi-perspective data will strengthen to decide upon. It enhances the capability of crisis manager to integrate and synthesis inflow of feed-back to make it believable and negate the propaganda. For this purpose CPB concept is introduced in the meta-model. It also facilitates the rating of configuration and planning of current crisis manager by comparing it with database w.r.t. intensity, exposure and type. A repository containing snapshots of planning with rating gives model a knowledge bank, distinguishing CRIPS to evaluate and evolve. With all these contributions, it also has limits in the form of data storage schema for repository, will be a challenging task itself due to unstructured inflow data. Also it gives only one aspect of Contingency Management System i.e. Response. In future, work will be carried out in defining schema for repository, creating protocols for integration and synthesizing inflow feedback.

6   Conclusion and Future Work

Interactive Contingency environment creation and Response Planning System (CRIPS) is a meta-model which facilitates the crisis manager to create contingency with its diverse ingredients and interactive response actions to cater it with comprehensive feedback mechanism and centroid wholesome picture board which are the essential components of any response mechanism Additionally it comprises of intelligent rating and repository to rate the planning of the crisis manager and save this whole contingency environment creation and planning snapshot for analysis and self-learning by model itself. In the case study section, the validity of proposed meta-model is demonstrated through a real terrorists attack on Army Public School (APS) Peshawar/Pakistan. The results prove that proposed model is capable of modeling simple as well as complex scenarios and allows a crisis manager to effectively model a response in order to deter the catastrophic effects. In future, work will be carried out in defining schema for repository, creating protocols for integration and synthesizing inflow feedback. A prototype tool for contingency planning, based upon the proposed meta-model, is also a milestone to be achieved.

References

1. Sell, C., Braun, I.: Using a workflow management system to manage emergency plans. In: Proceedings of the 6th International ISCRAM Conference, vol. 41 (2009)
2. Model, Business Process. Notation (bpmn) version 2.0. OMG Specification, Object Management Group, pp. 22–31 (2011)
3. Object Management Group (OMG). Welcome To UML Web Site
4. International Telecommunication Union (ITU). Recommendation Z.151 (10/12): User Requirements Notation (URN) - Language definition (2012)
5. Chinosi, M., Trombetta, A.: BPMN: an introduction to the standard. Comput. Stand. Inter. **34**(1), 124–134 (2012)
6. Braun, R., Esswein, W.: Classification of domain-specific BPMN extensions. In: Frank, U., Loucopoulos, P., Pastor, Ó., Petrounias, I. (eds.) PoEM 2014. LNBIP, vol. 197, pp. 42–57. Springer, Heidelberg (2014). https://doi.org/10.1007/978-3-662-45501-2_4
7. Braun, R., et al.: Extending a Business Process Modeling Language for Domain-Specific Adaptation in Healthcare. Wirtschafts informatik (2015)

8. Kyng, M., Nielsen, E.T., Kristensen, M.: Challenges in designing interactive systems for emergency response. In: Proceedings of the 6th Conference on Designing Interactive Systems. ACM (2006)
9. Uhr, C., Johansson, H., Fredholm, L.: Analysing emergency response systems. J. Contingencies Crisis Manag. **16**(2), 80–90 (2008)
10. García-Magariño, I., Gutiérrez, C., Fuentes-Fernández, R.: The INGENIAS development kit: a practical application for crisis-management. In: Cabestany, J., Sandoval, F., Prieto, A., Corchado, Juan M. (eds.) IWANN 2009. LNCS, vol. 5517, pp. 537–544. Springer, Heidelberg (2009). https://doi.org/10.1007/978-3-642-02478-8_68
11. Pavón, J., Gómez-Sanz, J.J., Fuentes, R.: The INGENIAS methodology and tools. Agent-oriented methodologies, pp. 236–276. IGI Global (2005)
12. García-Magariño, I., Gutiérrez, C.: Agent-oriented modeling and development of a system for crisis management. Expert Syst. Appl. **40**(16), 6580–6592 (2013)
13. Xie, J., Yang, T.: Using social media data to enhance disaster response and community service. In: 2018 International Workshop on Big Geospatial Data and Data Science (BGDDS). IEEE (2018)
14. Anwar, M.W., et al.: A model-driven framework for design and verification of embedded systems through SystemVerilog. Des. Autom. Embedd. Syst. **23**(3–4), 179–223 (2019). https://doi.org/10.1007/s10617-019-09229-y
15. Rasheed, Y., et al.: A model-driven approach for creating storyboards of web based user interfaces. In: Proceedings of the 2019 7th International Conference on Computer and Communications Management. ACM (2019)
16. Mustapha, K., Mcheick, H., Mellouli, S.: Modeling and simulation agent-based of natural disaster complex systems. Procedia Comput. Sci. **21**, 148–155 (2013)
17. Gascueña, J.M., Navarro, E., Fernández-Caballero, A.: Model-driven engineering techniques for the development of multi-agent systems. Eng. Appl. Artif. Intell. **25**(1), 159–173 (2012)
18. Khzam, N.B., Mussbacher, G.: Domain-specific software language for crisis management systems. In: 2018 IEEE 8th International Model-Driven Requirements Engineering Workshop (MoDRE). IEEE (2018)
19. Dong, R., et al.: Information diffusion on social media during natural disasters. IEEE Trans. Comput. Soc. Syst. **5**(1), 265–276 (2018)
20. Llesol, M., Intal, G.L.: Development and validation of a disaster response decision metamodel in the philippine context. In: 20019 IEEE 6th International Conference on Industrial Engineering and Applications (ICIEA). IEEE (2019)
21. HoseinDoost, S., et al.: A model-driven framework for developing multi-agent systems in emergency response environments. Softw. Syst. Model. **18**(3), 1985–2012 (2019). https://doi.org/10.1007/s10270-017-0627-4
22. HoseinDoost, S., Fatemi, A., Zamani, B.: Towards a model-driven framework for simulating interactive emergency response environments. J. Comput. Secur. **5**(1), 35–49 (2018)

Multi-criteria Differential Evolution for Optimization of Virtual Machine Resources in Smart City Cloud

Jerzy Balicki[1]([✉]) [ID], Honorata Balicka[2] [ID], Piotr Dryja[3] [ID],
and Maciej Tyszka[3]

[1] Faculty of Mathematics and Information Science, Warsaw University of
Technology, Koszykowa Street 75, 00-662 Warsaw, Poland
j.balicki@mini.pw.edu.pl
[2] Sopot University of Applied Sciences, Rzemieślnicza Street 5, 81-855 Sopot,
Poland
hbalicka@ssw.sopot.com
[3] Faculty of Telecommunications, Electronics and Informatics, Gdańsk
University of Technology, Narutowicza Street 11/12, 80-233 Gdańsk, Poland
piodryja@pg.gda.pl, tyszka.maciej@gmail.com

Abstract. In a smart city, artificial intelligence tools support citizens and urban services. From the user point of view, smart applications should bring computing to the edge of the cloud, closer to citizens with short latency. However, from the cloud designer point of view, the trade-off between cost, energy and time criteria requires the Pareto solutions. Therefore, the proposed multi-criteria differential evolution can optimize virtual machine resources in smart city clouds to find compromises between preferences of citizens and designers. In this class of distributed computer systems, smart mobile devices share computing workload with the set of virtual machines that can be migrated among the nodes of the cloud. Finally, some numerical results are studied for the laboratory cloud GUT-WUT.

Keywords: Artificial intelligence · Differential evolution · Smart city

1 Introduction

Artificial intelligence with metaheuristics and deep learning algorithms supports cloud computations in the city cloud by providing the high quality solutions for adequate NP-hard issues or for prediction, classification, regression, and clustering based on Big Data (BD) streams. Besides, the mobile devices can perform the selected AI algorithms combining an aggregation of data from sensors in real time and using low-energy consuming processors to achieve results without the extensive interactions with the cloud. Thousands sensors are able to send data about the state of physical world to the core of the cloud via the Internet of Things (IoT), and then decision making can be efficient.

© Springer Nature Switzerland AG 2020
K. Saeed and J. Dvorský (Eds.): CISIM 2020, LNCS 12133, pp. 332–344, 2020.
https://doi.org/10.1007/978-3-030-47679-3_28

In smart city, numerous smart applications on mobile devices can use Big Data files to support city services and citizens' expectations. However, to object detection or video classification, deep learning artificial neural networks are supposed to be trained. Although, there are accessed some high-quality pertained networks, the intensive learning is required on the workstations with graphical processors or supercomputers to achieve the accepted value of accuracy, precision, F1-score, or Area Under Curve (AUC). For example, 3D convolutional neural networks are able to recognize numerous human actions [2]. After long last training on the cloud server, they can be moved to the smart phones of citizens, and then quickly detect some critical actions.

Because of limited processing power and memory related to edge servers that are physically local to citizens, the larger cloud resources are required [3]. Besides, the large applications can be divided on many modules performed both at the edge device and at the core cloud servers. The high wireless transmission capacity 10 Gb/s offered by the communication 5G can support numerous smart agents that can move among hosts to optimize some different aspects of workload.

The Internet of Things (IoT), the fifth generation of wireless communications technologies supporting cellular data networks 5G and several AI-driven algorithms permit on knowledge mining about some phenomena and features of city dynamics [6]. In results, the deep artificial neural networks provide tools for better understanding citizens' behaviors or for inhibition some crisis situations. The Convolutional Neural Networks (CNNs) can be trained to anticipate the annual expenditure budget to satisfy a community expectation [5]. Also, numerous complaints of citizens about disadvantages of the city infrastructure can be efficiently detected and fixed. For example, 3D CNNs are able to recognize numerous human actions [13]. Moreover, the Long Short-Term Memory artificial networks (LSTMs) support an urban planning and monitoring the states of buildings [11]. Big Data processing is the main goal of deep learning algorithms. LSTMs serve nearly 30% of the artificial neural network inference workload in Google datacenters [33]. Deep learning is an emergence technology that can reduce the capacities of Big Data streams by data fusion, too.

Experiments carried out with smart city systems like the New York information network "City 24/7" [2], SmartSantander [32], Masdar city [2], FixMyStreet [16], Dublin public transport [17], and Tsukuba Science City [2] confirmed that the robust computer infrastructure is crucial to support fields related to citizens, living, and education. Correspondingly, ecological environment, research, and governance require an efficient cloud. Some authors indicate healthcare, economy, and employment opportunities as areas for development by new models and technologies, too. Mobility, energy management, city infrastructures, and technology for citizens should be integrated in the dynamic and smart city. Above key areas can be supported by smart human capital [31].

City applications can be updated if they implement deep machine learning algorithms. However, retraining of such algorithms is time-consuming for large datasets. It could last many days on PC computer with the newest CPU processors. That is why, the workstations with modern GPU processors are necessary. In the cloud of hosts, virtual machines (VMs) by migration can reduce the number of active physical machines [10]. The virtual layer between hardware and operating system supports that different applications can run on any host. The cloud virtualization software can manage VMs on one host. The active state of VM can be transferred from one host to another machine, what permits the possibility to move workload among several servers. We propose the multi-objective differential evolution to handle with VM migration.

To order many important issues in this paper, related work is described in Sect. 2. Then, the artificial neural networks for edge computing at smart city are characterized in Sect. 3. Next, Sect. 4 presents some studies under Tweeter's blogs for smart city apps. The design principles of a differential evolution for the smart city are analyzed in Sect. 5. Finally, some findings are presented in Sect. 6.

2 Related Work

Multi-criteria differential evolution for optimization of virtual machine resources is important for the complex computer systems like a smart city cloud [2]. By definition, a smart city is considered as an "environment which involves many technologies and multiple agents collecting data from sensors scattered around the whole city" [28]. Computing clouds provide large computational resources which are ready to use from anywhere, anytime on request that is compatible with the concept of smart city.

Sutar, Mali and More propose system providing dynamic and energy efficient live VM migration approach to reduce wastage of power by initiating sleep mode of idle hosts for energy saving [33]. Their task allocator determines overloaded servers, and then optimizer analyses load on physical machine using the ant colony optimization algorithm. Besides, Local Migration Agent select appropriate physical servers to migrate VMs. Finally, Migration Orchestrator moves the VM load to the hosts, and Energy Manager initiates sleep mode for idle physical machine [33]. In our work, we propose to extend our previous OpenStack based management model [2] by adding above dynamic saving energy procedure. Besides, we suggest using differential evolution for solving this new multi-criteria optimization problem.

Veeravalli and He consider economics and market mechanisms for computing cloud as an emerging computing market where cloud providers and users as the players share, trade and consume computing resources [36]. Furthermore, economic mechanisms (such as auctions and tiered pricing) can implement a diversifying pay-as-you-go paradigm. Besides, they study Cloud of Clouds wherein the computational and data infrastructure for handling scientific, business and enterprise applications span across multiple clouds and Data-Centers [36].

Agarwal and Raina study live migration of virtual machines in cloud [1]. Migration of VMs is live, because the original VM is running, while the migration is in progress to reduce the VM downtime for the order of milliseconds. They analyze the load balancing among the hosts regarding the processor usage or the IO usage subject to the limited virtual machines downtime. The control from virtual machines is converted to the management of services in Red Hat Cluster Suite, and then cluster services provide mechanisms for VM migration [1].

OpenStack is the suited cloud software for supporting live migration of VMs using [24]. Biswas et al. consider live migration of virtual machine using high network interfaces with transmission capacity 10 Gb/s. They use three physical machines Dell PowerEdge R815 (AMD Opteron 6366HE@ 3,6 GHz, 128 GB RAM) located in

Coleraine, Dublin, and Halifax. The first host provides resources for 10 Gb/s network interfaces supported software Openstack Icehouse with QEMU 1.2.1. The second server is a compute node running with Nova 2.18.1 and *nova-network services*. The third host works as both a compute node and the controller node providing also the other management services [7].

The most advantageous mechanism for managing resources in cloud are studied in many works [26]. Patel et al. recommend green cloud mechanism for resources management [30]. Yang et al. (2015) mentioned libvirt function to establish cloud infrastructure using KVM hypervisor to introduce green cloud computing by shutting down idle cloud resources [38].

Dhanoa and Khurmi analyze the effect of size of virtual machine and bandwidth of network on total migration time required to migrate VM. The purpose is saving energy to support green cloud computing [15]. Kumar and Prashar propose algorithms to equal load in cloud computing environment [23]. Wang et al. introduce mechanism to increase profit for placing VM in datacenter and diminish number of VM migrations [37].

Metaheuristics such as differential evolution or ant colony algorithms can be used for solving problems of cloud resource management. The differential evolution is able to solve the Chebychev Polynomial fitting problem. Besides, it produces high-quality solutions for non-convex optimization problem, multi-modal and non-linear functions, good multi-variable solving function, easy programming and simple operations [2].

Differential evolution uses a perturbation of two members as the vector. A new vector is obtained by adding to third member previous vector. At the crossover operation, the algorithm with certain rules mixes the new vector with the predefined parameters to produce test vectors. At the next step, the test vector of function is compared to the target function. If the test vector is less than target function, the test vector instead of the target is in the next generation. The same number of competitors in next generation is produce by final choice of the operation on all members of the population [30].

Differential evolution can reduce the received power and increase received data error rate for designing wireless communication network among high buildings. The algorithm evaluates different structures of the antenna arrays. It is investigated three different shapes of antenna the L shape, the Y shape and the circular shape. The test simulation was preform in New Taipei city [18].

Virtual machines can train deep learning algorithms to provide the best algorithms for decision making in smart city. Moreover, differential algorithm can optimize live migration of virtual machines with deep artificial networks dedicated for city tasks. Deep learning belongs to the most exciting and efficient approaches for designing artificial neural networks. Achievements and approaches related to development artificial neural networks for supporting mobile devices regarding edge and core cloud computing are presented in [20, 21, 28].

3   Deep Artificial Neural Networks for Smart City Applications

Areas of a smart city that related to deep machine learning are traffic issues, web clip recognition, waste management, energy supply and demand governance. For instance, sensors embedded in bins can indicate when to pick them up or the CNNs recognize objects to a waste segregation. Regarding the traffic solutions, such tasks like the smart parking management, a time control of road lights, and the alternative road recommendations are reinforced by deep learning algorithms [19].

To obtain the accuracy 99%, an elapsed time of the CNN training at the virtual machine image on Linux Fedora Server version 30 (the CPU Intel Core i7, 27 GHz, RAM 16 GB) is approximately 1 min for the German Traffic Sign Detection Benchmark Dataset. The CNN identifies the traffic signs from images taken at German roads. Each sign is represented by the matrix with 28×28 selected features. The dataset was divided on 570 training signs and 330 test ones [22].

Much more elapsed time consuming issue is the training of the LSTM for video classification that is important to detect some danger situations from the web camera monitoring system. We trained the Cityscapes Dataset that encloses 25,000 stereo videos about the street scenes from 50 cities – details of this dataset are presented in [13]. For this case, the above virtual machine is too slow for the practical using. Figure 1 shows the progress of training after 4,200 min (70 h) of the elapsed time of the LSTM implemented by the Matlab R2019b on single CPU Intel Core i7 with Windows 10. The validation of the training process is made after each 377 iterations per epoch.

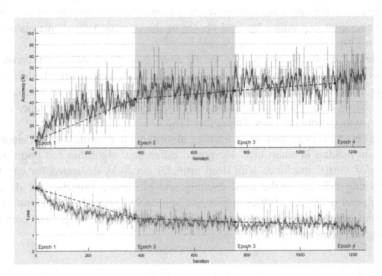

Fig. 1. The accuracy and the loss function improvement during training of the LSTM neural network

Migration of the virtual machine with the pre-trained LSTM to the more powerful workstation with the GPU can be done by using protocols HTTPS and WebSocket. Besides, micro-services exchange data with format JSON.

We constructed the computing cloud called GUT-WUT (Gdańsk University of Technology – Warsaw University of Technology) based on OpenStack software to confirm possibilities of live migration and resource management optimization. The GUT-WUT is the cloud version of the *Comcute* grid that was constructed at the Gdańsk University of Technology [3, 12].

Especially, the following instruction is developed for moving the VM to the target node in the cloud:

```
openstack server migrate 'id VM' 'target node'
```

Alternatively, it can be used the web API with the method POST and url: /servers/{id VM}/action, and then we can send data in JSON format by the following instruction:

```
{"migrate": {"host": "target node"}}.
```

Another useful dataset is the Human Motion DataBase (HMDB) with 6,849 clips divided into 51 action categories, each containing a minimum of 101 clips. For instance, the trained LSTM can detect smoking and drinking in the forbidden areas, pedestrian falling on the floor, or the shot gun. These automatic alarm classifications can allow counteracting many extreme situations on city streets [22].

Estimating road congestion can be developed by using some indirect information about mobile cell network load. This approach provides alarms about the unusual situations on the roads like traffic jams. The other example is the smart analysis of RFID tags applied to mark objects in stores and warehouses. These tags can be placed to any city objects in purpose to provide information. Artificial neural networks can be adapted to analyze RFID tags, too [27]. They can serve as a helper during navigation in warehouse and looking for the selected RFID tag. Besides, they can be applied for car navigation in a parking, or looking for car parts in the mechanical store.

Deep learning is used at ubiquitous IoT to support end devices and smart sensors [29]. Autonomous systems such as vehicles are supported by the CNNs, too. Besides, artificial neural networks are developed for detecting the flood attack by computer viruses. The robust intruder detection systems are applied for multi-hop wireless mesh networks. Those networks are integrated with other networks via the special gateways that provide a common interface to the system. These gateways are resistant against malicious attacks. One of the attack is the distributed denial of service attack (DDOS). The aim of the LSTM is to recognize the pattern of the network traffic, which might be caused by the ongoing flood attack [29].

A ubiquity of smart devices brings another challenge – it is expected that all devices like fridges, watches, and cars would have their own IP addresses. Therefore, there are various directions in research related to deep learning and the Internet of Things what force the intensive progress in the smart city and Industry 4.0 [25].

4 Agents and Social Media

Live migration of virtual machines with intelligent applications among computers in city cloud develops the idea of transferring much more autonomy to software agents. For example, the system JAnEAT for the decentralized traffic control is based on the messaging infrastructure JADE and the NEAT neuro-evolutionary learning algorithm for controlling the traffic lights [2]. Agents are divided into classes (adaptive agents, observer agents). Main agents are responsible for routing in the IoT network (including device detection). Adaptive agents performs neuro-evolutionary computations – effectively meaning preparing new versions of observer agents. Observer agents change lights in particular road intersections.

Healthy food for residents plays a key role in the urban community. This approach is effective even in food problem domains like detection of the decaying fruit [2]. The set of sensors retrieves information concerning the conditions in which fruits are stored. The other sensors are calibrated to spot out symptoms of early decay of fruits. Those pieces of information combined conclude an input to a neuro-evolutionary algorithm, which adjusts not only the size of fruit deliveries but also warns users of a pre-decay situation in particular fruit storage facilities. By selection an appropriate host for the smart VM with the application for detection the decaying fruit, the shorter time for training the CNN with higher accuracy can be achieved.

Soft sensors are called information from social media. They can be used for analyzing expectations of citizens. For example, to share information with greater audience we can use Twitter as a strong and easy tool [35]. Also, deep learning helps citizens in analyzing sentiments related to many issues. On the other hand, IoT is based on autonomous interaction not only between humans, but also between devices equipped with appropriate sensors. IoT can show up its power when the 3 'I's are combined together: Instrumented, Interconnected, and Intelligent. The first user account was created for the plant sensors allowing posting a tweet when water is needed. Besides, a drone controlled in a city exclusively by tweets was tested in 2014. In the UK, the authorities decided to use about 3,000 sensors installed in the rivers with the communication via Twitter accounts [14].

Social media can protect against crisis situations in smart city. Department of Communications in Roanoke, Virginia, U.S. launched a social media, when a heavy snowstorm hit in the winter of 2014. News streams from Facebook and Twitter are visible by citizens as well as some view posts from a selected agency. It is permit to browse and see the city's videos and news releases via some social media platforms, such as Instagram and Flickr. Social Media Center works in more effective way than a phone hotline 311 regarding some criteria related to query, complain or ask for help [9].

From Roanoke case study, we can observe that social media can support a smart city as a model of metropolis, which integrates a number the domain information systems inside the computer cloud with the Internet of Things to efficient management of urban resources.

5   Differential Evolution in Smart City

Differential evolution is an efficient technique for determination the positions of the transmitting antennas on buildings in a city area [18]. The results show that differential evolution gives the best minimization of the cost function than genetic algorithms. Many other examples from literature and our extensive experiments convinced us to optimize virtual machine migrations by differential evolution [18].

Let us consider a scenario with the control task of the smog distribution in the city. We use computer resources of the GUT-WUT laboratory cloud that includes the *Comcute* grid [12]. This laboratory cloud is used for smart city and educational purposes to verify the theoretical approaches.

Firstly, we have positively checked the migration ability of the virtual machines in the GUT-WUT as described in Sect. 3. Secondly, the resource optimization have been curried out with using migration of virtual machines controlled by differential evolution.

We determine the memory capacities of the virtual machines by simulation in the cloud. The virtual machine with data management agent requires 4 GB RAM and 5 GB HDD, and the VM with web cooperation agent needs 1.5 GB RAM and 0.5 GB HDD. We used the streams of 100,000 input data packages from 100 sensors to intelligent servers with LSTM and CNN neural networks via the part of the experimental cloud containing two VMs with data management agents and four VMs with web cooperation agents. The elapsed CPU time was estimated at 320 s for the web cooperation agent on DELL E5640 computer equipped with 2xIntel 4-core Xeon E-5640 2.66 GHz. Besides, the elapsed CPU time was estimated at 306 s for the data management agent.

The virtual machines with artificial neural networks migrate among servers with GPUs. On the other hand, the virtual machines with data management agents and web cooperation agents migrate among communication servers. We show the results of resource optimization by migration of virtual machines with data management agents and web cooperation agents migrate among communication servers.

Decision makers can select criteria to define the goal of optimization by Multi-criteria Differential Evolution [8]. We propose four criteria to establish some preferences of decision makers. These criteria should be minimised. Let $\hat{Z}_{max}$ be processing workload of the bottleneck CPU [s] and let $\tilde{Z}_{max}$ be communication capacity of the bottleneck node [s]. Moreover, we can consider E - electric power of the cloud [watt] as well as $\varXi$ - cost of hosts [money unit, i.e. USD] [4]. Besides, these criteria have the upper constraints that have to guarantee project requirements like maximum of computer load $\hat{Z}_{sup}$ or an upper limit of node transmission $\tilde{Z}_{sup}$. To save energy, let ξ_{max} be limit of electric power for the cloud. A budget constraint for project can be denoted as ε_{max}.

Furthermore, we consider the rebuilding of the experimental cloud WUT-GUT by increasing the number of computers from 6 to 15 [12]. It is assumed that maximal investment cost is \$87,500. Besides, the upper limit on the power of electricity is 22 kW. We can apply computers from 12 types of servers like Dell, Fujitsu, IBM and HP ProLiant. This instance is characterized by 855 binary decision variables, and binary searched space contains $2.4 * 10^{257}$ items. The number of integer decision variables is 60, and the number of all VMs migrations - $1.3 * 10^{69}$. The lowest cost configuration consists of 4 servers *Dell v1*, 5 computers *Dell v2*, and 6 servers *Infotronik* ATX i5-4430 with the cost \$6,942. On the other hand, electric power 10.5 kW can be reduced to 6 kW by using 15 servers *Infotronik* ATX i5-4430 to satisfy green energy target.

If we consider four criteria, there are six evaluation cuts with two criteria: $(\hat{Z}_{max}, \tilde{Z}_{max})$, $(\hat{Z}_{max}, \Xi)$, $(\hat{Z}_{max}, E)$, $(\tilde{Z}_{max}, \Xi)$, $(\tilde{Z}_{max}, E)$, and (Ξ, E). Figure 2 shows the representation of Pareto-suboptimal solutions $\{P_1, P_2, \ldots, P_{200}\}$ obtained for four criteria in the criteria space cut $(\hat{Z}_{max}, \tilde{Z}_{max})$. Although, points $P_5 = (448; 25,952; 82,626; 19,640)$ and $P_6 = (587; 25,221; 78,010; 20,300)$ are still non-dominated due to criteria $\hat{Z}_{max}$, $\tilde{Z}_{max}$, we can select the representation of Pareto evaluation points with 200 elements. There is the other non-dominated point P_7 for the cut $(\hat{Z}_{max}, \tilde{Z}_{max})$.

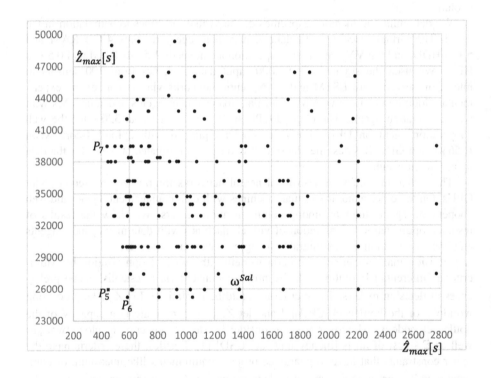

Fig. 2. Pareto front of the cloud resource evaluations for two criteria $(\hat{Z}_{max}, \tilde{Z}_{max})$

Differential evolution founds the compromise solution (Table 1) characterised by $\omega^{Sal} = (1,240; 25,952; 10,244; 11,630)$ and the smaller distance 0.49 to an ideal point $P^{inf} = (442; 25,221; 6,942; 6,750)$ in the four criteria space than P_5 with the distance

1,32. The normalised distance is $\sqrt{2}$ between P^{inf} and the nadir point $N^* = (2,764;$ 49,346; 87,359; 20,740).

Table 1. Specification of the compromise solution ω^{Sal}

Node number i	Computer type number j	Computer name	Virtual machine index v
1.	8	IBM x3650 M4	7, 17, 25, 29, 31
2.	1	DELL R520 E5640 v1	22, 26, 34, 39
3.	4	Infotronik ATX i7-4790	18, 38, 44
4.	2	DELL R520 E5640 v2	5, 14, 16, 42
5.	1	DELL R520 E5640 v1	2, 6, 19, 41
6.	3	Infotronik ATX i5-4430	12, 23
7.	3	Infotronik ATX i5-4430	4, 28
8.	2	DELL R520 E5640 v2	10, 24, 32, 37
9.	3	Infotronik ATX i5-4430	20, 33
10.	1	DELL R520 E5640 v1	9, 13, 35
11.	2	DELL R520 E5640 v2	15, 21, 45
12.	2	DELL R520 E5640 v2	3 (city task), 27
13.	4	Infotronik ATX i7-4790	8
14.	3	Infotronik ATX i5-4430	40, 43
15.	7	Fujitsu Primergy RX300S8	1, 11, 30, 36

Table 1 presents the specification of the compromise solution. There are preserved 3 servers DELL R520 E5640 v1 and 4 servers DELL R520 E5640 v2 from the current laboratory cloud.

However, 4 computers Infotronik ATX i5-4430, 2 servers Infotronik ATX i7-4790, one Fujitsu Primergy RX300S8, and one IBM x3650 M4 have to be buy and install in the specified nodes at the designed cloud (Table 1).

Virtual machines are assigned to dedicated computers in purpose to minimise workload of the bottleneck computer and communication traffic in the bottleneck node. This solution can be described by two row vectors, as follows: $X^{\alpha} = [15, 5, 12, 7, 4, 5, 1, 13, 10, 8, 15, 6, 10, 4, 11, 4, 1, 3, 5, 9, 11, 2, 6, 8, 1, 2, 12, 7, 1, 15, 1, 8, 9, 2, 10, 15, 8, 3, 2, 14, 5, 4, 14, 3, 11]$ – the vector of virtual machines' assignment, $X^{\beta} = [8, 1, 4, 2, 1, 3, 3, 2, 3, 1, 2, 2, 4, 3, 7]$ – the vector of computer assignment. If $X^{\alpha}(2) = 5$, the second virtual machine migrates to the computer at the fifth node. If $X^{\beta}(3) = 4$, the fourth computer type (Table 1) is located at the third node.

Workload of the bottleneck computer is 1,240.78 s that indicates a relative balance of CPUs in the cloud, too. Moreover, communication load of the bottleneck node is 25,952.49 s that approaching to almost optimal value. Cost of servers 10,244 $ is well below budget and electric power consumption 11,630 W gives perspective to the further rebuilding this cloud. As an additional result of increasing the number of modules up to 45, it is expected that ten city tasks can be served with the similar intensity as one task in the current cloud. In addition, the performance according to the CPU Mark test will be increased by at least 4.5 times while meeting essential requirements.

Above optimization can be done iteratively with the given period of time, i.e. 5 min. During this time all virtual machines can be autonomously supported by differential evolution to determine selected aspects of their destinations. Besides, resource management can be optimized by selection the compromise set of resources regarding four criteria. We can state that multi-criteria optimization by differential evolution can have a major impact on smart city cloud design and action. These efforts can increase the decision-making aid by allowing it to make intelligent and effective decisions at the appropriate time.

6   Concluding Remarks and Future Work

Multi-criteria differential evolution is an efficient technique for optimization of virtual machine resources in smart city cloud smart. This cloud can share city task workload that permits on efficient development machine learning applications, too. Solvers based on differential evolution can search Pareto-optimal allocations of virtual machines to optimize computer resource management. The compromise solution for parameter $p = 2$ from the set of Pareto-optimal ones is recommended regarding cloud optimization. Due to our experimental validation of Pareto solutions by differential evolution, the higher quality performance of the cloud is achieved than performance obtained by solutions from the well-known algorithms like the genetic algorithm or the ant colony optimization algorithm.

In our future work, we are going to study other assignment techniques for migration of the virtual machines with the extended set of optimization criteria. Some improvements of the agent cooperation are supposed to be discussed. Besides, we will focus on testing the different artificial intelligence algorithms for smart city infrastructures. Finally, the other cloud architectures will be studied towards smart city development.

References

1. Agarwal, A., Raina, S.: Live migration of virtual machines in cloud. Int. J. Sci. Res. Publ. **2** (6), 1–5 (2012)
2. Balicki, J., Balicka, H., Dryja, P., Tyszka, M.: Big data and the internet of things in edge computing for smart city. In: Saeed, K., Chaki, R., Janev, V. (eds.) CISIM 2019. LNCS, vol. 11703, pp. 99–109. Springer, Cham (2019). https://doi.org/10.1007/978-3-030-28957-7_9
3. Balicki, J., Korłub, W., Szymanski, J., Zakidalski, M.: Big data paradigm developed in volunteer grid system with genetic programming scheduler. In: Rutkowski, L., Korytkowski, M., Scherer, R., Tadeusiewicz, R., Zadeh, L.A., Zurada, J.M. (eds.) ICAISC 2014. LNCS (LNAI), vol. 8467, pp. 771–782. Springer, Cham (2014). https://doi.org/10.1007/978-3-319-07173-2_66

4. Balicki, J.: An adaptive quantum-based multiobjective evolutionary algorithm for efficient task assignment in distributed systems. In: Mastorakis, N., et al. (eds.) Recent Advances in Computer Engineering, 13th International Conference on Computers, Rhodes, Greece, pp. 417–422. WSEAS, Athens (2009)

5. Banerjee, S., Agarwal, N.: Analyzing collective behavior from blogs using swarm intelligence. Knowl. Inf. Syst. **33**(3), 523–547 (2012). https://doi.org/10.1007/s10115-012-0512-y

6. Batty, M., Axhausen, K., Giannotti, F., Pozdnoukhov, A., Bazzani, A., Wachowicz, M., et al.: Smart cities of the future. Eur. Phys. J. **214**(1), 481–518 (2012). https://doi.org/10.1140/epjst/e2012-01703-3

7. Biswas, M.I., Parr, G., McClean, S., Morrow, P., Scotney, B.: A practical evaluation in OpenStack live migration of VMs using 10 Gb/s interfaces. In: 2016 IEEE Symposium on Service-Oriented System Engineering (SOSE), pp. 346–351. IEEE (2016)

8. Błażewicz, J., Kovalyov, M., Węglarz, J.: Preemptable malleable task scheduling problem. IEEE Trans. Comput. **55**(4), 486–490 (2006)

9. Caragliu, A.: Del Bo, C., Nijkamp P.: Smart cities in Europe. Series Research Memoranda 0048. VU University Amsterdam, Faculty of Economics, Business Administration and Econometrics, Amsterdam (2009)

10. Cardoso, L.P., Mattos, D.M., Ferraz, L.H.G., Duarte, O.C.M., Pujolley, G.: An efficient energy-aware mechanism for virtual machine migration. In: Global Information Infrastructure and Networking Symposium, pp. 1–6. IEEE (2015)

11. Clohessy, T., Acton, T., Morgan L.: Smart city as a service (SCaaS): a future roadmap for e–government smart city cloud computing initiatives. In: 7th International Proceedings on Utility and Cloud Computing, pp. 836–841. IEEE/ACM (2014)

12. Comcute grid. http://comcute.eti.pg.gda.pl/. Accessed 17 Feb 2020

13. Cordts, M., et al.: The cityscapes dataset for semantic urban scene understanding. In: Proceedings of the Conference on Computer Vision and Pattern Recognition, pp. 1–12. IEEE (2016)

14. Curtis, S.: How Twitter will power the Internet of Things. http://www.telegraph.co.uk/technology/twitter/11181609/How-Twitter-will-power-the-Internet-of-Things.html. Accessed 17 Feb 2020

15. Dhanoa, I.S., Khurmi, S.S.: Analyzing energy consumption during VM live migration. In: International Conference on Computing, Communication & Automation, pp. 584–588 (2015)

16. FixMyStreet Project. https://www.mysociety.org/projects/. Accessed 17 Feb 2020

17. Galligan, S.D., O'Keeffe, J.: Big Data Helps City of Dublin Improves its Public Bus Transportation Network and Reduce Congestion. IBM Press (2013)

18. Gao-Yang, L., Ming-Guang L.: The summary of differential evolution algorithm and its improvements. In: 3rd International Proceedings on Advanced Computer Theory and Engineering (iCACTE), pp. 153–156 (2010)

19. Gea, T., Paradells, J., Lamarca, M., Roldan, D.: Smart cities as an application of internet of things: experiences and lessons learnt in Barcelona. In: 7th International Proceedings on Innovative Mobile and Internet Services in Ubiquitous Computing, pp. 552–557 (2013)

20. Guojun, L., Ming, Z., Fei, Y.: Large-scale social network analysis based on MapReduce. In: International Proceedings on Computational Aspects of Social Networks, pp. 487–490 (2010)

21. Komninos, N., Pallot, M., Schaffers, H.: Special issue on smart cities and the future internet in Europe. J. Knowl. Econ. **4**, 1–134 (2013). https://doi.org/10.1007/s13132-012-0083-x

22. Kuehne, H., Jhuang, H., Garrote, E., Poggio, T., Serre, T.: HMDB: a large video database for human motion recognition. In: Proceedings of the International Conference on Computer Vision, pp. 2556–2563. IEEE, Barcelona (2011)
23. Kumar, R., Prashar, T.: Performance analysis of load balancing algorithms in cloud computing. Int. J. Comput. Appl. **120**(7) (2015)
24. Kwak, J., Kim, Y., Lee, J., Chong, S.: DREAM: dynamic resource and task allocation for energy minimization in mobile cloud systems. IEEE J. Sel. Areas Commun. **33**(12), 2510–2523 (2015)
25. Macmanus, R.: The Tweeting House: Twitter + Internet of Things. http://readwrite.com/2009/07/20/the_tweeting_house_twitter_internet_of_things. Accessed 17 Feb 2020
26. Mashayekhy, L., Nejad, M.M., Grosu, D.: Physical machine resource management in clouds: a mechanism design approach. IEEE Trans. Cloud Comput. **3**, 247–260 (2014)
27. Nam, T., Pardo, T.A.: Conceptualizing smart city with dimensions of technology, people, and institutions. In: 12th International Proceedings on Digital Government Innovation in Challenging Times, pp. 282–291 (2011)
28. Naphade, M., Banavar, G., Harrison, C., Paraszczak, J., Morris, R.: Smarter cities and their innovation challenges. Computer **44**(6), 32–39 (2011)
29. Ning, H., Wang, Z.: Future internet of things architecture: like mankind neural system or social organization framework. IEEE Commun. Lett. **15**(4), 461–463 (2011)
30. Patel, V.J. Bheda, H.A.: An advanced survey on research issues of energy management in cloud computing. Int. J. Adv. Res. Comput. Sci. Softw. Eng. **4**(1) (2014)
31. Schaffers, H., Komninos, N., Pallot, M.: Smart Cities as Innovation Ecosystems Sustained by the Future Internet. Fireball White Paper (2012)
32. Smartsantander. Future Internet Research & Experimentation. http://www.smartsantander.eu/. Accessed 17 Feb 2020
33. Sutar, S.G., Mali, P.J., More, A.: Resource utilization enhancement through live virtual machine migration in cloud using ant colony optimization algorithm. Int. J. Speech Technol. **23**, 79–85 (2020). https://doi.org/10.1007/s10772-020-09682-2
34. Twardowski, B., Ryzko, D.: Multi–agent architecture for real–time big data processing. In: International Proceedings on Web Intelligence and Intelligent Agent Technologies, vol. 3, pp. 333–337 (2014)
35. Twitter. https://about.twitter.com/company. Accessed 17 Feb 2020
36. Veeravalli, B., He, B.: Guest editors' introduction: special issue on economics and market mechanisms for cloud computing. IEEE Trans. Cloud Comput. **3**(3), 245–246 (2015)
37. Wang, X., Yuen, C., Hassan, N. U., Wang, W., & Chen, T.: Migration-aware virtual machine placement for cloud data centers. In: Workshop on Cloud Computing Systems, Networks, and Applications. IEEE (2015)
38. Yang, C.T., Chuang, C.L., Liu, J.C., Chen, C.C., Chu, W.C.: Implementation of cloud infrastructure monitor platform with power saving method. In: 29th International Conference on Advanced Information Networking and Applications Workshops, pp. 223–228. IEEE (2015)

Dynamic Ensemble Selection – Application to Classification of Cutting Tools

Paulina Heda[1], Izabela Rojek[2] (ID), and Robert Burduk[1][(✉)] (ID)

[1] Faculty of Electronic, Wroclaw University of Science and Technology,
Wybrzeze Wyspianskiego 27, 50-370 Wroclaw, Poland
`robert.burduk@pwr.wroc.pl`
[2] Institute of Computer Science, Kazimierz Wielki University,
Chodkiewicza 30, 85-064 Bydgoszcz, Poland
`izarojek@ukw.edu.pl`

Abstract. In order to improve pattern recognition performance of an individual classifier an ensemble of classifiers is used. One of the phases of creating the multiple classifier system is the selection of base classifiers which are used as the original set of classifiers. In this paper we propose the algorithm of the dynamic ensemble selection that uses median and quartile of correctly classified objects. The resulting values are used to define the decision schemes, which are used in the selection of the base classifiers process. The proposed algorithm has been verified on a real dataset regarding the classification of cutting tools. The obtained results clearly indicate that the proposed algorithm improves the classification measure. The improvement concerns the comparison with the ensemble of classifiers method without the selection.

Keywords: Ensemble of classifiers · Ensemble selection · Cutting tool

1 Introduction

Supervised learning is one of the three main trends in machine learning [1]. In general, the individual supervised classification algorithm maps the feature space into a set of class labels. The prediction performance of a single machine learning model can be improved using a committee of classifiers, which is widely known as Multiple Classifier System (MCS) or Ensemble of Classifiers (EoC) [2]. The construction of EoC and finally using different predictions of individual supervised classification algorithms reduces the risk of choosing an incorrect hypothesis and therefore, improves the overall predictive performance [3].

In general, the procedure of creating EoC can be divided into three major steps [4,5]: (I) generation – a phase where base classifiers are trained and the pool of base classifiers is created, (II) selection – an optional phase where only several

© Springer Nature Switzerland AG 2020
K. Saeed and J. Dvorský (Eds.): CISIM 2020, LNCS 12133, pp. 345–354, 2020.
https://doi.org/10.1007/978-3-030-47679-3_29

models from the committee are taken to the next phase and (III) integration – a process of combining outputs of multiple classifiers to obtain one, integrated model classification. This step is optional if the selection phase results only a single classification model.

The selection phase is related to the choice of one or a set of base classifiers available after the generation phase. If one classifier is selected from the entire pool of base classifiers then we are talking about the selection of classifier. If the set of classifiers after the selection is larger, then we are talking about the ensemble selection [5].

In this work we propose a new algorithm of the ensemble selection that uses median of correctly classified objects from a learning set. In detail, we propose the method based on the analysis of decision profiles, which represent outputs of all base classifiers.

The article by Tan et al. [6] discussed the carbide-tool selection expert system for the purposes of CNC lathe. The aim of this study was to develop an optimum system for selecting a tool chuck, a cutting tool, and a plate, along with their machining parameters (i.e. feed and cutting speeds) by using decision rules. Igari et al. [7] proposed an optimum selection model for processing tools and parameters based on decision rules generated by decision trees. Also, earlier authors' articles show the selection of tools using individual classifiers in the form of single decision trees [8] and neural networks [9].

Given the above, the objectives of this work are the following:

- A proposal of a new dynamic ensemble selection algorithm that uses median of correctly classified objects to define the decision scheme.
- Experimental research to compare the proposed method of the dynamic ensemble selection with base classifiers on the real classification problem regarding the classification of cutting tools.

The paper is structured as follows: In the next section the proposed algorithm is presented Sect. 2. In Sect. 3 the experiments that were carried out are presented, while the results and the discussion are presented in Sect. 4. Finally, we draw conclusions and propose some future works in Sect. 5.

2   Proposed Algorithm of Dynamic Ensemble Selection

2.1   Basic Notation

In general, the classifier maps the feature space into a set of class labels $\Psi : \mathbb{X} \mapsto \Omega$, where each object from the input space $x \in \mathbb{X}$ belongs to one of class labels $\omega_c \in \Omega$, while $\Omega = \{\omega_1, \ldots, \omega_C\}$ is a set of possible class labels. This is a general classifier decision that does not take into account other indirect information regarding the classification process. Considering various types of information returned by base classifiers, we can highlight three cases [4].

- The abstract level – the classifier Ψ assigns the unique class label ω_c to a given recognized object x. The output of each base classifier indicates uniquely the class label [10,11].

- The rank level – in this case for each recognized object x, each classifier produces an integer rank array. Each element within this array corresponds to one of the defined class labels [12]. The array is usually sorted and the label at the top being the first choice.
- The measurement level – the output of a classifier is represented by a score function that addresses the degree of assigning the class label to the given recognized object x. An example of such a representation of the output is a posteriori probability returned by Bayes classifier [13,14].

Let $\Psi = \{\Psi_1, \ldots, \Psi_K\}$ be an ensemble of base classifiers. The majority voting is one of the simplest and most-used methods for combining the outputs of base classifiers. In general, the majority voting method allows counting base classifiers outputs as a vote for a class and assigns the input pattern to the class with the greatest count of votes. It is defined as follows:

$$\Psi_{SUM} = \arg\max_{\omega_i} \sum_{k=1}^{K} I(\Psi_k(x), \omega_i), \tag{1}$$

where $I(\cdot)$ is the indicator function with the value 1 in the case of the correct classification of the object described by the feature vector x, i.e. when $\Psi_k(x) = \omega_i$.

This means that having a pool of arbitrary classifiers and an object to classify, we assign to the object a label that is indicated by most of the models in the pool of base classifiers and each of the individual classifiers takes an equal part in building the ensemble of classifiers. As we mentioned earlier, the ensemble of classifiers using majority voting will be used as the reference classifier.

2.2 Score Function in Ensemble of Classifiers

In this work we consider the situation when each base classifier returns a score function. That is, we will not consider using for the process of classifier selection the abstract rank or level. However, a ensemble of classifiers using an abstract level will be used as a reference method.

For an object x that is recognized, each base classifier Ψ_k determines the value of the score functions (a posteriori probability functions) $[p_{k1}(x), \ldots, p_{kC}(x)]$. These functions are organized into a vector which is the output of the individual classifier. The K individual classifiers outputs for an object x can be represented as the matrix:

$$DP(x) = \begin{bmatrix} p_{11}(x) & \cdots & p_{1C}(x) \\ \vdots & \ddots & \vdots \\ p_{K1}(x) & \cdots & p_{KC}(x) \end{bmatrix}, \tag{2}$$

which is called the decision profile. The values in column j are the individual score functions for class label ω_j and values in row k are output of Ψ_k base classifier.

The previous work of authors [15] presents another algorithm that uses modification of score function to create an ensemble of classifiers. This article uses the median instead of the average, and the process of classifiers selection is performed, i.e. not all base classifiers are used to create the final ensemble of classifiers.

2.3 Proposed Algorithm

Now we present a novel algorithm for the dynamic ensemble selection. The dynamic ensemble selection means that a different set of base classifiers can be selected for each recognized object.

During the generation phase of the proposed algorithm we obtain N decision profiles, where N is the number of objects from the learning set. Using objects from the learning set that has been correctly classified, two decision schemes are defined according to Algorithm 1.

The classification of the new object x is the operational phase of the proposed algorithm. In this phase, the decision schemes calculated in the generation phase are used. The dynamic ensemble selection is carried out according to Algorithm 2. In the process of combining outputs of base classifiers after the selection we use maximum of the sum rule.

In the paper [16] we presented an algorithm that uses the average of the correctly classified objects to determine decision schemes. In the proposed approach we use the median and quartile of the correctly classified objects to define the decision schemes and the proposed algorithm has several additional parameters. In addition, the number of class labels in this paper is not limited to two labels.

3 Experiment Setup

3.1 Dataset of Cutting Tools

The dataset of the cutting tool examples was collected at a real company providing a wide range of products. The first stage of data acquisition involved identifying the organizational structure of the company, and its production process. The scope of data acquisition also depends on the type of production. Depending on the produced components, there are unit, small-scale, medium-scale and large-scale production types. Production typically involves a large number of variants of a batch or unit products. The diversity of these variants leads to a rather low degree of standardization. The analysis covered the products, machinery, tools, instrumentation and semifinished products. Account was taken of the existing technological processes for manufactured products. The manufacturing process planning is divided into several stages. The first stage involves selecting the semi-finished products. This is followed by designing the technological-process structure, i.e. the sequence of technological procedures and operations. Then, workpiece instrumentation, machine tools, cutting tools, tooling and machining parameters are selected for each technological procedure and operation [17].

This article presents classification models for one of the elements of technological process planning, namely the selection of cutting tools for technological operations.

Algorithm 1: Generation phase of dynamic ensemble selection algorithm

---

Input: Sequence of N labeled instance – learning set (LS), Set of base
      classifiers $\Psi_1, \ldots, \Psi_K$, Parameters γ and λ of the algorithm

Output: Decision schemes DS^{max}, DS^{min}

1 Train a base classifier $\Psi_1, \ldots, \Psi_K$ using LS.

2 Compute decision scheme: $DS = \begin{bmatrix} ds_{11} & \cdots & ds_{1C} \\ \vdots & \ddots & \vdots \\ ds_{K1} & \cdots & ds_{KC} \end{bmatrix}$, where

$$ds_{k\omega} = Med\left(\sum_{n=1}^{N} I(\Psi_k(x_n) = \omega_n) p_k(\omega_n | x_n) \right).$$

3 **return** *Decision scheme DS^{max}*:

$$DS^{max} = \begin{bmatrix} ds_{11}^{max} & \cdots & ds_{1C}^{max} \\ \vdots & \ddots & \vdots \\ ds_{K1}^{max} & \cdots & ds_{KC}^{max} \end{bmatrix},$$

where
$$ds_{k\omega}^{max} = ds_{k\omega} + \gamma * Q_1$$

Q_1 is a first quartile of $\left(\sum_{n=1}^{N} I(\Psi_k(x_n) = \omega_n) p_k(\omega_n | x_n) \right)$.

4 **return** *Decision scheme DS^{min}*:

$$DS^{min} = \begin{bmatrix} ds_{11}^{min} & \cdots & ds_{1C}^{min} \\ \vdots & \ddots & \vdots \\ ds_{K1}^{min} & \cdots & ds_{KC}^{min} \end{bmatrix},$$

where
$$ds_{k\omega}^{min} = ds_{k\omega} - \lambda * Q_1.$$

The cutting tools dataset contains 564 learning objects ($N = 564$) and 17 class labels ($C = 17$). The majority of information obtained from databases was raw, and incomplete. In order for such data to become useful for mining purposes, they need to be pre-processed, i.e. cleaned and transformed. Data cleaning entails the unification of records, the supplementation of missing entries, or the identification of extreme points. In turn, data transformation involves normalisation or coding.

The features of the object are : type of machining surface (e.g. shape machining), material symbol (e.g. EN-AW 5754 aluminum), demanded surface roughness (e.g. 6.3), milling tool structure (e.g. monolithic milling cutter), the kind of clamping tool milling (e.g. arbor), dimension (e.g. 16), milling cutter shape (e.g. cylindrical Weldon), tooth number (e.g. 3), total length of milling tool (e.g. 32), min. cutting speed vc (e.g. 250), max. cutting speed vc (e.g. 500), cutting depth ap

---

Algorithm 2: Operation phase of dynamic ensemble selection algorithm

---

Input: Decision schemes DS^{max}, DS^{min}, Set of base classifiers $\Psi_1, \ldots, \Psi_K$,
Parameters α and β of the algorithm, recognized object $- x$

Output: Ensemble decision after selection Ψ^S_{Med}

1 Set a decision profile:

$$DP(x) = \begin{bmatrix} p(x)_{11} & \cdots & p(x)_{1C} \\ \vdots & \ddots & \vdots \\ p(x)_{K1} & \cdots & p(x)_{KC} \end{bmatrix}.$$

2 Compute $DP^{sel}(x)$:

$$p^{sel}(x)_{k\omega} = \begin{cases} 0, & \text{if} \quad p(x)_{k\omega} < ds^{min}_{k\omega} \\ \beta p(x)_{k\omega}, & \text{if} \quad ds^{min}_{k\omega} \leq p(x)_{k\omega} \leq ds^{max}_{k\omega} \\ \alpha p(x)_{k\omega}, & \text{if} \quad p(x)_{k\omega} > ds^{max}_{k\omega} \end{cases}.$$

3 **return** *Ensemble decision* Ψ^S_{Med}:

$$\Psi^S_{Med}(x) = \max_{\omega} \sum_{k=1}^{K} p^{sel}(x)_{k\omega}.$$

(e.g. 9), milling width ae (e.g. 1), cutting feed f (e.g. 796), operating cost (e.g. 120). The class labels are the milling tool symbols (e.g. Fi16W).

3.2   Parameters of Ensemble Selection Algorithm

During the experiment 10 base classifiers were used. Three of them are k-NN classifiers: Ψ_1 is 13-NN, Ψ_2 is 15-NN and Ψ_3 is 19-NN classifier. The other seven base classifiers are decision trees.

The classifiers implemented in SAS 9.4 environment were used. In particular, we used DISCRIM procedure for k-NN base classifier and HPSPLIT procedure for decision tree base classifiers. Table 1 presents the parameters of these base classifiers.

Table 2 presents parameters $\alpha, \beta, \gamma, \lambda$ used in the proposed ensemble selection algorithm.

In addition, the nominal features were not used in the learning and testing process. SAS 9.4 does not allow the use of nominal values during the classification process because this data type has not been predefined for the variable list. The presented results are obtained via 10-fold-cross-validation method.

3.3   Classification Measures

To evaluate the proposed methods the following classification measures are used: average accuracy (AA), micro-averaged F_1 score ($F_1\mu$) and macro-averaged F_1

Table 1. Decision tree parameters – base classifiers $\Psi_4, \ldots, \Psi_{10}$ used in the experiment

Classifier	Maximum tree depth	Maximum number of tree branches	Partition fraction
Ψ_4	7	2	0.2
Ψ_5	15	2	0.3
Ψ_6	15	4	0.3
Ψ_7	5	3	0.4
Ψ_8	15	4	0.6
Ψ_9	10	2	0.3
Ψ_{10}	10	4	0.4

Table 2. Parameter sets used in the experiment – parameters of selection algorithm

Parameter set	α	β	γ	λ
PS_1	2	1.5	0.5	1
PS_2	2	0.5	0.5	1
PS_3	2	1.5	0	2
PS_4	2	0.5	0	2
PS_5	1.5	0.5	2.5	1
PS_6	2	1.5	2.5	1
PS_7	2	0.5	2.5	1
PS_8	2	1.5	1	0
PS_9	2	0.5	1	0
PS_{10}	2	1.5	1	−0.5
PS_{11}	2	0.5	1	−0.5
PS_{12}	1.5	0.5	1	−0.5

score (F_1M). Micro-averaged F_1 score represents the relations between data's positive labels and those given by a classifier based on sums of a per-class decisions while macro-averaged F_1 score represents the relations between data's positive labels and those given by a classifier based on a per-class average [18]. Macro and micro averaged measures were used to assess the performance for the majority and minority classes. This is because the macro-averaged measures are more sensitive to the performance for minority classes.

4   Results and Discussion

Experimental studies were carried out for various sets of parameters $\alpha, \beta, \gamma, \lambda$ of the proposed section ensemble algorithm. For all quality measures, the best set of parameters is as follows: $\alpha = 2, \beta = 1.5, \gamma = 0.5, \lambda = 1$. The ensemble selection algorithm Ψ^S_{Med} with this set of parameters was compared with all base classifiers $\Psi_1, \ldots, \Psi_{10}$ and EoC method without selection Ψ_{SUM}. Table 3 shows the results for three classification measures AA, $F_1\mu$ and F_1M.

Table 3. Results of experimental research for three classification measures

Classifier	AA	F_1M	$F_1\mu$
Ψ_1	0.913	0.721	**0.953**
Ψ_2	0.913	0.721	**0.953**
Ψ_3	0.908	0.715	0.951
Ψ_4	0.740	0.497	0.747
Ψ_5	0.798	0.530	0.805
Ψ_6	0.653	0.421	0.659
Ψ_7	0.743	0.498	0.749
Ψ_8	0.719	0.491	0.725
Ψ_9	0.798	0.530	0.805
Ψ_{10}	0.698	0.462	0.704
Ψ^S_{Med}	**0.942**	**0.785**	0.950
Ψ_{SUM}	0.927	0.754	0.935

The obtained results indicate that the proposed dynamic ensemble selection method significantly improves the quality of the cutting tools classification. In the case of two classification measures AA and F_1M the proposed approach is better than any base classifier and EoC without selection. For $F_1\mu$ measure the proposed ensemble selection is better than EoC without selection and slightly worse than three base classifiers. In this case, the deterioration in the classification quality does not exceed 0.3%.

5   Conclusions

The paper presents the new dynamic ensemble selection algorithm. The proposed algorithm uses the values of the base classifier outputs for calculation of decision schemes. In particular, we proposed an approach that uses median and quartile of correctly classified objects to define the decision schemes.

The discussed approach to the ensemble selection has been tested on a real dataset. This dataset represents real tool selections performed during the design of manufacturing processes. The obtained results, for three classification measures, indicate clearly that the classification quality of cutting tools can be improved by approx. 3%. The improvement concerns the comparison with EoC method without selection.

Acknowledgment. This work was supported by the statutory funds of the Department of Systems and Computer Networks, Wroclaw University of Science and Technology and Institute of Computer Science, Kazimierz Wielki University.

References

1. Bishop, C.M.: Pattern Recognition and Machine Learning (Information Science and Statistics). Springer, New York (2006)
2. Ulaş, A., Semerci, M., Yıldız, O.T., Alpaydın, E.: Incremental construction of classifier and discriminant ensembles. Inf. Sci. **179**(9), 1298–1318 (2009)
3. Sagi, O., Rokach, L.: Ensemble learning: a survey. Wiley Interdisc. Rev. Data Min. Knowl. Disc. **8**(4), e1249 (2018)
4. Kuncheva, L.I.: Combining Pattern Classifiers. Wiley, New York (2014). https://doi.org/10.1002/9781118914564
5. Britto Jr., A.S., Sabourin, R., Oliveira, L.E.: Dynamic selection of classifiers–a comprehensive review. Pattern Recogn. **47**(11), 3665–3680 (2014)
6. Tan, C., Ranjit, S.: An expert carbide cutting tools selection system for CNC lathe machine. Int. Rev. Mech. Eng. **6**(7), 1402–1405 (2012)
7. Igari, S., Tanaka, F., Onosato, M.: Customization of a micro process planning system for an actual machine tool based on updating a machining database and generating a database oriented planning algorithm. In: ASME/ISCIE 2012 International Symposium on Flexible Automation, pp. 35–42. American Society of Mechanical Engineers Digital Collection (2012)
8. Rojek, I.: Classifier models in intelligent CAPP systems. In: Cyran, K.A., Kozielski, S., Peters, J.F., Stańczyk, U., Wakulicz-Deja, A. (eds.) Man-Machine Interactions. AINSC 2009, vol. 59, pp. 311–319. Springer, Heidelberg (2009). https://doi.org/10.1007/978-3-642-00563-3_32
9. Rojek, I.: Technological process planning by the use of neural networks. AI EDAM **31**(1), 1–15 (2017)
10. Dey, A., Shaikh, S.H., Saeed, K., Chaki, N.: Modified majority voting algorithm towards creating reference image for binarization. In: Kumar Kundu, M., Mohapatra, D.P., Konar, A., Chakraborty, A. (eds.) Advanced Computing, Networking and Informatics - Volume 1. SIST, vol. 27, pp. 221–227. Springer, Cham (2014). https://doi.org/10.1007/978-3-319-07353-8_26
11. Ruta, D., Gabrys, B.: Classifier selection for majority voting. Inf. Fus. **6**(1), 63–81 (2005)
12. Przybyła-Kasperek, M., Wakulicz-Deja, A.: Comparison of fusion methods from the abstract level and the rank level in a dispersed decision-making system. Int. J. Gen Syst **46**(4), 386–413 (2017)
13. Bloch, I.: Information combination operators for data fusion: a comparative review with classification. IEEE Trans. Syst. Man Cybern. Part A Syst. Hum. **26**(1), 52–67 (1996)
14. Ho, T.K., Hull, J.J., Srihari, S.N.: Decision combination in multiple classifier systems. IEEE Trans. Pattern Anal. Mach. Intell. **16**(1), 66–75 (1994)
15. Burduk, R., Baczyńska, P.: Ensemble of classifiers with modification of confidence values. In: Saeed, K., Homenda, W. (eds.) CISIM 2016. LNCS, vol. 9842, pp. 473–480. Springer, Cham (2016). https://doi.org/10.1007/978-3-319-45378-1_42

16. Baczyńska, P., Burduk, R.: Classifier selection uses decision profiles in binary classification task. In: Choraś, R.S. (ed.) Image Processing and Communications Challenges 7. AISC, vol. 389, pp. 3–10. Springer, Cham (2016). https://doi.org/10.1007/978-3-319-23814-2_1

17. Black, J.T., Kohser, R.A.: DeGarmo's Materials and Processes in Manufacturing. Wiley, Chichester (2017)

18. Sokolova, M., Lapalme, G.: A systematic analysis of performance measures for classification tasks. Inf. Process. Mana. 45(4), 427–437 (2009). https://doi.org/10.1016/j.ipm.2009.03.002

Stochastic Model of the Simple Cyber Kill Chain: Cyber Attack Process as a Regenerative Process

Romuald Hoffmann[(✉)]

Institute of Computer and Information Systems, Faculty of Cybernetics,
Military University of Technology,
ul. gen. Sylwestra Kaliskiego 2, 46, 00-908 Warszawa, Poland
romuald.hoffmann@wat.edu.pl

Abstract. The proposed model extends the Markov model of the simple cyber kill chain already published in the literature by assumption of any continuous probability distribution of adversaries' and defenders' activity time. The description of the chain is based on the cyber kill chain concept initially introduced by Lockheed Martin's researchers as the intrusion kill chain. The model includes the assumption of repeatability of cyber-attacks. On this basis, the stationary probabilities of staying of the attack process in individual phases were determined.

Keywords: Cyber-attack process · Stochastic model · Cyber-attack life cycle · Regenerative stochastic process · Stationary probability

1 Introduction

1.1 Background

Today's progress of information technologies is one of the most important factors in the development of the modern world. This development affects many social, economic and military processes. This impact should be perceived in both a positive and negative sense. One of the undesirable effects of the impact is growing cybercrime [1–3]. Unfortunately, in today's world, the group of individual and/or organized cybercriminals has been joined by countries for which attacks in the cyberspace have become an element of aggressive economic and military policy, thus provoking the defensive reactions of others (e.g. [4]) Just now, one can already see a certain kind of arms race in cyberspace, which in the author's opinion in the near future will become the norm, leading in consequence to serious military conflicts. We will not be mistaken in claiming that the competition of countries in cyberspace has become a fact and this phenomenon is now rapidly increasing. Analyzing existing cases of most serious cyber-attacks, we can see a certain systematics - the attacks were carried out according to a certain pattern that can be described as a process of cyber-attack. Even today, despite the fact that there are often talks and writings about cyber-attacks, many organizations perceive a cyber-attack as a short-lived event that can hardly be resisted. In reality, a

K. Saeed and J. Dvorský (Eds.): CISIM 2020, LNCS 12133, pp. 355–365, 2020.
https://doi.org/10.1007/978-3-030-47679-3_30

serious cyber-attack is not a momentary act, but a process with a set of activities that have to be performed in the right order and which have their duration and place [4–9]. Depending on a purpose of an attack, these activities are combined into logical groups and implemented in stages, thus creating a cyber-attack process. This process has a finite duration and is called a cyber-attack life cycle [4] or a cyber kill chain [9, 10].

The cyber-attacks life cycles are practical models to describe cyber-attack processes that consist of different intrusion stages related to network security and information system security [7]. For instance, Mandiant in [4] its analytical report on the activities of Chinese cybercriminal units published a description of APT attack processes. Mandiant claims that this cycle was used by Chinese cyber-espionage units to penetrate the resources of many governments and corporations at that time.

Knowledge of the cycles may allow defenders, for example, to estimate: the probability of an cyber-attack over the time, the average duration of an attack or the average time to compromise the IT system (time-to-compromise) and ultimately the cost of an attack. Knowing these and other characteristics of the processes, we can try to answer the questions, e.g. when the probable cyber-attack occurs, what phase of the attack we are subject to and with what probability.

1.2 Related Work

In research literature the cyber-attack processes which are divided into phases are known as cyber kill chains [6, 9, 10] or cyber-attack life cycles [4, 8]. Cyber-attack life cycle phases are variously named, defined and described by researches. For instance, according to [12] a cycle consists of five stages: reconnaissance, scanning, system access, malicious activity and exploitation. In [1, 6, 11] a cyber-attack process is named as an intrusion kill chain and defined as the sequence of seven stages: reconnaissance, weaponization, delivery, exploitation, installation, command and control (C2), act on objectives (for description of stages see Table 1). This kill chain is also described by many researchers, e.g. in [10, 13].

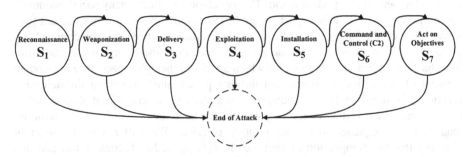

Fig. 1. Graph of transitions between the phases of the simple cyber kill chain [8].

In [6] there are indicated seven stages: initial compromise, establish foothold, escalate privileges, internal reconnaissance, move laterally, maintain presence, complete mission. Other researchers [14] point out six stages. These authors indicate that an attack on critical infrastructure should be considered as a sequence of six phases: reconnaissance, weaponization, delivery, cyber execution, control perturbation, physical objective realization.

In various cyber security papers, the cyber kill chain proposed in [1, 6] is a very popular conceptual model usually describing cyber-attack processes, e.g. [7].

Table 1. Phases of cyber kill chain/cyber-attack life cycle [1, 6–9].

Phase (Stage)	Description and examples of actions in the phase
Reconnaissance (S_1)	Identification of adversary/aggressor/attacker goals e.g. business, political, etc. Selection/profiling of attack (technical) targets by identifying the target environment e.g. TCP port scanning, indexing websites, conference materials, lists of email addresses, social networks, information about the technologies used (specific), sociotechnical phishing, etc.
Weaponization (S_2)	Preparation of cyberweapons, i.e. special malicious software, e.g. integrating Trojan horses into another malicious code (exploit) to create a deliverable payload using an automatic weaponizer. If we do not need to build or figure a software package, the stage reduced to just collecting required cyber weapons
Delivery (S_3)	Delivering cyberweapons to a target environment, e.g. use of the most common delivery methods (e.g. APT attacks) which, for example, are infected attachments to emails, crafted or maliciously modified website software (e.g. applets, links), SQL code, infected data drivers connected to USB ports
Exploitation (S_4)	Running malicious code (after providing a cyberweapon to the target environment), e.g. as a result of the use of personal/software vulnerabilities in an application or tampering users of the target system
Installation (S_5)	Installation of additional malicious code, e.g. Remote Access Trojan (RAT) horses, placement of backdoor in the target system in order to set up a permanent communication channel of the infected internal environment of the victim with the center (external environment) of command and control software malicious
Command and Control (S_6)	Command and control of an infected environment and (optionally) additional malicious action, e.g. escalation or additional system privileges, installation of remaining or additional malicious code (e.g. back-doors/trojans/rootkits), modification of file systems, browsing or modification of system databases
Act on Objectives (S_7)	Undertaking actions aimed at achieving the original goals, e.g. copying data, violating the integrity and/or availability of data, gaining access to the victim's e-mail in order to use it to deeper penetration of the hidden infrastructure or using e-mail to further spread the attack. In this phase, physical infrastructure degradation of the organization is not excluded

The Markov models of a simple cyber kill chain presented in [8, 15] are homogeneous continuous time Markov chains (CTMCs). In these models the basic concept of the cyber kill chain is understood as in [1, 6, 10]. In other words, a cyber-attack may pass sequentially through the stages from "reconnaissance" to "act on objectives", i.e. without any possibilities of skipping any of the chain stages or returning to the previous ones, however, cyber-attacks can be stopped or abandoned at any stage at any time (see Fig. 1). An illustration of the possible transitions between the phases of the simple cyber-kill chain is shown in Fig. 1.

Two other Markov models of cyber kill chains as CTMCs are considered in [8], where transitions between the chain's phases are in some cases permissible. These models are called models of cyber kill chains with iterations and refer to cases where recurring cyber-attacks take place in the same organizations that were previously targeted. These cases occur after correcting, completing or abandoning recent attacks. This also happens when many repetitive or new cyber-attacks are carried out in order to multiply their impact of on the organization and then impede forensic analysis. Table 1 consists of the description of the cyber kill chain for which above stochastic models are considered.

In all available approaches to description of cyber-attack life cycles there are not specified stages such as an initiation and a termination. Therefore, a generalized cyber-attack life cycle and its Markovian model has been proposed in [16] which includes these two additional phases. The first stage is an identification of the attacker's needs. The last stage of the cyber-attack process is a termination of the attack combined with removing traces of malicious activities by the adversary.

2   The Model

As mentioned in previous section the cyber kill chain has been modeled using Markov chains with a continuous time parameter. The source models assume exponential probability distributions of the durations of the cyberattack cycle phases, which may limit the use of the models in practice. Therefore, in this section we propose a stochastic model of the chain assuming any continuous probability distribution of the phases' durations.

In practice, most serious cyberattacks on an organization (enterprise, institution, state) do not end in one single attack cycle. The cyber-attacks are repeated hour by hour, day by day, months by month. Even if attackers got expected effects, they continue attacks by setting new targets and starting the cycle again. Therefore, it is assumed in this paper that cyber-attacks can be repeated, and the termination of the current attack immediately triggers an attacker's new cyber-attack cycle. Therefore, our model considers repetition of an attack, manifested in the repetition of the simple cyber kill chain. We show that the cyber-attack process can be modeled as a regenerative process created by sequence of simple cyber kill chain cycles.

2.1   Basic Assumptions

The basis of the stochastic model proposed in this section is the cyber-attack cycle understood as the simple cyber kill chain [8, 15] described in previous section (see Table 1 and Fig. 1 as well). For purpose of our model, we still assume that it is not possible to return from the current phase to the previous phases and skip any of the subsequent phases to successfully continue the current attack cycle (Fig. 1). However, the model is to take into account the dynamics of the attack process, assuming that the current attack cycle can be stopped or interrupted (eventually terminated) at any of its phases and at any time from the beginning of the current attack phase. Figure 1 depicts in the form of a directed graph possible transitions between individual phases of one simple cyber-attack cycle.

Our model is formally based on the following assumptions.

Let $T_n \in [0, +\infty)$ denote the time necessary to complete successfully the phase $S_n (n = 1, 2, \ldots, 7)$. Let $T_1, \ldots, T_7$ be independent random variables. We will call these times the necessary durations of cycle phases. We assume that T_n is a random variable with the continuous distribution function $F_n(t) = \Pr\{T_n < t\}$ and has the finite expected value ET_n, $ET_n = \int_0^{+\infty} t dF_n(t) < +\infty$.

Let $\tau_n \in [0, +\infty)$ denote the time after which the attack may be stopped, interrupted, or abandoned from the start of the phase S_n $(n = 1, 2, \ldots, 7)$. Let τ_n be a random variable with the continues distribution function $G_n(t) = \Pr\{\tau_n < t\}$ and has the finite expected value $E\tau_n$, $E\tau_n = \int_0^{+\infty} t dG_n(t) < +\infty$. We assume that the random variables $\tau_1, \ldots, \tau_7$ are independent.

In addition, we assume stochastic independence of the random variables T_n and τ_n, $n = 1, 2, \ldots, 7$.

From the definition of random variables τ_n and T_n one can conclude that the holding time in the phase S_n is equal to the random variable $\beta_n = \min\{T_n, \tau_n\}$ $(n = 1, 2, \ldots, 7)$.

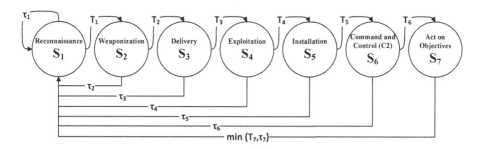

Fig. 2. Graph of state transitions of the cyber-attack cycle with the values τ_n and T_n.

As a consequence of above assumptions made, the graph of transitions between states is shown in Fig. 2. The return to the state of S_1 symbolizes the start of a new cycle of the simple cyber kill chain. The arches of the graph illustrating the transitions between the cycle's phases are described by random variables of duration after which a transition to the next phase may follow or the end (interruption) of current attack cycle.

2.2 Duration of a Single Cyber-Attack Cycle (a Simple Cyber Kill Chain)

Let $\alpha_n \in \{0, 1\}$ for each $n = 1, 2, \ldots 7$ be a binary random variable[1] such that $\alpha_n = 1$ when an event $\{T_n < \tau_n\}$ occurs, and $\alpha_n = 0$ when an event $\{T_n \geq \tau_n\}$ occurs.

Let $\Theta_1 \in [0, +\infty)$ denote the duration of a single cyber-attack cycle. The random variable Θ_1 can be expressed by the following formula

$$\Theta_1 = \beta_1 + \alpha_1 \cdot (\beta_2 + \alpha_2 \cdot (\beta_3 + \alpha_3 \cdot (\beta_4 + \alpha_4 \cdot (\beta_5 + \alpha_5 \cdot (\beta_6 + \alpha_6 \cdot \beta_7))))) \quad (1)$$

where $\beta_n = \min\{T_n, \tau_n\}$, $n = 1, 2, \ldots 7$.

Finally, the duration of one single cyber-attack cycle (1) is

$$\Theta_1 = \beta_1 + \alpha_1 \cdot \beta_2 + \alpha_1 \cdot \alpha_2 \cdot \beta_3 + \alpha_1 \cdot \alpha_2 \cdot \alpha_3 \cdot \beta_4 + \ldots + \alpha_1 \cdot \alpha_2 \cdot \ldots \cdot \alpha_6 \cdot \beta_7 \quad (2)$$

Formula (2) shows that the duration of one single cyber-attack cycle Θ_1 is the sum of dependent random variables: $\beta_1, \alpha_1 \cdot \beta_2, \alpha_1 \cdot \alpha_2 \cdot \beta_3, \ldots, \alpha_1 \cdot \alpha_2 \cdot \ldots \cdot \alpha_6 \cdot \beta_7$. It means that the duration of next phase depends on the behavior of the attack process in the phases preceding the phase.

Let $\overline{F_n}(t) = 1 - F_n(t)$ and $\overline{G_n}(t) = 1 - G_n(t)$. Since the random variables T_n and τ_n are independent by assumption thus the probability distribution of the random variable $\beta_n = \min\{T_n, \tau_n\}$ for each $n = 1, 2, .., 7$ is as follows

$$B_n(t) = \Pr\{\beta_n < t\} = Pr\{\min\{T_n, \tau_n\} < t\} = 1 - \overline{F_n}(t) \cdot \overline{G_n}(t)$$

Since $E\beta_n = \int_0^{+\infty} (1 - B_n(t))dt$ thus the expected value of random variable β_n is

$$E\beta_n = \int_0^{+\infty} \overline{F_n}(t) \cdot \overline{G_n}(t)dt \quad (3)$$

The expected value of the random variable α_n is

$$E\alpha_n = 1 \cdot Pr\{T_n < \tau_n\} + 0 \cdot Pr\{T_n \geq \tau_n\}$$

Therefore,

$$E\alpha_n = \int_0^{+\infty} F_n(t)dG_n(t) = \int_0^{+\infty} \overline{G_n}(t)dF_n(t) \quad (4)$$

From property of expected values of random variables [20] it appears that the expected value of duration of one cyber-attack cycle Θ_1 (2) is

---

[1] event indicator, binary indicator function

$$E\Theta_1 = E\beta_1 + E\alpha_1\beta_2 + E\alpha_1\alpha_2\beta_3 + \ldots + E\alpha_1\alpha_2 \cdot \ldots \cdot \alpha_6\beta_7$$

Then, based on assumptions made on the independence of random variables T_n i τ_n the expected duration of one cyber-attack cycle Θ_1 is expressed by

$$E\Theta_1 = E\beta_1 + E\alpha_1 \cdot E\beta_2 + E\alpha_1 \cdot E\alpha_2 \cdot E\beta_3 + \ldots + E\alpha_1 \cdot E\alpha_2 \cdot \ldots \cdot E\alpha_6 \cdot E\beta_7 \quad (5)$$

where $E\beta_n$, $n = 1, 2, \ldots, 7$ is given by formula (3), $E\alpha_n$ – by formula (4).

2.3   Cyber Attack Process as a Regenerative Process

Let $\{X(t)\}_{t \geq 0} \equiv \{X(t), t \in [0, +\infty)\}$ and $\{X_k(t)\}_{t \geq 0} \equiv \{X_k(t), t \in [0, +\infty)\}$ be stochastic processes with state space $S = \{1, 2, \ldots, 7\}$. The elements of the space $S = \{1, 2, \ldots, 7\}$ are indexes of the phases respectively $S_1, S_2, \ldots, S_7$. An index $k = 1, 2, \ldots$ in second process denote a serial number of an attack cycle. The stochastic processes $\{X(t)\}_{t \geq 0}$ and $\{X_k(t)\}_{t \geq 0}$ describes the dynamics of cyber-attacks conducted according the simple cyber kill chain.

Notice that the random variable Θ_1 (the duration of a single cyber-attack cycle) is the moment when the process $\{X_1(t), 0 \leq t < \Theta_1\}$ is stopped.

Let the cycles Θ_k, $k = 2, 3, \ldots$, be defined in the same manner as the random variable Θ_1 (see (1) and (2)), i.e. for each $k = 1, 2, 3, \ldots$

$$\Theta_k = \beta_1 + \alpha_1 \cdot \beta_2 + \alpha_1 \cdot \alpha_2 \cdot \beta_3 + \alpha_1 \cdot \alpha_2 \cdot \alpha_3 \cdot \beta_4 + \ldots + \alpha_1 \cdot \alpha_2 \cdot \ldots \cdot \alpha_6 \cdot \beta_7$$

By assumption different cycles are independent and all governed by the same probability low. Thus, the sequence $\{\Theta_k\}_{k \geq 1} \equiv \{\Theta_k, k = 1, 2, \ldots\}$ is the sequence of the independent, identically distributed random variables, with expected value $E\Theta_1 < +\infty$, i.e. for each $k = 1, 2, 3, \ldots$ the expected value $E\Theta_k = E\Theta_1$.

Now we define the process $\{X(t)\}_{t \geq 0}$ as follow

$$X(t) = \begin{cases} X_1(t) & \text{if} \quad 0 \leq t < \Theta_1 \\ X_2(t - \Theta_1) & \text{if} \quad \Theta_1 \leq t < t_2 \\ \cdots & \cdots \\ X_k(t - t_{k-1}) & \text{if } t_{k-1} \leq t < t_k \\ \cdots & \cdots \end{cases} \quad (6)$$

where $t_0 = 0$, $t_k = \Theta_1 + \Theta_2 + \ldots + \Theta_k$, $k \geq 1$.

The time $t_1 = \Theta_1$ is the first regenerative point when the process $\{X(t)\}_{t \geq 0}$ starts a new cycle completely from scratch independently of the past, the time .. is the second regenerative point, and etc. Thus, the sequence of the stochastic cyber-attack cycles $\{(\Theta_k, X_k(t))\}_{k \geq 1}$ are independent and identically distributed. The process $\{X(t)\}_{t \geq 0}$ as the model of cyclic cyber-attack processes is a regenerative process with the regeneration points t_k, $k \geq 1$ [21].

It should be noted here that the process $\{X(t)\}_{t \geq 0}$ may not go through all phases of the cycles Θ_k, $k = 1, 2, \ldots$. It follows from possibilities of interrupting a cyber-attack at any stage of a cyber kill chain. In other words, it is immediate consequence of construction of the duration of a single cyber-attack cycle Θ_1 (generally $\Theta_k, k = 1, 2, \ldots$).

2.4   Stationary Probability Distribution of an Attack Process

Our primary task is to find the stationary distribution of the regenerative process $\{X(t)\}_{t \geq 0}$ specified by (6). The task is to calculate the limit

$$\lim_{t \to \infty} \Pr\{X(t) = n\} = P_n \tag{7}$$

where n is the number of cyber-attack phase S_n, $n = 1, 2, \ldots, 7$.

From the key renewal theorem, Smith's theorem [19], [21] it appears that the limit (7) can be determined from following dependence

$$\lim_{t \to \infty} \Pr\{X(t) = n\} = \frac{1}{E\Theta_1} \int_0^{+\infty} \Pr\{\{X_k(t) = n\} \cap \{\Theta_k \geq t\}\} dt \tag{8}$$

The limit (8) can be calculated for any cycle of the sequence $\{(\Theta_k, X_k(t))\}_{k \geq 1}$ because the cycles are independent and identically distributed.

We shell firstly determine the probability of the event that at moment $t \geq 0$ the cyber-attack process is in the phase S_n, and has not yet ended, i.e.

$$\Pr\{\{X_k(t) = n\} \cap \{\Theta_k \geq t\}\} = Pr\left\{\left\{\sum_{i=1}^{n-1} d_i < t\right\} \cap \left\{\sum_{i=1}^{n} d_i \geq t\right\}\right\} \tag{9}$$

where $\Theta_k = d_1 + \ldots + d_7$, $d_1 = \beta_1$, $d_i = \alpha_1 \cdot \ldots \cdot \alpha_{i-1} \cdot \beta_i$ for $i = 2, \ldots, 7$.

By de Morgan Laws the right side of (9) can be written as

$$Pr\left\{\left\{\sum_{i=1}^{n-1} d_i < t\right\} \cap \left\{\sum_{i=1}^{n} d_i \geq t\right\}\right\} = 1 - Pr\left\{\left\{\sum_{i=1}^{n-1} d_i \geq t\right\} \cup \left\{\sum_{i=1}^{n} d_i < t\right\}\right\} \tag{10}$$

Because $\left\{\sum_{i=1}^{n-1} d_i \geq t\right\} \cap \left\{\sum_{i=1}^{n} d_i < t\right\} = \emptyset$ then

$$Pr\left\{\left\{\sum_{i=1}^{n-1} d_i \geq t\right\} \cup \left\{\sum_{i=1}^{n} d_i < t\right\}\right\} = Pr\left\{\sum_{i=1}^{n} d_i < t\right\} + Pr\left\{\sum_{i=1}^{n-1} d_i \geq t\right\} \tag{11}$$

Based on (10) and (11) the probability (9) is as follows

$$\Pr\{\{X_k(t) = n\} \cap \{\Theta_k \geq t\}\} = Pr\left\{\sum_{i=1}^{n} d_i \geq t\right\} - Pr\left\{\sum_{i=1}^{n-1} d_i \geq t\right\} \tag{12}$$

From (12) the integral $\int_0^{+\infty} \Pr\{\{X_k(t) = n\} \cap \{\Theta_k \geq t\}\} dt$ is as follows

$$\int_0^{+\infty} \Pr\{\{X_k(t) = n\} \cap \{\Theta_k \geq t\}\}dt = \int_0^{+\infty} Pr\left\{\sum_{i=1}^{n} d_i \geq t\right\} - Pr\left\{\sum_{i=1}^{n-1} d_i \geq t\right\}dt$$

(13)

where $\Theta_1 = d_1 + \ldots + d_7$, $d_1 = \beta_1$, $d_i = \alpha_1 \cdot \ldots \cdot \alpha_{i-1} \cdot \beta_i$ for $i = 2, \ldots, 7$.

In the above expression (13), the integral $\int_0^{+\infty} Pr\{\sum_{i=1}^{n} d_i \geq t\}dt$ determines the expected value of the random variable $\sum_{i=1}^{n} d_i$ [19]. From the properties of expected values [20] it follows that $E\{\sum_{i=1}^{n} d_i\} = \sum_{i=1}^{n} Ed_i$. Hence, we can write

$$\int_0^{+\infty} \Pr\{\{X_k(t) = n\} \cap \{\Theta_k \geq t\}\} = E\left\{\sum_{i=1}^{n} d_i\right\} - E\left\{\sum_{i=1}^{n-1} d_i\right\} = Ed_n \quad (14)$$

where $Ed_1 = E\beta_1$, $Ed_n = E\alpha_1 \cdot \ldots \cdot \alpha_{n-1} \cdot \beta_n$ dla $n \geq 2$.

In view of the above, and on the basis of (5) finally, the limit (8) is as follows for each $n = 1, 2, \ldots, 7$

$$\lim_{t \to \infty} \Pr\{X(t) = n\} = P_n = \frac{Ed_n}{E\Theta_1} \quad (15)$$

Finally, using (15) the stationary probabilities $P_1, \ldots, P_2$ defined by (7) are as follows

$$P_1 = \frac{E\beta_1}{E\Theta_1}, P_n = \frac{E\alpha_1 \cdot \ldots \cdot E\alpha_{n-1} \cdot E\beta_n}{E\Theta_1} \text{ for } n = 2, 3, \ldots, 7 \quad (16)$$

where $E\alpha_{n-1}, E\beta_n (n = 1, \ldots, 7)$ and $E\Theta_1$ are specified respectively by (3), (4) and (5).

3 Probabilistic Risk Assessment Based on the Proposed Model - Illustration

Traditional risk assessment quantifies risk as the product of the probability of an undesirable event leading to specific consequences and a measure of the negative impact on the organization due to this undesirable event (probabilistic risk assessment) [17] or as a triplet of threat, vulnerability, and consequences [18]. In this section we use probabilistic risk assessment to quantify cyber risks. To do this, we should first calculate the probability of each phase of the cyber kill chain, which can be determined using proposed model of the simple cyber kill chain "with iterations".

We can calculate "risk" traditionally as a product of likelihood of threats and their impacts on the assets of an organizations. To illustrate our approach simply let's assume that $A = [A_1, \ldots, A_7]$ is a vector of monetary values of the organization's assets calculated at each stage of the simple cyber kill chain. Then, "risk score" represented as $R(t)$ can be calculated using the following equation:

$$R(t) = P(t) \cdot A^T \tag{17}$$

where $P(t) = [P_1(t), \ldots, P_7(t)]$, $P_n(t) = P\{X(t) = n\}$ $(n = 1, \ldots, 7)$.

It is important to notice that we obtained (16) the vector of stationary probabilities P_k which can help us to calculate the risk score $R(t) \xrightarrow[t \to +\infty]{} R$ given by Eq. (17) as follows

$$R = P \cdot A^T$$

where $P = [P_1, \ldots, P_7]$ is given by (16).

In order to calculate risks at each stage of the simple cyber kill chain the stochastic model has to be parameterized. To estimate the stationary probabilities, it is necessary and enough to know the expected values $E\alpha_1, \ldots, E\alpha_{n-1}, \ldots, E\alpha_6, E\beta_1, \ldots, E\beta_n, \ldots,$ $E\beta_7$ and $E\Theta_1$. The most popular and straight-forward solution is to ask experts in cyber security domain assess the values $E\alpha_{n-1}, E\beta_n$ $(n = 1, \ldots, 7)$ and to base on their opinion, or to analyze existed empirical data, or a combination of both. The process of assessing the expected values is crucial, but it is not the primary focus of this article.

4   Conclusion

Due to the fact that until now in the available sources ([8, 15, 16]) there has not been published the stochastic model of cyber-attacks processes (based on the simple cyber kill chain) where has been assumed of any continuous probability distribution of adversaries' and defenders' activities time, this article fills this gap.

Like in [8, 15], it is assumed that the attack may be stopped while the aggressor is in any phase of the attack. In practice, an attack may be terminated not only by the will of the cybercriminals, but also, and perhaps primarily because of the operating cyber defense system.

It should be noted here that from a practical point of view to estimate the stationary probabilities, it is necessary and enough to know the expected values of individual times and to estimate the probabilities of success of completing each phase of the attack cycle.

On the basis of the presented model, obtained stochastic characteristics, such as: stationary probabilities of the attack process being in individual phases, the expected values of the duration of individual phases can be used for the needs of the risk assessment process and the security management of organizations and e-services provided.

References

1. McAfee: McAfee Labs Threats Report, August 2019. https://www.mcafee.com/enterprise/en-us/assets/reports/rp-quarterly-threats-aug-2019.pdf (2019), last accessed 2019/11/01
2. https://www.fireeye.com/cyber-map/threat-map.html. Accessed 02 Jan 2020
3. https://threatmap.checkpoint.com. Accessed 02 Dec 2019

4. Mandiant, APT1: Exposing One of China's Cyber Espionage Units. Mandiant (2013). www. fireeye.com/content/dam/fireeye-www/services/pdfs/mandiant-apt1-report.pdf. Accessed 31 Jan 2020
5. Cloppert M.: Security Intelligence: Attacking the Kill Chain. http://computer-forensics.sans. org/blog/2009/10/14/security-intelligence-attacking-the-kill-chain/. Accessed 02 Nov 2019
6. Hutchins, E.M., Cloppert, M.J., Amin, R.M.: Intelligence-driven computer network defense informed by analysis of adversary campaigns and intrusion kill chains, Leading Issues in Information Warfare and Security Research, vol. 1, pp. 78–104. Academic Publishing International Ltd, Reading, UK (2011)
7. ENISA: ENISA Threat Landscape Report 2018 (2019). https://www.enisa.europa.eu/ publications/enisa-threat-landscape-report-2018. Accessed 02 Nov 2019
8. Hoffmann, R.: Markov models of cyber kill chains with iterations. In: 2019 International Conference on Military Communications and Information Systems (ICMCIS) (2019). https://doi.org/10.1109/icmcis.2019.8842810
9. Lockheed Martin: Seven Ways to Apply the Cyber Kill Chain with a Threat Intelligence Platform. Lockheed Martin Corporation (2015). https://www.lockheedmartin.com/content/ dam/lockheed-martin/rms/documents/cyber/Seven_Ways_to_Apply_the_Cyber_Kill_ Chain_with_a_Threat_Intelligence_Platform.pdf. Accessed 02 Nov 2019
10. Yadav, T., Rao, A.M.: Technical aspects of cyber kill chain. In: Abawajy, Jemal H., Mukherjea, S., Thampi, Sabu M., Ruiz-Martínez, A. (eds.) SSCC 2015. CCIS, vol. 536, pp. 438–452. Springer, Cham (2015). https://doi.org/10.1007/978-3-319-22915-7_40
11. Spring, J.M., Hatleback, E.: Thinking about intrusion kill chains as mechanisms. J. Cybersecur. 3(3), 185–197 (2017)
12. Coleman, K.G.J.: Aggression in cyberspace. conflict and cooperation in the global commons: a comprehensive approach for international security, pp. 105–119. Georgetown University Press, Washington DC (2012)
13. Khan, M.S., Siddiqui, S., Ferens, K.: A cognitive and concurrent cyber kill chain model. In: Daimi, K. (ed.) Computer and Network Security Essentials, pp. 585–602. Springer, Cham (2018). https://doi.org/10.1007/978-3-319-58424-9_34
14. Hahn, A., Thomas, R.K., Lozano, I., Cardenas, A.: A multi-layered and kill-chain based security analysis framework for cyber-physical systems. Int. J. Crit. Infrastruct. Prot. 11, 39–50 (2015)
15. Hoffmann, R.: Markowowskie modele cykli życia ataku cybernetycznego. Roczniki Kolegium Analiz Ekonomicznych SGH, z 54, 303–317 (2019)
16. Hoffmann, R.: Ogólny cykl życia ataku cybernetycznego i jego markowowski model. Ekonomiczne Problemy Usług 131(2/1), 121–130 (2018)
17. Keller, W., Modarres, M.: A historical overview of probabilistic risk assessment development and its use in the nuclear power industry: a tribute to the late Professor Norman Carl Rasmussen. Reliab. Eng. Syst. Saf. 89(3), 271–285 (2005)
18. Kaplan, S., Garrick, B.J.: On the quantitative definition of risk. Risk Anal. 1(1), 11–27 (1981)
19. Klimow, G.P.: Procesy obsługi masowej. WNT, Warszawa (1979)
20. Beichelt, F.: Stochastic Processes in Science, Engineering and Finance. Taylor & Francis Group, LLC, New York (2006)
21. Kowalenko, I.N., Kuzniecow, N.J., Szurienkow, W.M.: Procesy stochastyczne. Poradnik. PWN, Warszawa (1989)

Genetic Algorithm for Generation Multistage Tourist Route of Electrical Vehicle

Joanna Karbowska-Chilinska(✉) and Kacper Chociej

Bialystok University of Technology, Wiejska Street 45A, 15-351 Bialystok, Poland
j.karbowska@pb.edu.pl

Abstract. The problem of selection points of interest which tourist wants to visit the most in the case when tourist travels by electric vehicle (EV) is examined in this paper. Furthermore the battery capacity of EVs is limited so charging stations are selected to the route, in order to a tourist could recharge the battery and move on to the next stage of the route. The genetic algorithm is proposed and tests for the different limitations of EVs batteries are conducted on realistic database. The experimental results show that the proposed genetic algorithm can successfully be used in this context.

Keywords: Tourist Trip Design Problem · Electric Vehicle Routing Tour Planning · Genetic algorithm

1 Introduction

When tourists visit a region for a short time and rent electric vehicles (EVs) they consider where they can travel using these vehicles. One of the essential problems to solve is to select from the set of points of interest (POIs) the most important attractions, taking into account the limitation of the driving range of battery and to plan places to charge the battery in order to continue the trip. In the literature the problem is known as an Electric Vehicle Routing Tour Planning (EVRTP) [1] and is an extension of Tourist Trip Design Problem (TTDP) [2]. The main objective of TTDP is to select POIs which maximize tourist satisfaction, taking into consideration limitations (e.g. the length of the route, the visiting time of each POI, the cost of hotels, entrance fees). The simplest model of TTDP is Orienteering Problem (OP) [3–5]. Orienteering is a sort of game: a competitor starts at given point and he tries to visit as many points with assigned score and back to the final point without exceeding the given time limit. The game goal is to maximize total collected score. This approach was applied to the tourist planning problem by Vansteenwegen et al. [6]. Extensions of OP are Team OP (TOP) [7], OP with Time Windows (OPTW) [8] or Time Dependent OP (TDOP) [9]. Generally consideration of opening hours, time travel between POIs and their time of visiting allows better modeling tourist trip planning problem [7,9–11].

© Springer Nature Switzerland AG 2020
K. Saeed and J. Dvorský (Eds.): CISIM 2020, LNCS 12133, pp. 366–376, 2020.
https://doi.org/10.1007/978-3-030-47679-3_31

The electric vehicles need recharging stops or swapping due to their limited ranges of batteries. For this reason the recharging schedule is an important issue in the tourist trip planning problem. Literature describes some problems such as Electric Vehicle Scheduling Problem (E-VSP) [12], Electric Vehicles Routing problem with Time Windows and Recharging Stations (E-VRPTW) [13] and Electric Vehicle Routing Tour Planning (EVRTP) [1]. In EVRTP the aim is to design a tour from the given starting to the given ending point, so the tourist can visit the most important POIs from the consideration of battery recharging or swapping and limitation of total trip time. Wang et al. [1] extends the model of OPWT and proposes model which simulate the change in batteries state of charge at each point along the route. The authors in [1] use the aforementioned model in heuristic algorithm to obtain a solution of EVRTP.

In this article we use the other model than in [1]. We also use an extension of OP, the Orienteering Problem with Hotel Selection (OPwHS) [14]. In OPwHS a set of hotels, a set of POIs with profits, D-the number of tour stages are given. The goal is to determine a tour that maximizes the total collected profit of visited POIs. The tour is composed of D+1 connected trips. Every trip starts and ends in one of hotels and each trip has the same limitation T_{max}. In our work we adapt OPwHS model- instead of hotels we consider charging stations. We assume that the starting point is a charging station and at this point the tourist fully charged batteries. The number of kilometers that the car can travel on a single battery charge is also known (it depends on the car brand). This value corresponds to T_{max}- the limitation length of each trip.

The contribution of this article is a genetic algorithm which generates the sequences of POIs and also chooses the stations when the batteries are recharged after driving T_{max} kilometers. In the stations the car is fully recharged and then the driver visits other POIs selected by the algorithm and so on until the ending point of a tour (located in the given charging station) is reached. The generated tour maximizes the total attractiveness from the visited POIs. In this work we do not consider the time span connected with the opening hours and visiting time of POIs or the driving time between POIs. The genetic algorithm is tested on realistic database for the different limitations of EVs batteries. Results and the execution time of the algorithm show that the presented solution could be use in practice.

The paper is organized as follows. Section 2 presents a formal definition of the regarded problem. Section 3 proposes a genetic algorithm which gives the tourist a tour composed of connected trips (between the trips will be recharging stations). The results of computational experiments conducted on realistic database are discussed in Sect. 4. Section 5 presents the conclusion of this work.

2   Problem Description

We define the described problem as a graph optimization problem in the same way as in the previous work [15]. The set A of n POIs and a set B of m EV charging stations will be the vertices of the graph. The distance t_{ij} between each adjacent POIs or POIs and EV charging stations are known. Each POI has a profit p bigger than zero, interpreted as an attractiveness. Each charging station has profit equal to zero. T_{max} denotes the number of kilometers that the car can travel on a single battery charging. The number of brakes D in a tour, in order to charge batteries is also given. The starting and the ending points, denoted by S and E respectively, are selected from the set B of EV charging stations. An ordered set of POIs between two EV charging stations is called a trip. The whole tour is composed of D+1 trips. In output the algorithm gives D+1 connected trips of visiting POIs and the charging stations at most once. Moreover the sum of collected profits is maximized and the total distance of each trip does not exceed the constraint T_{max}. The defined problem is the same as the Orienteering Problem with Hotel Selection [14].

3   Genetic Algorithm

In [15] for the presented problem we used the greedy method with local search operators. In this approach the genetic algorithm is used. The genetic algorithm gives us a chance to find more optimal solutions by changing all population by genetic operators.

First of all the initial population having P_{size} tours is generated. Every tour is composed of D+1 connected trips built randomly. Next each initial trip of the tour is shortened by standard 2opt operator [16]. When the whole initial tour is complete N new generations of the algorithm are created. Each generation is created as follows. First the selection process of the best individuals among T_{size} random individuals is performed. New population is built of the collected individuals having the same size as the previous population. Next $C_p/2$ random pairs of tours for the crossover process are collected, where C_p means the probability of taking pair of tours to try crossover operator on them (e.g. when $P_{size} = 60$ and $C_p = 0.5$, 30 tours are taken randomly (15 pairs) to crossover). After that the crossover process on every collected pair is performed. Next with probability M_p random individuals for the mutation process are collected where M_p means the probability of taking tours to try mutation operator on them (e.g. when $P_{size} = 60$ and $M_p = 0.5$, 30 tours are taken randomly for the mutation operator). After that the mutation process on the collected individual is performed. There are two types of mutation - insert and removal mutation. When the algorithm processes next generations there is a moment when the algorithm reaches the tour with some profit value and there is no improvement in further iterations. In order to eliminate unnecessary time consumption, the following stop condition is used: the same or lower value of profit is repeated for at least k generations the algorithm stops. The result of the algorithm is a tour with the highest value among all processed generations. The basic structure of the proposed genetic algorithm is given as follows.

```
population = generateInitialPopulation(P_size)

bestParents = [getBestParent(population)]
```

```
for(i = 0; i < N; i += 1) {
    population = selectBestFromPopulation(population, T_size)

    parentsToCrossover = getParentsToCrossover(population, C_p)

    for((parentA, parentB) in parentsToCrossover) {
        makeCrossover(parentA, parentB)
}

    parentsToMutation = getParentsToMutation(population, M_p)

    for(parent in parentsToMutation) {
        makeMutation(M_ratio, parent)
    }

    bestParents.push(getBestParent(population))

    if(checkStopCondition(bestParents, i + 1) == true) break
}
```

The details of the algorithm will be described in the next subsections.

3.1 Initial Population

At the beginning of the algorithm the initial population is generated. For this purpose P_{size} random tours are generated as follows. For one tour D+1 trips are created where the first trip starts with the chosen charging station S, the last trip ends with the chosen charging station E. Every trip (except first) starts with the last charging station of the previous trip. For the actual point of the trip algorithm inserts not visited POI in a random way. Next the algorithm finds the nearest not visited charging station which can finish building the actual trip. If the last trip is built, charging station E is the last point in the trip. Next algorithm checks if the total length of the actual trip does not exceed the T_{max} limitation. For this purpose following condition is checked:

$$actualTripDistance + distanceLeft + distanceRight <= T_{max} \qquad (1)$$

where $actualTripDistance$ - actual length of trip, $distanceLeft$ - distance between the last POI of an actual trip and the candidate for next POI, $distanceRight$ - distance between the candidate for next POI and the candidate for the charging station being the last point of the trip. If the condition is satisfied, the candidate for POI is added to the trip and the candidate for charging station is saved. Example of an insertion of the next POI and charging station to the trip is shown in Fig. 1. On the actual trip 2opt operator is executed. If T_{max} is exceeded the algorithm ends building this trip. The last point of the trip is the saved charging station.

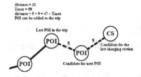

Fig. 1. Insertion new POI and charging station to the trip.

3.2 Selection

In this approach tournament selection is used. First algorithm takes randomly T_{size} tours from the current population. Next the tour with the highest sum of POIs profits is returned. This process is repeated until the desired amount of population is satisfied (P_{size} times). In Fig. 2 the example of the selection ($T_{size}= 3$) is presented.

3.3 Crossover

The crossover operation is intended to create a stronger individual through crossing genotype of two parents. In this approach algorithm tries to cross two different tours in point which is common for them. First the common point being charging station is found. Next two new tours are created. First new tour is composed of the first part of the first tour and the second part of the second tour. Second new tour is composed of the first part of the second tour and the second part of the first tour (see the Fig. 3). In the next step new tours are validated. If the new tour is built with exactly D+1 trips (contains D+2 charging stations), does not have any repetitions of POIs and each trip does not exceed T_{max} limitation the tour can be a candidate for the new tour. Otherwise the tour cannot be a candidate and is omitted. Both tours are validated this way. In the end a parent with the lowest sum of POIs profits is replaced with a new tour with the highest sum of profits (if new tour exists and its sum of profits is higher than parent's with the lowest sum of profits).

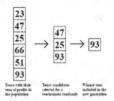

Fig. 2. Tournament selection ($T_{size} = 3$)

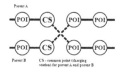

Fig. 3. Crossover operation

3.4 Mutation

There are two types of mutation in the algorithm - insertion and removal. Probability of insertion mutation is M_{ratio} and probability of removal mutation is $1 - M_{ratio}$.

3.5 Insert Mutation

In the insert type of mutation algorithm gets not visited POI in a random way. Next the algorithm searches the tour and its positions where the following ratio is the best:

$$\frac{candidate.profit^2}{distanceLeft + distanceRight} \qquad (2)$$

where variable *distanceLeft* means distance between the POI being before the candidate and the POI candidate for insertion, *distanceRight* - distance between the POI candidate for insertion and the POI being after the candidate. If the limitation T_{max} is not exceeded after potential POI's insertion- if the condition is satisfied:

$$distance - distanceRemoved + distanceLeft + distanceRight <= T_{max} \quad (3)$$

the new POI is added to the trip on the best possible position (see the example in the Fig. 4). In the end 2opt operator is performed.

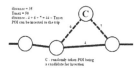

Fig. 4. Insertion mutation - C is included to the tour

3.6 Removal Mutation

In the removal type of mutation algorithm tries to eliminate the POI with the lowest profit so the following ratio is calculated:

$$\frac{poi.profit^2}{distanceLeft + distanceRight} \qquad (4)$$

where *distanceLeft* - distance between previous point and currently checked POI, *distanceRight* - distance between currently checked POI and next point. If the POI is found, the algorithm removes this POI from the tour. The Fig. 5 shows the example of removal mutation.

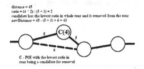

Fig. 5. Removal mutation - the elimination of C (profit $p_c = 4$)

3.7   Time Complexity

The time complexity of the algorithm depends on the number of operations during the subsequent steps of iterations of the genetic algorithm. Let s denotes the average numbers of POIs in a route i.e. (the sum of all POIs in the routes)/ P_{size}. The number of operation in the selection is $\theta(P_{size} \cdot s)$. The complexity of the crossover is $\theta((C_p/2) \cdot P_{size} \cdot s)$. The number of operation in the removal and insertion mutation is equal to $\theta(M_p \cdot P_{size} \cdot s)$ and $\theta(M_p \cdot P_{size} \cdot (s + s^2)$ respectively. In the last dependency function s^2 relates to the complexity of the 2 opt algorithm.

4   Experiments and Results

The presented algorithm was implemented in Python 3.6.6 version. All experiments were performed on data containing 303 POIs and 21 EV charging stations in the Silesian region (Poland). POIs are specified with coordinates (latitude and longitude), place name and profit (random number between 1 and 10).

4.1   Parameters

There were many tests with various parameters values performed. The number of iterations of the genetic algorithm and the number without improving results are set to $N = 100$ and $k = 10$ respectively. Initial experiments were performed on $P_{size} = 300$ but it turns out that only $P_{size} = 50$ gives values as good as the higher P_{size}. The smaller P_{size} value saves a lot of time wasted on processing. However the experiment for $P_{size} = 25$ gave visibly worse results. The calibration of parameters T_{size} (number of individuals in tournament selection) and C_p (probability of two random parents crossover) did not make significant changes in result value. In the algorithm $T_{size} = 5$ and $C_p = 0.4$ were used. Mutation process had the strongest effect on final value. For this approach $M_p = 0.4$ was used and gives the best results. For higher value of M_p parameter (e.g. $M_p = 0.8$) results were only a bit worse but the algorithm ends very often in earlier iteration (very often 35–50th generation). For lower value of parameter (e.g.$M_p = 0.2$) sums of POIs profits were visibly worse than for $M_p = 0.4$ but execution time was up to 6 times faster than for $M_p = 0.4$. In experiments also probability insertion and removal mutation was calibrated. For $M_{ratio} = 0.5$ algorithm was very unstable and the increase of the result was not so visible. The best results for $M_{ratio} = 0.9$ noted.

4.2   Comparison of the Results

The proposed genetic algorithm was tested on the various T_{max} lengths (distance that EV can drive on a single battery charge) - 120, 150, 170, 210 and 420 km. The results were compared to the greedy algorithm enriched by the local search operators such as 2opt and insert [15] (the algorithm was marked as GR+2opt+in). We decide to compare results to the mention method because both of them are based on the same model the Orienteering Problem with Hotel Selection. In the author's opinion it is worth to use this model to problem of determination of multistage tourist route of EVs. In the literature, we did not find publications that treat the problem as OPwHS, therefore we compare the results to the algorithm [15].

The comparison of results for D = 0, 1 is presented in Table 1–Table 2. The abbreviation of GA is used to the genetic algorithm. In tables the length is given in kilometers, execution time of the algorithm in seconds. The column Profit denotes sums of profits the best tour generated by the algorithm. The column

Table 1. Comparisons of the results GR+2opt+ in and GA (D = 0)

GR+2opt+ in				GA				
Tmax	Length	Profit	Time	Length	Profit	Time	Iterations	% gap between profits
120	119.73	132.80	13.66	119.02	342.80	2.14	87	258.1%
150	149.21	180.90	16.99	149.37	467.20	3.28	71	258.2%
170	168.81	198.20	17.51	168.80	525.20	4.00	59	265.0%
210	209.85	246.70	21.79	209.73	658.30	13.25	100	266.8%
420	419.91	818.30	63.84	419.79	1066.20	33.44	75	130.3%

Table 2. Comparisons of the results GR+2opt+ in and GA (D = 1)

GR+2opt+ in				GA				
Tmax	Length	Profit	Time	Length	Profit	Time	Iterations	% gap between profits
120	239.71	447.20	39.35	238.07	611.70	4.53	95	136.8%
150	299.63	642.00	49.63	299.79	782.30	8.19	68	121.9%
170	339.85	467.40	41.97	338.63	762.10	6.87	79	163.1%
210	419.12	798.90	57.38	419.88	930.50	11.98	77	116.5%
420	839.75	1224.80	62.47	839.29	1394.4	40.50	73	113.8%

Iterations contains the number of iteration to the stopping conditions. The column gap denotes % gap between the result profit of GA and GR+2opt+in.

For each value of T_{max} the algorithm GA returns much better results than GR+2opt+in. The very high increase of profit for the short trips is observed. In case D=0, when T_{max} is equal 120–210 km GA returns about 2,6 times better results than GR+2opt+ in. Tours are built much more effectively and more valuable POIs can be included in one tour when GA is used. Furthermore the execution time of GA is much lower than GR+2opt+ in. Figures 6–7 illustrate the example routes marked on map generated by GR+2opt+in and GA. In figures the points shaped as the plugs are charging stations and the other points are attractions. In comparison to greedy algorithm tours for GA are built more effectively. Experiments showed that genetic algorithm focuses more points in a bigger group (cluster) - often it is more profitable collecting more points being nearby each other and having lower profit than collecting points with higher profit but

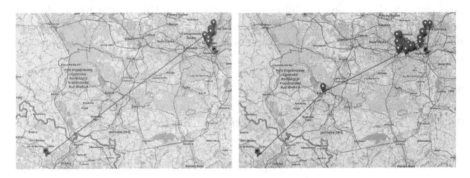

Fig. 6. The example of route generated by GR+2opt+in and GA (D $=0$, $T_{max} = 120$)

Fig. 7. The example of route generated by GR+2opt+in and GA (D $=1$, $T_{max} = 80$)

being a bit further. Furthermore in a genetic algorithm internal charging stations are different than in a greedy approach so it changes tour overall.

5   Conclusions and Further Work

In the paper the tourist trip planning problem of electric vehicle is examined. We include an information about limitation of batteries capacity using T_{max} - trip length limitation. The proposed genetic algorithm gives tourists the sequence of the most attractive POIs and the stations where they need recharge batteries. The tests on real data shows that in comparison to the greedy method, the population algorithm gives most attractive routes and its execution time is much shorter.

In the future in our approach we will include to this model time factor: opening hours and time visiting of POIs, travel time between POIs and also the charging batteries time. The enriched model will reflect more accurately the real situations.

Acknowledgements. The research was carried out as a part of research work nr WZ/WI/1/19 at Bialystok University of Technology financed from a subsidy of the Minister of Science and Higher Education.

References

1. Wang, Y.-W., Lin, C.-C., Lee, T.-J.: Electric vehicle tour planning. Transp. Res. Part D: Transp. Environ. **63**, 121–1361 (2018). https://doi.org/10.1016/j.trd.2018.04.016
2. Gavalas, D., Konstantopoulos, C., Mastakas, K., Pantziou, G.: A survey on algorithmic approaches for solving tourist trip design problems. J. Heuristics **20**(3), 291–328 (2014). https://doi.org/10.1007/s10732-014-9242-5
3. Vansteenwegen, P., Gunawan, A.: Orienteering Problems: Models and Algorithms for Vehicle Routing Problems with Profits. EATOR. Springer, Cham (2019). https://doi.org/10.1007/978-3-030-29746-6
4. Gunawan, A., Lau, H.C., Vansteenwegen, P.: Orienteering Problem: a survey of recent variants, solution approaches and applications. Eur. J. Oper. Res. **255**(2), 315–332 (2016). https://doi.org/10.1016/j.ejor.2016.04.059
5. Ostrowski, K., Karbowska-Chilinska, J., Koszelew, J., Zabielski, P.: Evolution-inspired local improvement algorithm solving orienteering problem. Ann. Oper. Res. **253**(1), 519–543 (2016). https://doi.org/10.1007/s10479-016-2278-1
6. Vansteenwegen, P., Souffriau, W., Vanden Berghe, G., Oudheusden, D.: The city trip planner: an expert system for tourists. Expert Systems with Applications **38**(6), 6540–6546 (2011). https://doi.org/10.1016/j.eswa.2010.11.085
7. Gunawan, A., Lau, H., Vansteenwegen, P., Lu, K.: Well-tuned algorithms for the team orienteering problem with time windows. J. Oper. Res. Soc. **68**(8), 861–876 (2017). https://doi.org/10.1057/s41274-017-0244-1
8. Karbowska-Chilinska, J., Zabielski, P.: Genetic algorithm with path relinking for the orienteering problem with time widows. Fundamenta Informaticae **135**(4), 419–431 (2014). https://doi.org/10.3233/FI-2014-1132

9. Ostrowski, K.: An effective metaheuristic for tourist trip planning in public transport networks. Appl. Comput. Sci. **14**(2), 5–19 (2018). https://doi.org/10.23743/acs-2018-09

10. Karbowska-Chilinska, J., Zabielski, P.: Maximization of attractiveness EV tourist routes. In: Saeed, K., Homenda, W., Chaki, R. (eds.) CISIM 2017. LNCS, vol. 10244, pp. 514–525. Springer, Cham (2017). https://doi.org/10.1007/978-3-319-59105-6_44

11. Gavalas, D., Konstantopoulos, C., Mastakas, K., Pantziou, G., Vathis, N.: Heuristics for the time dependent team orienteering problem: application to tourist route planning. Comput. Oper. Res. **62**, 36–50 (2015). https://doi.org/10.1016/j.cor.2015.03.016

12. Wen, M., Linde, E., Ropke, S., Mirchandani, P., Larsen, A.: An adaptive large neighborhood search heuristic for the Electric Vehicle Scheduling Problem. Comput. Oper. Res. **76**, 73–83 (2016). https://doi.org/10.1016/j.cor.2016.06.013

13. Erdelić, T., Carić, T., Erdelić, M., Tišiljarić, L.: Electric vehicle routing problem with single or multiple recharges. Transp. Res. Procedia **40**, 217–224 (2019). https://doi.org/10.1016/j.trpro.2019.07.033

14. Divsalar, A., Vansteenwegen, P., Chitsaz, M., Sörensen, K., Cattrysse, D.: Personalized multi-day trips to touristic regions: a hybrid GA-VND approach. In: Blum, C., Ochoa, G. (eds.) EvoCOP 2014. LNCS, vol. 8600, pp. 194–205. Springer, Heidelberg (2014). https://doi.org/10.1007/978-3-662-44320-0_17

15. Karbowska-Chilinska, J., Chociej, K.: Optimization of multistage tourist route for electric vehicle. In: Silhavy, R. (ed.) CSOC2018 2018. AISC, vol. 764, pp. 186–196. Springer, Cham (2019). https://doi.org/10.1007/978-3-319-91189-2_19

16. Tsiligirides, T.: Heuristic methods applied to orienteering. J. Oper. Res. Soc. **35**(9), 797–809 (1984). https://doi.org/10.1057/jors.1984.162

Event Ordering Using Graphical Notation for Event-B Models

Rahul Karmakar[1]([✉]) [iD], Bidyut Biman Sarkar[2], and Nabendu Chaki[3] [iD]

[1] The University of Burdwan, Burdwan, India
rkarmakar@cs.buruniv.ac.in
[2] Techno International, Rajarhat, Kolkata, India
bidyutbiman@gmail.com
[3] University of Calcutta, Kolkata, India
nabendu@ieee.org

Abstract. System requirements are sometimes either too complex or undefined. Event-B is a formal modeling method and is being used increasingly to model various systems. Event-B models support atomicity decomposition and are quite useful for complex refinement structures. However, neither a Event-B model represents any explicit control flows among the events, nor does it support links between the new events during refinements. This work aims to model the Stop and Wait mechanism for an Automatic Repeat Request (ARQ) protocol to analyze the complexities due to communication errors during data re-transmissions. The limitation is the lack of control flows among the events during successive refinements. This has been graphically represented in this work and embedded with Event-B notations for the atomicity decomposition of the model. Finally, the successive refinements presented using an Event-B model, has been validated using the Rodin tool. This leads to a successful ARQ model.

Keywords: Event-B · Formal modeling · Stop and wait ARQ · RODIN tool · Atomicity decomposition · ERS diagram

1 Introduction

Model-based verification techniques describe the system behaviour in a mathematical and unambiguous fashion [1]. Event-B modeling language is devised as an extension of classical B methods, which has different applications in diverse domains. It is a step by step process of system development. We start with an abstract model and refine the model successively to meet the requirements. An Event-B model has a static component called context, where we declare all the sets, constants and axioms. The dynamic part is the machine that sees the context. A machine has variables and invariants. The state changes of a machine are defined by guards and actions, which is called event [2]. Rodin [3] is a tool support for the validation of an Event-B model. The control flow between events cannot be handled explicitly in Event-B as it does not accept ordering of the events.

© Springer Nature Switzerland AG 2020
K. Saeed and J. Dvorský (Eds.): CISIM 2020, LNCS 12133, pp. 377–389, 2020.
https://doi.org/10.1007/978-3-030-47679-3_32

However, it can be managed implicitly by Event-B. Event-B modeling allows to refine a model incrementally. When a new event is introduced in a refinement stage, we could not link externally between the new and the abstract event. Atomicity decomposition of a model is supported by Event-B. The relationship between the events (atomicity) is a very important aspect to maintain, when we design a large and complex system. A case study on stop and wait for the ARQ technique with atomicity decomposition of the model is presented in this paper. It is a layer 2 flow control mechanism for data communication. Similar applications are found using Event-B [4] and Petri Net [5]. The basic concept of atomicity decomposition and model decomposition are explained below.

The atomicity decomposition technique is described with a brief overview. Figure 1 represents the explicit relationship among the events A to F. Event A is the abstract event and three events B, C and D are the children of A in the tree like structure. The event B does not refine the event A and represented by the dashed line. Events C and D refine the event A and the refinement relationship is represented by the solid lines [6]. Another aspect of the diagram is that, it implies event B will execute before event C and C will execute before event D. Thus, the ordering among the events (B, C, D) is also represented by the diagram. This ordering can be established by Event-B notations. The * between A and C signifies multiple instances of C, which means event C can execute multiple times. Different constructors can be used to represent event relationships. We get some of the representations from Fig. 1. Event E and F are exclusive to each other, which is represented by XOR constructor. Constructors like AND, OR can also be used to represent the relationships between Events. We can decompose the model with graphical notations and then implement the explicit ordering using Event-B notations.

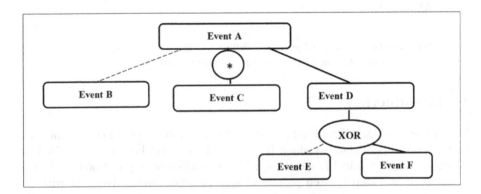

Fig. 1. Atomicity decomposition diagram.

Model decomposition technique decreases the complexity of large system and represented in Fig. 2. The large model can be divided into sub-models and refined individually. The events and variables are shared among the sub models for distributed and concurrent systems [7]. The shared event and variable are represented in Fig. 2. There are 3 events A, B, and C. Events X and Y share the variables A and event Y and Z share the event B. The machine M is divided into two submachine M1 and M2. Both the submachines share the event Y. This paper present an Event-B modeling approach using a graphical notation introduced as in Jackson System Development JSD [6]. This decomposition structure is explained in [8,9]. We could incorporate explicit ordering between the events using JSD graphical notations. This ordering has improved and enriched the conventional Event-B modeling. Refinements are also represented by the graphical structure. These two approaches help us to design the ARQ system in a more flexible manner, especially when re-transmission takes place due to frame loss or acknowledgement message loss. The paper is structured as follows, in Sect. 2 we present an overview of stop and wait ARQ technique. Section 3 presents a brief survey of the related works. Section 4 presents the ERS of stop and wait for ARQ in detail followed by some observations in Sect. 5. The concluding remarks highlighting the potential of the proposed work in shaping up paths for future research directions are presented in Sect. 6.

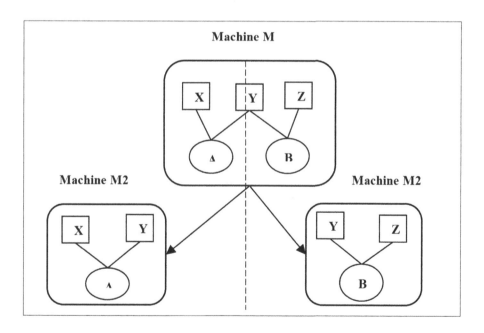

Fig. 2. Model decomposition

2   An Overview of ARQ Protocol and Requirements

Flow control is one of the important design issues in the data link layer when the speed mismatch happens between the sender and receiver. It controls the rate of the frames transmitted to the receiver. Figure 3(a) represents the normal operation of Stop and Wait for ARQ mechanism. The sender sends a packet and waits for the acknowledgment from the receiver within a specified time to ensure the successful transmission. The control variables S and R have the current value of the frame either 0 or 1. Timeout is an important aspect of this technique. Suppose the acknowledgment is not received within the specified time due to delayed acknowledgment showed in Fig. 3(b) or lost acknowledgment showed in Fig. 3(d) then the duplicate copy of the frame is sent by the sender after the timeout. The duplicate frame is then discarded at the receiving end in case of delayed acknowledgment showed in Fig. 3(b) and the duplicate acknowledgment is discarded at the sender's side in case of acknowledgment lost showed in Fig. 3(d). The duplication is identified by the control variable values. Figure 3(c) represents how transmission takes place when a frame is lost. The sender retransmits the frame after the timeout and accepted by the receiver [10]. The whole operation is represented using ERS in Sect. 4.

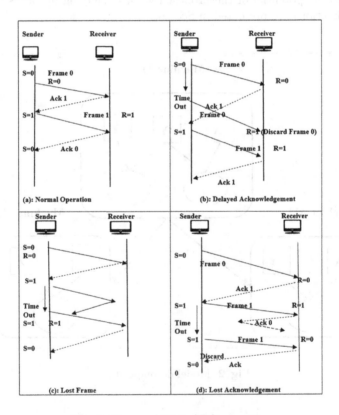

Fig. 3. Stop and Wait ARQ operation

3   Related Work

There have been works [8] proposing additional structuring to augment Event-B notation for atomicity decomposition of a complex refinement. Model decomposition is also presented using some case studies. Event Refinement Structure (ERS) was proposed by Butler [8]. It is a graphical notation based on JSD [6]. The relationships between the events are presented graphically to implement the Event-B notations. A technique is proposed to decompose a machine into sub-machines and those are refined independently. The paper addresses two important aspects of system development using Event-B i.e., atomicity decomposition and model decomposition using graphical notations. These are explained in the section of this manuscript. Atomicity decomposition of the Multimedia Protocol using Event-B is addressed in the paper [11]. It represents a protocol of media Channel System that establishes, modify and closes the channel by the communication parties. They also compared the model with the spin model checker. Further decompositions are performed using guards and events instead of sequential decomposition, which will be more useful for a complex system. No automatic model builder is used in this work. The main goal of the BepiColombo mission [7] had been to explore the planet Mercury. The whole system is controlled by the Mission-Critical-Software (MCS). The MCS controls the earth and also the device. It checks the Telecommand (TC) received from the earth and then validates the TC. TC is a control message and there are many types of TCs in the system. The atomicity, decomposition, and model decomposition are implemented using Event-B to handle this complex system. They validate the model using the RODIN tool. It shows how atomicity decomposition and model decomposition are effectively used to handle the control behavior manually. A plug-in tool Event Refinement Structure (ERS) is developed [9]. This tool automatically constructs the consistent Event-B model with control flows and refinement relationships. A context-free grammar notation Augmented Backus Normal Form (ABNF) is introduced to describe the ERS language syntax. Some of the ABNF features are flow, par, child, constructor, etc. Altogether 19 translation rules from ERS to Event-B are formed. All the constructors like a loop, logical XOR, replicator are also described. The development architecture is like; they define the ERS language specification in the Eclipse Modelling Framework (EMF) Meta-model. The source Meta-model is then transformed into the Event-B EMF target meta-model. The transformation from the target meta-model to Event-B is performed by a rule-based model-to-model transformation language called Epsilon Transformation Language (ETL). The ERS tool is then compared against the BepiColombo system [9] and the Multimedia Protocol [11]. It is found that the total proofs from the systems in [11] and [7] are substantially reduced. Here, the ERS tool does not provide a graphical environment of the ERS diagrams. The ERS diagrams are represented as an EMF model and manipulated by the EMF structure editor. More translations rules can be implemented for the ERS language. The improved version of the ERS tool provides a graphical environment for the ERS [12]. The Generic Diagram Extension Framework approach is proposed to transform ERS to Event-B. It has graphical and validation support for the model whereas the ERS tool needs another tool for model validation. It is a Java-based tool useful for complex case studies and validation of models. Authors claimed that more translation rules can be added in the future to add application-specific guards, actions, and invariants in the ERS environment without switching to Event-B editor. Another work [4], describes the stop and wait (SAW) technique is implemented by Event-B and verified by the UPPAAL model checker. They provide the mapping between Event-B to UPPAAL. The SAW model is implemented manually by Event-B and checked using RODIN then verifies the model using UPPAAL. The authors handle the complexity of the protocol with a single machine. Different cases of retransmission of message and acknowledgment are not presented clearly. These cases can be represented by the refinement steps. We find ERS as a very useful approach to model a system requirement using Event-B, from many of the existing works that we studied. The atomicity decomposition and model decomposition can be applied to a communication protocol. Event-B is a formal method for system-level modeling and analysis. This is often used to represent a system at different abstraction levels and for formal verification of consistency between refinement levels. However, there is no formal graphical representation for complex system refinements. This paper represents the stop and wait technique with ERS and establish the explicit ordering between events. Subsequently, these graphical notations are translated to Event-B notations. This helps to handle the complex behavior of the system effectively. The comparison of the complexity and flexibility of the approach with the non-ERS design-based approach has been done for justification.

4   Atomicity Decompositions of Stop and Wait Protocol

4.1   Abstract Specification: Basic Operations of SAW

This is the abstract representation of the Stop and Wait operations. The requirement goal is to establish sequencing among the events and the refinement relationships. The graphical representation of the abstract system is represented in Fig. 4 with five events. As discussed in Sect. 1, *Sender_Send_Request*, *Receiver_Receive_Request*, *Receiver_Send_Ack* and *Sender_Receiver_Ack* refine the abstract event *Stop_and_Wait_ARQ* and these are in the sequence to complete the basic Stop and Wait ARQ operation i.e. *Sender_Send_Request* is followed by *Receiver_Receive_Request*, followed by *Receiver_Send_Ack*. The requirement properties can be established with the rules given below. The ordering and refinement relationship between two events are established by using subset property. The relation between set variable of the abstract event and the concrete variable of the refined event can be represented as:

Preceding Abstract Event (variable name is same as the event name) ⊆ Succeeding Refined Event (variable name is same as the event name) It may be defined in Event-B by giving the same name of the variables with the events and established as an invariant property [7]. These properties are held while designing the whole system. The sequencing and refinement relationships among the events are showed in Fig. 4. This can be described as the Event-B properties 1 to 4 in Table 1. Property 1 ensures that the *Sender_Send_Request* can send multiple request. Hence, the variable *Sender_Send_Request* is a subset of the set Request. Four other scenarios are shown in Fig. 5(a), (b), and 6(a), (b). The successful receiving of the request by the receiver is checked. This decomposition is represented in Fig. 5(a) and in 5(b). The *Receiver_Receive_Fail* event is used for the purpose. The event relationship can be represented by the Event-B properties 5 and 7 in Table 1. The event *Sender_Receive_Fail* is used when the sender did not receive the acknowledgment sent by the receiver represented in Fig. 6(a) and (b). This relation can be represented by Event-B properties 6 and 8 in Table 1. These two events refine the abstract event *Stop_and_Wait_ARQ* and are represented by solid lines between them. The disjoint relationship between *Receiver_Receive_Request* and *Receiver_Receive_Fail* can be established using the intersection property Event X ∩ Event Y = ∅ and given as Event-B property 7. Property 8 established the same relation between *Sender_Receive_Ack* and *Sender_Receive_Fail*. The four events in Fig. 4, *Sender_Send_Request*, *Receiver_Receive_Request*, *Receiver_Send_Ack* and *Sender_Receiver_Ack* refine the event Stop_ and_ Wait _ARQ. The refinement relationship between the Stop_ and_ Wait _ARQ and the *Sender_Send_Request* showed in Fig. 5(a) and 5(b) for successful sending of request from sender to receiver can be modelled using Event-B machine given below.

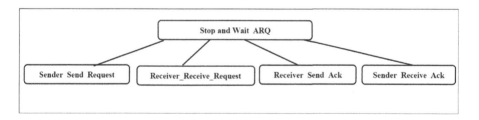

Fig. 4. Normal operations of stop and wait ARQ.

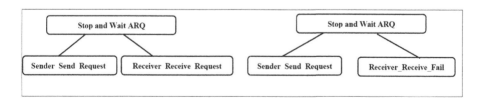

Fig. 5. (a)(b) Sender receives request or Sender receives fail.

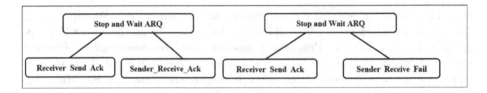

Fig. 6. (a)(b)Sender receives ACK or Sender receives ACK

Table 1. Variables and Properties

Variables	Event relationship Properties (P)
$Sender\_Send\_Request$	P1:$Sender\_Send\_Request \subseteq$ Request
$Receiver\_Receive\_Request$	P2:$Receiver\_Receive\_Request \subseteq Sender\_Send\_Request$
$Receiver\_Send\_Ack$	P3:$Receiver\_Send\_Ack \subseteq Receiver\_Receive\_Request$
$Sender\_Receive\_Ack$	P4:$Sender\_Receive\_Ack \subseteq Receiver\_Send\_Ack$
$Receiver\_Receive\_Fail$	P5:$Receiver\_Receive\_Fail \subseteq Sender\_Send\_Request$
$Sender\_Receive\_Fail$	P6:$Sender\_Receive\_Fail \subseteq Receiver\_Send\_Ack$
	P7:$Receiver\_Receive\_Request \cap Receiver\_Receive\_Fail = \emptyset$
	P8:$Sender\_Receive\_Ack \cap Sender\_Receive\_Fail = \emptyset$

$Sender\_Send\_Request$
refines
$Stop\_and\_Wait\_ARQ$
ANY
$Request$
WHERE
$grd1:$ $Req \in$ Request $\setminus$ $Sender\_Send\_Request$
THEN
$act1:$ $Sender\_Send\_Request :=$ $Sender\_Send\_Request \cup$ Request
END

The $Sender\_Send\_Request$ event refines $Stop\_and\_Wait\_ARQ$ event. The Stop and Wait ARQ system can send and receive multiple requests and acknowledgements so Request and ACK are considered as sets and Request and Ack are used as a parameter. The guard ensures that the event $Sender\_Send\_Request$ has not occurred with the current Request. The action of the event defines the inclusion of the new Request to the variable $Receiver\_Receive\_Request$. The Event relationship between $Receiver\_Receive\_Request$ or $Receiver\_Request\_Fail$ with Stop_ and _Wait _ARQ showed in Fig. 5(a) and (b) can be represented below.

$Receiver\_Receive\_Request$
refines
$Stop\_and\_Wait\_ARQ$
ANY

Request
WHERE
grd1: $Request \in Sender\_Send\_Request \setminus (Sender\_Send\_Request$
$\cup Sender\_Receive\_Fail)$
THEN
act1: $Receiver\_Receive\_Request :\,= Receiver\_Receive\_Request \cup Request$
END

The guard ensured the event *Sender_Send_Request* is executed before *Sender _Send_Request* or *Sender_Receive_Fail* event. The events can be modelled towards successful receiving of the acknowledgment showed in Fig. 6(a) and 6(b) from Receiver to Sender. This is given below.

Receiver_Send_Ack
refines
Stop_and_Wait_ARQ
ANY
Ack
WHERE
grd1: Ack $\in Receiver\_Receive\_Request \setminus Receiver\_Send\_Ack$
THEN
act1: $Receiver\_Send\_Ack:\,= Receiver\_Send\_Ack \cup Ack$
END

The *Sender_Receive_Ack* event can be modelled given below with Event-B notations.

Sender_Receive_Ack
Refines
Stop_and_Wait_ARQ
ANY
Ack
WHERE
grd1: Ack $\in Receiver\_Send\_Ack \setminus (Receiver\_Send\_Ack \cup Receiver\_Receive\_Fail)$
THEN
act1: $Sender\_Receive\_Ack :\,= Sender\_Receive\_Ack \cup Ack$
END

Confirm sending and receiving by the Sender and Receiver: Four new events *Sender_Ready*, *Timer_Set*, *Req_Seq_No* and *Wait_Ack* are introduced to complete the operation *Sender_Send_Request* depicted in Fig. 7. These events will not refine the event *Sender_Send_Request* and represented with dashed lines. The orderings among the events ensure the successful sending the request from sender's side to receiver. The *Receiver_Receive_Request*, *Receiver_Send_ACK* and *Sender_Receive_ACK* events are represented in

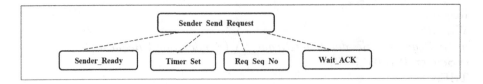

Fig. 7. Sender sends request to the receiver.

Fig. 8(a)(b) and 9. All the new events skip the refinements and their sequencing ensures the successful receiving the request by the receiver and Ack by the sender.

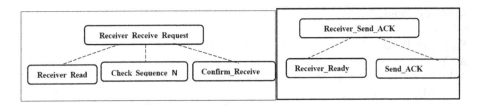

Fig. 8. (a)(b)Receiver receives the request and Receiver sends the Ack

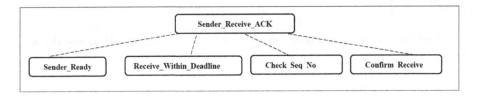

Fig. 9. Resend duplicate frame when ACK is lost (Regiment 1)

4.2   Refinement 1: Resending Request/Ack

As discussed in Sect. 2 all the resending scenarios of Request and Acknowledgement are represented with the ERS diagram as a refinement of the previous abstract model. Figure 10 represents resending of the old frame with three events *Sender_Ready*, *Not_Receive_Within_Time* and *Resend_Old_Request*. Only the *Resend_Old_Request* event refines the *Sender_Receive_Request*. This relationship between the events can be implemented by the subset relation between the concrete variable *Resend_old_Request* with the set variable *Sender_Receive_Ack*. The resending of duplicate frame, when the acknowledgment is lost from receiver to sender, is represented by the ERS notations in Fig. 11. There are four events *Sender_Ready*, *Not_Receive_Within_Time*, *Receive_Old_ACK*, and *Resend_Old_Request* which skip the refinement and the *Resend_Old_Request* refines the abstract event *Sender_Receive_ACK*. The refinement relationship between *Resend_Old_Request* and *Sender_Receive_ACK* can be represented by the property *Resend_Old_equest* ⊆ *Sender_Receive_ACK*. The resending of duplicate acknowledgment operation was discussed in Sect. 2. When a particular frame is not received in time or the duplicate frame is received, then the corresponding acknowledgment is resent. The ERS in Fig. 12 showed the operation and the relationship between events. *Receiver_Ready* and *Not_Receive_Request* or *Receive_Old_Request* events did not refine the abstract event *Receiver_Receive_Ready* and represented by dashed lines. The *Resend_ACK* event refines *Receiver_Receive_Request* and represented by the solid line. The property is represented as *Resend_ACK* ⊆ *Receiver_Receiver_Request*.

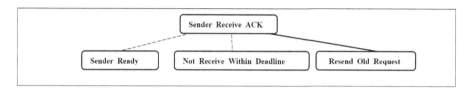

Fig. 10. Resend duplicate frame when ACK is lost (Regiment 1)

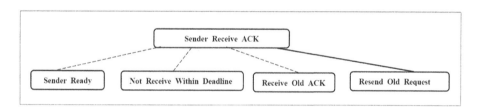

Fig. 11. Resend duplicate frame when ACK is delayed (Regiment 1)

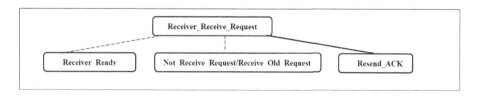

Fig. 12. Resend duplicate ACK when the frame is lost

5   Critical Observations

The atomicity decomposition is done at the stop and wait for the ARQ system in the proposed approach. The system design represents all the events and their relationship with graphical notations. This is the major contribution in this work that has eased the process of elicitation and helps to understand the system more unambiguously. This eventually helps efficient usage of formal notations to represent and design a complex system. The orderings of events are represented explicitly in this proposed methodology. All the refinement relationships between events represented before system design. Besides, in the proposed extension of the Event-B model, this technique helps to design the resending operations of frames and acknowledgments more effectively. As for example, in Fig. 10, the *Sender_Ready* event does not refine the abstract event *Sender_Receive_ACK*. The relationship is represented by the dashed lines and helps to covert the relations into Event-B notations. Relationships between events are represented by invariant properties. All the ERS structures shown in Sect. 4 are converted into Event-B notations. The model may be validated by the RODIN model checker. This technique is more flexible and less complex than the ARQ protocol designed using Petri Net because of the graphical representations of the events with the order [5]. All the Event-B notations described above will make SAW system modeling unambiguous and flexible.

6   Conclusions

The model, under evaluation, may further be decomposed into sub-models. The events and variables could be decomposed accordingly. The ERS structure not only establishes the atomicity decomposition of an Event-B model but is found to be quite useful to detect wrong atomicity decompositions. It can be detected using the proposed methodology because the system is represented graphically. An appropriate tool-support may also be developed to provide a comprehensive system to verify the event refinement structure automatically. The conversion of the ERS notations into Event-B is done manually in the proposed work. A tool may be developed in future for the automatic conversion of ERS into Event-B for complex system modeling. The translation rules can be designed effectively for the automatic development starting from a semi-formal requirements description using the graphical notation used.

References

1. Baier, C., Katoen, J.-P.: Principles of model checking. MIT Press, Cambridge (2008)
2. Abrial, J.-R.: Modeling in Event-B: System and Software Engineering. Cambridge University Press, Cambridge (2010)
3. Jastram, M., Butler, M.: Rodin User's Handbook: Covers Rodin v.2.8. CreateSpace, Scotts Valley (2014)

4. Filali, R., Bouhdadi, M.: Formal modeling and verification of time-constrained ARQ protocols with Event-B. Int. J. Eng. Technol. **8**, 1807–1816 (2016)
5. Best, E., Devillers, R., Koutny, M.: Petri Net Algebra. Springer, Heidelberg (2013). https://doi.org/10.1007/978-3-662-04457-5
6. Floyd, C.: A comparative evaluation of system development methods. In: Proceedings of the IFIP WG 8.1 Working Conference on Information Systems Design Methodologies: Improving the Practice, pp. 19–54. North-Holland Publishing Co. (1986)
7. Salehi Fathabadi, A., Rezazadeh, A., Butler, M.: Applying atomicity and model decomposition to a space craft system in Event-B. In: Bobaru, M., Havelund, K., Holzmann, G.J., Joshi, R. (eds.) NFM 2011. LNCS, vol. 6617, pp. 328–342. Springer, Heidelberg (2011). https://doi.org/10.1007/978-3-642-20398-5_24
8. Butler, M.: Decomposition structures for Event-B. In: Leuschel, M., Wehrheim, H. (eds.) IFM 2009. LNCS, vol. 5423, pp. 20–38. Springer, Heidelberg (2009). https://doi.org/10.1007/978-3-642-00255-7_2
9. Salehi Fathabadi, A., Butler, M., Rezazadeh, A.: Language and tool support for event refinement structures in Event-B. Formal Aspects Comput. **27**(3), 499–523 (2014). https://doi.org/10.1007/s00165-014-0311-1
10. Forouzan, A.B.: Data Communications & Networking (sie). Tata McGraw-Hill Education, New York (2007)
11. Salehi Fathabadi, A., Butler, M.: Applying Event-B atomicity decomposition to a multi media protocol. In: de Boer, F.S., Bonsangue, M.M., Hallerstede, S., Leuschel, M. (eds.) FMCO 2009. LNCS, vol. 6286, pp. 89–104. Springer, Heidelberg (2010). https://doi.org/10.1007/978-3-642-17071-3_5
12. Dghaym, D., Trindade, M.G., Butler, M., Fathabadi, A.S.: A graphical tool for event refinement structures in Event-B. In: Butler, M., Schewe, K.-D., Mashkoor, A., Biro, M. (eds.) ABZ 2016. LNCS, vol. 9675, pp. 269–274. Springer, Cham (2016). https://doi.org/10.1007/978-3-319-33600-8_20

Intraday Patterns in Trading Volume. Evidence from High Frequency Data on the Polish Stock Market

Joanna Olbryś[(✉)] and Adrian Oleszczak

Faculty of Computer Science, Bialystok University of Technology,
Wiejska 45A, 15-351 Bialystok, Poland
j.olbrys@pb.edu.pl

Abstract. According to the literature, there are some possible shapes of intraday patterns in stock market characteristics such as trading volume, transaction costs, order flows, depths, spreads, price returns, stock market resiliency, etc. Empirical investigation and visualization of intraday patterns may be a useful tool for investment decision–making process and can help an analyst to state how particular characteristics vary over a session. In this paper, intraday patterns in trading volume based on high frequency data, are investigated. The data set is large, and it contains transaction data rounded to the nearest second for 10 companies traded in the Warsaw Stock Exchange (WSE). The whole sample covers the long period from January 2005 to December 2018. Extensive studies document various hour-of-the-day patterns in volume on the stock markets in the world. The findings of empirical experiments for real-data from the WSE are in general consistent with the literature and they confirm that intraday trading volume reveals U-similar or M-similar patterns in the case of all investigated equities, for all analyzed periods.

Keywords: High frequency data · Stock market · Trading volume · Intraday patterns

1 Introduction

High frequency data offers various potential insights into the microstructure of financial markets. The intraday data availability provides researchers with the opportunity of being able to investigate stock market phenomena at the finest level of data [6]. A number of studies use transactions data to identify intraday behaviour of several stock market characteristics. Trading volume is one of them. Goodhart and O'Hara [7] stress that a fundamental property of high frequency data is that observations can occur at varying time intervals. Therefore, trades are not equally spaced over the day, which may result in intraday 'seasonal' patterns in the volume of trade.

Supported by the grant WZ/WI/1/2019 from Bialystok University of Technology and founded by the Ministry of Science and Higher Education.

K. Saeed and J. Dvorský (Eds.): CISIM 2020, LNCS 12133, pp. 390–401, 2020.
https://doi.org/10.1007/978-3-030-47679-3_33

The aim of this paper is a real-data investigation of intraday patterns in trading volume based on high frequency data from the Warsaw Stock Exchange (WSE). Furthermore, an analysis of robustness of the empirical findings to the choice of a sub-period is provided. Four time periods are explored in the study: (1) the whole sample period from January 2005 to December 2018, (2) the pre-crisis period, (3) the Global Financial Crisis (GFC) period, and (4) the post-crisis period. The GFC period is formally established based on the paper [15]. The goal is to assess whether the results during the investigated periods significantly differ between each other. To the best of the authors' knowledge, the empirical findings presented here are novel and have not been reported in the literature thus far.

The main contribution of this research is twofold:

1. deep investigation of investors' daily activity based on hourly trading volume,
2. robustness analyses with respect for various time periods including Global Financial Crisis.

The rest of the paper is organized as follows. Section 2 presents brief literature review concerning intraday patterns in trading volume on various stock markets throughout the world. Section 3 describes the database and provides the results of empirical experiments for high frequency data from the Warsaw Stock Exchange. Section 4 contains concluding remarks.

2 Intraday Patterns in Trading Volume

According to the literature, there are some possible shapes of intraday patterns in stock market characteristics such as trading volume, transaction costs, order flows, depths, spreads, price returns, stock market resiliency, etc. (see e.g. [17] and the references therein). Empirical investigation and visualization of intraday patterns may be a useful tool for investment decision–making process and can help an analyst to state how different characteristics vary over a session. It is rather not surprising that perfectly shaped visual patterns occurrence is scarce, but there are several attributes that help to differentiate and point out the most important features [17]:

- The M-pattern shows little values during the beginning and the ending of a session with the highest values slightly after the beginning and before the end. It is also marked by distinctively low value in the middle of a session.
- The W-pattern is the exact opposite of the M-pattern. The W-pattern reveals the highest values during the opening and the closing phases of a session with little values after the opening and before the ending phases. Visible peaks occur during the middle of a session.
- The U-pattern is a slight modification of the W-pattern. The only peculiar difference is lack of a peak during the middle of a trading session.
- The inverted U-pattern is the opposite of the U-pattern. It depicts little values during the beginning and the ending of a session with the highest values in the middle of a session.

– The J-pattern is a modification of the U-pattern. The only difference is that it unveils little values at the beginning of a session.

In this research, intraday patterns in trading volume are explored, therefore the analysis of previous literature is focused mostly on the studies related to trading volume investigation. Trading volume is treated as a proxy of stock liquidity and it is defined as a number of shares traded over a particular unit of time. The empirical findings concerning intraday patterns in trading volume on various stock markets in the world are different. In their seminal and frequently cited paper, Jain and Joh [9] find the U-shaped pattern in volume over the trading day on the New York Stock Exchange (NYSE), which is a hybrid market. They document that volume is highest during the first hour, declines until the fourth hour, and then increases during the last two hours. The authors emphasize that average volume traded reveals significant differences across trading hours of the day. Admati and Pfleiderer [2] provide a partial theoretical explanation of this phenomenon. They point out that the intraday patterns emerge as consequences of the interacting strategic decisions of informed and liquidity traders. Using transaction data for all stocks traded on the Toronto Stock Exchange, McInish and Wood [11] show that number of shares traded (i.e. volume) has a U-shaped intraday pattern. McInish and Wood [12] also report the presence of hour-of-the-day patterns in volume on the NYSE. They confirm that mean value differs significantly by hour of the trading day. The largest value occurs during the first hour. Atkins and Basu [4] try to explain the intraday U-shaped pattern in trading volume with the fact that announcements of new information can affect the trading volume of common stocks. Hamao and Hasbrouck [8] investigate the behaviour of intraday trades and quotes or individual stocks on the Tokyo Stock Exchange. They stress that as in the U.S. data, most market statistics exhibit a marked intradaily pattern. The volume tends to be elevated at the beginning and end of trading sessions. Abhyankar et al. [1] explore intraday patterns of various market characteristics on the London Stock Exchange (LSE), which is a dealership market. Their findings do not unveil a U-shaped pattern in volume, similar to that reported in other stock markets. They find a double-humped pattern with highs at 9.30 a.m. and then at 4.00 p.m. prior to the close. Cai et al. [6] also present empirical results of intraday spreads, trading volume, and volatility for high frequency data from the LSE. They confirm that the general pattern in volume for the UK is two-humped. Ahn and Cheung [3] investigate the Stock Exchange of Hong Kong (SEHK) and they find the U-shaped pattern in trading volume. It is important to note, that the SEHK is a pure electronic order-driven market without market makers. Lee et al. [10] analyse the most active stocks in the Taiwan Stock Exchange (TWSE). They report that although almost all previous studies find a U-shaped pattern in daily trading volume, their results reveal a J-shaped rather than a U-shaped pattern. It means that volume at the open is not significantly different from those of the other time intervals, excepting the last trading interval. Ranaldo [18] studies intraday market liquidity on the Swiss Stock Exchange (SWX). The empirical results show that the most characteristic feature of the SWX trading day is a triple-U-shape with the three peaks during the afternoon. Będowska-Sójka [5] conducts research concerning mainly intraday stealth trading on the Warsaw Stock Exchange (WSE) but she also recognizes the U-shaped pattern in trading activity within the day. Miłobędzki and Nowak [13] investigate intraday patterns in spreads and volumes on the WSE and they find that volumes are U-shaped.

3 Empirical Experiments for High Frequency Data from the Warsaw Stock Exchange

This section presents results of empirical experiments for high frequency intraday data from the Warsaw Stock Exchange (WSE). The database is large. It contains 21 381 230 records in total (see Table 2). Therefore, computations have been conducted with a dedicated programme. The programme has been made using Python language with the library providing data analysis tools called Pandas.

3.1 The WSE Structure and Real Data Description

The WSE is an order-driven market with an electronic order book. It means that liquidity is provided only by limit orders submitted by individual investors, and there are no market makers who support liquidity. There might be observed three main phases of exchange session on the WSE: opening phase, active trading, and closing phase with play-off. Table 1 presents short market trading schedule on the WSE. It often happens that the first and the last phase are those with the most extensive trading activity while the middle trading phase is usually more balanced [17].

Table 1. Market trading schedule on the WSE equities - continuous trading system

Market phase	Time
Opening call	8:30 am–9:00 am
Opening auction	9:00 am
Continuous trading	9:00 am–4:50 pm
Closing call	4:50 pm–5:00 pm
Closing auction	5:00 pm
Trading at last	5:00 pm–5:05 pm

Source: The WSE website (https:// gpw.pl/session-details).

In this study, 10 WSE-traded companies are investigated. The equities are included in the WIG 30 index. The choice of the companies was determined by data availability during the long time period (14 years). Table 2 gives a summary information about them.

High frequency data rounded to the nearest second is utilized. The data comes from the Bank Ochrony Środowiska (BOS, Bank for Environmental Protection) brokerage house (available at http://bossa.pl/notowania). The data set contains the opening, high, low, and closing prices, and volume, for each security over one unit of time. The whole sample covers the period from January 2, 2005 to December 31, 2018 (3491 trading days). Moreover, to verify the stability of the empirical findings, the calculations are conducted both for the whole sample and over three consecutive sub-samples, each of equal size (436 trading days) [16]:

1. the pre-crisis period from September 6, 2005 to May 31, 2007,
2. the Global Financial Crisis (GFC) period from June 1, 2007 to February 27, 2009,
3. the post-crisis period from March 2, 2009 to November 19, 2010.

The GFC period on the Polish stock exchange is specified based on the paper [15], in which the formal statistical procedure for the direct identification of market states is utilized.

Table 2. Basic information about the 10 WSE-traded companies.

	Company (PLN m.)	Trading value in 2018	Average no. of transactions per session in 2018	No. of records in the database
1	PKO	22674.9	4292	4415095
2	PKN	20140.6	4609	3457711
3	PEO	19273.4	3143	2733078
4	KGH	13890.3	3844	5380209
5	CCC	6038.2	1768	650044
6	LPP	5600.1	969	460286
7	OPL	2282.8	1152	2297101
8	EUR	2125.9	1564	1029346
9	MIL	1985.1	657	683410
10	ING	558.3	214	274950
	Total	94569.6	–	21381230

Notes: The 10 WSE-traded companies are labelled by ticker symbols and presented in decreasing order of the total trading value (in PLN million) in 2018.
Source: The WSE website and own calculations.

3.2   Intraday Patterns in Trading Volume on the WSE

To explore intraday patterns in trading volume, the average hourly values of aggregate volume and standard deviations of these values are calculated for each stock during the whole sample period and three sub-samples. Table 3 contains the results concerning the whole sample period. Due to the space restriction, remaining calculations are available upon a request. The 10 WSE-listed companies are labelled by ticker symbols and presented in decreasing order of their total trading value in 2018 (see Table 2). The empirical findings are illustrated in some figures.

Figure 1 exemplifies various patterns of hourly trading volume for eight companies, namely PEO, KGH, CCC, LPP, OPL, EUR, MIL, and ING, during the whole sample period January 2005–December 2016. All presented patterns are U-similar or M-similar. They are marked by distinctively low value of trading volume in the middle of a session.

Figures 2 and 3 present hourly patterns in trading volume within the whole sample (WS) and three sub-samples (P1, P2, P3) for the most liquid two companies, namely PKO and PKN, respectively (see Table 2). All figures unveil pronounced U-similar or M-similar patterns.

Table 3. The averaged hourly stock trading volume (in items) within the whole sample period from January 2, 2005 to December 31, 2018 (3491 trading days)

Stock	H1	H2	H3	H4	H5	H6	H7	H8
PKO	322806	275724	232967	202166	197044	242096	330741	362789
	(368200)	(291546)	(244104)	(249339)	(204074)	(230834)	(293771)	(335958)
PKN	128482	138803	118114	101398	98090	126418	171027	176060
	(132300)	(138799)	(111217)	(100254)	(89994)	(116014)	(150979)	(157388)
PEO	56596	52869	45607	40033	38743	47873	65784	73144
	(99091)	(71719)	(57499)	(56370)	(53062)	(60761)	(78587)	(80865)
KGH	150803	115976	90466	78398	78197	99643	131690	127307
	(137081)	(108458)	(80099)	(73068)	(71704)	(85415)	(104893)	(97516)
CCC	5383	6797	6090	6026	5314	7137	7163	7213
	(15094)	(16670)	(14578)	(27191)	(15299)	(36354)	(15237)	(16418)
LPP	198	190	178	161	144	173	263	251
	(1249)	(503)	(470)	(434)	(401)	(392)	(866)	(540)
OPL	361867	364175	296637	257544	248720	300915	420493	447791
	(720324)	(543315)	(541863)	(350839)	(321147)	(416522)	(508816)	(509328)
EUR	34847	41335	35310	31195	30615	36762	43656	41845
	(83837)	(116445)	(78790)	(73676)	(74342)	(83853)	(117422)	(77501)
MIL	112028	116860	113297	88535	76331	93586	120874	123496
	(359240)	(273380)	(289927)	(276636)	(170849)	(182997)	(237017)	(326022)
ING	1866	2097	1901	1527	1665	1668	2068	2377
	(8321)	(8540)	(7505)	(9666)	(11825)	(6176)	(6877)	(11711)

Notes: The hours:
H1 9:00am–10:00am; H2 10:00am–11:00am; H3 11:00am–12:00am; H4 12:00am–1:00pm;
H5 1:00pm–2:00pm; H6 2:00pm–3:00pm; H7 3:00pm–4:00pm; H8 4:00pm–5:00pm.
Standard deviations are given in brackets. Remaining notation like in Table 2.

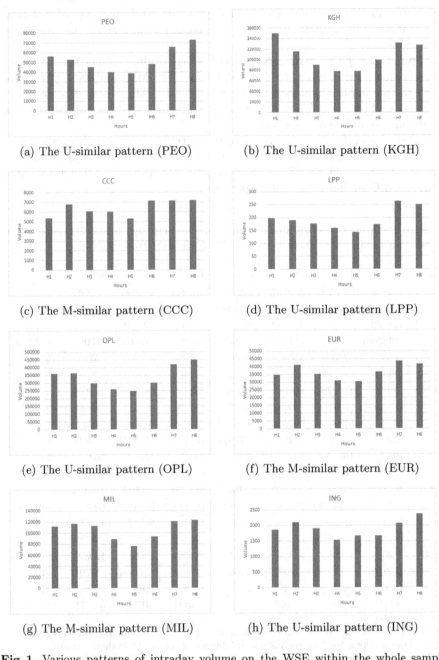

(a) The U-similar pattern (PEO)

(b) The U-similar pattern (KGH)

(c) The M-similar pattern (CCC)

(d) The U-similar pattern (LPP)

(e) The U-similar pattern (OPL)

(f) The M-similar pattern (EUR)

(g) The M-similar pattern (MIL)

(h) The U-similar pattern (ING)

Fig. 1. Various patterns of intraday volume on the WSE within the whole sample period January 2005–December 2016. The averages of hourly trading volume for stocks are reported in Table 3.

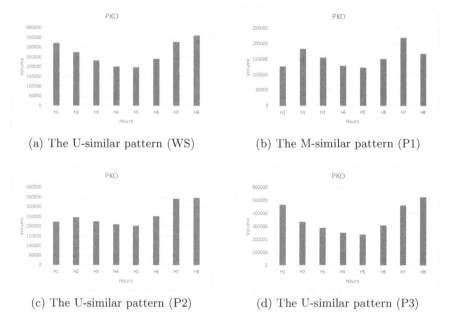

Fig. 2. Hourly patterns in trading volume within the whole sample (WS) and three sub-samples (P1, P2, P3) for the PKO equity. Notation like in Table 5

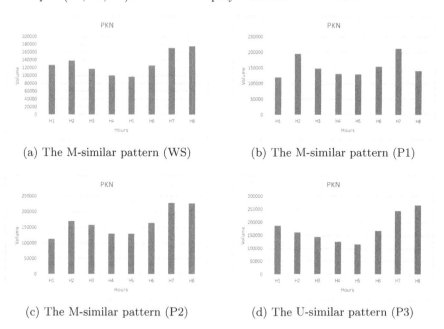

Fig. 3. Hourly patterns in trading volume within the whole sample (WS) and three sub-samples (P1, P2, P3) for the PKN equity. Notation like in Table 5

Table 4 reports summarized results of hourly patterns in trading volume within the whole sample (WS) and three sub-samples (P1, P2, P3) for 10 WSE-traded companies which are investigated in this study. Due to the space restriction, only selected patterns are showed in Figs. 1, 2 and 3. One can observe (in Table 4) that the M-similar pattern dominates (18 cases), the U-similar pattern appears slightly less often (16 cases). Only in six cases, other not perfectly shaped visual pattern occurs.

An additional goal of the real-data analysis is to assess whether the mean results of stock intraday trading volume within the Global Financial Crisis period significantly differ compared to the other investigated periods. To address this issue, the Z-statistic for independent large sample means is utilized:

$$Z = \frac{\overline{x_1} - \overline{x_2}}{\sqrt{\frac{s_1^2}{n_1} + \frac{s_2^2}{n_2}}}, \tag{1}$$

where $\overline{x_1}$ and $\overline{x_1}$ are sample means, s_1^2 and s_2^2 are sample variances, and n_1, n_2 denote sample size, respectively ($n_1 = 3491$ for the whole sample and $n_2 = 436$ for each sub-sample). The following two-tailed hypothesis is tested:

$$\begin{aligned} H_0 &: \mu_1 = \mu_2 \\ H_1 &: \mu_1 \neq \mu_2, \end{aligned} \tag{2}$$

where μ_1 and μ_2 are the expected values of trading volume for each stock during the compared periods, and the null hypothesis states that two expected values are equal.

Table 5 contains the summarized findings of the test for difference between two means of hourly trading volume for the whole group of equities. The null hypothesis H_0 (2) is rejected when $|Z| > 1.96$ (the critical value of the Z-statistic (1) at 5% significance level is equal to 1.96). The obtained results are not homogenous and do not explicitly confirm or deny the stability of average volumes within analyzed periods. No reason has been found to reject the null hypothesis H_0 for 210 times, while the hypothesis H_1 has been the result of calculations for 270 times. Furthermore, the findings seem to depend on the hour of the day. Specifically, the hypothesis H_0 is outweighed by the hypothesis H_1 in the case of the first two and last three hours within a session, i.e. H1 (9:00am–10:00am), H2 (10:00am–11:00am), H6 (2:00pm–3:00pm), H7 (3:00pm–4:00pm), and H8 (4:00pm–5:00pm). Moreover, one can observe that intraday trading volume does not reveal any unusual patterns occurring within the Global Financial Crisis period (which is labelled by P2 in Table 5). This evidence is consistent with the literature concerning other characteristics of the Polish market liquidity, e.g. stock market resiliency [16, 17].

Table 4. Summarized results of hourly patterns in trading volume for 10 WSE-traded companies.

	Company	WS	P1	P2	P3
1	PKO	U-similar	M-similar	U-similar	U-similar
2	PKN	M-similar	M-similar	M-similar	U-similar
3	PEO	U-similar	M-similar	U-similar	U-similar
4	KGH	U-similar	M-similar	M-similar	U-similar
5	CCC	M-similar	Other	Other	Other
6	LPP	U-similar	Other	M-similar	M-similar
7	OPL	U-similar	M-similar	M-similar	U-similar
8	EUR	M-similar	U-similar	Other	M-similar
9	MIL	M-similar	Other	M-similar	U-similar
10	ING	U-similar	M-similar	U-similar	M-similar

Notes: Notation like in Tables 2 and 5.
Source: Own calculations.

Table 5. Summarized results of the significance test for difference between two means of hourly trading volume for 10 companies

Hour	H1		H2		H3		H4		H5		H6		H7		H8	
Hypothesis	H_0	H_1	H_0	H_1	H_0	H_1	H_0	H_1	H_0	H_1	H_0	H_1	H_0	H_1	H_0	H_1
WS/P1	3	7	4	6	4	6	4	6	4	6	3	7	4	6	2	8
WS/P2	1	9	4	6	4	6	6	4	4	6	6	4	4	6	5	5
WS/P3	1	9	4	6	4	6	6	4	4	6	6	4	4	6	5	5
P2/P1	5	5	4	6	6	4	4	6	5	5	5	5	5	5	4	6
P2/P3	3	7	7	3	8	2	7	3	7	3	4	6	6	4	3	7
P1/P3	3	7	4	6	6	4	5	5	6	4	4	6	1	9	2	8
Sum	16	44	27	33	32	28	32	28	30	30	28	32	24	36	21	39

Notes:
WS - the whole sample period 2.01.2005–31.12.2018,
P1 - the pre-crisis period 6.09.2005–31.05.2007,
P2 - the crisis period 1.06.2007–27.02.2009,
P3 -the post-crisis period 2.03.2009–19.11.2010.
The critical value of Z statistic at 5% significance level is equal to 1.96. Remaining notation like in Table 3.

4   Conclusion

This paper contributes to the literature by deep investigation of investors' activity based on hourly trading volume. Intraday trading patterns have been explored for high-frequency real-data from the Polish stock exchange. We have utilized the data rounded to the nearest second for 10 stocks included in the WIG 30 index from the long time period January 2005–December 2018. The real-data experiments have revealed visible U-shaped or M-shaped patterns in the case of all investigated equities. These patterns mean that trading volume firstly falls. Nextly, during the middle of a trading session it becomes low and quite stable, and it finally rises. The findings are in line with the literature concerning the microstructure of stock markets in the world.

One of potential avenues for further research is to investigate hour-of-the-day patterns in trading volume across days of the week, as Nowak and Olbryś [14] document day-of-the-week effects in liquidity on the WSE. They use daily turnover as a liquidity proxy. The results indicate that liquidity on the WSE tends to be significantly lower on Mondays and higher on Wednesdays in comparison with the other days of the week. Moreover, Miłobedzki and Nowak [13] show that volumes exhibit either the day-of-the-week or the hour-of-the-day effect or both on the WSE. However, the authors utilize a shorter data sample from April 2013 to December 2016, and therefore they do not assess potential effects of the Global Financial Crisis on the WSE.

References

1. Abhyankar, A., Ghosh, D., Levin, E., Limmack, R.J.: Bid-ask spreads, trading volume and volatility: intra-day evidence from the London Stock Exchange. J. Bus. Finan. Account. **24**(3&4), 343–362 (1997)
2. Admati, A.R., Pfleiderer, P.: A theory of intraday patterns: volume and price variability. Rev. Finan. Stud. **1**(1), 3–40 (1988)
3. Ahn, H.-J., Cheung, Y.-L.: The intraday patterns of the spread and depth in a market without market makers: the Stock Exchange of Hong Kong. Pac.-Basin Finan. J. **7**, 539–556 (1999)
4. Atkins, A.B., Basu, S.: The effect of after-hours announcements on the intraday U-shaped volume pattern. J. Bus. Finan. Account. **22**(6), 789–809 (1995)
5. Będowska-Sójka, B.: Intraday stealth trading. Evidence from the Warsaw Stock Exchange. Poznan Univ. Econ. Rev. **14**(1), 5–19 (2014)
6. Cai, C.X., Hudson, R., Keasey, K.: Intra day bid-ask spreads, trading volume and volatility: recent empirical evidence from the London Stock Exchange. J. Bus. Finan. Account. **31**(5&6), 647–676 (2004)
7. Goodhart, C.A.E., O'Hara, M.: High frequency data in financial markets: issues and applications. J. Empirical Finan. **4**(2&3), 73–114 (1997)
8. Hamao, Y., Hasbrouck, J.: Securities trading in the absence of dealers: trades, and quotes on the Tokyo Stock Exchange. Rev. Finan. Stud. **8**(3), 849–878 (1995)
9. Jain, C., Joh, G.-H.: The dependence between hourly prices and trading volume. J. Finan. Quan. Anal. **23**(3), 269–283 (1988)
10. Lee, Y.-T., Fok, R.C.W., Liu, Y.-J.: Explaining intraday pattern of trading volume from the order flow data. J. Bus. Finan. Account. **28**(1&2), 199–230 (2001)
11. McInish, T.H., Wood, R.A.: An analysis of transactions data for Toronto Stock Exchange: return patterns and end-of-the-day effect. J. Bank. Finan. **14**(2&3), 441–458 (1990)
12. McInish, T.H., Wood, R.A.: Hourly returns, volume, trade size, and number of trades. J. Finan. Res. **14**(4), 303–315 (1991)

13. Miłobędzki, P., Nowak, S.: Intraday trading patterns on the Warsaw Stock Exchange. In: Jajuga, K., Locarek-Junge, H., Orlowski, L.T. (eds.) Contemporary Trends and Challenges in Finance. SPBE, pp. 55–66. Springer, Cham (2018). https://doi.org/10.1007/978-3-319-76228-9_6

14. Nowak, S., Olbryś, J.: Day-of-the-week effects in liquidity on the Warsaw Stock Exchange. Dyn. Econ. Models **15**, 49–69 (2015)

15. Olbryś, J., Majewska, E.: Bear market periods during the 2007–2009 financial crisis: direct evidence from the Visegrad countries. Acta Oeconomica **65**(4), 547–565 (2015)

16. Olbrys, J., Mursztyn, M.: Measuring stock market resiliency with Discrete Fourier Transform for high frequency data. Phys. A **513**, 248–256 (2019)

17. Olbrys, J., Mursztyn, M.: Estimation of intraday stock market resiliency: Short-Time Fourier Transform approach. Phys. A **535**, 122413 (2019)

18. Ranaldo, A.: Intraday market liquidity on the Swiss Stock Exchange. Finan. Mark. Portfolio Manag. **15**(3), 309–327 (2001)

An Efficient Metaheuristic for the Time-Dependent Team Orienteering Problem with Time Windows

Krzysztof Ostrowski[(✉)] [iD]

Faculty of Computer Science and Telecommunication,
Bialystok University of Technology, ul. Wiejska 45A, 15-001 Bialystok, Poland
k.ostrowski@pb.edu.pl

Abstract. The Time-Dependent Team Orienteering Problem with Time Windows (TDTOPTW) is a combinatorial optimization problem defined on graphs. The goal is to find most profitable set of paths in time-dependent graphs, where travel times (weights) between vertices varies with time. Its real life applications include tourist trip planning in transport networks. The paper presents an evolutionary algorithm with local search operators solving the problem. The algorithm was tested on public transport network of Athens and clearly outperformed other published methods achieving results close to optimal in short execution times.

Keywords: Time-Dependent Team Orienteering Problem with Time Windows · Evolutionary algorithm · Local search · Public transport network

1 Introduction

The Time-Dependent Team Orienteering Problem with Time Windows (TDTO PTW) belongs to the family of the Orienteering Problem (OP). The classic OP is defined on a weighted graph with nonnegative profits associated to vertices and nonnegative costs associated to edges. The goal of the OP is to find a path between two given vertices, limited by total cost of visited edges and maximizing total profit of visited vertices. The OP solution does not have to contain all vertices (usually it is impossible because of total cost constraint) and each vertex can be visited only once.

The Time-Dependent Orienteering Problem (TDOP) [12] is a generalization of the classic OP and is defined on time-dependent graphs. In such graphs edge costs (weights) are identified with travel times between vertices. More importantly, travel time between vertices depends on a moment of travel start (weights

The research was carried out as part of the research work number WZ/WI/1/19 at the Bialystok University of Technology, financed from a subsidy provided by the Minister of Science and Higher Education.

ⓒ Springer Nature Switzerland AG 2020
K. Saeed and J. Dvorský (Eds.): CISIM 2020, LNCS 12133, pp. 402–414, 2020.
https://doi.org/10.1007/978-3-030-47679-3_34

are functions of time). Public transport networks are good examples of time-dependent graphs (travel time determined by time-table). The purpose of the TDOP is to find most profitable path between given two vertices (starting at a given time) limited by total travel time.

The TDTOPTW [7] is an extension of the TDOP. The goal of the TDTOPTW is to find a set of paths (of a given cardinality) maximizing total collected profit. Each path has the same limit of total travel time and the same start/end vertices. Any vertex can be included only once in a multi-path solution. Additionally each vertex has some visit time as well as time window, which determines when a given vertex can be visited. Arriving too late makes it impossible to visit a vertex while arriving too early means waiting for its opening. It should be noted that time of edge traversals as well as time needed to visit vertices and waiting time are all included in total travel time.

Problems from the OP family have many practical applications including tourist trip planning [6,17], transport logistics and even DNA sequencing [3]. In tourist trip planning each attraction (point of interest - POI) has some profit (dependent on its popularity) and a time-window (opening hours). Finding an attractive multi-day tour (of a limited duration) in a time-dependent transport network (time-dependency: timetables and traffic) is equivalent to solving the TDTOPTW.

1.1   Literature Review

Problems from the OP family are NP-hard [9] and exact algorithms can be very time-consuming for larger graphs. For this reason most papers are devoted to metaheuristics. Various approaches for the OP were based i.a. on greedy and randomized construction of solutions [2], local search methods [4,20], tabu search [8], ant-colony optimization [18] and genetic algorithms [19].

Most papers about the Time-Dependent versions of the OP are associated with practical applications of the problem (trip planning in public transport networks). LI [12] published the first article about the classic TDOP and solved it with an exact dynamic programming algorithm. He obtained results for small test instances.

Garcia et al. [6] presented the first paper describing application of the TDTOPTW in POI and public transport network of San Sebastian. To solve the problem the authors proposed Iterated Local Search method (ILS). However, they performed computations on average daily travel times and assumed periodicity of public transport timetables.

Gavalas et al. [7] proposed an approach to the TDTOPTW which uses real time-dependent travel times in a transport network of Athens. The authors introduced two fast heuristics (TD_CSCR and TDSlCSCR), which based on ILS and vertex clustering, and made comparisons of a few methods. The authors created 20 topologies and 100 tourist preferences combining into 2000 different test cases.

Verbeeck et al. [21] developed new benchmark instances for the TDOP, which model street traffic. The authors proposed an ant-colony approach, which achieved high quality results in a short execution time. Gunawan et al. [10] modified Verbeeck's benchmarks (discretization of time) and compared a few approaches (adaptive ILS proved to be the most effective of them).

The author's previous papers were devoted to metaheuristics for various problems from the OP family. Methods developed by the author (composition of evolutionary algorithms and local search heuristics) proved successful on the OP [14,15] as well as on TDOP benchmark instances [16] and on TDOPTW public transport and POI network of Bialystok [17]. The algorithms achieved results close to optimal and outperformed other methods: GRASP [2] and GLS [20] (OP), ACS [21] and Adaptive ILS [10] (TDOP).

Recently a metaheuristic (tabu search+nonlinear programming) was proposed to tackle a new variant of the problem: Orienteering Problem with Service Time-Dependent Profits (OPSTP) [22]. In this variant vertex profits are not constant (as in all previous problems) but change with time (in a non-linear way), which is another aspect of time-dependency.

2   Problem Definition

Let $G = (V, E)$ be a directed, weighted graph. In time-dependent graphs each weight between vertices i and j $(i, j \in V)$ is identified with travel time between these vertices and is a function $w_{ij}(t)$ dependent on the moment of travel start t. Each vertex i has a nonnegative profit p_i, nonnegative visit time τ_i and time windows (opening time to_i and closing time tc_i) indicating its period of availability.

Given the time-dependent graph, moment of start t_0, time limit T_{max}, start and end vertices (s and e) the purpose of Time-Dependent Team Orienteering Problem with Time-Windows (TDTOPTW) is to find a set of m paths from s to e starting at time t_0 which maximizes total profit of visited vertices and total travel time of each path is limited by T_{max}. Each vertex (except s and e) can be included only once and only in one path. For simplicity let's assume that vertices s and e (start and end point of the tour) have no profits, no visit time and no windows (this is usually the case in practical applications).

TDTOPTW can be formulated as Mixed Integer Programming (MIP) problem. Let x_{ij} is 1 iff a solution contains direct travel from i to j and 0 otherwise. Let ta_i and tl_i be arrival and leave time at/from vertex i included in a solution. The purpose of TDOP is to maximize formula 1 while satisfying Eqs. 2–7:

$$max \sum_{i \in V} \sum_{j \in V} (p_i \cdot x_{ij}) \tag{1}$$

$$\sum_{i \in V} x_{si} = \sum_{i \in V} x_{ie} = m \tag{2}$$

$$\underset{i \in V \setminus \{s,e\}}{\forall} (\sum_{j \in V} x_{ij} = \sum_{j \in V} x_{ji} \leq 1) \tag{3}$$

$$tl_s = t_0 \tag{4}$$

$$\underset{i \in V, j \in V \setminus \{e\}}{\forall} (x_{ij} \cdot ta_j = x_{ij} \cdot (tl_i + w_{ij}(tl_i))) \tag{5}$$

$$\underset{i \in V}{\forall} (x_{ie} \cdot (tl_i + w_{ie}(tl_i)) \le x_{ie} \cdot (t_0 + T_{max})) \tag{6}$$

$$\underset{i \in V \setminus \{s,e\}, j \in V}{\forall} (x_{ij} \cdot ta_i \le x_{ij} \cdot tc_i \wedge x_{ij} \cdot tl_i = x_{ij} \cdot (max(ta_i, to_i) + \tau_i)) \tag{7}$$

Equation 2 guarantees that every path in a solution starts at vertex s and ends at vertex e. Formula 3 indicates each vertex can be visited at most once and no path ends in vertices other than s and e. Equation 4 guarantees that every path start at time t_0 while formula 5 guarantees that leave/arrival times of subsequent vertices are consistent with time-dependent weights. Constraint 6 means that total travel time of every path cannot be more than T_{max} while constraint 7 guarantees that time-windows are not violated.

3   Algorithm Description

To solve the TDTOPTW the author proposed an evolutionary algorithm with local search methods embedded, which was developed from the method solving the TDOP [16]. It uses both random and local search operators, 2-point heuristic crossover, disturb operator and deterministic crowding as selection mechanism. Path representation was used: subsequent genes indicate vertices visited. The algorithm starts with a random population of feasible solutions.

3.1   Evaluation

Fitness function of a solution is the sum of profits of its paths. Contrary to the author's TDOP algorithm, infeasible solutions are not present in the population. This is due to nature of the problem: additional time-window constraints and presence of multiple paths in a single solution. Genetic operators don't allow solutions to violate time-windows and T_{max} constraints. In the future the author may propose a fitness function for infeasible solutions, which takes into account TDTOPTW specificity.

3.2   Crossover

In each iteration $c_p \cdot P_{size}$ individuals are selected and arranged in random pairs (c_p - crossover probability, P_{size} - population size). The basic procedure of heuristic 2-point crossover (working on two single paths) as well as random parents selection was based on TDOP algorithm. Heuristic crossover tries to exchange fragments between successive common vertices of both paths in order to maximize fitness of the better child.

Each TDTOPTW solution contains m paths (instead of one) and for this reason an adaptation was needed. For each path from parent A the algorithm applies crossover with each path from parent B (m^2 single-path crossovers in total). After each procedure the crossed paths are inserted back into parents and duplicate vertices are removed (if needed). From all options the algorithm chooses the one which maximizes fitness of the better of two modified solutions (children). The procedure is explained in Fig. 1. On the other hand, random crossover version selects sub-paths and exchanged fragments randomly. Crossover specificity is determined by c_h parameter - it's the probability of using heuristic crossover (with $1 - c_h$ the probability of random version).

parent A: **parent B:**
A1: (1, 5, 4, 3, 8, 10, 2, 12, 7)   B1: (1, 17, 16, 3, 14, 5, 10, 17, 7)
A2: (1, 6, 9, 14, 16, 11, 7)   B2: (1, 8, 11, 15, 6, 20, 19, 7)

single crossover (A1 and B1):
A1: (**1**, 5, 4, **3**, 8, **10**, 2, 12, **7**)
B1: (**1**, 17, 16, **3**, 14, 5, **10**, 18, **7**)

exemplary result (1...3 fragments exchange):
An: (**1**, 17, ~~16~~, **3**, 8, **10**, 2, 12, **7**)
Bn: (**1**, 5, 4, **3**, 14, ~~5~~, **10**, 18, **7**)

child A: **child B:**
An: (1, 17, 3, 8, 10, 2, 12, 7)   Bn: (1, 5, 4, 3, 14, 10, 18, 7)
A2: (1, 6, 9, 14, 16, 11, 7)   B2: (1, 8, 11, 15, 6, 20, 19, 7)

Fig. 1. Exemplary crossover of two solutions (each consists of $m = 2$ paths). In the example first path of parent A (A1) is crossed with first path of parent B (B1). There are three possible fragment exchanges: between 1 and 3, between 3 and 10, between 10 and 7. First of them is presented and new paths (An and Bn) are created. Vertices 5 and 16 are removed from newly created paths due to vertices duplicates in newly created solutions. Afterwards new paths are inserted into original parents and children are formed. Heuristic crossover checks all crossing combinations between various paths and all fragments exchanges.

3.3 Selection

After crossover children compete with their own parents for a place in the population (survivor selection in the form of deterministic crowding [13]). For $m = 1$ edit distance function used to determine pairs is the same as in [16]. For $m > 1$ a modified solution compete with its original version (they have $m - 1$ paths in common and are similar). Such selection approach maintains diverse population for longer enabling a more effective search before convergence.

3.4 Mutation

In each iteration $m_p \cdot P_{size}$ individuals are chosen for mutation (m_p - mutation probability). Mutation include a few operators. Compared to TDOP solution, new operator was added (*move*) and others were modified to operate on multi-path solutions. Initially all m paths of the selected individual undergo 2-opt procedure, which exchanges two edges (connections) present in a path with another two edges in order to reduce total travel time (Fig. 2). Afterwards *move* operator tries to move a single vertex from one path to another (within the chosen solution) in order to reduce the total travel time of the solution as much as possible. All vertices are considered by *move* and if no option reduces the travel time the solution is not modified. This operator helps to balance routes and is presented in Fig. 3. Afterwards a vertex insertion or vertex deletion (heuristic or random) is carried out: (probability of insertion/deletion is 0.5 each). The goal of heuristic insert is to find a non-included vertex maximizing profit to travel time increase ratio: all vertices, all insertion places and all sub-paths are considered. An example of heuristic insert is given in Fig. 4. Heuristic deletion works a bit analogically: it deletes a vertex minimizing profit to travel time decrease ratio. Specificity of operators (random/heuristic) is steered by parameter m_h, which is equal to the probability of usage heuristic insertion/deletion (analogically to c_h in crossover).

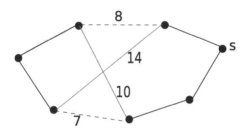

Fig. 2. Exemplary 2-opt operator on Euclidean plane. A solution is a cycle which starts and ends in vertex s. Elimination of edge intersection in the cycle (exchanging red edges for dashed edges) will reduce total length of the cycle by 9. The proposed algorithm operates on time-dependent travel times (not distances) but this example was used for simplicity. (Color figure online)

3.5 Disturb

In each iteration $d_p \cdot P_{size}$ individuals are chosen for disturb (d_p - disturb probability). Disturb is a different kind of mutation. It is executed way less often than standard mutation but it can cause larger changes in individuals. The operator removes a path fragment (consisted of up to 10% of vertices) in a random or heuristic way (it is steered by d_h parameter). The operator is destructive in nature and should be used rarely but it's usage can help escape local optima and slightly improve results.

3.6 Operators Optimization

In time-dependent paths each modification (i.e. insertion of new vertex) requires travel time recalculation for a path fragment after the modification point.

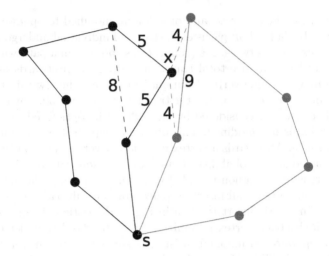

Fig. 3. Exemplary move operator on Euclidean plane. A solution contains two paths (black and red), both start and end in vertex s. Edge distances are marked near edges. It can be seen that moving vertex x from black path to red path results in reduction of total distance of two paths (reduction in black path (2) is larger than increase in red path (1)). The proposed algorithm operates on time-dependent travel times (not distances) but this example was used for simplicity. (Color figure online)

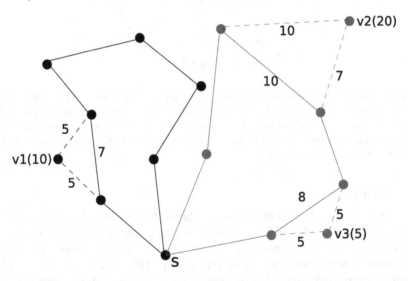

Fig. 4. Exemplary heuristic insertion operator on Euclidean plane. The presented solution consists of two paths (black and red) starting and ending in s. Vertex profits are in parenthesis and edge costs are marked near edges. The goal is to find new vertex and insertion point (in any path), which maximize ratio of profit to cost increase. For $v1$ the ratio is 10/3, for $v2$ it's 20/7 and for $v3$ it's 5/2. Vertex $v1$ will be included in the black path by the operator. The proposed algorithm operates on time-dependent travel times (not distances) but this example was used for simplicity. (Color figure online)

In case of insert operator (which checks all possible insertion places for a new vertex) naive searching would require n^2 time complexity (n - path size). However, determining best insertion place (in terms of travel time increase) for a given vertex can be done in linear time with just one loop over a path. Let's assume that the path includes vertices $1, 2, ..., n$ and let $t_1, t_2, ..., t_n$ be vertex arrival times for this path. We want to insert new vertex x in the best place (minimizing total travel time after insertion). This problem is equivalent of finding earliest arrival time to vertex n (assuming that x was included). This can be calculated recursively. Let $EAT(k)$ be earliest arrival time to vertex k assuming that vertex x was included in a path somewhere before k. Here are formulas (for their simplicity there are no time-windows and visit times):

$$EAT(2) = t_1 + w_{1x}(t_1) + w_{x2}(t_1 + w_{1x}(t_1)) \tag{8}$$

$$EAT(k) = MIN \left\{ \begin{array}{l} EAT(k-1) + w_{k-1k}(EAT(k-1)) \\ t_{k-1} + w_{k-1x}(t_{k-1}) + w_{xk}(t_{k-1} + w_{k-1x}(t_{k-1})) \end{array} \right\} \tag{9}$$

The optimal solution of insertion x before vertex k is either:
1) inserting x optimally before vertex $k - 1$ and connect $k - 1$ and k directly.
2) inserting x directly before vertex k (and after $k - 1$)
Out of these two options we choose one with earliest arrival time at vertex k. Calculating it in one loop (from vertex 2 to n) we can obtain $EAK(n)$ and optimal insertion point in linear execution time.

Additional optimizations were also performed on time-consuming 2-opt operator. Precomputation of reversed path fragments is done to estimate fast if a given 2-opt move is promising to perform. Details were presented in [16].

4   Experimental Results

Experiments were conducted on a computer with Intel Core i7 3.5 GHz processor and the algorithm was implemented in C++. The algorithm was tested on public transport and POI network of Athens (tests prepared by Gavalas et al. [7]). The authors created 20 topologies and 100 tourist preferences combining into 2000 different test cases. Results presented in this section (profits and execution times) are averaged over execution runs for all test cases. Gaps are expressed in percent and illustrate relative differences between profits of EVO100 and other methods. The author's metaheuristic is compared to:

- Time-dependent heuristics (TD_CSCR, TDSlCSCR) by Gavalas et al. [7] and their version working on average travel times (AvgCSCR).
- ILS algorithm working on average travel times (AvgILS) by Garcia et al. [6] and its time-dependent version (TD_ILS).
- Exact algorithm (based on branch-and-bound and dynamic programming techniques) implemented by the author (marked as OPT).

Two versions of the evolutionary algorithm (with different population sizes) were tested: EVO100 and EVO20. Parameter values were derived from the TDOP algorithm [16]. Originally parameter values were computed by automatic tuning procedure - ParamILS [11,14]. Parameters N_g and C_g for EVO20 were scaled down to 1000 and 100 accordingly. Version with reduced population size (20) was chosen because its execution times were similar to other compared methods. Parameters are given in Table 1.

Table 1. Parameters of the evolutionary algorithm

Param.	Value	Description	Param.	Value	Description
P_{size}	100/20	Population size	m_p	1	Mutation probab.
N_g	5000/1000	Max. generations number	c_p	1	Crossover probab.
C_g	500/100	Max. generations number	d_p	0.01	Disturb probab.
		Without improvement	m_h	1	Mutation heuristic coeff.
d_h	0.8	Disturb heuristic coeff	c_h	0.8	Crossover heuristic coeff

Table 2. Experimental results for different numbers of paths (m). Gaps are expressed in percent and illustrate relative differences between profits of EVO100 and other methods. Execution times are given in seconds. Max. trips duration: 5 hours, start at 10:00.

Method	m = 1		m = 2		m = 3		m = 4		Exec.
	Profit	Gap	Profit	Gap	Profit	Gap	Profit	Gap	Times
AvgILS	298.5	13.4	561.9	16.0	819.7	13.1	1078.7	14.5	0.02–0.26*
TD_ILS	326.3	5.3	641.4	4.1	939.8	3.4	1219.2	3.3	0.02–0.38*
AvgCSCR	332.0	3.6	643.3	3.9	933.7	4.1	1209.3	4.1	0.03–0.2*
TDSlCSCR	342.1	0.7	657.9	1.7	946.5	2.8	1219.1	3.3	0.05–0.32*
TD_CSCR	337.8	1.9	654.3	2.2	948.1	2.6	1225.5	2.8	0.04–0.26*
EVO20	343.6	0.3	666.5	0.4	966.3	0.7	1248.9	1.0	0.04–0.36
EVO100	344.5	0.0	669.2	0.0	973.3	0.0	1261.1	0.0	0.66–7.6
OPT	344.6	-0.1	−0.1/−0.4**						

*Other methods were executed on a different computer - times given for informative purposes.
**Because of long execution times optimal solutions for $m = 2$ were computed only for preference number 205 (and all 20 topologies). Given gaps are differences between OPT and both EVO versions.

In Table 2 experimental results for all methods are given. One can see that the proposed evolutionary algorithm (both versions) clearly outperforms all other methods. On average EVO20 is 1.5 and 1.8% better than the best of remaining metaheuristics (TDSlCSCR and TD_CSCR) and the difference is about 2% for $m > 2$. Gaps of other methods are even larger (3–16%). What is more, EVO20 achieves results close to optimal (average gap of 0.4% for $m \leq 2$) in a very short execution time. Differences between the best algorithms gradually grow as m is increased. Larger solution space (larger m) is explored more effectively by the stronger version of the evolutionary algorithm (EVO100) and gaps between these two versions rise from 0.3 to 1.0%. Better results by EVO100 (nearly optimal) are achieved at the cost of increased execution time.

Table 3. Additional results (profits) obtained by EVO for longer trips (maximum duration: 8 h, start time: 9:00).

Method	m = 1	m = 2	m = 3	m = 4	Exec. times
EVO20	560.2	1069.4	1529.9	1957.6	0.07–0.7
EVO100	561.4	1074.8	1542.7	1976.2	0.9–16
OPT	561.5				

In Table 3 results for longer trips are presented. For $m = 1$ both algorithm versions achieve results very close to exact algorithm. As m gets larger EVO100 gains advantage over EVO20 (up to 1%). No optimal solutions are known for $m > 1$ but these results are presented for future comparisons.

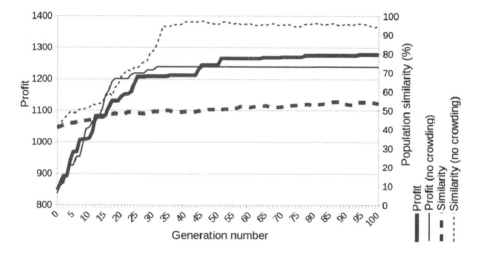

Fig. 5. Exemplary runs of two algorithm versions: with and without crowding. Profit of the best solution found so far and average population similarity are presented as a function of generation number. $P_{size} = 20$, $m = 4$, $topology = 1$, $preference = 408$.

In Fig. 5 there is a comparison of two algorithm runs. One of them uses deterministic crowding and the other uses random assignment of competition pairs during survivor selection. Population similarity (based on longest common subsequence metric) grows very fast without deterministic crowding - similarity close to 100% signals convergence around one solution. Deterministic crowding forces competition between more similar individuals, which enables to preserve population diversity for longer and improve results in later generations (as seen in the figure). Usage of crowding improved average results by 1.4% for $m = 4$. This method of selection proved to be very effective (compared to standard parent selection methods) in the classic OP as well [14].

Table 4. Comparison of average results for different values of parameters: heuristic crossover coefficient (c_h) and heuristic mutation coefficient (m_h). Relative profit losses (in percent) to the best configuration (in bold) are given. Popul. size was 20 and $m = 4$.

c_h	m_h					
	0.0	**0.2**	**0.4**	**0.6**	**0.8**	**1.0**
0.0	23.8	11.0	3.5	1.4	0.8	0.8
0.2	19.9	8.2	2.4	0.9	0.5	0.5
0.4	16.3	6.4	2.0	0.7	0.3	0.3
0.6	13.6	5.4	1.6	0.5	0.2	0.2
0.8	11.8	4.7	1.3	0.4	0.1	0.1
1.0	10.4	4.1	1.2	0.3	**0.0**	0.1

In Table 4 the algorithm performance for different values of parameters is presented. Heuristic crossover coefficient (c_h) is the probability that a given crossover will be heuristic (the probability of a random version of crossover is $1 - c_h$). Parameter m_h plays analogical role for mutation. It can be seen that usage of heuristic crossover and local search during mutation has a very good influence on results quality. The best profits are achieved for high values of m_h and medium/high values of c_h. The algorithm isn't very sensitive to changing parameter values: almost half of parameter configurations in the table is less than 1% worse than the best configuration.

5   Conclusions and Further Research

In this paper an effective algorithm solving the Time-Dependent Team Orienteering Problem with Time Windows was presented. The described method (evolutionary algorithm with local search heuristics) proved to be very effective compared to other metaheuristics. It was confirmed that the presented approach is efficient for various problems from the OP family. Test were conducted on public transport and POI network of Athens and high-quality solutions were achieved in a very short execution time. This signals potential application of the algorithm in e-tourism. Further research will concentrate on adaptation of the method to related problems: the Orienteering Problem with Time-Dependent Profits [22], the Orienteering Problem with Hotel Selection [5] and the Stochastic Orienteering Problem [1].

References

1. Campbell, A.M., Gendreau, M., Barrett, W.T.: The orienteering problem with stochastic travel and service times. Ann. Oper. Res. **186**(1), 61–81 (2011)
2. Campos, V., Marti, R., Sanchez-Oro, J., Duarte, A.: Grasp with path relinking for the orienteering problem. J. Oper. Res. Soc. **156**, 1–14 (2013)
3. Caserta, M., Voss, S.: A hybrid algorithm for the DNA sequencing problem. Discrete Appl. Math. **163**(1), 87–99 (2014)
4. Chao, I., Golden, B., Wasil, E.: Theory and methodology - a fast and effective heuristic for the orienteering problem. Eur. J. Oper. Res. **88**, 475–489 (1996)
5. Divsalar, A., Sorensen, K., Vansteenwegen, P., Cattrysse, D.: A memetic algorithm for the orienteering problem with hotel selection. Eur. J. Oper. Res. **237**(1), 29–49 (2014)
6. Garcia, A., Vansteenwegen, P., Arbelaitz, O., Souffriau, W., Linaza, M.T.: Integrating public transportation in personalised electronic tourist guides. Comput. Oper. Res. **40**(3), 758–774 (2013)
7. Gavalas, D., Konstantopoulos, C., Mastakas, K., Pantziou, G., Vathis, N.: Heuristics for the time dependent team orienteering problem: application to tourist route planning. Comput. Oper. Res. **62**, 36–50 (2015)
8. Gendreau, M., Laporte, G., Semet, F.: A tabu search heuristic for the undirected selective travelling salesman problem. Eur. J. Oper. Res. **106**, 539–545 (1998)
9. Golden, B., Levy, L., Vohra, R.: The orienteering problem. Naval Res. Logist. **34**, 307–318 (1987)
10. Gunawan, A., Yuan, Z., Lau, H.C.: A Mathematical Model and Metaheuristics for Time Dependent Orienteering Problem. Angewandte Mathematik und Optimierung Schriftenreihe AMOS 14 (2014)
11. Hutter, F., Hoos, H.H., Leyton-Brown, K., Stutzle, T.: ParamILS: an automatic algorithm configuration framework. J. Artif. Intell. Res. **36**, 267–306 (2009)
12. Li, J.: Model and algorithm for Time-Dependent Team Orienteering Problem. Commun. Comput. Inf. Sci. **175**, 1–7 (2011)
13. Mahfoud, S.W.: Crowding and preselection revisited. In: Proceedings of the 2nd International Conference on Parallel Problem Solving from Nature (PPSN II), Brussels, Belgium, pp. 27–36. Elsevier, Amsterdam (1992)
14. Ostrowski, K.: Parameters tuning of evolutionary algorithm for the orienteering problem. Adv. Comput. Sci. Res. **12**, 53–78 (2015)
15. Ostrowski, K., Karbowska-Chilinska, J., Koszelew, J., Zabielski, P.: Evolution-inspired local improvement algorithm solving orienteering problem. Ann. Oper. Res. **253**(1), 519–543 (2017)
16. Ostrowski, K.: Evolutionary algorithm for the Time-Dependent Orienteering Problem. Lect. Notes Comput. Sci. **10244**, 50–62 (2017)
17. Ostrowski, K.: An effective metaheuristic for tourist trip planning in public transport networks. Appl. Comput. Sci. **12**(2) (2018)
18. Schilde, M., Doerner, K., Hartl, R., Kiechle, G.: Metaheuristics for the biobjective orienteering problem. Swarm Intell. **3**, 179–201 (2009)
19. Tasgetiren, M.: A genetic algorithm with an adaptive penalty function for the orienteering problem. J. Econ. Soc. Res. **4**(2), 1–26 (2001)
20. Vansteenwegen, P., Souffriau, W., Vanden Berghe, G., Oudheusden, D.V.: A guided local search metaheuristic for the team orienteering problem. Eur. J. Oper. Res. **196**(1), 118–127 (2009)

21. Verbeeck, C., Sorensen, K., Aghezzaf, E.H., Vansteenwegena, P.: A fast solution method for the time-dependent orienteering problem. Eur. J. Oper. Res. **236**(2), 419–432 (2014)
22. Yu, Q., Fang, K., Zhu, N., Ma, S.: A matheuristic approach to the orienteering problem with service time dependent profits. Eur. J. Oper. Res. **273**(2), 488–503 (2019)

Measurement and Optimization Models of Risk Management System Usability

Tomasz Protasowicki[(✉)] [iD]

Military University of Technology, ul. gen. Sylwestra Kaliskiego 2,
00-908 Warsaw, Poland
tomasz.protasowicki@wat.edu.pl

Abstract. Systematization of issues related to the approach to defining and measuring the usability of a Risk Management System (RMS) is one of the key milestones on the way to understand it. A number of models and methods exists in this area. Author attempted to propose the model of the RMS usability and methodology of its application, as a results of in-depth studies carried out by himself in multiple organizations. The outcomes of research are aligned to the requirements that allow their use in any organization regardless of its field of operations, location, size and type of ownership. The model provides fundamentals that enables optimization of the RMS functional configuration in cases where it is possible to generate several acceptable its configuration's variants for specific organization.

Keywords: Usability model · Risk management system · Mathematical modeling · Multi-criteria optimization

1 Introduction

Adopting an active approach to measuring the usefulness of risk management systems in organizations allows them not only to avoid system failures but also potentially costly incidents. In addition it allows to reach viable benefits in terms of business performance. The same indicators that reveals if the risk is under control can often show whether business conditions have been optimized.

Each enterprise where a risk of failure or a risk in information security exists should consider implementation of the methodology for measuring the usability of the implemented Risk Management System (RMS). Reason of this expectation is that proper methodology of measuring the security status and usability of the organization's RMS is substantial for safety of its operations regardless of location, size, field of operations and nature or type of ownership.

The results achieved during research shows that the Risk Management Systems must have specific functional properties. That are determined by such attributes as: functionality, reliability, rationality, efficiency, effectiveness, and above all quality and security. Therefore it is substantial to elaborate proper models allowing one to quantitative measuring an overall usability of the RMS with use of different measures and measurement techniques combined together. The results obtained in the course of research work allows an author to propose the model for the measurement of the RMS

© Springer Nature Switzerland AG 2020
K. Saeed and J. Dvorský (Eds.): CISIM 2020, LNCS 12133, pp. 415–425, 2020.
https://doi.org/10.1007/978-3-030-47679-3_35

usability, as well as to formulate the problem of multi-criteria optimization of the RMS usability against its functional configuration.

2 Research Methodology

Methodology of research was developed by adoption of the generalized systems analysis. For purpose of this paper systems analysis is defined as set of methods and techniques aimed to analytics, evaluation and decision making to solve in rational way some systemic problem situations. Methodology of system analysis includes following steps:

- identification of objectives,
- exploration of options to achieve identified objectives with taking into account a new alternative solutions;
- assessment of all possible effects of each of the options taking into account the uncertainty and risk;
- comparative analysis of options according to selected criteria,
- presentation of results in a way that allows decision making.

2.1 Research Objectives

The main research objective addressed in following paper is to determine the possibilities of using existing models and measurement methods and to propose how to use them in evaluating the usability of the RMS. It requires to formulate a model for measuring current features, attributes or usability properties and justify the method of doing so, which ensures that the required level of usability of the RMS is maintained, and enables measurement and optimization of the organization's security level.

2.2 Research Questions

Research objectives has required to answer a set of research questions including, but not limited to:

- what are the key features of the RMS essential while its design and evaluation its implementation usability?
- which methods and models, that currently exists, might be adopted to measure quantities describing the usability of the RMS?
- what are the key outcome indicators and key activities indicators that might be used to measure the RMS usability and how to evaluate their values?
- what are expected properties of the RMS usability model, which will be developed?
- how to formulate problem for the multi-criteria optimization of the RMS functional configuration?

2.3   Methods

The research methods which were used: literature studies, analysis of standards, guidelines, directives and codes in field of the cyber and critical infrastructure and overall organizational security, critical analysis and assessment of currently used in Europe methods of measuring the risk and evaluating usefulness of a RMS, document research, interviews and surveys with professionals in the field of construction and implementation of a risk management systems.

3   Results

Most of modern organizations have systems in place for the areas of: business process management, quality management, business continuity management, project management etc. They are an excellent source of data for risk analysis, as well as for measurement of RMS usability. On the other hand RMS provides us with knowledge about threats, vulnerabilities and potentials in these areas of the organization's activities. RMS is closely coupled with other management systems within organization, and should not be analyzed separately.

3.1   Definition of the RMS Usability Measurement Model

For the purposes of this paper, the following definition of the RMS has been adopted: an organization's risk management system is a set of rules, policies, procedures, practices, human resources and technical resources (including, but not limited to, policies and instructions for identifying, measuring, monitoring and controlling risk) relating to risk analysis and review processes in organization (1).

RMS must fulfill specific usability (U) attributes including, but not limited to: functionality (F), reliability (N), quality (J), safety (B), innovation (I) and business continuity (C), under certain environmental conditions. Therefore U, as the primary characteristic of the RMS, is a complex function of individual usability properties or attributes (2).

Literature exploited during the research provides many models for measuring the usability of the system [1, 2, 6, 7, 9, 11]. Available sources include examples of simple, developed and complex models. These models has got various forms, e.g.:

- numerical models - written using mathematical formulas;
- graphic models - in the form of drawings, diagrams or constructed on the basis of aggregates (lists) of questions, or constructed on the basis of matrices, tables, maps;
- integrated (combined, mixed) models - resulting from a combination of above.

In this paper the RMS usability measurement has been defined by numerical model by the expression (3).

Proposed formulas for discussed models of: RMS (1), RMS usability (2) and RMS usability measurement (3) have been stated as follows:

$$\mathbf{RMS} = \langle \mathbb{E}^{RMS}, \mathbf{R}^{RMS} \subseteq \mathbb{E}^{RMS} \times \mathbb{E}^{RMS} \rangle \rightarrow \mathbf{U}^{RMS} \tag{1}$$

$$\mathbf{U^{RMS}} = f\left(w_i^{RMS} \in \mathbb{W}^{RMS}; i \in \mathbb{I}^{RMS}\right) \tag{2}$$

$$UMM = \langle ZP, \{U_i\}, F^{SZR}, MP \rangle \wedge i \in \mathbb{I}^{RMS} \tag{3}$$

where:

RMS is the Risk Management System, $\mathbb{E}^{RMS}$ is a set of elements of the RMS, $\mathbf{R}^{RMS}$ is a set of system relations, $\mathbf{U}^{RMS}$ is an usability of the RMS; $\mathbb{W}^{RMS}$ is a set of distinguished attributes of the RMS, w_i^{RMS} is an *i-th* attribute of the usability of the RMS, $\mathbb{I}^{RMS}$ is a set of numbers of distinguished usability attributes of the RMS; ZP is a set of key elements of RMS evaluated for determining the state of its usability; $\{U_i\}$ is a set of usability functions of RMS attributes, F^{SZR} is an RMS usability function determined on the set $\{U_i\}$, MP is an usability measurement methodology.

While considering the RMS usability model we also assumed that each usability attribute $w_i^{RMS} \in \mathbb{W}^{RMS}$ is identified by the number $i \in \mathbb{I}^{RMS}$ and is described by the set C_i^{RMS} of the names of the features. If all the differing sets of C_i^{RMS} features, which are described by individual usability attributes, are numbered with the variable $b = \overline{1, B}$ (called the type of the attribute), then two attributes are of the same type when they are described by identical sets of features. Sets of Q_i^{RMS} numbers of attributes describing the attribute $i \in \mathbb{I}^{RMS}$ and their corresponding sets of C_i^{RMS} names may not be null for each $i \in \mathbb{I}^{RMS}$. Set $\mathbb{E}^{RMS}$ elements of the RMS contain following subsets: $\mathbb{E}^{POL}$ of policies, human resources and general risk management procedures; $\mathbb{E}^{PRO}$ of general company procedures and practices; $\mathbb{E}^{TEC}$ of technical elements of organization; $\mathbb{E}^{BCP}$ of business continuity plans and procedures; $\mathbb{E}^{FAI}$ of detection, notification and investigation procedures for failures.

3.2 Methodology of RMS Usability Measurement

Analyzing various approaches to measuring usability in relation to RMS, the question arises if there is a possibility of creating a complete and consistent assessment methodology? Research outcomes shows that these approach must be based on internal and external factors related to RMS usability attributes. Moreover the methodology must allow for the full and unambiguous assessment of RMS's functional usability. Substantial requirement is to achieve possibility of application these methodology - not only on the theoretical field, but primarily in practice. Following figure presents methodology of RMS functional usability measurement, that is result of performed research activities (Fig. 1). It is a fundament for further optimization of RMS functional configuration and development within business organizations as described in following parts of paper.

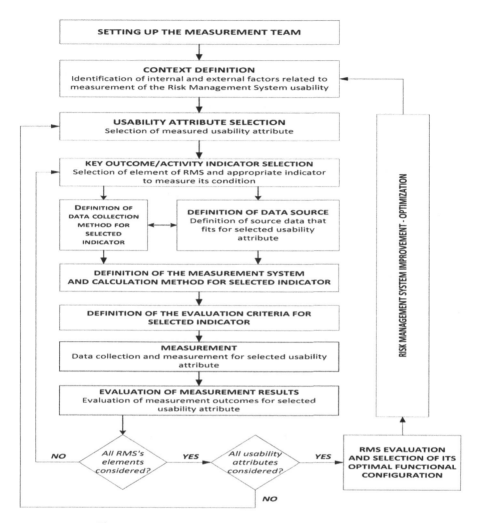

Fig. 1. Proposed RMS usability measurement methodology.

Generalized RMS usability is a complex function of individual attributes mentioned in previous Sect. (3.1). Therefore a critical step in the process of measuring the usability of the RMS is to determine:

- what should we measure (e.g. elements of RMS like: people, processes, activities, threats, policies, procedures, documentation, technical resources etc.)?
- what attributes, properties or usability features will be considered (e.g. security, business continuity, etc.)?
- how will the data be collected?
- what data collection techniques will be used? (e.g. testing, research, interviews, observations, instruments, combined methods)?

- what kind of measures best match selected elements and usability attributes? (i.e. binary measures, categories, ordered measurements, coefficient measurements, interval measurements)?
- which category of measuring system best fits the given measuring situation (e.g. descriptive, threshold or trend)?
- which measuring system and/or measure is best to use? (e.g. Likert scale, binary measure, quotient measure, interval measure)?

3.3   Key Outcome and Activity Indicators for the RMS Usability Attributes

The indicator is understood as an observed value which allows an insight into the usability properties of each attribute of RMS's and is difficult to measure directly (like safety, quality etc.).

Outcome indicators shows whether the expected results have been achieved. In other words are the expected results in terms of usability attributes achieved, or are not?

Activity indicators enable to determine if the result of actions to be taken by the organization will endeavor in reducing of risk. These indicators often measure the state of the attribute of usability (e.g. Security) against level of tolerance. This allows to discover if there is a difference between real and desired state at particular moment.

The combination of outcome indicators and actions indicators provides two perspectives on whether the basic elements of an RMS (such as policies, procedures, practices, personal resources and technical resources) are useful and works properly. Defining these indicators and theirs measurement systems is a recurring process for each RMS usability attribute and for each key element of an RMS's. Defining outcome indicators and activities indicators as well as processes of measurement and measurement systems is a repeatable process for each of RMS usability attribute and each of its key element. Each of indicator mentioned above is an element w_k of the set W^{RMS}, associated with a specific usability attribute w_i^{RMS}.

The measurement system is used to quantify the steady state of the RMS's usability attribute (e.g. security, quality, functionality). This system have to be defined for each indicator, with corresponding set of: techniques, methods or procedures for data collection, time regime for measurement, roles and responsibilities in collecting and reporting values and specific thresholds or tolerance levels, i.e. points at which deviations of indicator's value should initiate actions within organization.

Because of the vast set of both outcome and activity indicators exists, the examples of calculation formulas has been skipped as they are broadly available in literature: [2–5, 8, 10].

3.4   Quantitative Measurement of the RMS Usability

Having defined indicators in relation to the set $\{U_i\}$ of the RMS usability attributes, we can introduce a definition of partial and total usability of the Risk Management System.

RMS's partial usability is associated with a specific usability attribute w_i^{RMS} and is determined using a vector $\vec{R}_i$. This vector is a linear combination of the set of indicators

$w_k \in W^{RMS}$ (described in Sect. 3.3) with coefficients u_k reflecting the importance or share of this indicator in the usability of the attribute U_i defined as follows (4):

$$\vec{R}_i = \sum_{k \in K} w_k u_k \tag{4}$$

The modification of the coordinates of the vector $\vec{R}_i$ for the i-th usability attribute, where $i \in \mathbb{I}^{RMS}$ as in Eq. (2), to consider the impact of individual indicators on the final level of the system utility attribute is given by the following weighted vector $\mathcal{R}_i$:

$$\mathcal{R}_i = \overrightarrow{\mathcal{P}_i} \otimes \vec{R}_i \tag{5}$$

Vector $\overrightarrow{\mathcal{P}_i}$ in Eq. (5) is an coordinate priorities vector of the i-th usability attribute vector $\vec{R}_i$ reflecting the weights of the impact of individual indicators on the total value of this attribute, defined as follows:

$$\overrightarrow{\mathcal{P}_i} = [p_{q,v}]_{Q \times V}; \bigwedge q \in Q \wedge v \in V p_{q,v} \geq 0, 1; \sum_q^Q \sum_v^V p_{q,v} = \mathcal{N}^i = |W^{RMS}| \tag{6}$$

The constraint defined in formula (6) guarantees that each element w_k of the set of key outcome and activity indicators W^{RMS} with cardinality $\mathcal{N}^i$ has a non-zero effect on the final level for the i-th utility attribute, $i \in \mathbb{I}^{RMS}$. The measure of partial usability for the RMS's attribute U_i is called the number $R_i \in R$ equal to the length of the vector $\mathcal{R}_i$:

$$R_i = \left\| \mathcal{R}_i \right\| \tag{7}$$

Having defined partial usability for a set of RMS's usability attributes, we can finally introduce a definition of its total usability as the number $U \in R$ equal to the length of the vector $\vec{U}_i$:

$$U = \left\| \vec{U}_i \right\| \tag{8}$$

Similarly to the $\vec{R}_i$ vector (4), $\vec{U}_i$ is a weighted vector of partial usability $\vec{U}_i$ where $i \in \mathbb{I}^{RMS}$ defined as follows (9):

$$\vec{U}_i = \mathcal{B}_i \otimes \mathcal{R}_i \tag{9}$$

where:

$$\mathcal{B}_i = [b_{t,s}]_{T \times S}; \bigwedge t \in T; s \in S; b_{t,s} \geq 0, 1; \sum_t^T \sum_s^S b_{t,s} = \mathcal{I} = |\mathbb{I}^{RMS}| \tag{10}$$

Taking above into consideration the value of U is quantitative measure of total RMS's usability, which is the result of application the methodology of measuring the usefulness of the Risk Management System proposed in Sect. 3.2 of this paper.

3.5 Optimization of the RMS Usability Against Its Functional Configuration

Formulating the problem of multi-criteria optimization of the RMS usability against its functional configuration makes sense only in cases where it is possible to generate several acceptable variants of its configuration for specific organization. These configurations must be based on a set of key RMS elements, which are subject to measurement and constitute the basis for determining the state of RMS usability. This set includes the most important policies, procedures and practices applied in the organization in the field of risk management as well as all available human resources and technical resources.

If there are many acceptable functional configurations with significantly different values of the distinguished outcome and activity indicators, there is usually a need to choose the best one. In this sense the best configuration is the most aligned to the requirements contained in the standards or specified at the stage of designing the risk management system for a given organization. This require to set up:

- a set KU^{dop} of decision variables $x = [x_1, x_2, \ldots, x_k,]$ related to acceptable RMS's functional configurations for which outcome indicators are known in relation to attributes describing the RMS's usability:

$$KU^{dop} = \left\{ KU_x, x = \overline{1, X} \right\} \tag{11}$$

- a function vector whose elements represent the objective function $f(x) = (f_1(x), f_1(x), \ldots, f_k(x))$, in the form of a vector criterion function $\underline{\bar{Q}}$:

$$\bar{Q}(KU_x) = (Q_1(KU_x), Q_2(KU_x), Q_3(KU_x), Q_4(KU_x), Q_5(KU_x)) \tag{12}$$

where:

$Q_1(KU_x)$ - defines the degree to which evaluated functional configuration KU_x meets the requirements of the organization's codes, standards and practices,

$Q_2(KU_x)$ - defines the generation time for functional configuration KU_x, which reflects the velocity of change management in the organization,

$Q_3(KU_x)$ - defines the degree to which the functional configuration KU_x is susceptible to the occurrence of incidents resulting from the improper application of procedures in the organization,

$Q_4(KU_x)$ - defines the degree of component redundancy in functional configuration KU_x,

$Q_5(KU_x)$ - defined ratio of functional configuration KU_x support costs to overall risk management costs in the organization;

- constraints to decision variables $g_i(x) \geq 0, (i = 1, 2, \ldots, k)$, $h_j(x) = 0, (j = 1, 2, \ldots, l)$:

$$Q_1(KU_x) \geq 0, \ Q_2(KU_x) \leq T, \ Q_3(KU_x) \geq 0, \ Q_4(KU_x) \geq 0, \ Q_5(KU_x) \geq 0 \qquad (13)$$

where T is an allowable time to develop and implement any one of the acceptable functional configuration of the RMS;

- dominance relation in criterion space ($\geq$).

Taking above into consideration the problem of multi-criteria optimization of user configuration can be written as follows:

$$\left(\overline{KU}^{dop}, Q, \geq\right) \qquad (14)$$

The set of acceptable functional configurations has the following form:

$$\overline{KU}^{dop} = \left\{ \begin{array}{c} KU_x \in KU^{dop} : Q_1(KU_x) \geq 0 \wedge Q_2(KU_x) \leq T \wedge Q_3(KU_x) \geq 0 \\ \wedge Q_4(KU_x) \geq 0 \wedge Q_5(KU_x) \geq 0; x \in X \end{array} \right\} \qquad (15)$$

The vector criterion function is defined as follows:

$$Q : \overline{KU}^{dop} \rightarrow Y, Q(KU_x) = \underline{y} \qquad (16)$$

The criterion space is given by:

$$Y = \left\{ \underline{y} = \overline{Q}(KU_x) \in R^5 : KU_x \in \overline{KU}^{dop} \right\} \qquad (17)$$

where vector of ratings $\overline{Q}(KU_x)$ related to acceptable RMS's functional configuration KU_x is given by formula (12) and criteria for choosing the optimal functional configuration are defined as:

$$Q_1(KU_x) = y_1 = \frac{L_x^K + L_x^S + L_x^P}{L_x} 100\% \rightarrow max \qquad (18)$$

$$Q_2(KU_x) = y_2 = -\frac{1}{N_x} \sum_{i=1}^{N_x} t_i^x \rightarrow max \qquad (19)$$

$$Q_3(KU_x) = y_3 = -\frac{L_x^B + L_x^{NA} + L_x^{NP}}{L_x} 100\% \rightarrow max \qquad (20)$$

$$Q_4(KU_x) = y_4 = \frac{\overline{\overline{F}}_x - \overline{\overline{F}}_x^D}{\overline{\overline{F}}} 100\% \rightarrow max \qquad (21)$$

$$Q_5(KU_x) = y_5 = -\frac{K_x^{OS} + L_x^{KT}}{K_x} 100\% \rightarrow max \qquad (22)$$

where: L_x^K is number of codes, L_x^S is a number of standards, L_x^P is a number of practices, L_x is an overall number of requirement; N_x is a number of experiments related to functional configuration KU_x, t_i^x is a configuration generation time; L_x^B is a number of gaps in RMS's procedures implementation in organization, L_x^{NA} is a number of inadequate implementations or obsolete RMS's procedures in organization, L_x^{NP} is a number of procedures not respected by members of the organization, L_x is an overall number of procedures that should exists in organization; F_x is a set of functions available in specific functional configuration KU_x, F_x^D is a set of functions required to maintain ongoing execution of the RMS's processes, F is a set of RMS's functions fixed during its design phase, $\overline{\overline{F}}_x, \overline{\overline{F}}_x^D, \overline{\overline{F}}$ are cardinalities of sets F_x, F_x^D, F; K_x^{OS} is cost of human resources required to enable proper operation of RMS having functional configuration KU_x, K_x^{KT} is cost of technical resources required to enable proper operation of RMS in specific functional configuration KU_x, K_x is overall risk management costs in the organization.

The dominance relation $\geq$ in the criterion $\mathbf{Y}$ space is the dominance relation in the PARETO sense.

4 Conclusions

Designing a Risk Management Systems with focus on optimization of theirs usability is a new field of research that has appeared recently. One of the best practices for measuring RMS's usability is to develop an effective measurement method with an adequate measurement model and a properly selected set of key outcome and activity indicators. Solving the problem of multi-criteria optimization of the RMS's functional configuration is crucial when it is possible to implement in organization one of several designs of its configuration. Presented approach is a proposal of a partial solution for the measurement problem, which would allow ongoing assessment of the state of usability of the RMS in any organization. Application of research results presented herein in any organization should ensure the management that the required level of usability of the RMS is maintained, and enables current measurement and future optimization in this area of organization's operations.

References

1. Au, F., Baker, S., Warren, I., Dobbie, G.: Automated usability testing framework. In: Proceedings of the Ninth Conference on Australasian User Interface, AUIC 2008, vol. 76, pp. 55–64 (2008)
2. Bastien, J.: Usability testing: a review of some methodological and technical aspects of the method. Int. J. Med. Informatics **79**, 18–23 (2009)
3. Budworth, N.: Indicators of performance in safety management. Safety Health Pract. **14**, 23–29 (1996)
4. Campbell, D.J., et. al.: Performance measurement of process safety management systems. In: International Conference and Workshop in Reliability and Risk Management (1998)

5. Cunningham, S.M.: The major dimensions of perceived risk. In: Cox, D.F. (ed.) Risk Taking and Information Handling in Consumer Behavior. Harvard University Press (1967)
6. Fitzpatrick R.: Strategies for Evaluating Software Usability, Methods (1998)
7. Greatorex, M., Mitchell, V.: Developing the perceived risk concept. In: Emerging Issues in Marketing, Proceedings Marketing Education Group Conference, vol. 1, pp. 405–415 (1993)
8. Kiedrowicz, M., Napiórkowski, J., Stanik, J.: Model of automated control and monitoring system of the current level of information security. In: Proceedings of the 25th Anniversary Conference Geographic Information Systems Conference and Exhibition (2018)
9. Laskowski, M.: Czynniki zwiększające jakość użytkową interfejsów aplikacji internetowych. Logistyka **6**, 2191–2199 (2011)
10. Protasowicki, T., Stanik, J.: Metodyka kształtowania ryzyka w cyklu rozwojowym systemu informatycznego. In: Kosiuczenko, P., Śmiałek, M., Swacha, J. (eds.) Od procesów do oprogramowania: badania i praktyka, pp. 27–44 (2015)
11. Scott, J.E., Zikmund, W.G.: An investigation of the role of product characteristics in risk perception. Rev. Bus. Econ. Res. **13** (1977)

Development Methodology to Share Vehicles Optimizing the Variability of the Mileage

Luis E. Ramírez Polo[1] , Alcides R. Santander-Mercado[2] ,
and Miguel A. Jimenez-Barros[3(✉)]

[1] Universidad Autonoma del Caribe, Barranquilla, Colombia
luis.ramirez@uac.edu.co
[2] Universidad del Norte, Barranquilla, Colombia
asantand@uninorte.edu.co
[3] Universidad de la Costa, Barranquilla, Colombia
mjimenez47@cuc.edu.co

Abstract. A simulation is a tool used to visualize the behaviors of a system, which will later help make decisions regarding how to handle the variables involved in the system, as well as the specific changes that have to be made. This study shows a case of vehicle allocation for different people within a company, evaluating methodologies, vehicle rotation to reduce the variance of the mileage and eliminating penalties with rental agencies for exceeding the permitted mileage. The paper shows a literature review of allocation models and similar studies, and later displays a detailed description of the problem, the variables that was used, the composition of the simulation and the optimization model that were generated, the results of the simulation, and finally, the findings of the research.

Keywords: Simulation · Optimization · Assignment problem · Vehicle share · Mileage limit

1 Introduction

Assigning issues that occur in everyday situations, not knowing who is responsible for said issues, and ignoring a methodology where a decision was lightly taken and without an in-depth analysis may represent direct or indirect losses (missed opportunities), that later could have been converted into profits. Situations where contractors were assigned to projects, machine workers, and sales agent districts, among others, demand the making of a good decision, which is not always a simple one because of the various factors it must consider. Regarding vehicles, the focus of transportation planning has gradually shifted to infrastructural change, to improve the management of the existing one [1]. When the company has a transport fleet, the problem relies on knowing how to keep vehicles in certain mileages before the implied warranty expires. If the size of the fleet increases, data monitoring and collection become more operators that are problematic, especially if different are assigned the responsibility of driving the same vehicle [2].

K. Saeed and J. Dvorský (Eds.): CISIM 2020, LNCS 12133, pp. 426–435, 2020.
https://doi.org/10.1007/978-3-030-47679-3_36

This paper shows the development of a methodology to find the perfect moment to make a rotation of the vehicles, that was assigned by a company to specific regions and employers, was design using simulation and optimization techniques in order to optimize the efficiency in that time and at the same time reduce the variance of the mileage of all the vehicles and minimizing penalties for exceeding the mileage allowed by car.

2 Literature Review

The problem of allocation has become a focus of research papers in different scenarios. The idea is to act in the best way in every process where there are decisions to be made [3]. Provide a systematic review of selected publications that offer method-based solutions to the vehicle relocation issues in car-sharing networks. Part of the planning process, especially the process of managing your assets is related to creating the best possible configuration [4] and finding the best solution that comes close to reality [4]. In what regards the transportation issues, researches base their analysis on deciding who will lead an activity, proposing routes, points of entry and exit. In the history of research, [5] investigated the reasons behind traffic situation, government rules and optimal working hours in the rail system that allow the designing of a discrete simulation model to allocate equipment and its movements. The model allowed them to verify the impact of the changes that occur in the flow of trains, work rules and government regulations on the overall operational efficiency. The system helps evaluate changes to the current crew and allows them to test different allocation scenarios related to work schedules.

The platform used for the simulation was Trainsim; said the platform provides a replica of the allocation of the task force as a function of time and operational parameters, thus achieving a close prediction of the results of an assignment of the teamwork based on historical data. The information obtained can answer questions like "what if?" made in the simulation. Model inputs that determine system performance are variable. These inputs are:

a) Traffic on trains.
b) Regulations of the staff and costs associated.
c) Allocation scenarios including team schedules.

During the simulation, Trainsim reproduces the process of assigning working trains to teams based on traffic conditions. As a result, the output of the simulation model allowed us to acknowledge the total costs needed to operate under the rules set for work teams in different schedules [3]. Another common scenario where you can apply the assignment problem is the project-related scenario. This occurs in situations where several projects and decisions that must be made, regarding who the leader of each of the situations will be. This is of vital importance in project management. According to research done by [7], this problem has been named "Expert Assignment Problem". The authors formulated a mathematical model for this problem using genetic algorithms; however, there have been drawbacks in convergence rates when utilizing these algorithms.

For the former case, we propose to use optimization through the *ant colony algorithm*, which has greater abilities to solve such complex problems of discrete optimization and to solve the "Expert Assignment Problem" as well. In such problems, they had to consider three main criteria:

a) Should we assign a senior academic leader for each project?
b) The fewer the leaders, the better the draft allocation plan.
c) The number of projects that should be checked by every leader needs to be moderate.

Given these criteria, the mathematical model for the problem was formulated using a heuristic ant colony optimization, applied to meet the needs of the model and to perform the allocation of project leaders. This procedure resulted in making the model more efficient and, unlike genetic algorithms, in improving the convergence speed by finding the best solutions [4].

Another scenario where the assignment problem is the focus of the investigation is in a logistics park. An example is the Lianyungang Port Logistics Park, where the forecast traffic volume and traffic assignment are the focus of research by [9]. To forecast the volume, we implemented a TVNC method (for its acronym in English that stands for "Non-repeated Traffic Volume Method"). To assign traffic, we considered theoretical foundation-related models in equilibrium and disequilibrium models. In this research, a model was handled in balance for its effectiveness; the said model is known as System Optimum Assignment Model. Experimental study results made the logistics park more effective, which is reflected in significant social and economic benefits [9]. In a research conducted by [10], they developed a transport model frequently used to solve problems of physical distribution and location, which applies to typical situations, such as resource allocation, scheduling of vehicles, specialized cooperation and redistribution of the plant. However, the transport model had three disadvantages:

a) The ways to balance production - market and transportation costs are presented separately, which is a drawback for worktables.
b) Figures and tables are listed separately. This represents a problem for the calculation of different values.
c) Minimize the generation of Hamilton Circles.

These authors present some methods to overcome these drawbacks and a case study of the transport model applying the proposed methods [10].

Problems that have involved railways were researched by [10], who used a method that is the basis of the DAI (Distributed Artificial Intelligence), which is used for mapping the flow of passengers on railways and evaluating its performance through simulation. Passenger flow assignment is a key resource in the location, design, and scheduling of service vehicles, in this case, trains. In this study, the model of linear planning is considered for this method and is found distorted, due to the behavior of the passengers. This model seems to work in theory but not in practice. Therefore, the authors present the following method: The allocation of passenger flow, based on competition, cooperation and allocation of passenger flow through the decision making with multi-agents [7].

Airports are also environments where the allocation problem can happen in various ways. One is applied to the gate, which allows assigning an entry for arrivals and departures of flights to ensure that they are under the established schedules. In research conducted by [11], the approach was to allocate those gates, which are key to ensuring the efficiency of airports, with high rotation flights. This problem of assigning gates can be represented as a complex optimization problem. The study proposes a model of robust allocation to minimize the variability of productive time. According to the intrinsic characteristics of the formulated objective function, the authors implemented a search algorithm called TABU (Tabu Search Algorithm) and a meta-heuristic to solve the problem of gate assigning [8]. Furthermore, car-sharing services and analysis have become much more relevant in these times due to their environmental contribution and sustainable way of transportation [12]. Analyzed and classified s137 papers, covering the last fifteen years of research and deriving an insight of all the mainstream in this topic.

3   Problem Description

As has been mentioned by many authors, the projection of a good mobility system will be contemplating the key drivers like lower costs and lower environmental costs [11]. In this specific case, a company dedicated to providing services gives its employees a car to move to different places, depending on their role in the company; Technical, commercial, and managerial areas, among many others. After a period has passed, the distance traveled by the vehicles always varied greatly according to the position and the region where the employee was located. The company, who rents the vehicles, often had to pay large fines because some of the rented vehicles ran more mileage than what was agreed to, while others did not reach half of this value. The main propose is to rotate vehicles that use a higher amount of mileage with those that run below the established limit to obtain less of used average by vehicle. The distance traveled by each employee does not vary, the work area where they mobilize is the same, and however, depending on the velocity the time behaves like a random variable. Many companies that handle this type of rental systems incur in extra costs because users exceed the allowed mileage limit, and that results in renting a second vehicle for the person or extending the contract, while others stay within the mileage allowed in the stipulated time.

The problem lies in finding the ideal point in time for rotations, giving all the possible permutations that can be done, taking both the proximity of vehicles and the traveled mileage into consideration. In Fig. 1, a problem statement is observed for groups of cities; internal rotation, including a maximum and estimated rotations that the vehicle may have per period. Focusing on the company, in this case, helps maximize the utilization of vehicles and tries to reduce the use of GAP vehicles, decreasing unnecessary costs caused by penalties for exceeding the mileage, and delivering a continuous performance in the fleet car. This model contemplates that decide on the best option from a financial performance viewpoint, there may be overriding practical limitations, which dictate the ultimate choices made [12].

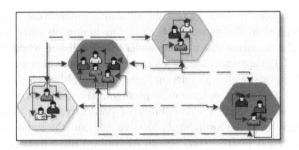

Fig. 1. Possible vehicle rotations.

4 Model Description

Considering all the variables that have an impact on decision making, the required data for the development of the model was collected and analyzed. Based on the employee's position and geographical location, the traveled mileage was analyzed as the main variable. According to this situation, the development of a dynamic model that combines simulation and optimization is necessary, where the best options will be found to optimize the time and distance traveled by car and therefore, we will be able to correct the rotation of the vehicles according to the proportion of distance traveled.

To estimate the parameters of the model was recognized that each employee could change the vehicle that they have a disposal that they can be managed as an entity at the simulation model. Was assign certain characteristics (attributes) to each of the employees or entities like position, location in the country and/or region where they are, and the random variable associated with the distance. This last parameter is considered, due to the stochastic nature of the collected data. For the calculation of the type of distribution and the values of the associated statistical parameters, according to the employee and the region, a test of goodness of fit and independence was used.

In the model this different sets and variables were considered:

Sets:

i: a set of short periods to analyze
j: a set of vehicles available
x: a set of employers

Variables:

D_{ij}: represents the cumulative distance traveled by vehicle j in time i
$Vkm(i)$: variance of the total traveled by all vehicles in time i

Parameters

$M(x)$: represents the distance traveled by the employer in a period

To simulate the process, the methodology applied in Fig. 2 was developed. First, each vehicle j is assigned an employee x by default and the D_{ij} measurements are assigned the value of $M(x)$, when the time is over, the mileage attributes of each vehicle are updated, considering that car temporally ownership is an important determinant of car usage, general travel behavior, and energy consumption [13].

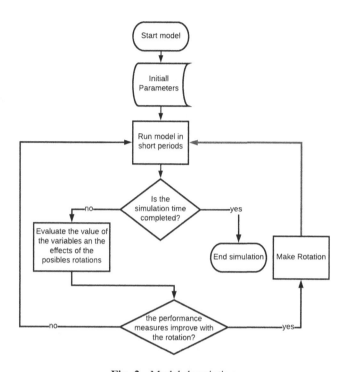

Fig. 2. Model description

To perform the exchange at D_{ij}, the vehicles are sorted from "low to high" according to their current mileage. The situation is evaluated if a change is necessary. To do this, we take into consideration the generated exchange zones and exchange pairs, which are those who are geographically closer to each other and at the same time can compensate the traveled mileage (Fig. 4). This is done to reduce the average mileage of vehicles (Fig. 3).

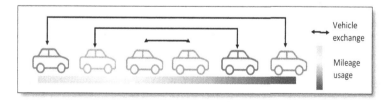

Fig. 3. Vehicle rotation logic

When the evaluation of changes is made, the model analyzes the variable $(Vkm(i))$ as an objective function to measure the difference between the vehicle's trajectory and the total average trajectory or the variance of the distance of all vehicles.

$$Vkm(i) = \frac{\sum_j (D_{i,j} - \bar{D}_{i,j})}{j} \qquad (1)$$

The main idea is to minimize this difference to reduce the variance. In such a case that a vehicle is changed with another that is very far away from the location where it is supposed to be working, a mileage corresponding penalty incurred as a consequence of the switching process from the current area to the one assigned on time $D_{i+1,j}$. Once the changes are made, the entity goes to the step to have delays and this procedure repeats while the total time to be analyzed is completed.

5 Model Results

To validate the methodology, a 24-month history of data from a local company was worked, in which data analysis was carried out to determine the behavior of the data for each of the employees x and to know through goodness-of-fit tests how was the probability function for the mileage traveled in a short time I. Once the data was processed, a simulation model was designed using the Arena 10.0 software to emulate the current and proposed situation. With this model, three scenarios were evaluated; in the first scenario, we recreated a model that represented the current situation to validate and compare the results obtained by the program with the real data analyzed. A second scenario was created with the proposed improvements with the new methodology without including the rotations of vehicles between cities and the tired model was design including an exchange between different cities.

Models 2 and 3 were created by adding an optimization model to determine the number of suitable changes to minimize the variation of the accumulated distances and handle asymmetry between them, all this under the following structure;

Minimize Max[**Vkm(1), Vkm(2),..., Vkm(i)**]
Subject to :   1 ≤ i ≤ 12

The model input information, as well as the final results, were validated through data analysis through sample sizes, their length and ideal run numbers, and hypothesis tests. With the first scenario, it was possible to validate the simulation model and with scanners 2 and 3, the best results were analyzed. Once the different scenarios were run, an improvement of more than 20% was obtained in the leveling of the used vehicles was evident, which has a positive direct and impact on the company's achievements, since by improving the use of these resources, the penalization imposed by exceeding the mileage of the vehicles are considerably reduced.

In Fig. 4, we can see some previous penalized for exceeding the mileage admitted in each one of the three cities analyzed, with the implementation of the new proven methodology through the simulation model, all these penalties disappear, since by minimizing the variance of the time traveled, the vehicles are used more in relation to the average mileage.

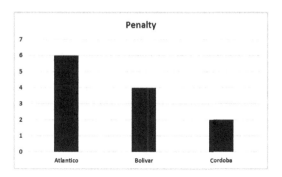

Fig. 4. Number of previous penalties by location

With respect to vehicle mileage, we can see that plans to share vehicles among employees reduce the variation in the average mileage used throughout the system. Figure 5 shows the result comparing the variations between the three cities analyzed. The line in bold shows the results under normal conditions and with the dotted line the results are shown applying the methodology. With the results obtained, the total amplitude of the mileage traveled by the vehicles by zone (the difference between the vehicles that had more use with the one that was used less) was reduced by 28% compared to the range that they currently drive, also, vehicles with less use increase their use by 55%, while those with more use reduce their activity by 44%, which balances their mileage. Also, in this case, analyzed in particular and due to the large distances between the cities, there were no differences between the results obtained between scenarios 2 and 3.

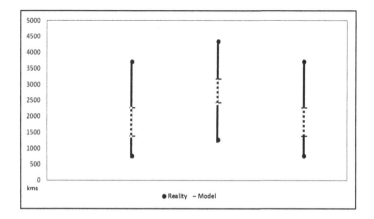

Fig. 5. Min and max distance drove before and after the optimization

6 Conclusions

To determine the best plan of action, a problem should be, in many cases, analyzed as a complete system, this approach allows comparisons between different cases. The lack of communication between those responsible for the process allowed these possibilities for improvement to go unnoticed.

The results show that an option that could go unnoticed can sometimes increase the productivity of a sector. In this particular case, it was possible to achieve measurable improvements of more than 20% in the use of vehicles and penalty reductions which has a significant impact on the company's finances and helps improve the environment. Because of this investigation, we can highlight the importance of linking simulation and optimization methods to evaluate, validate and improve the decision-making process.

Additionally, in the work was possible to develop a model that can predict the consequences of a specific decision, allowing companies to see that changes are not always necessary if the decision-making processes work well. Our research has some limitations, as we were not able to measure the financial and environmental advantages of the changes in the proposed method, the reason why we recommend that further analysis should be undertaken in those areas.

References

1. Sundaram, S., Koutsopoulos, H.N., Ben-Akiva, M., Antoniou, C., Balakrishna, R.: Simulation-based dynamic traffic assignment for short-term planning applications. Simul. Model. Pract. Theory 19(1), 450–462 (2011)
2. Lawrence, K.R., Tunke, J.A.: Analysis and profiling of vehicle fleet data. United States Patent 6505106 (1999)
3. Illgen, S., Höck, M.: Literature review of the vehicle relocation problem in one-way car sharing networks. Transp. Res. Part B: Methodol. 193(204), 120 (2019)
4. Herazo-Padilla, N., Montoya-Torres, J., Muñoz-V, A., Nieto Isaza, S., Ramirez Polo, L.: Coupling ant colony optimization and discrete-event simulation to solve a stochastic location-routing problem. In Proceedings of the 2013 Winter Simulation Conference: Simulation: Making Decisions in a Complex World, pp. 3352–3362, 2013
5. Ramirez Polo, L., Mendoza, F., Jimenez, M.: Simulation model to find the slack time for schedule of the transit operations in off-peak time on the main terminal of massive transport system. Espacios 38(13), 1 (2017)
6. Guttkuhn, R., Dawson, T., Trutschel, U., Walker, J., Moroz, M.: Simulation planning and rostering: a discrete event simulation for the crew assignment process in north american freight railroads. In: Proceedings of the 35th Conference on Winter Simulation: Driving Innovation (2003)
7. Guttkuhn, R., Dawson, T., Tmtschel, U., Walker, J., Moroz, M.: A discrete event simulation for the crew assignment process in North American freight railroads. In: Winter Simulation Conference, Washington, D.C (2013)
8. Li, N., Zhao, Z., Gu, J., Liu, B.: Ant colony optimization algorithm for expert assignment problem. In: Proceedings of the Seventh International Conference on Machine Learning and Cybernetics, Kuming (2008)

9. Xie, H., Ji, S., Shen, J., Han, X.: Research on traffic volume forecast and assignment of logistics park. In: 2008 IEEE, Vancouver, WA (2008)
10. Donghua, W., Xue, L.: A study on transportation problem model. In: International Conference on Management and Service Science, Wuhan (2009)
11. Jiang, X., Bao, Y.: A DAI-based method for rail traffic passenger flow assignment and simulation. In: International Conference on Artificial Intelligence and Computational Intelligence (2010)
12. Zheng, P., Hu, C., Zhang, C.: Airport gate assignments model and algorithm. In: International Conference on Artificial Intelligence and Computational Intelligence, Sanya, China (2010)
13. Ferrero, F., Perboli, G., Rosano, M., Vesco, A.: Car-sharing services: an annotated review. Sustain. Cities Soc. **37**, 501–518 (2018)
14. Hamilton, A., Waterson, B., Cherrett, T., Robinson, A., Snell, I.: The evolution of urban traffic control: changing policy and technology. Transp. Plann. Technol. **36**(1), 24–43 (2013)
15. Wright, S.: Designing flexible transport services: guidelines for choosing the vehicle type. Transp. Plann. Technol. **36**(1), 76–92 (2013)
16. Tanishita, M.: Life events, car transaction, and usage by car type: longitudinal data from Japan. J. Traffic Transp. Eng. **6**, 88–96 (2018)

Optimisation Model of Military Simulation System Maintenance

Wojciech Stecz[1,2(⌧)] and Tadeusz Nowicki[2]

[1] PIT-RADWAR, Warsaw, Poland
wojciech.stecz@pitradwar.com
[2] Faculty of Cybernetics, Military University of Technology, Warsaw, Poland

Abstract. We present logistic support architecture of real simulation training system for military troops preparing to conduct battle operations at a level up to a whole battalion. Due to the need of cost reduction of system maintenance, the system components were designed to be easily serviced with an integrated supply chain support. A framework of multi-period multi-level multi-item production planning models with open and closed loop supply chains is introduced in the paper. Open loop supply chain (OLSC) and closed loop supply chain (CLSC) support alternatively production planning (PP) processes. In CLSC, a PP process is supported by the products recovery processes made in a closed loop chain. A computational complexity involved in preparing production plans for a long planning horizon is important problem when solving these models, so the model optimised for quick solution finding is presented. Most elements of the presented models are common for the military and civil industry.

Keywords: MILP · Supply chain · Closed loop · Production planning

1 Introduction

A feasible production plan can be defined as one that satisfies the given demand over the planning horizon with no backorders or unsatisfied demands (resulting in the lost sales). A feasible production fulfils the specified production regulations for each product and does not violate any production constraints. Optimisation of a production plan means usually the minimisation of the total production costs or maximisation of the total income [2, 8]. In this paper, a feasible production model in complicated supply chain for military is presented. Great emphasis is put in the article on speeding up numerical calculations. That is why the presented model is quite complicated. The extension of the model is necessary due to the fact that calculations within the supply chain are performed repeatedly over a short period of time. That is why it is so important that the model is efficient and gives optimal solutions.

Production planning activities are executed in an open loop supply chain (OLSC) or a closed loop supply chain (CLSC). The planning activities for a predefined planning horizon prepare the results like: MPS (Master Production Schedule – a basis of the production tasks) and MRP (Material Resource Plan – a schedule for raw materials and components' purchase process). MPS is a production plan considering the most

K. Saeed and J. Dvorský (Eds.): CISIM 2020, LNCS 12133, pp. 436–450, 2020.
https://doi.org/10.1007/978-3-030-47679-3_37

important constraints for the final products. In the paper, a company that produces two final products described with Bill-of-Material is presented (simplified versions of individual laser system described in Fig. 1). BOM structures are shown in Fig. 2.

An OLSC forms a logistic construction which delivers the raw materials and components to the factories and distributes the final products to the clients. A CLSC is more complicated. A CLSC forms an additional network of the collection, recovery, and distribution centres which cooperate with clients who use some products that can be recovered and resold once again (see Fig. 3). This integrated structure was referred to as a CLSC in [3, 5]. An idea of CLSC behaviour is based upon the assumption that there exists the clients' willing to return used products to the manufacturers and get new ones instead [10, 11]. The returned goods are recovered by the manufacturers and sold once again [7, 9]. A CLSC framework described in [1, 7] is taken as a framework of CLSC in the paper.

We present logistic support architecture of real simulation training system for military troops preparing to conduct battle operations at a level of one soldier up to a whole battalion. Due to the need of cost reduction of system maintenance, all of the system components were designed to be easily serviced with an integrated supply chain support. These kinds of systems have been used in many NATO armies.

From the point of view of the company providing technical support, it is important to maintain minimum inventory of items that should be replaced. Therefore, the article presents a supply chain model that is used to calculate the demand for parts over a given period. The prepared model is of the MILP class. Models of this type allow analysts to quickly find optimal inventory that should be maintained.

There are some features that distinguish this paper from other papers of the subject. First is a longer planning horizon for the very complicated military environment. Next, the model presented is a multi-level multi-item PP with setup times and backlogs. This is important to underline that the longer a planning horizon and BOM complication, the more complicated a search process becomes. The model considers a multi-period multi-level multi-item production problem of a commodity with BOM.

The model presented in the paper uses the deterministic parameters, but our calculations are made for a planning horizon of more than 13 weeks (one period can describe a week). One should realise that CLSC models are the MILP (mixed integer linear programming) models and even an average complicated model consists of thousands of integer variables. Therefore, developing the right form of the model can speed up calculations by up to several dozen percent.

The rest of this article is organised as follows. In the next section, we describe the assumptions for the military training environment and origins of uncertainty for that environment. After this introduction, the mathematical models are proposed for the multi-period multi-level multi-item production planning with BOM constraints supported by CLSC. We introduce the modifications of models that take into consideration the possible disturbances of the parameters: volumes of demands and returns, and disturbances of efficiency for a few supply chain centres. In the last section, we present the optimisation results for the presented models which relate to the supply chain uncertainties.

2   Modern Military Training System Architecture

The integrated military training laser system is used by many armed forces around the world for training purposes. It uses electronic devices composed of lasers transmitters and detectors (receivers) to simulate an actual battle. Individual soldiers carry small laser receivers, scattered over their bodies, which detect when the soldier has been illuminated by a firearm's laser. Each laser transmitter is set to mimic the effective range of the weapon on which it is used. When a soldier is hit then the subsystem installed on his military harness informs him about his virtual life status (usually three statuses are used: killed, injured, operational). Dismounted soldiers (soldiers outside a vehicle) often wear a vest or harness with sensors as well as a set of sensors on their helmets.

A technical architecture of an individual soldier and hand weapon sets is presented in Fig. 1. For identifying a responder device, an interrogation device transmits a coded signal (each soldier laser beam is different) which is detected in the responder device and is converted into electrical signals which are supplied to a central unit on the receiving end for transmitting the identification messages back to the transmitting device in accordance with decisions made by this central unit.

Different versions of systems are available to armies. The capabilities of the individual systems can vary significantly but in general all modern systems carry information about the shooter, weapon, and ammunition in the laser (this information is presented at EXCON – headquarters for commanders that prepare and evaluate their subordinates).

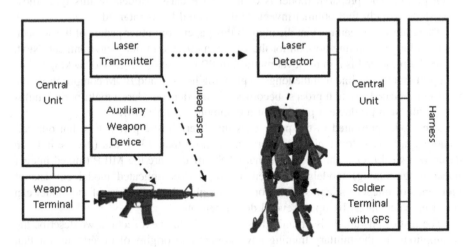

Fig. 1. A general structure of transmitter/receiver system mounted in soldier's harness and in vehicles

In addition to the complicated transmitter/receiver laser systems used by the soldiers and vehicles to support a proper cooperation among these elements, some data linking architecture is needed. Usually, this architecture consists of many complicated radio and GPS systems mounted on stationary antennas or on moveable tripods. Precise description of subsystems is beyond a scope of the paper but one has to understand that a producer of such installations has to take into account two main factors. First is that these subsystems must be resistant to extreme weather conditions, dirt, and dust. Second is that during the exercise all components of the system must be operational. This means that after any kind of malfunction, the subsystem must be serviced in maximum 8 h for the complicated system failure and at once for the individual soldier equipment.

Because there is no possibility to produce fully reliable electronic systems, the producers have to maintain the significant levels of reserve components or to design the architecture of the system in such a way to be able to easily and quickly replace broken elements and later remanufacture them. In that military systems, modality plays a crucial role whereas production of any component is very expensive.

We assume that a standard military training system consists of data (message) linking elements, a few hundred transmitter and receiver components for fighting vehicles and a few thousand components for individual soldiers. One may assume that such systems will be used by up to 15 years and modified during a life cycle of the system. One should assume also that 5%–10% of the electronic devices will be repaired each year due to intensity of use. Some components are more sensitive to damage than the others. Moreover, it can be assumed that subcomponents of a system, like laser receivers, can be bought from more than one producer what allows us to think about our system as having modular system architecture. Many identical components like laser beam receivers or fixings are used in different weapons.

3 Production Planning Maintenance Model for Military Supply Chains

3.1 Supply Chain with Production Planning Model

Our production planning model with open or closed loop support belongs to a level of multi-item multi-level multi-period lot-sizing capacitated problem. Multi-item problem means that there is more than one final product for which MPS (master production schedule) is prepared. Multi-level production planning means that a production process of a final product consists of a few phases in which raw materials are used to produce a semi-product and later these semi-products (components) are used for production of a final product depending on BOM structure (see Fig. 2).

The MPS is prepared for a planning horizon, so a planning process is multi-period. Lot-sized means that one takes into account a production lot sizes. Capacitated models are the models in which the capacity constraints upon the production facilities are put on.

CLSC models consist of a few components: collection, recovery, and redistribution centres. Sometimes other centres are added. One may notice that OLSC is a component of CLSC. So, in this paper, CLSC model is presented as the most complicated version of supply chain management. A classic CLSC model is expanded by the addition of the production components. A CLSC is used by the producer as a production support. In the paper, a production process is treated as a main business process (this is a proper assumption when one takes into consideration that clients usually prefer new products than recovered ones).

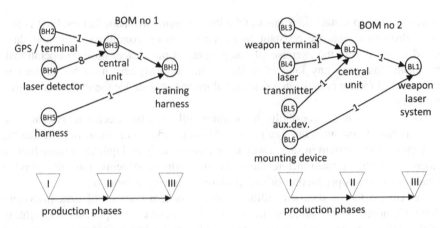

Fig. 2. BOMs for two equipment configuration for which the production plan is prepared

CLSC processes are the additional and supporting processes that allow a producer to minimise its costs. CLSC plays a role of a second producer. CLSC uses the returned products in the recovering processes and regenerates the returned products without dismantling them into parts and using these parts in the production. Sometimes, the product must be completely dismantled and many of its parts must be changed but we treat this product as a regenerated one. So, for example, this product retains the same part's number. We allow also the products to be dismantled and their parts used in the production of new products. Figure 3 describes our production and supply chain environment.

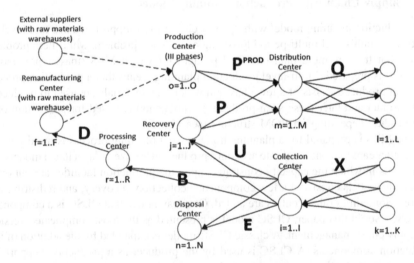

Fig. 3. Closed loop supply chain supporting a production process maintaining a military system

These products are collected in the I collection centres, recovered at the J recovery centres, and then moved to the M distribution centres. Then, some the L clients can purchase them. Simple CLSC model presented by Pishvaee [6] did not take into consideration the cases when there were shortages of the returned products. In these cases, the model estimates only the gap between demand and return, which is unacceptable in practice. Our CLSC supports a production planning process and a shortage may exist only when one decides to test a scenario in which there is no possibility to purchase raw materials. We modified the classic CLSC by introduction of a production process. Calculation efficiency of this model was improved by addition of some constraints. The model was optimised for planning purposes for a horizon longer than 12 planning periods. We omit the opening and closing costs of the facilities and use the operational costs of these units. These kinds of costs better reflect the CLSC operational maintenance.

3.2 Model Formulation

Indices

i - index of collection centre, $i = 1..I$; j - index of recovery centre, $j = 1..J$; m - index of redistribution centre, $m = 1..M$; n - index of disposal centre, $n = 1..N$; k - index of market customer zone (returns), $k = 1..K$; l - index of market customer zone (demands), $l = 1..L$; t - index of planning period, $t = 1..T$; r - index of processing centres, $r = 1..R$; f - index of manufacturing centres, $f = 1..F$; g - index of final product, $g = 1..G$; o – index of factory, $o = 1..O$; WB - number of BOM parts.

rp - square matrix of BOM structure. This structure is modelled as the directed acyclic graph $G^{BOM} = (V^{BOM}, A^{BOM})$. The nodes represent the product parts (and a final product). Associated arcs with the values $rp_{w1,w2 \in V^{BOM}} > 0$ indicate that $rp_{w1,w2 \in V^{BOM}}$ units of part $w1$ are needed in the production of each unit of part $w2$.

Parameters

d_{glt} - demand of the customer l for the recovered products g at the time t; r_{gkt} - return of the used products g of the customer k at the time t; c_i^{ColC} - capacity of the collection centre i (we assume a common warehouse area); c_j^{RecC} - capacity of the recovery centre j; c_m^{RedC} - capacity of the redistribution centre m; c_n^{DisC} - capacity of handling the scrapped products at the disposal centre n; f_i^{ColC} - fixed cost of maintenance for the collection centre i; f_j^{RecC} - fixed cost of maintenance for the recovery centre j; f_m^{RedC} - fixed cost of maintenance for the redistribution centre m; a_{gijt}^{ColRec} - shipping cost per unit of recoverable products from the collection centre i to the recovery centre j at the time t; a_{gjmt}^{RecRed} - shipping cost per unit of recovered products from the recovery centre j to the redistribution centre m at the time t; $a_{gmlt}^{RedCout}$ - shipping cost per unit of recovered products from the redistribution centre j to the customer zone l at the time t; a_{gint}^{ColDis} - shipping cost per unit of scrapped products from the collection centre i to the disposal centre n at the time t; a_{gkit}^{CinCol} - shipping cost per unit of returned products from the customer zone k to the collection center i at the time t; a_{girt}^{ColPro} - shipping cost per unit of products from the collection centre i to the processing centre r at the time t; a_{grft}^{ProMan} - shipping cost per unit used in shipping among the processing and manufacturing centres;

a_{ogt}^{PROD} production costs of the product g at the o-th facility at the time t; b_{ogt}^{PROD} - other non-production costs of the product g at the o-th facility at the time t; a_{ogt}^{BLOG} - backlog production costs of the product g at the o-th facility at the time t; π_{gl} - penalty cost per unit of non-satisfied demand of the customer l for the product g; h_t^{CLSC} storage costs at the time t – the same for the CLSC elements; h_t^{PROD} - storage costs at the time t for the production warehouses – the same for the production factories, final products, and raw materials

Variables

X_{gkit}	quantity of returned products g from the zone k to the collection centre i at t
U_{gijt}	quantity of recoverable products from the centre i to the recovery centre j at t
P_{gjmt}	quantity of recovered products g shipped from the recovery centre j to the redistribution centre m at the time t
Q_{gmlt}	quantity of recovered products g shipped from the redistribution centre m to the customer zone l at the time t
E_{gint}	quantity of scrapped products g shipped from the collection centre i to the disposal centre n at the time t
B_{girt}	quantity of returned products g shipped from the collection centre i to the processing centre r at the time t
D_{grftw}	quantity of product parts shipped from the processing centre r to the manufacturing centre f that recovers the part w at the time t
δ_{glt}	quantity of non-satisfied demand of the customer at the time t
Y_{it}	1 if a collection centre is open at the location i at the time t; 0 otherwise
Z_{jt}	1 if a recovery center is open at the location j at the time t; 0 otherwise
W_{mt}	1 if a redistribution centre is open at the location m at the time t; 0 otherwise
s_{git}^{Col}	storage level in the collection centre i
s_{gjt}^{Rec}	storage level in the recovery centre j
s_{gmt}^{Red}	storage level in the redistribution centre m
s_{grwt}^{Proc}	storage level of the part w in the processing centre r

Elements of the production planning process:

X_{ogwt}^{PROD}	production at the time t in the production centre o for the part w of product g
Y_{ogwt}^{PROD}	1 if the production centre o is open at the time t for the part w
P_{ogmt}^{PROD}	demand for the produced items at the time t for the part w
s_{ogwt}^{PROD}	storage level in the production centre o of the part w at the time t
X_{ogt}^{BLOG}	production backlog (a volume of unsatisfied demand in t)
X_{ogwt}^{LACK}	volume of the raw materials w for product g that is deficiency in the factory o

The optimisation function consists of a few elements presented below:

$$FixColC = \sum_{i \in I, t \in T} f_i^{ColC} * Y_{it} \tag{1}$$

$$FixRecC = \sum_{j \in J, t \in T} f_j^{RecC} * Z_{jt} \tag{2}$$

$$FixRedC = \sum_{m \in M, t \in T} f_m^{RedC} * W_{mt} \tag{3}$$

The fixed cost of collection centre maintenance per time unit is shown in Eq. (1). The fixed cost of recovery centre maintenance per time unit is shown in Eq. (2) and the costs of redistribution centre are shown in Eq. (3).

Shipping costs of the products among the CLSC components (see Fig. 3):

$$ShipCostX = \sum_{g \in G, k \in K, i \in I, t \in T} a_{gkit}^{CinCol} * X_{gkit} \tag{4}$$

$$ShipCostU = \sum_{g \in G, i \in I, j \in J, t \in T} a_{gijt}^{ColRec} * U_{gijt} \tag{5}$$

$$ShipCostP = \sum_{g \in G, j \in J, m \in M, t \in T} a_{gjmt}^{RecRed} * P_{gjmt} \tag{6}$$

$$ShipCostQ = \sum_{g \in G, m \in M, l \in L, t \in T} a_{gmlt}^{RedCout} * Q_{gmlt} \tag{7}$$

$$ShipCostE = \sum_{g \in G, i \in I, n \in N, t \in T} a_{gint}^{ColDis} * E_{gint} \tag{8}$$

$$ShipCostB = \sum_{g \in G, i \in I, r \in R, t \in T} a_{girt}^{ColPro} * B_{girt} \tag{9}$$

$$ShipCostD = \sum_{g \in G, r \in R, f \in F, t \in T, w \in WB} a_{grft}^{ProMan} * D_{grftw} \tag{10}$$

These are the shipping costs of products among the CLSC components. Shipping costs among the customers' returning products to the collection centres are shown in Eq. (4). Other shipping costs are depicted in Eqs. (5)-(10).

$$PenaltyCost = \sum_{g \in G, l \in L, t \in T} \pi_{gl} * \delta_{glt} \tag{11}$$

$$ProductionCost = \sum_{o \in O, g \in G, w \in WB, t \in T} \left(a_{ogt}^{PROD} * X_{ogwt}^{PROD} + b_{ogt}^{PROD} Y_{ogwt}^{PROD} \right)$$
$$+ \sum_{o \in O, g \in G, t \in T} \left(a_{ogt}^{BLOG} * X_{ogt}^{BLOG} \right) \tag{12}$$

$$StorageCost = \sum_{g \in G, i \in I, t \in T} h_t^{CLSC} * s_{git}^{Col} + \sum_{g \in G, j \in J, t \in T} h_t^{CLSC} * s_{gjt}^{Rec}$$
$$+ \sum_{g \in G, m \in M, t \in T} h_t^{CLSC} * s_{gmt}^{Red} + \sum_{g \in G, w \in WB, o \in O, t \in T} h_t^{PROD} * s_{ogwt}^{PROD} \tag{13}$$

Penalty costs are greater than zero when demand is greater than the sum of production and recovery. This is shown in Eq. (11). Production costs are shown in Eq. (12). Storage costs of CLSC and PP facilities are shown in Eq. (13). Only production warehouses can have different costs during the planning period.

Optimisation goal:

minimise $FixColC + FixRecC + FixRedC$
$+ ShipCostX + ShipCostU + ShipCostP + ShipCostQ + ShipCostE$
$+ ShipCostB + ShipCostD + StorageCost + ProductionCost + PenaltyCost$

subject to constraints:

$$\sum_{m \in M} Q_{gmlt} = d_{glt} - \delta_{glt}, \forall g \in G, l \in L, t \in T \tag{14}$$

$$\sum_{i \in I} X_{gkit} = r_{gkt}, \forall g \in G, k \in K, t \in T \tag{15}$$

$$\sum_{J \in J} U_{gijt} + \sum_{n \in N} E_{gint} + \sum_{r \in R} B_{girt} + s_{git}^{Col} = \sum_{k \in K} X_{gkit} + s_{gi,t-1}^{Col} \forall g \in G, \\ \forall i \in I, \forall t \in T \tag{16}$$

$$\sum_{n \in N} E_{gint} \geq s^* \sum_{k \in K} X_{gkit}, \forall g \in G, i \in I, t \in T \tag{17}$$

$$\sum_{n \in N} E_{gint} \leq \sum_{k \in K} X_{gkit}, \forall g \in G, i \in I, t \in T \tag{18}$$

$$\sum_{j \in J} P_{gjmt} + s_{gm,t-1}^{Red} + \sum_{o \in O} P_{ogmt}^{PROD} = \sum_{l \in L} Q_{gmlt} + s_{gmt}^{Red}, \forall g \in G, m \in M, t \in T \tag{19}$$

$$X_{ogwt}^{PROD} \leq \sum_{l \in L, tt \in T: tt \geq t} d_{gl,tt} * Y_{ot}^{PROD}, \forall g \in G, w = 1, o \in O, t \in T \tag{20}$$

$$X_{ogwt}^{PROD} \leq \sum_{l \in L, tt \in T: tt \geq t} d_{gl,tt}, \forall g \in G, w = 1, o \in O, t \in T \tag{21}$$

$$\sum_{o \in O} X_{ogwt}^{PROD} \leq \sum_{l \in L, tt \in T: tt \geq t} d_{gl,tt}, \forall g \in G, w = 1, t \in T \tag{22}$$

$$\sum_{o \in O, m \in M} P_{ogmt}^{PROD} + \sum_{j \in J, m \in M} P_{gjmt} \leq \sum_{l \in L, tt \in T: tt \geq t} d_{gl,tt}, \forall g \in G, t \in T \tag{23}$$

$$\sum_{o \in O, m \in M} P_{ogmt}^{PROD} + \sum_{j \in J, m \in M} P_{gjmt} \geq \sum_{l \in L, tt \in T: tt \geq t} d_{gl,tt}, \forall g \in G, t \in T \tag{24}$$

$$\sum_{m \in M} P_{gjmt} + s_{gjt}^{Rec} = \sum_{i \in I} U_{gijt} + s_{gj,t-1}^{Rec}, \forall g \in G, j \in J, t \in T \tag{25}$$

$$\sum_{k \in K} X_{gkit} \leq Y_{it} * c_i^{ColC}, \forall g \in G, i \in I, t \in T \tag{26}$$

$$\sum_{i \in I} U_{gijt} \leq Z_{jt} * c_j^{RecC}, \forall g \in G, j \in J, t \in T \tag{27}$$

$$\sum_{j \in J} P_{gjmt} \leq W_{mt} * c_m^{RedC}, \forall g \in G, m \in M, t \in T \tag{28}$$

$$\sum_{i \in I} E_{gint} \leq c_n^{DisC}, \forall g \in G, n \in N, t \in T \tag{29}$$

$$\sum_{i \in I} \mu_{gw} B_{girt} + s_{grw,t-1}^{Proc} = \sum_{f \in F} D_{grftw} + s_{grwt}^{Proc}, \forall g \in G, r \in R, t \in T \tag{30}$$

$$s_{ogw,t-1}^{PROD} + X_{ogwt}^{PROD} = \sum_{m \in M} P_{ogmt}^{PROD} + s_{ogwt}^{PROD}, \forall g \in G, o \in O, w = \{1\}, t \in T \tag{31}$$

$$s_{ogw,t-1}^{PROD} + X_{ogwt}^{PROD} = \sum_{w1 \in WB} \left(rp_{w1,w} * X_{ogwt}^{PROD} \right) + s_{ogwt}^{PROD}, \forall g \in G, \forall o \in O,$$
$$w \in \{3\}, \forall t \in T \tag{32}$$

$$s_{ogw,t-1}^{PROD} + X_{ogwt}^{PROD} + X_{ogwt}^{LACK} = \sum_{w1 \in WB} \left(rp_{w1,w} * X_{ogwt}^{PROD} \right) + s_{ogwt}^{PROD}, \forall g \in G, o \in O,$$
$$w \in \{2, 4\}, t \in T \tag{33}$$

$$X_{ogwt}^{PROD} = \sum_{f \in F, r \in R} D_{grftw}, \forall g \in G, o = \{1\}, w \in \{2, 4\}, t \in T. \tag{34}$$

The number of products recovered and produced Eq. (14) is equal to the difference between demand and non-satisfied demand (for any time unit). The quantity of returned products is equal to the quantity of products sent to the collection centres Eq. (15).

A balance for a flow of products between the following periods is modelled by Eq. (16). In this equality, for the collection centres, the quantity of products collected in the collection centres is equal to the quantity of products sent later to the recovery, processing and the disposal centres (taking into account the storage levels of products in the collection centre warehouses at t-1 and t).

The quantity of products sent to the disposal centres from the collection centres is greater or equal to s * 100%. s is set up as an a-priori known parameter which is shown in Eq. (17). The quantity of products sent to the disposal centres from the collection centres is less than or equal to the quantity of products collected Eq. (18).

The sum of the production and recovery activities must be equal to the distribution, taking into account the warehouse levels, which is shown in Eq. (19). The quantity of recovered products shipped from the recovery centres to the redistribution centres plus the quantity of produced goods shipped to the redistribution centres is equal to the quantity of products sent to the clients.

Equality for the production centres is shown in Eq. (20). Production should be equal to demand, taking into account the storage levels of the production warehouses. Additional constraints can be added to tighten the model formulation.

At the time t, the quantity of produced goods is less than or equal to the quantity of demands for the remaining time units, which is shown in Eq. (20). Equation (21) is a technical inequality that improves the solver searching process. This inequality is almost the same as in Eq. (20), but the 0-1 variable was omitted. Equation (22) is a technical inequality that improves the solver searching process.

The sum of quantity of produced goods and the recovered goods at t is less than or equal to the demands for the remaining time periods, which is shown in Eq. (23). Equation (24) shows that the sum of the quantity of produced goods and recovered goods at t is greater than or equal to the demands at t.

Equation (25) means that a quantity of products recovered and sent to the distribution centres is equal to the quantity of products sent to the recovery centres, taking into account the storage levels at t-1 and t.

Equations (26)–(29) constrain the centre's capacities. These capacities are modelled as the spaces needed for collection, remanufacturing, and distribution activities, and are different to the storage capacities of these centres.

The decomposition process of a returned product into parts is presented in Eq. (30). μ_{gw} is the number of parts of the type w that can be removed from the product g. For reader convenience, we omit constraints that describe the capacity of the manufacturing and processing centres.

The production planning model consists of two parts: production planning constraints for the components and production planning constraints for the final product. Equation (31) shows the equality for the production centres – final product (index w=1 describes the final product). Equation (32) shows the equality for the production of components (formulation depends on BOM structure). In Eq. (33), the analogous constraint is presented, but it describes the lowest level of BOM. The variable X_{ogwt}^{LACK} is introduced that describes the purchases of raw materials from the external sources (see Fig. 4 and BOM structure in Fig. 3).

Equation (34) shows a situation where remanufactured materials are sent only to one predefined factory. In a general case, this should be modified but in the model presented it reflects a real business case we know.

4   Results and Discussion

Several numerical experiments were prepared in order to check the feasibility and performance of the presented models. The most important results are reported in this section. First, we present the numerical results for SCM model with 6 military training areas treated as the SCM clients. In the paper, a military training area plays a role of a client and depicts an installation of a laser training system. For example, in Poland there will be about 6 installations.

We assumed in our experiments different numbers of SCM components. In the paper we present results for the SCM system with 3 collection centres (CC), 3 distribution centres (DC), 1 recovery centre (RC), 2 disposal centres and the one set of processing centre (PC), remanufacturing centre (RF) and production centre (PR) (see Fig. 4). The locations of some SCM components were predefined according to the localisations of military grounds and the company's partners. In Fig. 4 we presented the results for 52 planning periods (weeks) what means that the model was calculated for one year for all military grounds. Results presented in Fig. 5 describe the product flows for BOM no 1 – training harness (final product BH1).

Because in presented SCM some harnesses could be dismantled and the components could be used in new products so the processing centre, remanufacturing centre and production centre are present. A flow of harness parts are presented in Fig. 3. Number of elements sent among SCM components depends on BOM structure (see Fig. 2).

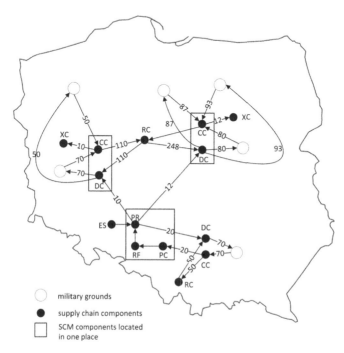

Fig. 4. Product (BH1) flow for 52 planning periods. Number of final product pieces (harness BH1) sent among the SCM components are placed on the arcs of SCM net.

Next, we calculated the production planning problem for the varying demands and the different time periods (13,26,52 weeks). First results are presented in Table 1. The most important factor influencing the results was the SCM structure, i.e. the numbers of collection, recovery, and distribution centres. A number of production facilities was also important for a computational efficiency, because the plans for these facilities were established in correspondence to BOMs. All the experiments were conducted with IBM CPLEX solver [4].

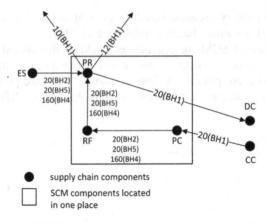

Fig. 5. Product (BH1) flow from a processing centre to a production centre.

In Table 1 we presented the optimisation results of PP model with CLSC for 6 military grounds and SCM elements presented in Fig. 4.

Table 1. Optimisation results for different PP with CLSC problems.

No.	Problem size $\|I\| \times \|J\| \times \|M\| \times \|N\| \times \|K\| \times \|L\| \times \|O\| \times \|R\| \times \|F\|$	Planning period	No of variables	Number of constraints	Optimisation time [s]
1	$3 \times 2 \times 3 \times 2 \times 6 \times 6 \times 1 \times 1 \times 1$	4	408	296	8
2	$3 \times 2 \times 3 \times 2 \times 6 \times 6 \times 1 \times 1 \times 1$	9	918	666	25
3	$3 \times 2 \times 3 \times 2 \times 6 \times 6 \times 1 \times 1 \times 1$	13	1326	962	40
4	$3 \times 2 \times 3 \times 2 \times 6 \times 6 \times 1 \times 1 \times 1$	26	2652	1924	140
5	$3 \times 2 \times 3 \times 2 \times 6 \times 6 \times 1 \times 1 \times 1$	39	3978	2886	320
6	$3 \times 2 \times 3 \times 2 \times 6 \times 6 \times 1 \times 1 \times 1$	52	5304	3848	763

One should notice that the extended formulation leads to more efficient calculations, which is the most important goal of the task. The best efficiency was obtained with the improved model when additional constraints were added to set up a lower bound of the production storage at the time t. This simple modification introduced only at the last phase of production let us to speed up the calculations by about 18% provided that there are differences in production costs in corresponding periods. One should also remember that this type of task is usually solved many times, hence the time to find a solution is critical.

5 Conclusions

In the paper, a production planning model supported by CLSC is presented for the military laser simulation system. The model describes a CLSC network that consists of the closed loop elements and the production elements (with the open loop components). In the paper, a recovery process in CLSC is a supporting activity. So, a production process is the leading element that decides whether customer demands are satisfied. Such assumption is justified because a client usually prefers new equipment than the used one. However, during a life cycle of the laser system, some devices must be remanufactured to reduce a cost of maintenance.

Our calculations have been performed on the models with the longer time horizon, which has had a great impact on the efficiency of a solution finding process. Our model is suitable for use in production planning with horizons up to 12 months (52-week time periods). We have used a CPLEX solver to solve the optimisation problem which has been based on an IP model.

We have described some modifications of the standard OLSC and CLSC models in to order to lower the computing time. The very important factor hampering the calculations was that the presented model was multi-item multi-level production. These additional constraints make the calculations complicated when one takes into account possible disturbances.

References

1. Amin, S.H., Zhang, G.: A multi-objective facility location model for closed-loop supply chain network under uncertain demand and return. Appl. Math. Model. **37**, 4165–4176 (2013)
2. Graves, S.C.: Manufacturing Planning and Control. Handbook of Applied Optimization, pp. 728–746. Oxford University Press, New York (2002)
3. Guide, V.D.R., Harrison, T.P., Van Wassenhove, L.N.: The challenge of closed loop supply chains. Interfaces **33**(6), 3–6 (2003)
4. IBM Homepage: IBM ILOG CPLEX User's manual for CPLEX (2016). https://www.ibm.com/support/knowledgecenter/SSSA5P_12.7.0/ilog.odms.studio.help/pdf/usrcplex.pdf
5. Krikke, H.R.R., Pappis, C.P., Tsoulfas, G.T., Bloemhof-Ruwaard, J.M.: Design principles for closed loop supply chains. Technical report ERS-2001-62-LIS ERIM (Erasmus Research Institute of Management) Report Series Reference (2001)
6. Pishvaee, M.S., Jolai, F., Razmi, J.: A stochastic optimization model for integrated forward/reverse logistics network design. J. Manuf. Syst. **28**(4), 107–114 (2009)
7. Pishvaee, M.S., Rabbani, M., Torabi, S.A.: A robust optimization approach to closed-loop supply chain network design under uncertainty. Appl. Math. Model. **35**, 637–649 (2011)
8. Shapiro, J.F.: Mathematical programming models and methods for production planning and scheduling. In: Handbooks in Operations Research and Management Science Logistics of Production and Inventory, vol 4, pp. 371–443. Elsevier Science Publishers, Amsterdam (1993)
9. Stecz, W.: Production optimization with closed loop supply chain support. Res. Logist. Prod. **6**(2), 117–128 (2016)

10. Fathollahi-Farda, A.M., Hajiaghaei-Keshtelia, M., Mirjalili, S.: Hybrid optimizers to solvea tri-level programming model for a tire closed-loop supply chain network designproblem. Appl. Soft Comput. **70**, 701–722 (2018)
11. Darestani, S.A., Hemmati, M.: Robust optimization of a bi-objective closed-loop supply chain network for perishable goods considering queue system. Comput. Ind. Eng. **136**, 277–292 (2019)

Imbalanced Data: Rough Set Methods in Approximation of Minority Classes

Jaroslaw Stepaniuk[(✉)]

Faculty of Computer Science, Bialystok University of Technology,
Wiejska 45A, 15-351 Bialystok, Poland
j.stepaniuk@pb.edu.pl

Abstract. The imbalanced data problem turned out to be one of the most important and challenging problems in artificial intelligence. We discuss an approach of minority class approximation based on rough set methods and three-way decision. This approach seems to be more general than the traditional one. However, it requires developing some new logical tools for reasoning based on rough sets and three-way decision, which is often expressed in natural language.

Keywords: Imbalanced data · Minority class approximation · Rough sets · Three-way decision · Granular computing

1 Introduction

More than twenty five years ago the imbalanced data problem turned out to be one of the most important and challenging problems [1]. Indeed, missing information about the minority class leads to a significant degradation in classifier performance. Moreover, comprehensive research has proved that there are certain factors increasing the problem's complexity. These additional difficulties are closely related to the data distribution over decision classes. In spite of numerous methods which have been proposed, the flexibility of existing solutions needs further improvement.

In this paper, we discuss an approach of the minority class approximation based on partial inclusion of granules as some kind of judgment [2]. In the judgment process, the arguments *for* and *against* the hypothesis about membership of the perceived case to a given concept are collected. In this way, the results of judgment clearly indicating that a given case belongs to one of the regions e.g., lower approximation, boundary region, or complement to the upper approximation are obtained (see also three-way decision [3]). We consider information granules as the computational building blocks that are necessary for cognition. Our approach is consistent with the opinion of L. Valiant [4]:

A fundamental question for artificial intelligence is to characterize the computational building blocks that are necessary for cognition. A specific challenge is to build on the success of machine learning so as to cover

K. Saeed and J. Dvorský (Eds.): CISIM 2020, LNCS 12133, pp. 451–460, 2020.
https://doi.org/10.1007/978-3-030-47679-3_38

broader issues in intelligence. This requires, in particular a reconciliation between two contradictory characteristics–the apparent logical nature of reasoning and the statistical nature of learning.

The paper is structured as follows. In Sect. 2, we present the rough set approach towards minority class approximation in the context of k-nearest neighbors approximation space. In Sect. 3, we discuss some aspects for the deeper reasoning about imbalanced data. In Sect. 4, we discuss judgment and three-way decision related to imbalanced data sets.

2   Minority Class Approximation in a k-Nearest Neighbors Approximation Space: Rough Set Approach

In the section, we present an approximation space based on a k-nearest neighbors classifier. We illustrate this notion with three-way decision strategies related to imbalanced data classification.

A k-nearest neighbors approximation space (more general cases are considered, *e.g.*, in articles [5,6]) can be defined by a tuple $AS_k = (U, NN_k, \nu)$, where U is a non-empty set of objects, NN_k is a function defined on U with values in the powerset $\mathcal{P}(U)$ of U ($NN_k(x)$ is the set of k-nearest neighbors of $x \in U$) and ν is the *inclusion function* defined on the Cartesian product $\mathcal{P}(U) \times \mathcal{P}(U)$ with values in the interval $[0, 1]$ measuring the degree of inclusion of sets [6]. For $X \subseteq U$ the lower and upper approximation operations can be defined in AS_k by

$$LOW_{AS_k}(X) = \{x \in U : \nu(NN_k(x), X) = 1\}, \tag{1}$$

$$UPP_{AS_k}(X) = \{x \in U : \nu(NN_k(x), X) > 0\}. \tag{2}$$

For $X, Y \subseteq U$ the standard rough inclusion relation ν_{SRI} is defined by

$$\nu_{SRI}(X, Y) = \begin{cases} \dfrac{card(X \cap Y)}{card(X)}, & \text{if } X \text{ is non} - \text{empty}, \\ 1, & \text{otherwise.} \end{cases} \tag{3}$$

Another example of rough inclusion function ν_t can be defined using the standard rough inclusion and a threshold $t \in (0, 0.5)$ using the following formula:

$$\nu_t(X, Y) = \begin{cases} 1 & \text{if } \nu_{SRI}(X, Y) \geq 1 - t \\ \frac{\nu_{SRI}(X,Y)-t}{1-2t} & \text{if } t \leq \nu_{SRI}(X, Y) < 1 - t \\ 0 & \text{if } \nu_{SRI}(X, Y) \leq t \end{cases} \tag{4}$$

This strategy can be described in natural language as follows. Due to uncertainty, it is not possible to perceive objects exactly. The objects are perceived using information about them represented by vectors of attribute values. These vectors define some k-nearest neighbors of objects. In making decision we use these k-nearest neighbors of objects to judge membership of a perceived object into a minority class. The result of judgment is based on the degree to which the k-nearest neighbors are included into the minority class.

Example 1. Let U be a set of objects and let $d : U \to V_d$ be a decision attribute with $V_d = \{+, -\}$ as the set of values. Let an approximation space $AS_{k,t}$ can be defined by a tuple $AS_{k,t} = (U, NN_k, \nu_t)$. For two decision classes minority class $X_{d=+} = \{x \in U : d(x) = +\}$, and majority class $X_{d=-} = \{x \in U : d(x) = -\}$, where $X_{d=+} \cup X_{d=-} = U$, we define

$$LOW_{AS_{k,t}}(X_{d=+}) = \{x \in U : \nu_t(NN_k(x), X_{d=+}) = 1\},$$

$$UPP_{AS_{k,t}}(X_{d=+}) = \{x \in U : \nu_t(NN_k(x), X_{d=+}) > 0\},$$

$$BOUNDARY_{AS_{k,t}}(X_{d=+}) = UPP_{AS_{k,t}}(X_{d=+}) - LOW_{AS_{k,t}}(X_{d=+}),$$

$$NEGATIVE_{AS_{k,t}}(X_{d=+}) = U - UPP_{AS_{k,t}}(X_{d=+}).$$

We can label objects $x \in U$ as in Table 1.

Table 1. The judgment about which label to assign to the object $x \in U$.

$Label_{k,t}(x)$	Argument "for" $\nu_t(NN_k(x), X_{d=+})$	Argument "against" $\nu_t(NN_k(x), X_{d=-})$
$LOWER$	$=1$	$\neq 1$
$BOUNDARY$	$\in (0,1)$	$\notin (0,1)$
$NEGATIVE$	$=0$	$\neq 0$

A similar strategy, enhanced by conflict resolution, is also widely used in inducing classifiers from decision tables (training samples).

3   Partition of Minority Class in Imbalanced Data

In this section we discuss an approximation (partition) of minority class in imbalanced data tables. We extend a strategy of labeling objects from the minority class introduced in [7,8].

The idea of formation of information granule facilitates splitting the problem into more feasible subtasks. Then, they can be easily managed by applying appropriate approaches, dedicated to specific types of entities. After defining groups of similar instances $NN_k(x)$ (namely minority class objects $x \in X_{d=+}$ and their k nearest neighbors $NN_k(x)$), the inclusion degree of each information granule $NN_k(x)$ in $X_{d=+}$ is examined. Based on this analysis, $Label(x)$, the labels are assigned to all positive examples $x \in X_{d=+}$.

We assume that evaluation of information granules is crucial for further processing. Before applying oversampling mechanism, each information granule, defined by $NN_k(x)$ having positive instance x as the anchor point, is labeled with one of the following etiquettes: $SAFE, BOUNDARY$ and $NOISE$.

The category of the individual entity is determined by the inclusion degree of $NN_k(x)$ in the information granule $X_{d=+}$ (the whole minority class).

Details of the proposed technique are presented in Eq. 5 and explained below.

$$Label_{k,t}(x) = \begin{cases} SAFE, & \text{if } \nu_t\left(NN_k(x), X_{d=+}\right) = 1 \\ BOUNDARY, & \text{if } 0 < \nu_t\left(NN_k(x), X_{d=+}\right) < 1 \\ NOISE, & \text{if } \nu_t\left(NN_k(x), X_{d=+}\right) = 0 \end{cases} \quad (5)$$

where $t \in (0, 0.5)$.

– Etiquette $Label_{k,t}(x) = SAFE$ for $x \in X_{d=+}$
A high inclusion degree indicates that the information granule $NN_k(x)$ is placed in a homogeneous area and therefore x can be considered as $SAFE$. The inclusion level is obtained by an analysis of granule characteristics, especially cardinalities of instances from both classes. The number of positive class representatives belonging to the analyzed entity (except the anchor example) i.e. $card(NN_k(x) \cap X_{d=+})$ is compared to the number of negative class instances i.e. $card(NN_k(x) \cap X_{d=-})$ (see Table 2 for $k = 7$ and $t = 0.3$).
– Etiquette $Label_{k,t}(x) = BOUNDARY$ for $x \in X_{d=+}$
A low inclusion degree is determined by a large representation of the majority class $X_{d=-}$ in the information granule $NN_k(x)$. These kind of entities are placed in the area surrounding class boundaries, where examples from both classes overlap (see Table 2 for $k = 7$ and $t = 0.3$).
– Etiquette $Label_{k,t}(x) = NOISE$ for $x \in X_{d=+}$
A very low inclusion degree of the information granule $NN_k(x)$ in the minority class $X_{d=+}$ is identified with the situation where the information granule is created around the rare individual placed in the area occupied by representatives of the negative class $X_{d=-}$. This case is considered in Table 2 for $k = 7$ and $t = 0.3$).

The example of labeling instances from the minority class is presented in Table 2. It shows all possible cases of the minority class instance's neighborhood.

Assuming that the k parameter is equal to 7, the second column presents the number of nearest neighbors belonging to the same class as the instance under consideration, and the third column shows the number of nearest neighbors representing the opposite class.

Example 2. Consider data consisting of 1000 objects, i.e. $card(U) = 1000$, where there are 950 objects in the majority class, i.e. $card(X_{d=-}) = 950$ and there are 50 objects in the minority class, i.e. $card(X_{d=+}) = 50$. Let $x, y \in X_{d=+}$ i.e. the decision $d(x) = d(y) = +$. Let $x_1, x_2, x_3, x_4, x_5, x_6, x_7$ be seven nearest neighbors of x i.e. $NN_7(x) = \{x_1, x_2, x_3, x_4, x_5, x_6, x_7\}$. Let us assume that $d(x_1) = -$, $d(x_2) = +, d(x_3) = +, d(x_4) = -, d(x_5) = -, d(x_6) = +$ and $d(x_7) = +$. We obtain $card(NN_7(x) \cap X_{d=+}) = card(\{x_1, x_2, x_3, x_4, x_5, x_6, x_7\} \cap X_{d=+}) = card(\{x_2, x_3, x_6, x_7\}) = 4$ and $card(NN_7(x) \cap X_{d=-}) = card(\{x_1, x_2, x_3, x_4, x_5, x_6, x_7\} \cap X_{d=-}) = card(\{x_1, x_4, x_5\}) = 3$. Hence

$$\nu_{SRI}\left(NN_7(x), X_{d=+}\right) = \frac{card(NN_7(x) \cap X_{d=+})}{card(NN_7(x))} = \frac{4}{7}$$

Table 2. Identification of the type of the minority class instance in case of $k = 7$ nearest neighbors and $t = 0.3$.

$Label_{7,0.3}(x)$	$card(NN_7(x) \cap X_{d=+})$	$card(NN_7(x) \cap X_{d=-})$	$\nu_{0.3}(NN_7(x), X_{d=+})$
$SAFE$	7	0	1
	6	1	1
	5	2	1
$BOUNDARY$	4	3	0.69
	3	4	0.32
$NOISE$	2	5	0
	1	6	0
	0	7	0

and

$$\nu_{0.3}(NN_7(x), X_{d=+}) = \frac{\nu_{SRI}(NN_7(x), X_{d=+}) - 0.3}{1 - 0.6} = \frac{\frac{4}{7} - 0.3}{1 - 0.6} = 0.69.$$

We conclude that the correct label for x is $Label_{7,0.3}(x) = BOUNDARY$. Let $NN_7(y) = \{x_1, x_2, x_3, x_6, x_7, x_8, x_9\}$, where $d(x_8) = +$ and $d(x_9) = +$. We obtain $card(NN_7(y) \cap X_{d=+}) = card(\{x_1, x_2, x_3, x_6, x_7, x_8, x_9\} \cap X_{d=+}) = card(\{x_2, x_3, x_6, x_7, x_8, x_9\}) = 6$ We conclude that the correct label for y is $Label_{7,0.3}(y) = SAFE$.

After categorizing instances from minority class $X_{d=+}$, the mode of algorithm for oversampling is obtained. Three methods are proposed in [7,8] to deal with various real–life data characteristics. They mainly depend on the number of information granules labeled as $BOUNDARY$. Assuming that a certain threshold value is one of the parameters of the algorithm , the complexity of the problem is defined based on this value and the number of granules recognized as $BOUNDARY$. The threshold indicates how many instances of the entire minority class should be placed in boundary regions to treat the problem as a complex one.

Having less $BOUNDARY$ entities, $i.e.$,

$$\frac{card(\{x \in X_{d=+} : Label_{k,t}(x) = BOUNDARY\})}{card(X_{d=+})} < complexity\_threshold$$

means that the problem is not complex and the following method of creating new instances can be applied:

Low Complexity mode (see also [7])
Low Complexity mode for obtaining balanced decision table $DT_{balanced}$ from the table DT: $DT \longmapsto^{LowComplexity} DT_{balanced}$

- $Label_{k,t}(x) = SAFE$: there is no need to significantly increase the number of instances in these safe areas. Only one new instance per existing minority

SAFE instance is generated. Numeric attributes are handled by the interpolation with one of the k nearest neighbors. For the nominal features, new sample has the same values of attributes as the instance under consideration.

- $Label_{k,t}(x) = BOUNDARY$: the most of synthetic samples are generated in these borderline areas, since numerous majority class representatives may have greater impact on the classifier learning, when there are not enough minority examples. Hence, many new examples are created closer to the instance x under consideration. One of the k nearest neighbors is chosen for each new sample when determining the value of numeric feature. Values of nominal attributes are obtained by the majority vote of k nearest neighbors' features.
- $Label_{k,t}(x) = NOISE$: no new samples are created.

Example 3. This didactic example is a continuation of the Example 2. Let us assume that *complexity_threshold* $= 0.2$ and there are eight information granules labeled as *BOUNDARY* i.e.

$$card(\{x \in X_{d=+} : Label_{7,0.3}(x) = BOUNDARY\}) = 8.$$

We obtain that the problem is not complex with respect to selected threshold:

$$\frac{card(\{x \in X_{d=+} : Label_{7,0.3}(x) = BOUNDARY\})}{card(X_{d=+})} = \frac{8}{50} = 0.16 < 0.2.$$

Let us consider two objects $x, y \in X_{d=+}$ as in Example 2. Only one new instance per existing minority *SAFE* instance y ($Label_{7,0.3}(y) = SAFE$) is generated. Numeric attributes are handled by the interpolation with one of the 7 nearest neighbors. One chooses one minority class sample among the 7 neighbors of y e.g. x_2. Finally, one generates the synthetic sample y_{new} by interpolating between x_2 and y as follows:

$$y_{new} = y + rand(0,1) \times (x_2 - y),$$

where $rand(0,1)$ refers to a random number between zero and one. For object x, where $Label_{7,0.3}(x) = BOUNDARY$ and

$$NN_7(x) \cap X_{d=+} = \{x_1, x_2, x_3, x_4, x_5, x_6, x_7\} \cap X_{d=+} = \{x_2, x_3, x_6, x_7\}$$

four synthetic objects are generated.

On the other hand, prevalence of *BOUNDARY* information granules, *i.e.*,

$$\frac{card(\{x \in X_{d=+} : Label_{k,t}(x) = BOUNDARY\})}{card(X_{d=+})} \geq complexity\_threshold$$

involves more complications during the learning process. Therefore, dedicated approach described below is chosen:

High Complexity mode (see also [7])
High Complexity mode for obtaining balanced decision table $DT_{balanced}$ from DT table: $DT \longmapsto^{HighComplexity} DT_{balanced}$

– $Label_{k,t}(x) = SAFE$: assuming that these concentrated instances provide specific and easy to learn patterns that enable proper recognition of minority samples, plenty of new data is created by interpolation between SAFE instance and one of its k nearest neighbors. Nominal attributes are determined by majority vote of k nearest neighbors' features.

– $Label_{k,t}(x) = BOUNDARY$: the number of instances is doubled by creating one new example along the line segment between half of the distance from $BOUNDARY$ instance and one of its k nearest neighbors. For nominal attributes values describing the instance under consideration are replicated.

– $Label_{k,t}(x) = NOISE$: new examples are not created.

If $\{x \in X_{d=+} : Label_{k,t}(x) = SAFE\} = \emptyset$, then the following method is applied:

No Safe mode (see also [7])

No Safe mode: $DT \longmapsto^{NoSafe} DT_{balanced}$

– $Label_{k,t}(x) = BOUNDARY$: all of the synthetic instances are created in the area surrounding class boundaries. This particular solution is selected in case of especially complex data distribution, which does not include any SAFE samples. Missing $SAFE$ elements indicate that most of the examples are labeled as $BOUNDARY$ (there are no homogeneous regions). Since only $BOUNDARY$ and $NOISE$ examples are available, only generating new instances in the neighborhood of $BOUNDARY$ objects would provide sufficient number of minority samples.

– $Label_{k,t}(x) = NOISE$: no new instances are created.

$NOISE$ granules are completely excluded from the preprocessing phase, since their anchor instances are erroneous examples or outliers. Therefore, they should not be removed, but they also should not be taken into consideration when creating new synthetic instances to avoid more inconsistencies.

Looking on the above formalization of labeling of objects from the minority class, one can find possible judgment strategy which can be easily expressed in a fragment of natural language. However, in real-life applications, the situation may be much more complex. For example, one can ask about the risk of making decision based on such modeling. For example, labels $SAFE$ or $NOISE$ are related to complex vague concepts. In the real-life medical applications, the judgment leading to decision about labeling by $SAFE$ and $NOISE$ may require interactions with domain knowledge, which can be the experience or recent discoveries in medicine reported in the literature, or can be additional testing of objects or performing some other actions on them. Some advanced judgments strategies should be used to obtain relevant information for making proper decision.

4 Judgment About Minority Class and Three-Way Decision

Three-way decision play a key role in everyday decision-making and have been widely used in many fields and disciplines [3].

Suppose *Universe* is a finite nonempty set and *Criteria* is a finite set of criteria. The problem of three-way decisions is to divide, based on the set of criteria *Criteria*, *Universe* into three pair-wise disjoint regions.

Example 4. In Sect. 2 we consider a universe U of all objects and three pairwise disjoint regions called the lower, boundary, and negative regions, respectively (see Fig. 1).

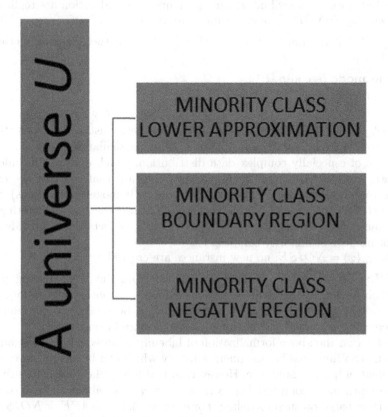

Fig. 1. Partition of a universe U into three pairwise disjoint regions

Example 5. In Sect. 3 we consider *Universe* as the set $X_{d=+}$ of all positive examples (the set of minority class objects). *Criteria* are based on the inclusion degree. Every minority class instance belongs to exactly one region called the safe $\{x \in X_{d=+} : Label_{k,t}(x) = SAFE\}$, boundary $\{x \in X_{d=+} : Label_{k,t}(x) = BOUNDARY\}$, and noise $\{x \in X_{d=+} : Label_{k,t}(x) = NOISE\}$ region, respectively (see Fig. 2).

Corresponding to the three regions, one may construct rules for three-way decision how to create new synthetic minority class objects.

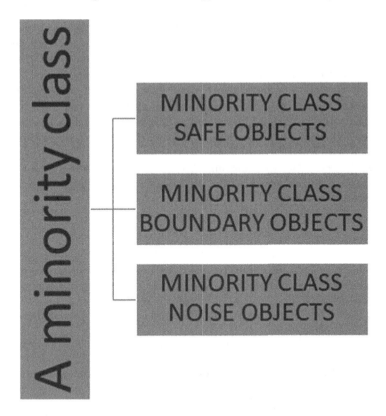

Fig. 2. Partition of a minority class $X_{d=+}$ into three pairwise disjoint regions

From the above discussion it follows that judgment is performed on the basis of information granules representing the current result of perception of the situation. On the basis of this result, the judgment resulting in selection of the most relevant action(s) (with respect to the current goals) in the perceived situation is performed. By performing actions the complex granule about the current situation is updated and again the judgment leads to the selection of next action(s). In this way computations on complex granules are controlled by actions aiming to preserve the required constraints. For example, the constraints related to imbalanced data classification tasks may be seen as reaching conclusion about the membership of perceived situation in the considered minority class.

5   Conclusions and Future Research

We discussed two approaches to minority class approximation. Some didactic illustrative examples are included. Both approach are pointing out the necessity of developing new methods for modeling of judgment in imbalanced data sets.

In neighborhoods different types of objects from the minority class such as *safe*, *borderline*, *rare examples*, and *outliers* are also considered [9]. In the future we plan to investigate in more detail the methods how to distinguish the outliers from the noise. In [9] it is emphasized that the results of the noise identification by filters are often identified by medical experts as valid outliers. Hence, it is visible that to provide a decision support system some more advanced reasoning tools, which we call judgment are required (see [10]). These reasoning tools should help the system to judge properly about such perceived cases. It should be noted that this judgment should be supported by the relevant information about perceived cases extracted from knowledge bases representing experience.

Acknowledgments. The work was supported by the grant from Bialystok University of Technology and funded with resources for research by the Ministry of Science and Higher Education in Poland.

References

1. Fernández, A., García, S., Galar, M., Prati, R.C., Krawczyk, B., Herrera, F.: Learning from Imbalanced Data Sets. Springer, Cham (2018). https://doi.org/10.1007/978-3-319-98074-4
2. Smith, B.C.: The Promise of Artificial Intelligence: Reckoning and Judgment. MIT Press, Cambridge (2019)
3. Yao, Y.Y.: Three-way decision and granular computing. Int. J. Approximate Reasoning **103**, 107–123 (2018)
4. Valiant, L.: https://people.seas.harvard.edu/valiant/researchinterests.htm
5. Skowron, A., Stepaniuk, J.: Approximation spaces in rough-granular computing. Fundamenta Informaticae **100**, 141–157 (2010)
6. Skowron, A., Stepaniuk, J., Swiniarski, R.: Modeling rough granular computing based on approximation spaces. Inf. Sci. **184**, 20–43 (2012)
7. Borowska, K., Stepaniuk, J.: Granular computing and parameters tuning in imbalanced data preprocessing. In: Saeed, K., Homenda, W. (eds.) CISIM 2018. LNCS, vol. 11127, pp. 233–245. Springer, Cham (2018). https://doi.org/10.1007/978-3-319-99954-8_20
8. Borowska, K., Stepaniuk, J.: A rough-granular approach to the imbalanced data classification problem. Appl. Soft Comput. **83**, 105607 (2019). https://doi.org/10.1016/j.asoc.2019.105607
9. Napierala, K., Stefanowski, J.: Types of minority class examples and their influence on learning classifiers from imbalanced data. J. Intell. Inf. Syst. **46**(3), 563–597 (2015). https://doi.org/10.1007/s10844-015-0368-1
10. Stepaniuk, J., Góra, G., Skowron, A.: Concept approximation based on rough sets and judgment. In: Mihálydeák, T., Min, F., Wang, G., Banerjee, M., Düntsch, I., Suraj, Z., Ciucci, D. (eds.) IJCRS 2019. LNCS (LNAI), vol. 11499, pp. 16–27. Springer, Cham (2019). https://doi.org/10.1007/978-3-030-22815-6_2

Run-Time Schedule Adaptation Methods for Sensor Networks Coverage Problem

Krzysztof Trojanowski⑩, Artur Mikitiuk$^{(\boxtimes)}$⑩, and Jakub A. Grzeszczak

Cardinal Stefan Wyszyński University in Warsaw, Warsaw, Poland
{k.trojanowski,a.mikitiuk}@uksw.edu.pl

Abstract. In this paper, we study run-time adaptation methods of a schedule of sensor activity generated for ideal temperature conditions. Such a schedule cannot be completed in low-temperature conditions due to a shorter lifetime of sensor batteries. We proposed several methods of selecting the next slot to be executed when the currently scheduled slot is unfeasible. Our experiments showed that in most cases, the best method is the one selecting the next slot based on computing the mean and standard deviation values of the battery load level for all the sensors active in a given slot.

Keywords: Maximum lifetime coverage problem · Schedule optimization · Adaptation to varying temperature

1 Introduction

Wireless sensor networks consist of devices deployed within the target area that gather and transfer information from the monitored field. In this research, we assume that devices are immobile and have the same sensing range. The devices are battery-powered, and the capacity of batteries defines network lifetime. Batteries are fully charged at the beginning of the monitoring. However, outdoor conditions are often far from laboratory ones. Notably, in the low temperature, battery performance drops significantly, and the lifetime of sensors shrinks. Thus, the network cannot guarantee that sensors cover an appropriate percentage of the monitored field (which is called the required level of coverage) over its entire lifetime. In many networks, the number of sensors is high, and sensing regions overlap. Monitoring redundancy creates an opportunity for optimization by turning off the redundant sensors. Typically, the optimization of the network activity schedule is computationally expensive. Therefore, it is executed before the network starts and for fixed working conditions. Further rearrangement of sensors' activity schedules performed in the network run-time and concerning the current temperature is desirable because it may enlarge significantly the period between activation of the network and the first moment when the level of coverage drops below the required threshold. The problem of a schedule adaptation to varying temperature conditions is the subject of the research in this paper.

© Springer Nature Switzerland AG 2020
K. Saeed and J. Dvorský (Eds.): CISIM 2020, LNCS 12133, pp. 461–471, 2020.
https://doi.org/10.1007/978-3-030-47679-3_39

In our research, the time is discrete, and thus every schedule defines sensors' activity for time units of the same length called slots. Slots represent short periods, so we can assign a temperature to every slot that is constant over the entire period. The temperature influences sensors' energy consumption, thus changes their lifetime. Input schedules define sensors activity in subsequent slots assuming constant temperature. The application of varying temperatures to slots verifies their feasibility due to the non-uniform battery discharge levels. In low-temperature conditions, when the battery performance drops, the last slots in the schedule cannot guarantee sufficient coverage of the area by active monitoring devices. We are interested in finding the most extended uninterrupted sequence of fully operational slots. Thus, we are interested in the rearrangement of the schedule content to maximize the length of this sequence. Methods of such rearrangement are the subject of our research.

The paper consists of five sections. The model of sensor network control is presented in Sect. 2. Section 3 discusses the process of schedule adaptation to the varying temperature conditions in the proposed model. Section 4 describes the experimental part of the research. Section 5 concludes the paper.

2   The Model of a Sensor Network

In our research, we work with the activity schedules of a network consisting of N_S immobile sensors. The sensors are randomly deployed over an area to monitor N_P points of interest (POI). All sensors have the same sensing range r_{sens} and the same finite battery capacity. We propose a model of this problem where some real-word properties are simplified. First, we assume that time is discrete, that is, consists of periods of the same length. During every period, a sensor can be active or sleep. A working sensor consumes one unit of energy per time unit for its activity. We assume that in the sleeping state, the energy consumption is negligible. Thus, assuming constant and ideal temperature in the surroundings, every sensor can be active during the same number of time steps (consecutive, or not).

In real life, effective battery capacity depends on various factors. For example, regular turnings on and off the battery shorten its lifetime. Thus, it does matter whether the battery is on in consecutive time steps or is in every time step turned on/off. Another issue concerns power consumption for communication activity between sensors. Depending on the communication topology, batteries of some sensors may discharge faster than the others. In our model, these issues are negligible except one, that is, the working temperature of sensors.

2.1   A Battery Discharge Model

For the modeling of a discharge process when batteries work in varying temperature conditions, we use rate discharge curves of Li-ion battery given in [5]. We assume that sensor batteries discharge linearly from its initial voltage to a threshold value. Thus, for different temperatures, we can estimate respective battery lifetimes in operating time units. Therefore, we measure the battery capacity in time units rather than in milliamps × hours. The maximum battery capacity T_{batt} represents the number of time units when the device can be in an active state running in the ideal temperature equal to 25 °C. A percentage amount of discharge in a unit of operating time equals 100% divided by the battery lifetime for the actual temperature.

This model is extremely simplified and does not represent real processes occurring in the battery under the influence of temperature changes. However, it allows approximating the battery discharge based on actual weather conditions. It must be stressed that the model refers neither to any particular battery nor the sensor device's real discharge curves. It takes into account neither costs of a transition from low power mode to high power mode nor the side effects of drawing current at a rate higher or lower than the discharge rate. Details of this model are described in [1].

2.2   Sensor Activity Schedule Representation

A schedule is a matrix H of 0s and 1s. The i-th row of this matrix defines the activity of the i-th sensor over time. The j-th column of H denoted H^j defines the state of all sensors during the j-th time unit. $H^j[i] = 1$ means that the i-th sensor is active at time j while $H^j[i] = 0$ means that the i-th sensor is at time j in a sleeping state. Each of the columns (called slots) guarantees the required level of coverage, that is, for every slot, all its active sensors cover the appropriate percentage of POIs. The number of slots is identical to the lifetime of the network because every slot takes precisely one unit of time. The number of 1s in a row of a schedule represents the working time of the corresponding sensor. It should not be higher than T_{batt}.

Schedules that are a subject of experiments are suboptimal. Therefore, we assume that turning off even a single sensor in a slot can make this slot incorrect. The schedules have been generated under the assumption that the amount of sensor battery discharge in every slot is the same and equals one operating time unit, that is, the network works at a constant temperature of 25 °C all the time. All schedules used in experiments described in the further text come from the algorithm based on the hypergraph model approach (for more details, the reader is referred to [2–4]).

2.3   Influence of the Temperature on the Schedule Execution

In the proposed model, every slot has its temperature, which is constant and can be different than in the neighboring slots. In the real world, the temperature may also differ slightly depending on the location of the sensor because, for example, objects in the monitored terrain may cast a shadow on some sensors. However, we consider a simplified model where the temperature is always precisely the same for all sensors.

Figure 1 shows an example schedule for the network consisting of 5 sensors. The network lifetime equals six time units. Two versions of the same schedule for two different operating temperatures contain the battery discharge percentages in the table cells. Zero means that the sensor is in the sleeping state, and non-zero—in an active state. All sensors are the same and have the same batteries. In the case of $t = 25\,°C$, the battery discharge during a time unit equals 20%. Let us assume that in the case of $t = -20\,°C$, the battery discharge is higher and equals, for example, 28.6%. The last column contains the sums of values in every row.

1	2	3	4	5	6	sum	1	2	3	4	5	6	sum
20	20	20	20	0	0	80	28.6	28.6	28.6	28.6	0	0	114.4
20	0	20	0	20	0	60	28.6	0	28.6	0	28.6	0	85.8
0	0	20	20	20	20	80	0	0	28.6	28.6	28.6	28.6	114.4
20	20	0	0	20	20	80	28.6	28.6	0	0	28.6	28.6	114.4
0	20	0	20	0	20	60	0	28.6	0	28.6	0	28.6	85.8

| (a) operating temperature: 25°C | (b) operating temperature: −20°C |

Fig. 1. Example schedule for 5 sensors; the network lifetime equals 6 time units. The table contains battery discharge percentages of sensors for $t = 25\,°C$ (left), and for $t = -20\,°C$ (right)

In the case of low temperature, the execution of a schedule discharges sensor batteries more than in the case of high temperature. Particularly, the schedule in Fig. 1(b) could not be executed for $-20\,°C$, because sensors no. 1, 3, and 4 discharge their batteries above the level of 100%. In this case, slots no. 4 and 6 are incorrect because of the lack of energy for the full working time of three sensors: sensor no. 1 in slot no. 4, and sensors no. 3 and 4 in slot no. 6. Such schedule can be denoted as $[C_1, C_2, C_3, I_4, C_5, I_6]$ (where C means a correct slot and I—an incorrect one) and its lifetime deteriorates to 3 because the third slot is the last one in the uninterrupted sequence of correct slots. Typically, in the case of the schedule application at a low temperature, slots located on the left end of the schedule are still fine, but in the slots located on the right end, some sensors have to be off. We assumed the worst-case scenario, where none of the active sensors in a slot can be off. Otherwise, the slot becomes incorrect and has to be removed from the schedule. If this is the case, there appears extra energy in all the remaining active sensors, which can be used by them in the slots standing in the schedule after the removed one. For example, when we remove slot no. 4, sensors no. 3 and 5 decrease their battery discharge.

Finally, we also assumed that we could not try to fix the coverage by changing the set of active sensors within a slot because we know neither the sensor nor POIs localization. The only option is to remove incorrect slots.

3   Methods of the Schedule Adjustment

In our approach, we try to fix sensor activity schedules corrupted by the need to work in low temperatures. First, we assume that the network uses a thermometer and knows the operating temperature for the current time unit. Therefore, it can also calculate the discharge for all batteries over time as well as their predicted discharge levels assuming they would have to work in the present temperature for a time unit. The network validates every slot in the input schedule H before execution. As far as the slots are feasible, that is, the battery levels for all sensors active in the current time unit are sufficiently high, the network executes slots one by one according to their order in H. Detection of the first infeasible slot H^k starts the repair procedure.

There is no prediction of the temperature, so the repair procedure searches for the best alternative slot just for the position k concerning the current temperature measured by the thermometer. The new slot comes from the remaining slots from H^{k+1} to H^n, where n represents the total number of slots in H.

One can divide these repair procedures into two groups. In the first group of methods, we verify slots from the queue containing slots from H^{k+1} to H^n. The first feasible slot, say H^m, that is, the slot which defines feasible activity control of the network respectively to the current temperature, is moved to position k, and the slots from H^k to H^{m-1} are shifted to the right by one. In the second group, for each slot deemed feasible, we calculate some slot's parameters and then compare it to the slot currently considered as best. After each feasible slot is checked, the winner is moved to position k. These groups of methods are described below.

3.1   Feasible Slot Selection

First-Fit (FF) verifies slots from the queue containing slots ordered from H^{k+1} to H^n. Finding the first feasible one stops the verification process.

Biased First-Fit (BFF) also verifies slots from the queue just like *First-Fit* but moves every negatively verified slot from the front to the end of the queue. This way, in the next call of this method, the order of slots in the queue is different. Now, the slots which were negatively verified recently, are to be verified last. This approach comes from the observation that the temperature seldom changes rapidly, and the slots which do not fit now, most probably will not fit in the nearest time step. Thus, we will find the first feasible slot sooner if we start looking elsewhere.

3.2   Slots Ranking

MinMax (MIMA) among the active sensors calculates mean for the two battery load levels: the highest and the lowest one. To calculate this, we have to find two sensors with extreme load levels in the slot. The slot with a higher mean value wins the comparisons.

Mean (MEAN) works the same way as MIMA except that we calculate the mean battery load level for all sensors which are active in the slot.

STDMean (STDM) compares slots in pairs to select the better one. First, for each of the slots, we calculate the mean μ and standard deviation σ values of the battery load level for all active sensors. Then, we attempt to compare slots to each other. When two slots H^i and H^j have disjoint intervals $[\mu - \sigma, \mu + \sigma]$, the winner is the one with the higher values in its interval, that is, $\mu - \sigma$ of the winner is greater than $\mu + \sigma$ of the loser. When the intervals overlap, we have to make some additional comparisons. Let us assume that H^i has lesser μ than H^j. In spite of this, H^i may win when: (1) $\sigma^i < \sigma^j$, (2) $\mu^i - \sigma^i$ is greater than $\mu^j - \sigma^j$, and (3) $\mu^i + \sigma^i$ is greater than μ^j. Otherwise, H^j wins.

Tournament (TOUR), in the beginning, makes sequences of active sensors for both compared slots. Each sequence consists of sensors active exclusively in its slot, that is, sensors active in both slots do not enter the sequences. Then, the sequences are sorted in ascending order of battery charge levels. Finally, we compare battery charge levels in sensors occupying the same positions in sequences, pair by pair. A point goes to the slot that activates the sensor with a higher battery charge level. In the case when one sequence is longer than the other one, the sensors without a pair bring points to none of the slots. The better slot is the one having a higher score. In the case of a tie, the better slot is the one with a shorter sequence, that is, the one that uses fewer sensors.

3.3   Computational Cost of the Repair Procedures

The methods from the first group have a linear complexity $O(sN_S)$, which depends on the number of sensors N_S and the number of remaining slots $s = n - k$. In the methods from the second group, the computational cost of finding the winning slot is also linear and equals $O(s)$. The cost of calculations in slots comparison methods equals $O(N_S)$ except the case of TOUR, where complexity depends on the sort algorithm, that is, equals $O(N_S \log(N_S))$.

4   Experiments

4.1   Benchmark and Plan of Experiments

Experiments were conducted for two outdoor temperature conditions: a colder weather (Series A in Fig. 2(a)) and a warmer one (Series B in Fig. 2(b)). Both series were measured every five minutes in a backyard of a house in a town near Warsaw for the weather typical for February and March. One can see that the temperature is always higher in a day with a peak in the early afternoon and decreases at night.

Input schedules are the best-found solutions for the SCP1 benchmark [2–4], which consists of 320 instances of problems (eight classes, 40 instances each). For each of the instances, we conducted experiments for five values of maximum battery capacity T_{batt}: 10, 15, 20, 25, and 30. Eventually, we obtained 1600 schedules in total.

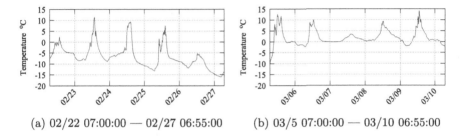

(a) 02/22 07:00:00 — 02/27 06:55:00   (b) 03/5 07:00:00 — 03/10 06:55:00

Fig. 2. Two series of outdoor temperatures measured every 5 min

The experiments consist of two groups, for the temperature measurement series A and B, respectively. In every group, the random search procedure searched for the most extended sequence of correct slots in each of 1600 schedules obtained from experiments with SCP1. In each of the groups, we get five mean lengths of final sequences because we defined five values of T_{batt}.

The length of an uninterrupted sequence of the correct slots is the primary output parameter of the experiments. However, the sequence lengths may differ significantly depending on T_{batt}, so the straight comparisons of the output sequence lengths may be misleading. Therefore, for each of the sequences, we calculated its percentage improvement respectively to the lengths of input sequences obtained when no improvement has been made. This normalization allows comparing the efficiency of methods over different classes of problems.

4.2   The Results

Tables 1, 2, 3, 4 and 5 present results of our experiments with methods described in Sect. 3 for eight classes of the SCP1 benchmark for values of T_{batt} from 10 to 30. The upper part of every table shows results for temperature from Series A, the lower part – for temperature from Series B. Numbers printed in bold represent the highest mean percentage improvement among the six methods for a given class of problems and given value of T_{batt}.

One can see in these tables that STDM gave the best mean percentage improvement in 59 out of 80 cases. In 7 cases, FF and BFF gave the same best percentage improvement. FF alone turned out to be the winner in 6 cases, BFF alone – in 4 cases. STDM and TOUR both gave the same best improvement twice. STDM and MEAN turned out to be both the winners once, and in one case, TOUR was the sole winner. MIMA never won the competition. However, FF and/or BFF were the winner for class 5 of the SCP1 benchmark in 8 out of 10 cases, and for class 4 of the SCP1 benchmark in 5 out of 10 cases. Thus, even if STDM seems to be the most effective method of run-time schedule adjustment, for some test cases, methods FF and BFF work better. Let us also notice that the increase of sensors battery capacity is accompanied by the performance drop of our schedule adjustment methods. We observe the most significant mean percentage improvements for $T_{batt} = 10$, in many cases, more than 200%.

Table 1. Mean percentage improvements of the best-found final sequences returned by the tested methods with respect to the lengths of input sequences for eight classes of the SCP1 benchmark for $T_{\text{batt}} = 10$; temperature from Series A: top part, Series B—bottom part

No.	FF	BFF	MIMA	MEAN	STDM	TOUR
1	183.1%	183.3%	190.9%	194.6%	**195.4%**	194.4%
2	234.5%	230.6%	228.5%	234.3%	**237.9%**	231.4%
3	212.6%	212.7%	212.1%	218.1%	**220.6%**	215.8%
4	**265.6%**	261.2%	243.6%	257.8%	261.4%	251.6%
5	150.4%	147.8%	161.7%	165.7%	**169.9%**	162.1%
6	184.4%	184.4%	193.4%	196.5%	**198.1%**	197.3%
7	231.0%	229.1%	225.7%	233.2%	**236.0%**	228.0%
8	183.1%	173.8%	202.8%	202.6%	**203.3%**	197.1%
1	171.8%	171.8%	180.5%	185.0%	**188.1%**	183.9%
2	223.9%	222.9%	217.1%	223.3%	**225.3%**	220.6%
3	180.4%	180.5%	179.7%	185.8%	**188.0%**	182.4%
4	266.4%	**267.2%**	247.6%	259.9%	263.0%	250.8%
5	201.5%	**202.6%**	190.6%	195.9%	197.5%	195.0%
6	176.6%	176.6%	186.4%	190.2%	**192.8%**	189.8%
7	212.6%	212.2%	208.4%	213.7%	**215.0%**	207.5%
8	215.1%	**217.3%**	208.1%	213.0%	213.7%	209.2%

Table 2. Mean percentage improvements of the best-found final sequences returned by the tested methods with respect to the lengths of input sequences for eight classes of the SCP1 benchmark for $T_{\text{batt}} = 15$; temperature from Series A: top part, Series B—bottom part

No.	FF	BFF	MIMA	MEAN	STDM	TOUR
1	115.7%	115.0%	133.2%	137.5%	**139.0%**	136.2%
2	133.1%	133.0%	130.8%	134.0%	**136.1%**	134.4%
3	127.5%	127.5%	132.7%	134.9%	**136.4%**	133.7%
4	**156.0%**	155.8%	149.8%	153.4%	155.0%	151.6%
5	**146.3%**	144.2%	140.6%	143.5%	144.9%	143.5%
6	108.8%	108.9%	119.7%	125.8%	**129.2%**	125.7%
7	134.8%	134.7%	133.8%	137.6%	**139.3%**	136.7%
8	**140.9%**	140.9%	135.8%	138.6%	140.1%	138.3%
1	119.0%	118.9%	128.9%	131.1%	**132.9%**	130.2%
2	134.4%	**134.5%**	129.7%	132.1%	134.4%	131.8%
3	113.5%	113.5%	117.6%	121.0%	**122.2%**	118.9%
4	**150.7%**	150.2%	142.4%	145.6%	146.7%	143.9%
5	128.1%	128.2%	128.4%	129.5%	**130.9%**	127.9%
6	115.6%	115.6%	124.2%	126.6%	**128.9%**	125.6%
7	133.0%	133.2%	129.7%	133.1%	**135.6%**	132.2%
8	127.2%	127.0%	125.8%	127.7%	**128.3%**	127.3%

Table 3. Mean percentage improvements of the best-found final sequences returned by the tested methods with respect to the lengths of input sequences for eight classes of the SCP1 benchmark for $T_{\mathrm{batt}} = 20$; temperature from Series A: top part, Series B—bottom part

No.	FF	BFF	MIMA	MEAN	STDM	TOUR
1	104.0%	104.3%	115.7%	117.8%	**119.2%**	116.8%
2	119.2%	119.2%	118.1%	121.3%	**124.0%**	121.9%
3	101.8%	101.8%	107.1%	109.5%	**110.8%**	109.5%
4	120.7%	120.7%	118.7%	121.3%	**122.1%**	119.4%
5	**95.0%**	**95.0%**	92.8%	93.6%	93.6%	93.4%
6	112.2%	112.4%	123.1%	125.1%	**126.7%**	124.4%
7	111.7%	111.7%	111.9%	114.5%	**115.7%**	114.2%
8	95.3%	95.3%	92.2%	94.6%	**95.6%**	95.2%
1	106.4%	106.4%	114.7%	116.9%	**118.0%**	116.1%
2	105.0%	105.0%	103.1%	106.3%	**108.6%**	107.7%
3	98.5%	98.5%	104.6%	106.8%	**108.6%**	105.7%
4	**109.0%**	**109.0%**	104.5%	107.0%	107.0%	106.1%
5	**96.3%**	96.1%	92.6%	94.6%	94.6%	94.0%
6	112.2%	112.1%	120.3%	122.6%	**123.8%**	121.6%
7	92.0%	91.9%	90.5%	93.2%	**94.5%**	93.4%
8	**92.6%**	**92.6%**	88.6%	90.7%	91.5%	91.5%

Table 4. Mean percentage improvements of the best-found final sequences returned by the tested methods with respect to the lengths of input sequences for eight classes of the SCP1 benchmark for $T_{\mathrm{batt}} = 25$; temperature from Series A: top part, Series B—bottom part

No.	FF	BFF	MIMA	MEAN	STDM	TOUR
1	105.4%	105.4%	117.3%	119.8%	**120.8%**	118.2%
2	104.0%	104.0%	103.8%	106.4%	108.2%	**108.3%**
3	98.6%	98.2%	105.4%	107.4%	**108.7%**	106.1%
4	108.4%	108.4%	108.3%	109.2%	**110.1%**	108.7%
5	**87.3%**	**87.3%**	85.0%	86.0%	87.0%	86.0%
6	103.0%	102.9%	112.1%	114.4%	**115.6%**	113.3%
7	98.1%	98.1%	99.1%	101.1%	**103.1%**	102.4%
8	91.6%	91.6%	89.3%	91.1%	**92.8%**	91.4%
1	101.6%	101.7%	110.2%	112.0%	**112.9%**	111.2%
2	99.4%	99.4%	97.4%	101.2%	**102.7%**	102.6%
3	91.3%	91.0%	97.5%	98.7%	**99.6%**	97.9%
4	100.0%	100.0%	98.1%	99.8%	**100.5%**	99.5%
5	**75.7%**	**75.7%**	73.6%	74.5%	75.0%	74.3%
6	102.5%	102.4%	110.3%	112.3%	**113.0%**	111.1%
7	88.6%	88.6%	88.5%	90.9%	**92.9%**	91.7%
8	75.0%	75.0%	72.8%	74.4%	**75.3%**	**75.3%**

Table 5. Mean percentage improvements of the best-found final sequences returned by the tested methods with respect to the lengths of input sequences for eight classes of the SCP1 benchmark for $T_{batt} = 30$; temperature from Series A: top part, Series B—bottom part

No.	FF	BFF	MIMA	MEAN	STDM	TOUR
1	86.8%	86.7%	96.0%	97.4%	**97.9%**	96.5%
2	96.0%	96.3%	95.6%	100.3%	**102.8%**	102.7%
3	91.8%	91.8%	100.9%	102.4%	**103.5%**	101.4%
4	102.5%	102.5%	103.1%	104.0%	**105.4%**	104.2%
5	**81.6%**	**81.6%**	80.0%	80.4%	81.3%	80.6%
6	98.4%	98.0%	107.4%	109.6%	**110.1%**	108.2%
7	89.6%	89.5%	90.1%	92.9%	**95.0%**	93.7%
8	85.1%	85.1%	83.1%	85.2%	**86.1%**	85.7%
1	78.2%	78.2%	85.8%	**87.3%**	**87.3%**	86.5%
2	93.3%	93.4%	92.9%	96.2%	**99.0%**	98.5%
3	89.3%	89.4%	94.9%	95.9%	**97.2%**	95.2%
4	94.3%	94.3%	94.0%	95.5%	**96.3%**	95.0%
5	**69.9%**	**69.9%**	68.9%	69.4%	69.7%	69.4%
6	89.9%	89.9%	97.6%	99.5%	**99.6%**	98.5%
7	91.2%	91.0%	91.2%	94.8%	**96.7%**	95.4%
8	76.0%	76.0%	74.1%	75.7%	**76.4%**	**76.4%**

On the other hand, for $T_{batt} = 30$, in most cases, improvements are below 100%. It is hard to indicate the easiest or the most challenging class of problems to solve. For $T_{batt} = 10$ or 15, all the methods cope with class 4 of the SCP1 benchmark better than with other classes. But it is not the case for higher values of T_{batt}. When we look at the least improvements, in the case $T_{batt} = 30$, class 5 is the most difficult to improve for all the methods. However, for $T_{batt} = 10$ class 5 is the most difficult just in the case of temperature from Series A. For the temperature from Series B, we observe the least improvements for class 1. Thus, we may say that the performance of our schedule adjustment methods depends both on the capacity of sensor batteries and the network.

5 Conclusions

In the presented research, we proposed some methods of run-time adaptation to changing temperature conditions of a sensor network schedule of activity. Our goal is to maximize the lifetime of a homogeneous network consisting of sensors with limited battery capacity. The starting point to our research are schedules of sensors activity generated under the assumption of an ideal temperature which does not affect sensor battery performance. Especially during the winter season, such schedules cannot be completed because low temperatures cause the early death of sensor batteries. To extend the lifetime of a schedule, we proposed several methods of run-time schedule slots rearrangement. Such a rearrangement verifies whether a slot to be executed next is feasible, that is, whether sensors active in this slot have enough energy to work under current temperature. If not, the schedule is searched for another slot possible to execute. One of our methods looks in the queue of slots to be executed later for the first feasible slot (First-Fit). We also considered a variation of this method called Biased First-Fit. Moreover, we proposed several methods ranking all feasible slots according to some slot parameters to select the best slot to be executed next. We call these methods MinMax, Mean, STDMean, and Tournament. Our experiments showed that in most cases the most effective one in terms of the mean percentage improvement of the length of schedule is STDMean, selecting next slot to be executed based on computing the mean and standard deviation values of the battery load level for all the sensors active in a given slot. However, for some of our test cases, methods First-Fit or Biased First-Fit worked better. Moreover, the experiments demonstrated that the performance of schedule adjustment methods also depends on the capacity of sensor batteries and the network.

In the future, we plan to study other approaches to schedule adaptation to low-temperature conditions.

References

1. Trojanowski, K., Mikitiuk, A.: Sensor network schedule adaptation for varying operating temperature. In: Palattella, M.R., Scanzio, S., Coleri Ergen, S. (eds.) ADHOC-NOW 2019. LNCS, vol. 11803, pp. 633–642. Springer, Cham (2019). https://doi.org/10.1007/978-3-030-31831-4_47
2. Trojanowski, K., Mikitiuk, A.: Local search approaches with different problem-specific steps for sensor network coverage optimization. In: Le Thi, H.A., Le, H.M., Pham Dinh, T. (eds.) WCGO 2019. AISC, vol. 991, pp. 407–416. Springer, Cham (2020). https://doi.org/10.1007/978-3-030-21803-4_41
3. Trojanowski, K., Mikitiuk, A.: Maximization of the sensor network lifetime by activity schedule heuristic optimization. Ad Hoc Netw. **96**, 101994 (2020). https://doi.org/10.1016/j.adhoc.2019.101994
4. Trojanowski, K., Mikitiuk, A., Kowalczyk, M.: Sensor network coverage problem: a hypergraph model approach. In: Nguyen, N.T., Papadopoulos, G.A., Jedrzejowicz, P., Trawiński, B., Vossen, G. (eds.) ICCCI 2017, Part I. LNCS (LNAI), vol. 10448, pp. 411–421. Springer, Cham (2017). https://doi.org/10.1007/978-3-319-67074-4_40
5. Wang, K.: Study on low temperature performance of Li ion battery. Open Access Libr. J. **4**(11), 1–12 (2017). https://doi.org/10.4236/oalib.1104036

Spectral Cluster Maps Versus
Spectral Clustering

Sławomir T. Wierzchoń[ID] and Mieczysław A. Kłopotek[(⊠)][ID]

Institute of Computer Science, Polish Academy of Sciences,
ul. Jana Kazimierza 5, 01-248 Warsaw, Poland
{slawomir.wierzchon,mieczyslaw.klopotek}@ipipan.waw.pl

Abstract. The paper investigates several notions of graph Laplacians and graph kernels from the perspective of understanding the graph clustering via the graph embedding into an Euclidean space. We propose hereby a unified view of spectral graph clustering and kernel clustering methods. The various embedding techniques are evaluated from the point of view of clustering stability (with respect to k-means that is the algorithm underpinning the spectral and kernel methods). It is shown that the choice of a fixed number of dimensions may result in clustering instability due to eigenvalue ties. Furthermore, it is shown that kernel methods are less sensitive to the number of used dimensions due to downgrading the impact of less discriminative dimensions.

Keywords: Graph clustering · Spectral clustering · Spectral embedding · Kernel clustering

1 Introduction

Spectral clustering found numerous applications in machine learning, exploratory data analysis, statistics, pattern recognition, entity resolution, protein sequencing, computer vision (e.g. text/image separation), and speech processing (including the hot topic of speech separation). The growing diversity of applications creates an urgent need for a deeper understanding of the intrinsic nature of spectral clustering algorithms in general and of their particular brands.

Spectral clustering has been studied for a long time as a method of approximating graph clustering according to criteria like *RCut* [5] or *NCut* [11,13].[1]

The respective clustering methods are explained in the tutorial [10] and in [9]. The structure of the algorithms of these types is as follows: Given the graph G of m nodes to cluster, compute its Laplacian matrix. Then compute $k < m$ eigenvectors, $\mathbf{v}_1, \ldots, \mathbf{v}_k$, associated with k lowest (nonzero) eigenvalues of that Laplacian. i-th element of each such vector corresponds to i-th node of the graph G. Construct a matrix $U = (\mathbf{v}_1, \ldots, \mathbf{v}_k)$ of m rows and k columns. One can eventually normalize the rows of U to be of unit (Euclidean) length. Then perform k-means clustering on the row vectors of this matrix to produce a clustering of m objects into k clusters. The used Laplacians are

---

[1] The graphs may originate from data embedded in Euclidean space, for which a graph was constructed based on thresholding similarities/distances between the objects [12].

ⓒ Springer Nature Switzerland AG 2020
K. Saeed and J. Dvorský (Eds.): CISIM 2020, LNCS 12133, pp. 472–484, 2020.
https://doi.org/10.1007/978-3-030-47679-3_40

- the combinatorial (or unnormalized) Laplacian – in this case the clustering will approximate the *RCut* criterion [5];
- the normalized Laplacian – the clustering approximates the *NCut* criterion [13];
- the normalized Laplacian with unit length correction – in this case the clustering will approximate random walk partition [11];
- the random walk Laplacian – approximating *NCut* [9].

In a forthcoming paper, [8], we show that spectral clustering fails to approximate *NCut* for non-uniform edge weights. This justifies search for another clustering interpretation. On the other hand, kernel methods [7] are quite popular in clustering sets of objects for which their vector description is unavailable, but their mutual similarities are available. The idea of kernel clustering consists in embedding the objects into a high dimensional space, called feature space, in such a way that the elements of similarity matrix correspond to dot products of the position vectors in that space. This similarity matrix can be easily converted into weighted graph. As already stressed by [4], both spectral and kernel clustering methods use or can be explained by usage of eigen-decomposition of the similarity matrix and the clustering in the space spanned by appropriately selected eigen vectors. This fact suggests that one could attempt to propose a unified view for both spectral clustering and kernel methods as a clustering of an embedding instead of (*Ratio/N*)-cut approximations or feature space mappings.

Hence, while the previous research concentrated on what the spectral clustering approximates, we want to concentrate here on the issue of what the spectral clustering actually produces. This is different from e.g. [3]. They wanted to find conditions when spectral/kernel clustering and graph cut clustering would converge to the same result. We want to make the differences between various spectral and kernel clustering methods explicit in that we explore the differences in the way how the graphs are embedded.

2 Similarity-Based 1D Embedding

Hall [6] proposed the following approach to graph embedding in one-dimensional Euclidean space. Consider a graph G with m nodes described by a similarity matrix S such that $s_{ii} = 0^2$ and $s_{ij} = s_{ji} \geq 0$ for $i, j = 1, \ldots, m$. The embedding $\mathbf{y} = (y_1, y_2, \ldots, y_m)^T$ should minimize

$$\mathcal{E}_H(\mathbf{y}; S) = \frac{1}{2} \sum_{i=1}^{m} \sum_{j=1}^{m} (y_i - y_j)^2 s_{ij} \tag{1}$$

The above condition expresses the intuition that nodes similar to one another should be placed close to one another in the embedding as their distance mode strongly increases $\mathcal{E}_H(\mathbf{y}; S)$. However, this minimization condition has two deficiencies: A constant vector $\mathbf{y}$ would produce a minimum equal to zero but we do not want all nodes of the graph to be placed at one point. So it has to be excluded. Also scaling $\mathbf{y}$ by some factor $0 < \gamma < 1$ would always decrease $\mathcal{E}_H$, hence some constraint has to be imposed. We require

$$\mathbf{y}^T \mathbf{y} = 1 \tag{2}$$

---

[2] $s_{ii} = 0$ is set for technical reasons. Self-similarity does not affect the basic formula, but the value 0 is useful when handling Laplacians.

the advantage of which will be obvious later. But

$$\frac{1}{2}\sum_{i=1}^{m}\sum_{j=1}^{m}(y_i-y_j)^2 s_{ij} = \frac{1}{2}\sum_{i=1}^{m}\sum_{j=1}^{m}(y_i^2 - 2y_iy_j + y_j^2)s_{ij}$$

$$= \sum_{i=1}^{m}(y_i^2\sum_{j=1}^{m}s_{ij}) - \sum_{i=1}^{m}\sum_{j=1}^{m}(y_iy_js_{ij}) = \mathbf{y}^T D\mathbf{y} - \mathbf{y}^T S\mathbf{y} = \mathbf{y}^T L\mathbf{y}$$

where $D = diag(S\mathbf{1})$ is a diagonal matrix[3] with elements $d_{ii} = \frac{1}{2}\left(\left(\sum_{j=1}^{m}s_{ij}\right) + \left(\sum_{j=1}^{m}s_{ji}\right)\right)$ and $L = D - S$ is the combinatorial Laplacian of the graph G. If $\lambda_1 \leq \cdots \leq \lambda_m$ are eigenvalues of L, and $\mathbf{v}_1, \ldots, \mathbf{v}_m$ are the corresponding unit length eigenvectors (which are hence orthogonal to each other), then each vector $\mathbf{y}$ can be represented as

$$\mathbf{y} = \sum_{i=1}^{m}\beta_i\mathbf{v}_i$$

From the assumption (2) if follows that $\sum_{i=1}^{m}\beta_i^2 = 1$ Therefore

$$\mathbf{y}^T L\mathbf{y} = (\sum_{i=1}^{m}\beta_i\mathbf{v}_i^T)L(\sum_{i=1}^{m}\beta_i\mathbf{v}_i) = (\sum_{i=1}^{m}\beta_i\mathbf{v}_i^T)(\sum_{i=1}^{m}\beta_i\lambda_i\mathbf{v}_i) = \sum_{i=1}^{m}\beta_i^2\lambda_i\mathbf{v}_i^T\mathbf{v}_i = \sum_{i=1}^{m}\beta_i^2\lambda_i$$

This weighted sum of eigenvalues with non-negative weights summing up to 1 reaches minimum when $\beta_i^2 = 1$ at the lowest λ_i. The lowest eigenvalue of L is equal to 0 and its eigenvector is a constant if the graph is connected which we excluded. So the minimum is equal to λ_2, and $\mathbf{v}_2$ is the required embedding.

Let us recall, however, that the eigenvalues do not need to be unique, in particular it may happen that $\lambda_2 = \lambda_3$. This means that there may exist different embeddings yielding the very same optimum of $\mathcal{E}_H(\mathbf{y};S)$.

This gives rise to the question whether these embeddings differ. One straight forward approach to clustering is to use the Fiedler eigenvector ($\mathbf{v}_2$). If we cluster the data into two clusters, one assigns the nodes with positive components of this eigenvector to one cluster and the remaining ones to the other. If two eigenvectors (with same eigenvalue) would yield the same clustering, then their dot product would differ from zero and hence they will not be orthogonal. Therefore Fieldler eigenvalue based embeddings would yield contradictory clusterings. One can ask whether the same applies to other clustering methods, like k-means, frequently used in spectral clustering. If $p+1$ consecutive eigenvalues starting with λ_q are identical, then any vector $\sum_{j=0}^{p}\gamma_{q+j}\mathbf{v}_{q+j}$ such that $\sum_{j=0}^{p}\gamma_{q+j}^2 = 1$ is a unit eigenvector of L. So there exist infinitely many optimal embeddings.

It may be demonstrated by example that under such conditions the clusterings may differ (see also the formal derivation below). Therefore it is recommended that in such a case the embedding shall be performed in the $q+p$ dimensional space. In such a case

---

[3] As the matrix S is symmetric, this definition is a bit too laborious, but later on we will refer to asymmetric matrices.

if there exists only one optimum in a given set of eigenvectors, it remains the same if a different set of eigenvectors for this very eigenvalue is taken.

To prove this consider two points d, e in the original space. They will have the coordinates $(v_{q,d}, v_{q+1,d}, \ldots, v_{q+p,d})^T$ and $(v_{q,e}, v_{q+1,e}, \ldots, v_{q+p,e})^T$ resp. Their distance will amount to $d(d,e)^2 = (v_{q,d} - v_{q,e})^2 + (v_{q+1,d} - v_{q+1,e})^2 + \cdots + (v_{q+p,d} - v_{q+p,e})^2$ Any different coordinate system will assign coordinates $(v'_{q,d}, v'_{q+1,d}, \ldots, v'_{q+p,d})^T$ and $(v'_{q,e}, v'_{q+1,e}, \ldots, v'_{q+p,e})^T$ resp. Their distance will amount to $d'(d,e)^2 = (v'_{q,d} - v'_{q,e})^2 + (v'_{q+1,d} - v'_{q+1,e})^2 + \cdots + (v'_{q+p,d} - v'_{q+p,e})^2$ As $\mathbf{v}_{q+j}$s and $\mathbf{v}'_{q+j}$s are points on a hypersphere, and are two orthogonal coordinate systems, then they must be obtained by rotation. But the rotation preserves distances. So $d'(d,e) = d(d,e)$. For this reason k-means clustering for any embedding taking the complete set of eigenvectors with identical eigenvalues yields the same result.

Same result is not guaranteed if we take arbitrary subsets of such coordinates. Let us concentrate on $\mathbf{v}_2, \mathbf{v}_3$ only assuming the corresponding $\lambda_2 = \lambda_3$. Consider the cost function $J((\mathbf{v}_2, \mathbf{v}_3); \mathcal{C})$ of k-means under such assumption, whereby $\mathcal{C} = \{C_1, \ldots, C_k\}$ is the set of clusters into which the points $1, \ldots, m$ are clustered, $\mu_{c,2}, \mu_{c,3}$ are cluster centers of cluster C_c in the dimension spanned by $\mathbf{v}_2, \mathbf{v}_3$ respectively.

$$J((\mathbf{v}_2, \mathbf{v}_3); \mathcal{C}) = \sum_{c=1}^{k} \sum_{j \in C_c} \left((v_{j,2} - \mu_{c,2})^2 + (v_{j,3} - \mu_{c,3})^2 \right)$$

$$= \sum_{c=1}^{k} \sum_{j \in C_c} (v_{j,2} - \mu_{c,2})^2 + \sum_{c=1}^{k} \sum_{j \in C_c} (v_{j,3} - \mu_{c,3})^2 = J(\mathbf{v}_2; \mathcal{C}) + J(\mathbf{v}_3; \mathcal{C})$$

which means that the cost of clustering for a clustering $\mathcal{C}$ in a two-dimensional space is the sum of respective costs in each of the one-dimensional sub-spaces. Let $\mathcal{C}_{o23}$ be the optimal k-means clustering in the two-dimensional space spanned by $\mathbf{v}_2, \mathbf{v}_3$, $\mathcal{C}_{o2}$ in $\mathbf{v}_2$ and $\mathcal{C}_{o3}$ in $\mathbf{v}_3$. If $\mathcal{C}_{o2} \neq \mathcal{C}_{o3}$, we are done - different eigenvectors for the same eigenvalue yield different clusterings. So assume now that $\mathcal{C}_{o2} = \mathcal{C}_{o3}$. Assume that $\mathcal{C}_{o23} \neq \mathcal{C}_{o3}$. In such a case we get a contradiction because $J(\mathbf{v}_2; \mathcal{C}_{o3}) + J(\mathbf{v}_3; \mathcal{C}_{o3}) < J(\mathbf{v}_3; \mathcal{C}_{o23}) = J((\mathbf{v}_2, \mathbf{v}_3); \mathcal{C}_{o23})$ because $\mathcal{C}_{o23}$ was assumed to be the optimal clustering in the $(\mathbf{v}_2, \mathbf{v}_3)$ space. So let $\mathcal{C}_{o23} = \mathcal{C}_{o2} = \mathcal{C}_{o3}$. But if $\lambda_2 = \lambda_3$ then for any δ, $\sin(\delta)\mathbf{v}_2 + \cos(\delta)\mathbf{v}_3, \cos(\delta)\mathbf{v}_2 - \sin(\delta)\mathbf{v}_3$ are also two mutually orthogonal unit length eigenvectors of the very same Laplacian. Let us discuss the k-means clusterings in the new (that is rotated) space.

$$J((\sin(\delta)\mathbf{v}_2 + \cos(\delta)\mathbf{v}_3, \cos(\delta)\mathbf{v}_2 - \sin(\delta)\mathbf{v}_3); \mathcal{C})$$

$$= \sum_{c=1}^{k} \sum_{j \in C_c} \left((\sin(\delta)v_{j,2} + \cos(\delta)v_{j,3} - \sin(\delta)\mu_{c,2} - \cos(\delta)\mu_{c,3})^2 \right.$$

$$\left. + (\cos(\delta)v_{j,2} - \sin(\delta)v_{j,3} - \cos(\delta)\mu_{c,2} + \sin(\delta)\mu_{c,3})^2 \right)$$

$$= \sum_{c=1}^{k} \sum_{j \in C_c} \left(v_{j,2}^2 + v_{j,3}^2 + \mu_{c,2}^2 + \mu_{c,3}^2 - 2v_{j,2}\mu_{c,2} - 2v_{j,3}\mu_{c,3} \right)$$

$$= \sum_{c=1}^{k} \sum_{j \in C_c} \left((v_{j,2} - \mu_{c,2})^2 + (v_{j,3} - \mu_{c,3})^2 \right)$$

$$= J((\mathbf{v}_2, \mathbf{v}_3); \mathfrak{C})$$

This result is not surprising: The value of the cost function in 2D does not depend on the rotation of the coordinate system. With $k = 2$ clusters, imagine rotating of the coordinate system in such a way that the rotated cluster centers $\sin(\delta)\mu_{c,2} + \cos(\delta)\mu_{c,3}$, for $c = 1,2$, being equal to zero each (it is always possible to find such δ).

A clustering with such centers would not be optimal for sure. So upon rotation we found that a clustering along one axis differs from the clustering along two axes. We conclude that in general, when multiple eigenvectors exist with identical eigenvalue, one shall not use for clustering a subset of them as the results may be arbitrary.

Let us recall at this point, that for L all the eigenvalues different from zero have corresponding eigenvectors where the sum of their elements is equal to zero. Just sum up the rows of the equation system $L\mathbf{v} = \lambda\mathbf{v}$. If only one eigenvalue is equal to zero, then its eigenvector has all values identical (this is assured by the orthogonality requirement). Hence, it is not important if we use the eigenvector $\mathbf{v}_1$ in k-means clustering or not. However, when more than one eigenvalue is equal to zero, the abovementioned effect occurs that the eigenvectors may be arbitrary. Under these circumstances it is necessary to use $\mathbf{v}_1$.

Consider a grid graph 4×4 (4 rows of nodes, 4 columns of nodes). In this case we have two different vectors corresponding to Fiedler eigenvalue (5.857864e-01) that k-means clusters either into: the nodes 1–7 and 9, and the nodes 8 and 10–16, or it clusters into: the nodes 2, 3, 4, 7, 8, 11, 12, 16, and 1, 5, 6, 9, 10, 13, 14, 15. If we use instead both eigenvectors as the space for $k = 2$, cluster 1 consists of nodes 1–8, cluster 2 of nodes 9–16.

Let us define the regularized combinatorial Laplacian $L_R(t) = I + tL$ with $t > 0$ and define the embedding $\mathbf{y}$ as one minimizing $\mathcal{E}_R(\mathbf{y}; S) = \mathbf{y}^T L_R \mathbf{y}$ subject to $\mathbf{y}^T \mathbf{y} = 1$. It is easily seen that the minimizing $\mathbf{y}$ is the same as in $\mathcal{E}_H(\mathbf{y}; S)$.

3   Density Corrected 1D Embedding

Belkin and Niyogi [2] proposed a different embedding style, though they follow the guideline of Eq. (1). The embedding $\mathbf{z} = (z_1, z_2, \dots, z_m)^T$ should minimize

$$\mathcal{E}_B(\mathbf{z}; S) = \frac{1}{2} \sum_{i=1}^{m} \sum_{j=1}^{m} \left(\frac{z_i}{\sqrt{d_{ii}}} - \frac{z_j}{\sqrt{d_{jj}}} \right)^2 s_{ij} \qquad (3)$$

(where d_{ii} is the element of the aforementioned diagonal matrix D) subject to constraint

$$\mathbf{z}^T \mathbf{z} = 1 \qquad (4)$$

The embedding implies a special treatment of nodes of high degree. Such nodes can be set far more apart than in case of lower degree nodes.

To see the similarity and difference to the Hall approach, let us introduce the vector $\mathbf{z} = D^{1/2}\mathbf{y}$. Then $\mathcal{E}_B(\mathbf{z}; S) = \mathcal{E}_B(D^{1/2}\mathbf{y}; S) = \mathcal{E}_H(\mathbf{y}; S)$. But instead of the constraint (2), we obtain

$$\mathbf{y}^T D\mathbf{y} = 1 \tag{5}$$

This allows us to write (1) as

$$\mathcal{E}_B(\mathbf{z}; S) = \mathbf{y}^T L\mathbf{y} = \mathbf{z}^T D^{-1/2}LD^{-1/2}\mathbf{z} = \mathbf{z}^T \mathcal{L}\mathbf{z}$$

where $\mathcal{L}$ is the normalized Laplacian of the graph G. This allows us to repeat an identical argument that $\mathbf{z}$ should be the second lowest eigenvalue but this time of normalized Laplacian, with the additional remark about the case of equal eigenvalues. The embedding that will be used is the vector $\mathbf{z}$. Hence k-means usage with spectral clustering based on normalized Laplacian is justified by the respective geometry.

Let us define here the Pageranked normalized Laplacian $\mathcal{L}_P(\alpha) = (1-\alpha)I + \alpha\mathcal{L}$ with $0 < \alpha < 1$ and define the embedding $\mathbf{z}$ as one minimizing $\mathcal{E}_P(\mathbf{z}; S) = \mathbf{z}^T \mathcal{L}_P\mathbf{z}$ subject to $\mathbf{z}^T\mathbf{z} = 1$. It is easily seen that the minimizing $\mathbf{z}$ is the same as in $\mathcal{E}_B(\mathbf{z}; S)$.

4   Random Walk 1D Embedding

Still another proposal consists in observing the random walk behaviour on the graph. It is assumed that a random walker goes to one of neighbouring nodes in the graph with probability proportional to the similarity between the nodes. So given it is now at node j, he goes to node i with probability $P(i|j) = s_{ij}/\sum_{l=1}^{m} s_{lj}$. The transition matrix P with $p_{ij} = P(i|j)$ may be expressed as $P = SD^{-1}$. Let us consider the embedding $\mathbf{u}$ minimizing

$$\mathcal{E}_{rws}(\mathbf{u}; P, \pi) = \frac{1}{2}\sum_{i=1}^{m}\sum_{j=1}^{m}(u_i - u_j)^2 p_{ij} \tag{6}$$

This resembles formally the embedding (1). However, P is an asymmetric matrix even in case of an undirected graph. Let us define $L = diag(P) - 0.5(P + P^T)$. Then

$$\mathcal{E}_{rws}(\mathbf{u}; P) = \mathbf{u}^T L\mathbf{u}$$

where L shall be called here the skewed random walk Laplacian. Again, we need a constraint on $\mathbf{u}$ and let it be $\mathbf{u}^T\mathbf{u} = 1$. Under these circumstances the optimal embedding $\mathbf{u}$ would be the eigenvector corresponding to the second lowest eigenvalue of L. Note that we have here a couple of further possibilities. We considered only one step random walks. We can instead play with t step random walks and accordingly consider embeddings minimizing $\mathcal{E}_{rws}(\mathbf{u}; P^t)$.

We assumed that the random walk distance is restricted to one step only. But it does not need to. We can consider the reachability probability in t or fewer steps instead. Define $P(t) = P(t-1)P$, and $P(1) = P$. Furthermore let $P_c(t) = P_c(t-1) + (1 - P_c(t-1)) \circ P(t)$, where $\circ$ indicates the element-wise multiplication of matrices, and let $P_c(1) = P$. Let us define $L(t) = diag(P_c(t)) - 0.5(P_c(t)^T + P_c(t))$. Then

$$\mathcal{E}_{rws}(\mathbf{u}; P, t) = \mathbf{u}^T L(t)\mathbf{u} \tag{7}$$

5   Multidimensional Extensions

Embedding into a one-dimensional space would have various advantages like simplicity of clustering criteria. However, it is more common to make at least two dimensional visualisation. Let us follow the pathway of reasoning suggested by [6]. The form of the target function for embedding is an analogy of Eq. (1), but now with embedding into a space spanned by r orthogonal vectors $\mathbf{y}_1 = (y_{11}, y_{21}, \ldots, y_{m1})^T$, $\ldots \mathbf{y}_r = (y_{1r}, y_{2r}, \ldots, y_{mr})^T$,

$$\mathcal{E}_H(\mathbf{y}_1, \ldots, \mathbf{y}_r; S) = \frac{1}{2} \sum_{i=1}^{m} \sum_{j=1}^{m} \sum_{d=1}^{r} (y_{id} - y_{jd})^2 s_{ij} \tag{8}$$

By imposing the criterion of [6] that $\mathbf{y}_j^T \mathbf{y}_j = 1$ for each $j = 1, \ldots, r$ we get

$$\mathcal{E}_H(\mathbf{y}_1, \ldots, \mathbf{y}_r; S) = \sum_{d=1}^{r} \mathbf{y}_d^T L \mathbf{y}_d$$

We require again that no dimension should be degenerate (that is coordinates of at least two points are different for that dimension). In this way we can conclude (taking into consideration the orthogonality requirement) that the coordinates should be the eigenvectors corresponding to $\lambda_2, \ldots, \lambda_{r+1}$. In Fig. 1 you can see the clustering based on the embedding into two dimensions.

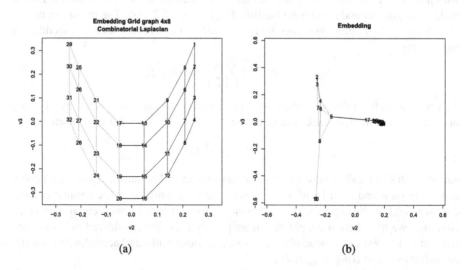

(a)   (b)

Fig. 1. (a) Hall embedding of (a) 4×8 grid graph, and (b) a graph consisting of two subgraphs connected by a single edge. Results of k-means clustering into two clusters by using both $\mathbf{v}_2, \mathbf{v}_3$ (green and blue; black edges separate clusters). (Color figure online)

By analogy we proceed with all the other embeddings. See Fig. 2(a) for Belkin embedding, 2(b) for skewed random walk embedding.

6 Kernel Embedding

In the kernel-based methods, including kernel k-means, a different type of embedding is considered, based on so-called kernel matrix K. It is assumed to be the matrix of dot products of vectors pointing from the origin of the coordinate system to the (embedded) data points in some high dimensional space. Let Q be a matrix representing this embedding in high dimensional space with rows representing the embeddings of objects, and let $\mathbf{q}_i, \mathbf{q}_j$ be two row vectors from Q representing objects i, j. Then $K_{ij} = K_{ji} = \mathbf{q}_i \mathbf{q}_j^T$, that is $K = QQ^T$. Under eigen-decomposition $K = V \Lambda V^T$ where Λ is the diagonal matrix of eigenvalues and V is the matrix with columns being the corresponding eigenvectors of $K = V \Lambda^{1/2} \Lambda^{1/2} V^T$. Hence, $Q = V \Lambda^{1/2}$. As K is well approximated when lower values of λ are ignored (set to zero), so also Q may be reduced to a lower dimensional space.

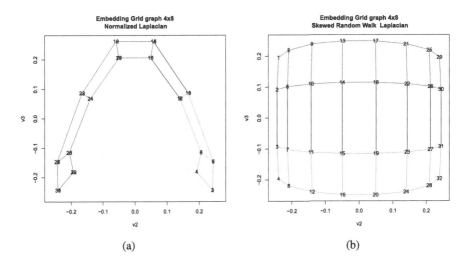

(a)   (b)

Fig. 2. Belkin embedding (a) vs. Skewed Random Walk embedding (b) of 4×8 grid graph. Results of k-means clustering into two clusters by using both $\mathbf{v}_2, \mathbf{v}_3$ (green and blue; black edges separate clusters) (Color figure online)

Let us consider now two of the types of Laplacian Kernels: the Regularized Laplacian Kernel $K_{RLK}(t) = (I + tL)^{-1}$ and the Modified Personalized PageRank Based Kernel $K_{MPPRK}(\alpha) = (D - \alpha S)^{-1}$, $0 < \alpha < 1$ [1]. A closer look at the definitions of the two kernels. $K_{RLK}(t)$ and $K_{MPPRK}(\alpha)$, reveals that both of them can be considered as approximated inverse of L. $K_{MPPRK}(\alpha)$ does so with $\alpha \to 1$, $K_{RLK}(t)$ with $t \to \infty$. $K_{RLK}(t)$ divided by t approximates it when $t \to \infty$. This means that their eigenvectors and inverted eigenvalues approximate those of L. Nonetheless the coordinates in the embeddings are distinct from those of Hall embedding, as they are multiplied by inverse square roots of eigenvalues.

Let us look more precisely at $K_{RLK}(t)$. We have already defined the regularized combinatorial Laplacian as $L_R(t) = I + tL$. Obviously, for any eigenvector $\mathbf{v}$ of L,

$L_R(t)\mathbf{v} = \mathbf{v} + t\lambda\mathbf{v} = (1 + t\lambda)\mathbf{v}$. So eigenvectors of L are also eigenvectors of $L_R(t)$, while for any eigenvalue λ of L, $(1 + t\lambda)$ is an eigenvalue of $L_R(t)$. As $K_{RLK}(t)$ is an inverse of $L_R(t)$, eigenvectors of L are eigenvectors of $K_{RLK}(t)$, and eigenvalues λ of L are transformed to $\frac{1}{1+t\lambda}$. For an example of embedding via $K_{RLK}(t)$ see Fig. 3(a).

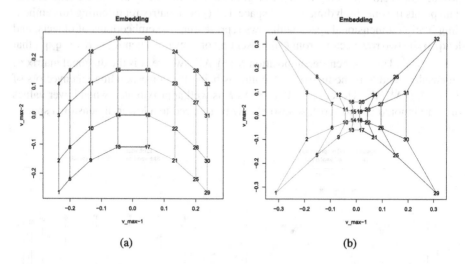

(a)   (b)

Fig. 3. (a) Regularized Laplacian Kernel embedding of 4×8 grid graph - k-means clustering into two clusters (green and blue; black edges separate clusters) (b) Modified Personal PageRank Based Kernel embedding of 4×8 grid graph - k-means clustering into two clusters (green and blue) (Color figure online)

As K_{MPPRK} is concerned, consider the following: $L_P(\alpha) = D - \alpha S = (1 - \alpha)D + \alpha L = D^{1/2}((1 - \alpha)I + \alpha\mathcal{L})D^{1/2} = D^{1/2}\mathcal{L}_P(\alpha)D^{1/2}$. $K_{MPPRK}(\alpha)$ is just an inverse of $L_P(\alpha)$. Eigenvectors of $\mathcal{L}_P(\alpha)$ are those of $\mathcal{L}$. Hence eigenvectors of $K_{MPPRK}(\alpha)$ are those of $\mathcal{L}$ multiplied with $D^{-1/2}$. Eigenvalues of $\mathcal{L}_P(\alpha)$ are $(1 - \alpha) + \alpha\lambda$, where λ is an eigenvalue of $\mathcal{L}$, and eigenvalues of $K_{MPPRK}(\alpha)$ are their reciprocals. Interestingly, still another view, a random walk view is applicable here. Consider the random walk matrix $P = SD^{-1}$ reflecting the closeness of nodes in terms of reachability within a single random walk step. Then P^t expresses the probability of reaching one node from the other within t steps. In order to express the overall reachability, we would sum these probability matrices, however being less interested in $t + 1$ steps by factor α than in t steps. So the reachability is expressed via $\sum_{t=0}^{\infty}(\alpha)^t P^t = (I - \alpha P)^{-1}$. If we multiply this with a particular initial "distribution" of random walkers D^{-1}, then we get $K_{MPPRK}(\alpha)$. For an example of embedding via $K_{MPPRK}(\alpha)$ see Fig. 3(b).

7 Subspace Choice

Spectral and kernel embeddings have two essential parameters: the targeted number of clusters and the subspace to be used for clustering. They are coupled by the algorithms, however we want to examine what will be the effect of decoupling them.

We evaluated them under two aspects: the unexplained variance (Figs. 4(a), 5(a), 6(a)) and the difference between the actual and "true" clustering, (Figs. 4(b), 5(b), 5(b)). The X axis shows the dimensionality for which the clustering was performed. The true clustering for k-means is that using exactly k dimensions. Let e.g. $\mathfrak{C} = \{C_1, C_2, ..., C_k\}$ be the true clustering and let $\mathfrak{C}' = \{C_1', C_2', ..., C_k'\}$ be the clustering obtained via k-means.

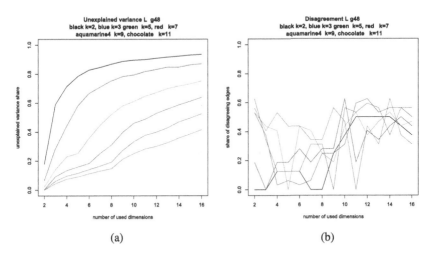

(a)   (b)

Fig. 4. (a) Unexplained variance of k-means on increase of the number of dimensions for a 4×8 grid graph. Hall embedding for the same graph for various values of k. (b) Disagreement between clusterings on increase of the number of dimensions for a 4×8 grid graph. Hall embedding for the same graph for various values of k for k-means.

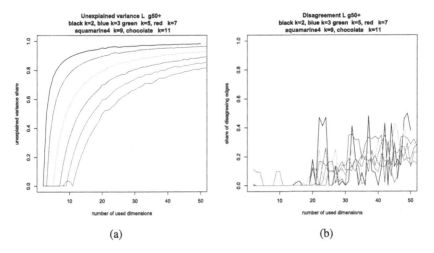

(a)   (b)

Fig. 5. (a) Unexplained variance of k-means for Hall embedding. Each curve for a different graph consisting of loosely coupled k-subgraphs. (b) Disagreement between the true clusterings and those from of k-means for Hall embedding. Each curve for a different graph consisting of loosely coupled k-subgraphs.

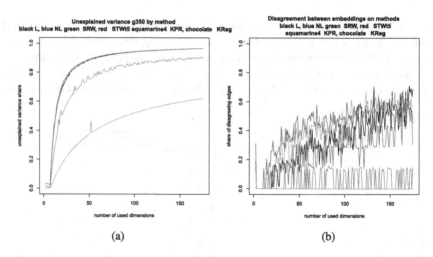

(a)   (b)

Fig. 6. (a) Unexplained variance when the number of dimensions is increased for a clear graph structure. k fixed at 7. Various embedding methods. (b) Disagreement between the trues clusterings and ones got from k-means when the number of dimensions is increased for a clear graph structure. k fixed at 7. Various embedding methods.

We define a unique mapping $m : \{1,\dots,k\} \to Z\{1,\dots,k\}$. If $\mathbf{x} \in C_i$ but $\mathbf{x} \notin C'_{m(i)}$, then $\mathbf{x}$ is misclassified. The mapping m is defined in such a way as to minimize the number of misclassified elements. The "disagreement" is the ratio of the number of misclassified elements to the number of all elements in the dataset.

We performed the experiments for two types of graphs: one without any structure in it (grid graph) and the other with a clear cluster structure. While k-means always produces k clusters, it is an important question whether or not the clusters reveal an intrinsic structure in the data. With these experiments, we wanted also to shed some light on the ongoing research issue how to decide if relevant clusters were discovered.

Figure 4(a) shows the dependence of unexplained variance percentage when clustering a structureless grid graph 4×8 for Hall embedding for various values of k. As expected, the higher k the lower the unexplained variance. In all cases adding more dimensions introduces more noise. Figure 5(a) shows that dependence for graphs with clear cluster structure. For each k a graph consisting of k loosely coupled 50-node subgraphs was generated. Each subgraph was generated as a loop connecting all nodes plus a number of random links so that 15% of all possible links are there. Coupling was generated by adding two random edges linking each next component with the former ones. Findings are as above except that the rapid variance growth starts at the number of dimensions corresponding to the number of "true" clusters. Figure 6(a) compares with one another various embedding methods studied in this paper, applied to the graph with 7 components of 50 nodes each. No embedding method seems to be immune to overshooting the number of dimensions though two of them give less drastic results. Regularization based kernel seems to be the best, followed by PR-based kernel. Other perform worse.

Figures 4(b), 5(b), and 5(b) present the respective differences between the "true" clusterings and the obtained ones in terms of misclassified elements. If the data is structureless, Fig. 4(b), the misclassification is chaotic. If there is a true structure in the data, Fig. 5(b), misclassifications are significantly less frequent. When comparing the methods, Fig. 5(b), we see important differences. Regularized kernel (KReg) seems to be nearly insensitive to the number of dimensions used. Pagerank based kernel (KPR) is the second best. Next is the Belkin (KNL) embedding comparable with our skewed random walk with 5 steps (SRWt5, formula (7)). Worst are the Hall embedding (L) and the skewed random walk with 1 step (SRW).

8 Concluding Remarks

In this paper we suggested a different approach to the issue of understanding spectral clustering algorithms. Instead of insisting on demonstrating some approximations to some graph clustering criteria like *NCut* or *RCut*, we propose to look at spectral clustering as based on graph embeddings close to similarity measures. The embedding techniques have been presented as approximations of similarity measures in terms of the inverse of squared distances. Hence, spectral clustering techniques invoking *k*-means can be considered as ones working in a space approximating inverted square-rooted similarities. Thus spectral clustering approximates *NCut/RCut* minimizing sum of square roots of weights of cut edges.

The experimental study showed also that the difference between kernel methods and spectral methods consisting in taking or not taking into account the eigenvalues (besides eigenvectors) may contribute to stability of clustering even if the number of dimensions is not accurately chosen, that is there is a mismatch between the number of dimensions and the number of clusters, due e.g. to ties or near-ties in the eigenvalue spectrum.

References

1. Avrachenkov, K., Chebotarev, P., Rubanov, D.: Kernels on graphs as proximity measures. In: Bonato, A., Chung Graham, F., Prałat, P. (eds.) WAW 2017. LNCS, vol. 10519, pp. 27–41. Springer, Cham (2017). https://doi.org/10.1007/978-3-319-67810-8_3
2. Belkin, M., Niyogi, P.: Laplacian eigenmaps for dimensionality reduction and data representation. Neural Comput. **15**(6), 1373–1396 (2003)
3. Dhillon, I., Guan, Y., Kulis, B.: A unified view of kernel k-means, spectral clustering and graph cuts. Technical report, UTCS Technical report, TR-04-25, February 2005
4. Filippone, M., Camastra, F., Masulli, F., Rovetta, S.: A survey of kernel and spectral methods for clustering. Pattern Recogn. **41**(1), 176–190 (2008)
5. Hagen, L., Kahng, A.: New spectral methods for ratio cut partitioning and clustering. IEEE Trans. Comput. Aided Des. Integr. Circuits Syst. **11**(9), 1074–1085 (1992)
6. Hall, K.M.: An r-dimensional quadratic placement problem. Manage. Sci. **17**(3), 219–229 (1970)
7. Hofmann, T., Schölkopf, B., Smola, A.: Kernel methods in machine learning. Ann. Stat. **36**(3) (2008)
8. Kłopotek, M., Wierzchoń, S., Kłopotek, R.: Weighted Laplacians of grids and their application for inspection of spectral graph clustering methods (2020, submitted)

9. Kurras, S.: Variants of the graph Laplacian with applications in machine learning. Ph.D. thesis, University of Hamburg, Germany (2017)
10. von Luxburg, U.: A tutorial on spectral clustering. Stat. Comput. **17**(4), 395–416 (2007)
11. Ng, A.Y., Jordan, M.I., Weiss, Y.: On spectral clustering: analysis and an algorithm. In: Advances in Neural Information Processing Systems, pp. 849–856. MIT Press (2002)
12. Schiebinger, G., Wainwright, M.J., Yu, B.: The geometry of kernelized spectral clustering. Ann. Statist. **43**(2), 819–846 (2015)
13. Shi, J., Malik, J.: Normalized cuts and image segmentation. IEEE Trans. Pattern Anal. Mach. Intell. **22**(8), 888–905 (2000)

Author Index

Printed in the United States
By Bookmasters